SVA: The Power of Assertions in SystemVerilog

Eduard Cerny • Surrendra Dudani • John Havlicek
Dmitry Korchemny

SVA: The Power of Assertions in SystemVerilog

Second Edition

Springer

Eduard Cerny
Synopsys, Inc.
Worcester
MA, USA

John Havlicek
Cadence Design Systems
Austin, TX, USA

Surrendra Dudani
Synopsys, Inc.
Newton, MA, USA

Dmitry Korchemny
Intel, Kfar Saba, Israel

ISBN 978-3-319-07138-1 ISBN 978-3-319-07139-8 (eBook)
DOI 10.1007/978-3-319-07139-8
Springer Cham Heidelberg New York Dordrecht London

Library of Congress Control Number: 2014942642

Printed on acid-free paper

Springer is part of Springer Science+Business Media (www.springer.com)

Preface

This book is the result of the deep involvement of the authors in the development of EDA tools, SystemVerilog Assertion standardization, and many years of practical experience. One of the goals of this book is to expose the oral knowhow circulated among design and verification engineers which has never been written down in its full extent. The book thus contains many practical examples and exercises illustrating the various concepts and semantics of the SystemVerilog assertion language. Much attention is given to discussing efficiency of assertion forms in simulation and formal verification. We did our best to validate all the examples, but there are hundreds of them and not all features could be validated since they have not yet been implemented in EDA tools. Therefore, we will be grateful to readers for pointing to us any needed corrections. The book is written in a way that we believe serves well both the users of SystemVerilog assertions in simulation and also those who practice formal verification (model checking). Compared to previous books covering SystemVerilog assertions we include in detail the most recent features that appeared in the IEEE 1800–2009 SystemVerilog Standard and were further improved and enhanced in the recent IEEE 1800–2012 Standard. In particular, it concerns the new encapsulation construct "checker" and checker libraries, Linear Temporal Logic operators, and semantics and usage in formal verification. However, for integral understanding we present the assertion language and its applications in full detail.

This second edition of the book not only introduces the enhancements and corrections present in the 2012 SystemVerilog standard, but also it is reorganized in a way that facilitates basic assertion understanding, initial deployment in simple forms, fully detailed exposition of the power of the language, and deployment in simulation and formal verification. Not to mention corrections to several errors and inconsistencies in the first edition.

The book is divided into six parts. These parts are organized from introductory to advanced, as well as separating as much as possible aspects related to simulation and formal verification. Part I is enough to get basic understanding of SystemVerilog

Assertions. By adding Part II, the reader should be able to write typical assertions and use them in simulation. Parts III and IV provide deep understanding of the assertion language. Part V is dedicated to formal verification and provides formal semantics of the language. Finally, Part VI deals with the design of checker libraries and the use of checkers in formal verification. The Appendix concisely describes sampling of variables, as it is different than in the previous 2009 Standard and thus the 1st edition of this book.

Part I, **Opening**, is an extended introduction to assertions, their use in simulation, formal verification and other tools, and their meaning in relation to the rest of the SystemVerilog language.

Chapter 1 introduces the concept of assertions, their place in history of design verification, and discusses the use of assertions in hardware design and verification flow.

Chapter 2 introduces minimal necessary concepts from the SystemVerilog language, other than assertions, that are useful for understanding assertions and their usage.

Chapter 3 provides the basics of SystemVerilog simulation semantics. It discusses how SystemVerilog processes are executed in the various scheduling regions and how assertions interact with the simulation of the design model.

Part II, **Basic Assertions**, goes into sufficient details of the assertion language to understand and write simple assertions.

Chapter 4 describes the different assertion statements that can be used to ascertain correctness, provide constraints and collect coverage, both in clocked concurrent and unclocked immediate (simple and deferred) forms.

Chapters 5 and 6 provide the basic information on how to write simple properties and sequences that form the operational core of assertions.

Chapter 7 exposes system functions that help to write assertions without having to resort to additional procedural code and introduces several system tasks for controlling assertion and action block execution. The main difference with the earlier version of the standard is the introduction of a more powerful function to detect the presence of `x` and `z` signal values and a unified task for controlling the execution of assertions.

Part III Metalanguage Constructs

Chapter 8 considers reusability of assertion bodies by showing how Boolean expressions, sequences, and properties can be defined and parameterized for later reuse.

Chapter 9 provides a detailed exposition of the encapsulation construct "checker." This construct is the basis for forming what could be called a super assertion, that is, an assertion entity that combines procedural code, modeling variables, variable assignments, coverage, assertion and assumption statements into one reusable parameterized unit. Checkers now allow different kinds of procedures as well as continuous assignments which were illegal previously. Sampling as described in the Appendix has also changed in checkers.

Part IV Advanced Assertions

Chapters 10 and 11 delve into the full intricacies of property and sequence operators. The former chapter also defines precisely the notions of vacuous and non-vacuous evaluations of assertions.

Chapter 12 describes sampling clocks, clock flow through assertions, and multiclocked assertions.

Chapter 13 provides information on the ways synchronous property evaluation can be terminated with success or failure using asynchronous and synchronous abort operators.

Chapter 14 shows how to use concurrent assertions inside always procedures, and how the leading clock is inferred. It also describes how evaluation attempts are started depending on the conditional and looping statements inside procedures.

Chapter 15 apologizes for local variables, but in fact shows how local variables provide much flexibility to assertions, especially in simulation.

Chapter 16 exposes the various forms of local variable declarations and rules of deployment, including special local variable arguments to properties and sequences.

Chapter 17 shows another facet of SystemVerilog assertions, that of recursive properties. They provide an alternate and succinct form for expressing complex properties.

Chapter 18 discusses coverage collection that is needed to measure the verification progress. Two forms are described, using assertion cover statements alone and in combination with test bench `covergroups` to form powerful data collection constructs.

Chapter 19 briefly introduces some techniques for debugging assertions, independently of services provided by specific EDA tools, and then discusses the efficiency of various assertion forms in simulation and formal verification.

Part V Formal Verification

Chapter 20 provides an introduction to the treatment of assertions in formal verification by discussing the different ways formal verification can proceed and its role in the verification process.

Chapter 21 exposes details of the models and algorithms used in formal verification, in particular, *model checking*.

Chapter 22 gives the theoretical base for full and precise understanding of the meaning of assertions. This chapter is particularly important to anyone who implements some form of an assertion verification engine, simulation or formal.

Part VI, Advanced Checkers, is primarily concerned with developing effective reusable verification objects.

Chapter 23 shows how checkers can be used effectively in formal verification. The chapter also provides deeper understanding of the behavior of checker variables.

Chapter 24 discusses how to create libraries of verification statements based on assertions, from simple **`let`** or **`property`** based forms, to the complex ones using **`checker`** encapsulation.

We did our best to verify and compile each and every example and verify the text, however, not all SystemVerilog constructs introduced in the 2009 and 2012 standards are supported by commercial tools. In spite of the great improvements in the 2012 SystemVerilog LRM there are still some incomplete or ambiguous points in the language. We do not guarantee correctness and do not assume any liability and responsibility for issues resulting from applying the techniques described in the book.

Acknowledgments

The authors wish to express their gratefulness to many people who reviewed sections of the draft of the first edition of the book which served as the basis for this improved edition, in particular, to Shalom Bresticker (Intel), Lena Korchemny (Inango), Jacob Katz (Intel), Scott Little (Freescale), Zeljko Zilic (McGill University), Christian Berthet (ST Microelectronics), Chris Spear (Synopsys), and Erik Seligman (Intel). Furthermore, the book was mostly written over the weekends, vacations, and evenings, hence we are thankful to our spouses and families for their patience and understanding. We also acknowledge the support of our employers in this endeavor.

Worcester, MA, USA — Eduard Cerny
Newton, MA, USA — Surrendra Dudani
Austin, TX, USA — John Havlicek
Kfar Saba, Israel — Dmitry Korchemny

Contents

Acronyms

ABV	Assertion-based verification
ALU	Arithmetic logic unit
DUT	Device under test
FPGA	Field-programmable gate arrays
FSM	Finite state machine
FV	Formal verification
HDL	Hardware description language
HDVL	Hardware description and verification language
IP	Intellectual property
LRM	Language reference manual
LSB	Least significant bit
LTL	Linear temporal logic
MSB	Most significant bit
NBA	Non-blocking assignment `<=`
OVL	Open verification library
OVM	Open verification methodology
PLI	Programming language interface
RTL	Register transfer level
SoC	System on chip
SSA	Static single assignment form
SV	SystemVerilog
SVA	SystemVerilog assertions
SVF	Sampled value function
SVTB	SystemVerilog testbench
UVM	Universal verification methodology
VMM	Verification methodology manual for systemverilog
VPI	Verification programming interface

Part I
Opening

Chapter 1
Introduction

Πασῶν τῶν τεχνῶν ἀρχὴ χαλεπή.

The beginning of all arts is difficult.

In comparison with the total chip development effort, the portion of effort spent in design verification is growing at a faster rate and thus consuming a significantly larger portion of the development cost. Despite more automation of various processes and new techniques, the cost containment for verification continues to be a challenge. There are at least two important cost motivations behind the increased effort. One is the damaging effects of a late discovery of a bug in the design flow on project schedules, which ultimately results in product delays. The other is the enormous manufacturing cost of a chip re-spin due to revelation of a design flaw after the initial prototype of the chip.

Consequently, there is a strong belief that investing more in developing new design verification techniques and, correspondingly, increased effort to uncover design bugs early in the design flow are worthwhile. We now have new techniques, such as constrained random simulation, verification coverage closure, and assertion checking, employed by many major organizations with the aim of speeding up creation of testbenches and uncovering design errors.

SystemVerilog [8] is an extension of Verilog [2],[1] a well-known Hardware Description Language (HDL), to support new verification techniques that have already shown promising results in various organizations. Whereas Verilog was oriented primarily to design and test at the Register Transfer Level (RTL) and gate level, SystemVerilog added means for describing testbenches (SystemVerilog Testbench, SVTB), defining functional coverage, and specifying assertions

[1]Until 2009, Verilog and SystemVerilog had separate standards. In 2009 both standards were merged into the SystemVerilog standard.

E. Cerny et al., *SVA: The Power of Assertions in SystemVerilog*,
DOI 10.1007/978-3-319-07139-8_1

(SystemVerilog Assertions, SVA).[2] By virtue of many new enhancements geared toward testing and checking, SystemVerilog surpassed Verilog as an HDL to become an HDVL—Hardware Description and Verification Language.

The assertions technology was developed on the premise that writing specifications formally plays a critical role in detecting design errors. This is because tools can automatically detect errors based on the implemented design and its specification [49]. Keeping this goal in mind, SVA was designed as an integral part of SystemVerilog.[3]

This book is dedicated to SystemVerilog Assertions. It teaches how to write assertions, how to design and use assertion libraries, and it discusses assertion applications and checking. To read and to understand this book, very basic knowledge of SystemVerilog is sufficient. A tour of important features of SystemVerilog is presented in Chap. 2. For an introduction to the basic Verilog layer of SystemVerilog, see [61].

In this chapter, we introduce SystemVerilog assertions informally, before their systematic treatment in the subsequent parts. For developing intuition on assertions, their meaning is explained by way of simple examples. The reader does not have to understand all the details at this stage, but only to grasp the concepts behind the examples. This chapter also includes an informal introduction of SVA language features that are handled in great detail in the rest of this book.

1.1 The Concept of Assertion

An assertion is a positive statement about a property of a design. It is positive in the sense that, should the statement be found as false, it indicates an error. Designers place assertions to express the intended behavior as specifications that can be interpreted and analyzed by tools. Since the property only states the behavior, it is often used to ensure that the design implementation of the behavior matches the assertion.

The use of assertions in contemporary hardware design methodologies has become widespread and matured over the past years. In programming languages, assertions have had a longer history of use, primarily because the assertions tend to be simpler, embedded in the code to check Boolean properties. HDLs model behavior over explicit time domains, with properties synchronous to clocks as well as asynchronous with specific time delays. This aspect of HDL modeling pushed forward the development of language techniques for expressing complex temporal behavior in the form of assertions and algorithms to interpret temporal assertions. As a result, several commercial languages have emerged to support the growing

[2]SystemVerilog also introduced many important object-oriented enhancements to Verilog, such as aggregate data types, classes, and interfaces [5].

[3]The history of the SVA standardization is described in [25].

```
1  module m(input logic c, clk);
2    logic a = 1'b0;
3    logic b = 1'b1;
4    always @(posedge clk) begin
5      a <= c;
6      b <= !c;
7    end
8    // assertion
9    a1: assert property (@(posedge clk) a != b)
10     else $error("a != b does not hold");
11 endmodule : m
```

Fig. 1.1 A simple assertion

needs of designers to perform design verification with the assistance of assertions, sometimes as the central groundwork in monitoring the progress of a design project.

By assertions, we mean statements that express properties to be true in a more general sense, without implying any specific intention or application. Although the most common application is for checking a design in order to detect bugs, other uses include making assumptions about the environment and tracking test scenarios for functional or behavioral coverage. Later in this chapter, we introduce various forms of applications that are provided by SVA.

SystemVerilog provides assertion features that are declarative. That is, assertions do not describe how to check, but only what to check. Figure 1.1 shows a simple example of an assertion in SVA.

Consider module `m`. Its functionality is implemented in Lines 4–7 using an **`always`** procedure[4] and nonblocking assignments.

Assertion `a1` checks at each rising edge of clock `clk` that `a != b`. If the assertion does not hold, an error message `a != b does not hold` is issued. This assertion may be checked in simulation or by formal verification tools.

The experience of using assertions has shown important benefits described below.

Implementing Checks in Verilog Is Difficult

One could ask how RTL correctness was checked in Verilog before the invention of SystemVerilog. Checks can be obtained by implementing Verilog code that is equivalent to writing assertions.

To appreciate why assertion implementation in RTL code is nontrivial and error prone, consider checking that `grant` should be asserted four clock cycles after `req`. For simplicity we ignore here the reset issue (see Exercise 1.2). A natural solution

[4]The reader may be used to the term "**`always`** block", but according to the SystemVerilog 2012 standard, this construct is called "**`always`** procedure".

```
module reqgranted1(input logic req, grant, clk);
  bit [2:0] ctr = '0;
  always @(posedge clk) begin
    if (req) ctr <= 1;
    else if (ctr > 0 && ctr < 4) ctr <= ctr + 1;
    else if (ctr == 4) begin
      if (!grant) $display("Request not granted.");
      ctr <= '0;
    end
  end
endmodule : reqgranted1
```

Fig. 1.2 Checking nonoverlapping evaluations

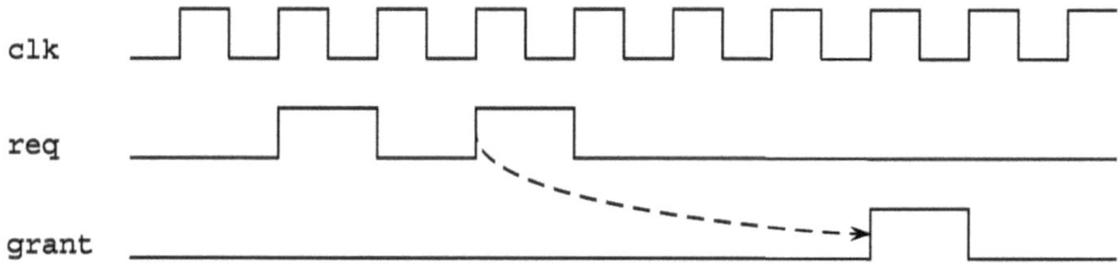

Fig. 1.3 Two requests before grant

```
module reqgranted2(input logic req, grant, clk);
  bit [3:0] sreg = '0;
  always @(posedge clk) begin
    sreg <= {sreg, req};
    if (sreg[3] && !grant) $display("Request not granted.");
  end
endmodule : reqgranted2
```

Fig. 1.4 Checking overlapping evaluations

is to activate a counter when `req` is detected, and when the counter value becomes 4, check for `grant`, as shown in Fig. 1.2.

What happens if there are two `req` issued before the first `grant` is seen, as shown in Fig. 1.3? We would expect our checker to fail because the first request is not granted. But, instead, the checker will pass. When the second `req` comes, we reset the counter and start counting anew—when we wrote the checker we did not think about overlapping evaluations!

To take overlapping evaluations into account we can use a shift register instead of a counter, as shown in Fig. 1.4. When `req` is asserted it is fed into the shift register `sreg`. The Most Significant Bit (MSB) of `sreg` is set into 1 when there was a `req` four cycles ago, and therefore `grant` must be asserted in this case.

The same intent may be expressed with a single assertion:

```
assert property(@(posedge clk) req |-> nexttime[4] grant);
```

Readers who are not yet convinced by this example can carry out Exercise 1.3.

Assertions Formally Express Design Intent

SVA is a *Formal Specification Language*. It is used to describe design properties unambiguously and precisely. Usually properties are written as part of the high level design specifications in a text document. But writing specification in a natural language is ambiguous.

Consider the following typical property specification: *Each request should be granted in four clock cycles*. This specification is ambiguous:

- Do we count four clock cycles starting from the cycle when the request was issued, or from the next cycle?
- Do we require that the grant is issued during the first four cycles or exactly at the fourth cycle?
- May two requests be served by the same grant or should they be served by two separate grants?

The same specification written in SVA is unambiguous:

```
assert property (@(posedge clk) request |-> nexttime[4] grant);
```

This specification defines a clocked, or concurrent assertion, and it reads: when `request` is issued, it should be followed by `grant` in the fourth clock cycle measured from the clock cycle when `request` was issued.

Because of the formal nature of SVA, specifications can be interpreted by tools, and what is more important, understood by humans. When the specifications are formally defined, there is no place for misunderstanding.

Assertions Improve Bug Detection

Assertions promote systematic methodologies by tapping into several flexible ways of inserting design checks. Once a methodology is set up to accommodate the needs of a project and assertion libraries are established for the design style, the effort to craft assertions becomes on par with writing ordinary code. The use of assertions proliferates, within the design code and at the interfaces of design units, and thus the design scrutiny is raised to trap errors. The checks remain in place from test to test, without expending any additional effort. In a way, writing assertions turns into something as simple as inserting comments. As a matter of course, bugs get detected early and efficiently because of the widespread and comprehensive set of assertion probes.

Assertions Promote Faster Root Cause Analysis

Because assertions can represent temporal behavior, a failure of an assertion that stretches out over multiple clock cycles detects a bug and concisely isolates it to the assertion expression. Now, one only needs to examine and analyze the temporal expression of the assertion to determine the root cause of the failure. Several modern debug tools support such an analysis to speed up the bug-fixing process. For example, by providing a precise window of time in which the failure occurred along with the cycle-by-cycle values of assertion signals, engineers can direct the analysis in the immediate design area of the failure. This is highly valuable and efficient.

A similar case with traditional techniques requires engineers to detect the failure by examining the output results, perhaps a mismatch of a value many cycles after the occurrence of the failure. A manual trace of values cycle by cycle through the design, without having specific clues about the behavior that actually caused the failure, is what the engineer must typically follow. As such, much of the difficult debugging is confined to the experts who retain intricate knowledge of the design.

Assertions Can Use Simulation and Formal Checking

The essential SVA features that exhibit temporality and clocking are strictly based on a mathematical framework, well understood and studied in academia. The availability of formalism made it possible to adopt proven algorithms for simulation and formal analysis. This enables taking the same assertion and applying both methods, one in a simulation environment for a set of tests, and the other in formal verification to conduct thorough proof analysis. Barring the limitations imposed by the fundamental differences between the two methods, both methods are applied to the same assertions for benefits within their individual processes.

The great benefit from this symmetry of use is that engineers need to write assertions only once. In most cases, a failure in simulation can be reproduced in formal analysis, and vice versa. A good methodology provides sufficient adjustments to take care of the small practical differences and limitations between the two methods. Besides being efficient, using the same assertions expels the major impediment to ensuring that the same behavior is checked by simulation and formal methods. Formal methods would check the critical blocks, and simulation would concentrate on system integration and the remaining blocks within it.

Assertions Are Part of Design Documentation

Traditionally, design documentation at any level consists of two parts. One is a stand-alone specification in a natural language, outside the domain of SystemVerilog. The other is a set of comments spread throughout the code in SystemVerilog. Neither of these specifications has any executable impact on the behavior of the design. That is, the behavior shown by the interpretation of SystemVerilog code does not necessarily correlate with the documented comments or outside specifications. Another commonly experienced pitfall is the enduring burden to keep the documentation up to date as the SystemVerilog design code matures and evolves over the life of a project.

Assertions, on the contrary, are executable statements that check the behavior of the design, no matter at what stage the design is being interpreted. This enforces the stipulation that assertions must change in accordance with the changes in the design, so there is no additional undertaking. As the design progresses, the assertions get updated to comply with the changes, thereby maintaining the documentation aspect of the design as well. Any laxness in the process gets caught by an assertion failure, which must be corrected to move forward. Writing formal specifications in SVA is not always an easy task, but it usually pays off because it leads to better understanding of the design, better design quality, and better documentation.

1.2 Assertions in Design Methodology

Assertions are an important part of the design and verification flow. This section discusses the use of assertions at various stages of the flow.

Figure 1.5 depicts a block diagram of a typical design and verification flow. The stages in the flow less relevant to assertions are omitted from this block diagram.

Usually, at the project inception hardware designers or architects write a product specification in a natural language, for example, in English. Based on the product specification, a high-level architectural model [43] is created in languages such as SystemVerilog or SystemC [7]. This architectural model may not be anchored with an accurate model of the design clock cycle; rather, it models functionality at a high level. The objective is to develop an efficient architecture by performing tradeoff analysis for time and area estimates, physical and power domain partitioning, input/output port definitions, etc.

Once an architecture is determined from the high level analysis, the architectural model is taken as the basis for developing a clock cycle accurate RTL model. Ultimately, this model evolves into a stable base for driving synthesis and physical design. In many organizations, the RTL model is considered the golden reference for the design. This dictates that the model be maintained and updated with any changes to the design. The RTL model is then synthesized automatically or manually into a gate or a transistor level with a netlist that specifies the connectivity [56]. This synthesized netlist is processed further by place and route tools to physically place and connect the gates on a chip for manufacturing.

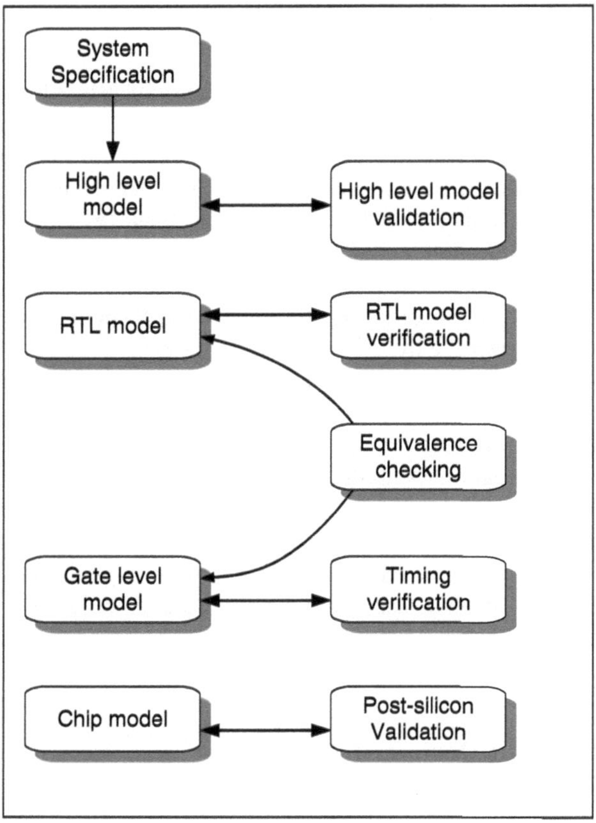

Fig. 1.5 High-level design and validation flow

For each design stage, there is a related verification stage which checks the design correctness at the corresponding level of abstraction. Below we discuss the role of assertions at the relevant design and verification stages.

1.2.1 Using Assertions for High Level Model

The design specification is a document usually written in a natural language describing its architecture and functionality. This document normally includes the main design components, data formats and communication protocols. Below is a typical example of a specification:

> The system consists of a transmitter and a receiver connected by a point-to-point duplex channel. The transmitter sends to the receiver packets and gets an acknowledgment from the receiver upon the packet receipt. The packet contains a header and a body. The header consists of 8 bits, and the two most significant bits contain information about the transaction type: *data* (10), *control* (01), or *void* (00). The remaining 6 bits of the header contain the transaction tag in case of a data transaction, and are 0 in case of a control transaction. For void packets the tag field may contain any value. The packet body consists of three bytes; these bytes contain raw data for data transactions and commands for control transactions ...
>
> Upon receipt of a data or a control packet the receiver sends back to the transmitter an acknowledgment signal. The acknowledgment consists of 7 bits: the most significant bit is set to 1, and the remaining 6 bits contain the tag of the received packet. If a void packet is received, its contents are ignored and no acknowledgment is sent ...
>
> The transmitter is not allowed to send a new packet before an acknowledgment is received. If timeout is reached, the transmitter sends the same packet again. If after three retries it does not get an acknowledgment, it asserts the error signal and requires a manual reset.

Specifications written in natural languages are ambiguous, and cannot be processed by tools. If we rewrite the properties from this specification in SVA as shown in Fig. 1.6, tools are then able to verify model compliance to its specification. We do not give the complete specification here (for example, we do not specify how timeout is set, and how a packet is resent), but only a fragment to illustrate the concept.

We briefly describe the SVA code in Fig. 1.6, derived from the specification, to get an intuitive idea of how assertions can be extracted from a specification and used for ensuring model compliance. Lines 1–7 define new types. Line 1 defines an enumeration type giving names to specific integral values. Types `tag_t` and `data_t` define new names for logic arrays of corresponding bounds. Lines 6 and 7 assign names to the combined pieces of data.

Lines 8–38 define the specification as a **`checker`**. A **`checker`** is a special verification unit containing assertions and their related code. By default, assertions `tx_packet_legal`, `rx_packet_legal`, and `no_ack` use the clock declared in Line 17. We discuss the **`default clocking`** statement in Sect. 12.2.2. Similarly, Line 18 describes a default reset for all concurrent assertions in this checker. When `rst` is asserted, no concurrent assertion is checked. This is carried out to disable assertion checking during the reset sequence. Handling clocks and resets is a very important topic, but we postpone the discussion on the related SVA features until Chaps. 12 and 13.

Lines 20, 22, and 24 define aliases for different conditions used in the assertions.

Sequence `no_ack_thrice` defined in Lines 32–34 models the relationship between `right_ack` and `timeout` from the specification: "The transmitter is not allowed to send a new packet before an acknowledgment is received. If timeout is reached, the transmitter sends the same packet again. If after three retries it does not get an acknowledgment, it asserts the error signal and requires a manual reset." Thus, the whole sequence

```
(!right_ack[+] ##1 timeout)[*3]
```

```
typedef enum logic[1:0] {
    txa_data = 2'b10, txa_control = 2'b01,
    txa_void = 2'b00, txa_forbid = 2'b11 } txa_t;
typedef logic [5:0] tag_t;
typedef logic [23:0] data_t;
typedef struct packed { txa_t txa; tag_t tag; data_t data; }
    packet_t;
typedef struct packed { logic ack_received; tag_t tag; } ack_t;
checker spec (
  packet_t tx_packet, // Packet to be transmitted
  packet_t rx_packet, // Last received packet
  logic sent,         // Packet sent
  ack_t ack,          // Acknowledge
  logic timeout,      // Timeout active
  logic err,          // Error signal
  event clk,          // System clock
  logic rst);         // Reset
default clocking @clk; endclocking
default disable iff rst;
// Legal transaction type
let legal_txa(txa) = txa != txa_forbid;
// Non-void packet sent
let packet_sent = sent && tx_packet.txa != txa_void;
// Right acknowledgment received
let right_ack = ack.ack_received && ack.tag == tx_packet.tag;
// Transmitted packet is always legal
tx_packet_legal: assert property (legal_txa(tx_packet.txa))
    else $error("Transmitted packet is malformed");
// Received packet is always legal
rx_packet_legal: assert property (legal_txa(rx_packet.txa))
    else $error("Received packet is malformed");
// No acknowledgment thrice
sequence no_ack_thrice;
  (!right_ack[+] ##1 timeout)[*3];
endsequence
// Despair - raise error flag
no_ack: assert property (packet_sent ##1 no_ack_thrice |->
  always err) else $error("Err indication does not persist");
endchecker : spec
```

Fig. 1.6 System specification

detects a situation when signal `timeout` is activated three times while awaiting acknowledgment.[5] We discuss sequences in Chap. 6.

[5]This specification does not provide details about how the `timeout` condition is formed. Of course, a complete specification should provide them.

Assertion `no_ack` in Line 36 states that if upon sending the packet (`packet_sent`) there is no acknowledgment (`no_ack_thrice`), then the error flag `err` should be asserted forever, that is, until the system reset occurs.

The above example shows that assertions can be inferred from a system level description for expressing them in SVA. These assertions can then be applied to a high level model in SystemVerilog and checked in simulation or proven formally.

Using SVA for system level has some difficulties. The main issue is clocks. The system description is not always clock accurate, and is often formulated in terms of transactions and real time. The signal activities are not expressed in terms of transitions synchronized by design clocks. In contrast, SVA requires an exact clock specification for every assertion. For example, the SVA specification in Fig. 1.6 is less abstract than its verbal counterpart with respect to the specification of the clock, reset, and timeout signals.

However, SVA has an important advantage that the same assertions may be used directly or after some refinement as checkers in RTL verification, after the high level model is refined to an RTL model of the design. As long as the high level model embodies an approximate notion of a clock, it may still be possible to describe assertions for the model. SVA provides several means for managing abstractions. For example, we can define clock and reset only once using **`default clocking`** and **`default disable iff`** statements. Specifying the exact clocks and resets requires changing these statements only. The assertion building blocks are encapsulated in **`let`**, **`sequence`**, and **`property`** statements. System refinement can thus be handled using appropriate abstraction means. In our example, if the timeout mechanism needs to be refined, it is enough to modify sequence `no_ack_thrice`. The mechanics of attaching or substituting one model with another in the refinement process is further supported by a configuration mechanism of SystemVerilog.[6] This is an iterative step which continues to refine the architecture and to adjust the assertions according to the modified requirements.

The situation gets more complicated if the language of the model is not SystemVerilog. For example, high level models of Systems on Chip (SoCs) are often written in SystemC, as this language can be natively integrated with the C/C++ code of software components. Unfortunately, SystemC does not yet have its own formal specification subset for assertions or provide a standard integration with external formal specification languages [62]. Therefore, using SVA specifications with SystemC models is tool dependent. There are other languages, such as TLA [48] designed especially for high level system specification. These languages are more abstract than SVA, and therefore better suited for this purpose. Detailed discussion about high level model validation is out of the scope of this book.

There are attempts in the academia and in the industry [21, 55, 68] to synthesize an RTL model directly from its formal specification, and some promising results have been demonstrated. Nevertheless, there is still a long way to go for this approach to become practical, and we do not discuss it in this book.

---

[6]Configurations were introduced in Verilog 2001 [1].

1.2.2 Using Assertions for RTL Models

The methodology of using assertions for RTL design verification is commonly known as Assertion-Based Verification (ABV), [19, 37, 44, 67]. The idea is to instrument RTL code with assertions to capture the design intent and to check the local correctness of the design. In the former case, assertions are written for checking the behavior at the interfaces. In the latter case, assertions are embedded in the design units, interspersed through the code as needed to check the local correctness.

Assertions on Interfaces

In this method, assertions are written to express the behavior as seen at an interface. Customarily, verification engineers write such assertions as they do not require intimate knowledge of the design details. This method of verification where the design units are viewed as black boxes, meaning without the knowledge of internal design details, is called *black-box verification.*

Verification engineers examine the high level specification of a design unit to infer rules and properties that must be satisfied by the design unit. Each rule may translate to one or more assertions. Once written and corrected, these assertions tend to remain unaffected by the changes in the internal design unit code. They are nonintrusive to the design units and can be retained physically outside the design units as well. SystemVerilog provides means of attaching checkers to the design units whenever needed using the `bind` statement, without actually modifying the source code of the design units (see Sect. 9.3.3).

Some examples of the functionality checked by the interface assertions are:

- Bus communication protocols
- Memory transactions
- Data transformations
- Transaction arbitration

Another pivotal use of these assertions is to detect errors when various design units are assembled into a larger unit. As integration issues emerge, they are effectively captured by these assertions. By maintaining the consistency of interfaces, individual design units are effectively freed from outside considerations, at least for verification purposes.

Embedding Assertions Within Design

Most often designers attend to their design units for local correctness. Within the scope of the design unit, they write assertions as they develop code to ensure signal

```
1  module shreg (input logic clk, rst, set, logic [7:0] val,
2    output logic [7:0] shift_reg);
3    always @(posedge clk or posedge rst) begin
4      if (rst) shift_reg <= 0;
5      else begin
6        if (set) shift_reg <= val;
7        else begin
8          shift_reg <= shift_reg << 1;
9          shift_reg[0] <= shift_reg[7];
10       end
11     end
12   end
13   check_shift: assert property (@(posedge clk)disable iff (rst)
14     set or
15     nexttime shift_reg == $past({shift_reg[6:0], shift_reg[7]}));
16   check_rst: assert final (rst -> shift_reg == '0);
17 endmodule : shreg
```

Fig. 1.7 Checks for a shift register

relationships. When assertions are written over internal signals of a design unit for performing local checks, the verification process is termed as *white-box verification*.

A typical example is shown in Fig. 1.7.

Module `shreg` in Fig. 1.7 implements a shift register `shift_reg` in RTL code (Lines 3–12). Two assertions `check_shift` and `check_rst` verify the implementation correctness of the code.

The first assertion `check_shift` checks that the new value of `shift_reg` is obtained by left rotation of its old value unless the new value was set explicitly when `set` was asserted. The system function `$past` used in this assertion returns the value of its argument evaluated at the previous clock cycle (see Sect. 7.2.1.2). Note the operator **`disable iff`**, which disables the assertion check when the reset signal is active.

The second assertion `check_rst` checks that `shift_reg` is reset correctly. We use a final assertion which is not clocked, and not a concurrent one here because the register reset is asynchronous and we need to check it at each simulation step, and not only at the clock cycles. Section 1.3 and Chap. 4 explain the difference between deferred and concurrent assertions in more detail.

Below are some typical items to check in white-box verification:

- Compliance of interface
- Finite State Machine (FSM) transitions
- Memory access correctness
- Stack and queue overflow and underflow
- Arithmetic overflow
- Signal stability

The complete list depends on a specific methodology; see [19, 37, 44, 67] for suggestions.

Although local assertions do not completely verify the design, their advantage is huge. They make design debugging more effective—a bug is detected and caught close to its origin. Thus, in Fig. 1.7 an incorrect implementation of the shift register will be immediately detected, and a failure of assertion `check_shift` or `check_rst` will point to the problematic code because these assertions are physically adjacent to it. Without these assertions, an implementation bug could manifest itself in another part of the design and probably several clock cycles later. One can imagine the difficulty of debugging that error.

Assertion Coverage

When administering a large verification project, one needs to know whether the intended functionality has been verified in its full scope, covering all functional scenarios of interest and all corner cases. Clearly, just making assertions a part of design flow does not adequately provide confidence in judging that the verification is complete or even comprehensive. Therefore, assertion coverage plays a critical part in decision making and tracking progress of the design verification project, keyed to the inquiry—"Are there enough assertions?"

The question is, what kind of coverage can be obtained from assertions to provide substantive indications? We note that, generally, there are two ways to approach this question. In the first approach, inquiries about the functionality are the central focus. In this regard, behavioral fragments expressed by various assertions must be matched against the specifications to determine the extent of the functionality included in the umbrella of the assertions. In the second approach, structural aspects of the design form the criteria. For example, the number of design elements (signals, registers, etc.) included in the assertion checks, the number of input and output signals included in assertions, and the number of assertions relative to the design code size. Both approaches are useful indicators that provide meaningful guidance in determining the required level of verification effort.

Coverage-Based Verification

A complementary approach to assertion-based verification is *coverage based verification*. Coverage-based verification starts by taking functional scenarios (*coverage points*) from the test plan and then collecting coverage of these scenarios on available tests. The goal is to refine tests so that all coverage points are hit. The main problem with this approach is its practical infeasibility: some scenarios are extremely difficult to cover, and some of them are even impossible. Usually, the first few tests hit many coverage points, and up to 60 % coverage is quickly reached. The additional tests cover fewer and fewer new coverage points, while reaching 80 % coverage or higher becomes increasingly challenging [19, 59] and unlikely in practice. Verification managers usually empirically set the desired coverage percentage, called the *coverage goal*. We discuss SVA tools for checking coverage in Sect. 4.7 and in Chap. 18.

1.2.3 *Using Assertions Beyond RTL*

Although assertions are most frequently used at the RT level, other areas of development later in the design phase can also benefit from their specification. Some analysis tools have already been developed, while others have been explored to take advantage of the expressibility of the assertion features. We discuss three important areas here.

Equivalence Verification

Equivalence [45] of two models usually means checking that the synthesized gate-level netlist is equivalent to the golden RTL model (Fig. 1.5). Equivalence checking is also needed when local changes are made in the design to improve its performance or power consumption.

Equivalence checking is usually done by formal verification because comparing model behaviors in simulation cannot provide good confidence in the correctness of design transformations. At first, it seems that there is no need for assertions in equivalence verification, but it turns out that the role of assertions is quite significant because two models are equivalent only under some assumptions on their inputs. For example, the input signal `go` of an RTL model may correspond to two input signals of the synthesized model: `go` and its negation `ngo`. To prove equivalence the following assumption about signal inversion should be supplied to the tool:

```
assume final (go ^ ngo);
```

In addition, assumptions about internal signals are used as hints for formal equivalence verification. To maintain correctness of the proof of equivalence, these assumptions must be proven as assertions in the corresponding blocks (see Sect. 20.5). Assumptions written for formal equivalence verifications are usually nontemporal; therefore, they are best represented with final assumptions having the syntax **`assume final`** as shown above.

Timing Verification

When an RTL model is synthesized into a gate-level model, a critical step is to verify its timing to ascertain correct functioning of the circuit [27]. Even though timing verification significantly differs from RTL verification, assertions are used there, although for different purposes. For example, RTL assertions are used to characterize signal paths, as in the following cases:

False Path Elimination Circuit performance is limited by the delay of the longest combinatorial path. Given the circuit configuration, if the actual signal transmission along this path is not possible then this path should be ignored for critical path and performance analysis [15, 28].

Clock Domain Crossing When data are transferred from a state element controlled by one clock to a state element controlled by another clock the data should be stable long enough to guarantee that it be sampled by the second clock [54].

Multicycle Path A multicycle path is a path between two state elements having a delay greater than one clock cycle. A multicycle path permits the sum of the delays of its combinatorial logic elements to be greater than one clock cycle. In this case, the second state element should be stable during the corresponding number of clock cycles [15].

We also mention the need for analog assertions [52] to specify the timing behavior of electrical components and interconnections. This type of assertion is specific to analog circuit analysis and performance verification, and it is currently not part of SVA. We do not discuss it further in this book.

Post-Silicon Validation

The advantages of RTL verification, on the one hand, are flexibility and high observability—all signal values at any time may be observed in simulation. On the other hand, it is very slow, and does not allow checking many important global scenarios. With post-silicon validation (and to a great extent in emulation) the situation is the opposite: chip speed is very high, but signal observability is low [11,57,65].

Postsilicon debugging is challenging because a bug can remain unobserved for millions of cycles after its actual occurrence. ABV may help coping with this problem. For example, the most critical RTL assertions may be synthesized into the chip. Assertions fire immediately upon detecting an error, thus making bug detection and debugging much more efficient.

1.3 Assertions in SystemVerilog

There are three kinds of assertions in SystemVerilog[7]:

- Immediate assertions
- Deferred assertions
- Concurrent assertions

---

[7]Our terminology for the kinds of assertions matches what we hear customarily in practice, but it differs from the terminology in the SystemVerilog 2012 LRM. What we call "immediate" is called "simple immediate" in the LRM, and what we call "deferred" is called "deferred immediate" or just "deferred" in the LRM. The LRM uses "immediate" to mean either "simple immediate" or "deferred immediate", but we will use the phrase "immediate or deferred" for this union. We find our terminology clear and less verbose than the LRM, and there is usually no confusion in context about the kind or kinds of assertions being discussed.

```
1  module m2(input logic c, d, clk);
2    logic a, b;
3    always_comb begin
4      a = c & d; b = c | d;
5      // Immediate assertion
6      a1: assert (a -> b);
7      // Deferred assertion
8      a2: assert final (a -> b);
9      // Concurrent assertion
10     a3: assert property (@clk a != b);
11   end
12   // Deferred assertion
13   a4: assert final (a -> b);
14   // Concurrent assertion
15   a5: assert property (@clk a != b);
16 endmodule : m2
```

Fig. 1.8 Kinds of assertions

The simplest assertions are *immediate assertions*. They act as procedural **if** statements and are legal in any place where procedural **if** statements may appear. Immediate assertions are nontemporal and are executed when the control flow reaches them. The main advantage of immediate assertions is that they have unrestricted applicability in various kinds of designs, synchronous and asynchronous, and in testbenches. Their ease of use makes them appealing, but the limited expressiveness lends their efficacy to detecting only simple bugs. In some cases, they are also prone to producing spurious failures due to simulation races. This is explained in Sect. 4.2.

Deferred assertions are an improvement over immediate assertions. They are similar to immediate assertions in that they are nontemporal and unrestricted in their use. Two important differences make them immensely useful over immediate assertions: they do not produce spurious failures, and they can be placed both inside and outside procedural code. Deferred assertions are further subdivided into observed and final (sometimes called simply "final"). They are explained in Sect. 4.3.

The most interesting and complex assertions are *concurrent assertions*. They are temporal and can describe design behaviors over time. For example, a concurrent assertion can state that a request should be granted in two clock cycles. Concurrent assertions are always clocked. Assertion `a1` in Fig. 1.1 is a simple example of a concurrent assertion. This assertion is Boolean and checked immediately before the rising edge of `clk`. Concurrent assertions may appear both inside and outside procedural code, but they cannot be placed in functions and tasks. The description of concurrent assertions occupies the major part of this book.

Figure 1.8 illustrates the use of all three kinds of assertions. Here, assertion `a1` is an immediate assertion, `a2` and `a4` are final deferred assertions, and `a3` and `a5` are concurrent assertions. Operator `->` used in assertions `a1`, `a2`, and `a4` is an implication operator. **`always_comb`** used in Line 3 is explained in Sect. 2.2. Notice that deferred assertions (e.g., `a4`) are legal outside procedural code, but immediate assertions would be illegal there. Immediate assertion `a1` and deferred assertions `a2` and `a4` are checked whenever either `a` or `b` changes value, while concurrent assertions `a3` and `a5` are checked only at the clock event `clk`, which occurs when `clk` changes value.

Concurrent assertions use *sampled* values of their variables. For design signals, the sampled value is the value of the variable at the beginning of the simulation step, before any values change in the time step. Therefore, in waveforms it looks as if concurrent assertions used past values of design signals from the preceding time steps. This is described in detail in Sect. 4.4.3.

Assertion Statements

Assertion statements specify properties on the behavior of signals. There are three major assertion statements in SVA: assertions (the keyword **`assert`**), assumptions (the keyword **`assume`**), and cover statements (the keyword **`cover`**). Each type of statement directs what to do with the specified property. These statements may be immediate, deferred, or concurrent.

An **`assert`** statement makes sure that the design behaves as the properties in the statement prescribe. Its purpose is to check the correctness of the system. An **`assume`** statement states assumptions, specifying properties that should hold to enable the proper functioning of the system. Its purpose is to ensure that the checking is conducted under a system or environment that complies with the stated conditions. A **`cover`** statement checks that the behavior it specifies is actually exhibited while testing the system. Unlike **`assert`** statements, the interest lies in detecting only selective cases from all possible cases of valid behavior.

In the following example, the same condition is used in all three types of assertions (`a1`,`m1`,`c1`) to depict the difference in motivation.

```
bit ok;
// ...
a1: assert property (@clk ok);
m1: assume property (@clk ok);
c1: cover property (@clk ok);
```

The difference between `a1` and `m1` is as follows: `a1` states that `ok` must be `1'b1` for correct behavior. It is a property that the design is obliged to satisfy. `m1` states that it can be taken for granted that `ok` is `1'b1`. Assertions are checked while taking into account assumptions. Typically, but not necessarily, assumptions are written on primary inputs of the design, characterizing the behavior of the environment of the design.

The **`cover`** statement `c1` states that there exists some valid system behavior where `ok` gets value `1'b1`. This does not prevent the system from having other valid behaviors where this condition does not hold. Cover statements are usually written to ascertain that there exist tests exercising specific scenarios.

SVA assertion statements are further discussed in Chaps. 4 and 18.

1.4 Checking Assertions

This section introduces how assertions are checked in different verification environments:

- Simulation.
- Emulation.
- Formal analysis.

1.4.1 Checking Assertions in Simulation

Simulation [61] is modeling the behavior on a sequence of input stimuli, called a (simulation) test. Simulation is the most popular method for checking assertions. All major SystemVerilog simulators, for example, VCS®, Questa®, and Incisive®, support SVA and can check assertions, assumptions, and cover statements.

Of course, in simulation it is only possible to check whether an assertion is violated in a given test case. If it is violated, we find a problem. But the absence of violation does not mean that the design is correct—the same assertion may be violated in another test case, or it may even be violated later in simulation if the same test case is extended.

Although in theory no reasonable number of test cases is sufficient to exhaustively check correctness of real designs, in practice simulation with coverage measurements does provide significant confidence in system correctness if no assertion violations are detected.

Typical simulators can report not only assertion violations, but also individual transaction completions. By a transaction, we mean an individual case of assertion evaluation. Figure 1.9 illustrates transaction completions for the following assertion:

```
req_ack: assert property (@(posedge clk) req |-> ##[1:3] ack);
```

This assertion states that each request `req` receives an acknowledgment `ack` in one to three clock cycles from the moment when `req` was asserted. Figure 1.9 shows two transactions, *trans1* and *trans2*. Transaction *trans1* starts at time 20 and completes at time 40, while transaction *trans2* starts at time 90 but does not complete before the end of simulation. An incomplete transaction does not necessarily indicate a correctness problem because, had simulation lasted longer, the transaction might complete as expected. However, this situation requires further analysis. When crafting tests (directed and random) it is desirable to leave no transactions *pending* (incomplete). Time points where there are no pending transactions are called *quiescent points*. Although it is a good practice to ensure that the simulation always ends at quiescent points, in reality it is hard to attain such a state for all assertions at the same quiescent point. In general, it is necessary to analyze incomplete transactions for any unexpected behavior.

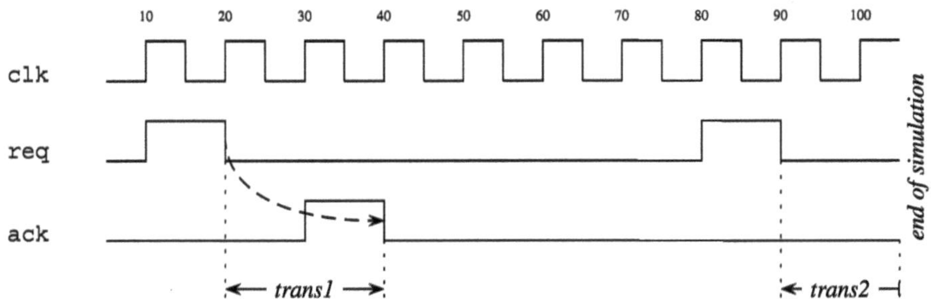

Fig. 1.9 Transactions of assertion `req_ack`

In Fig. 1.9 the transactions are delayed by one clock cycle with respect to the times at which `req` and `ack` rise. This may seem strange. For example, why does *trans1* last from time 20 until 40, and not from 10 until 30? Concurrent assertions use sampled values of their variables, that is, the values that these variables have at the beginning of a simulation step (Sect. 1.3). At the beginning of the simulation step corresponding to time 10, the sampled value of `req` is still 0. Assertion `req_ack` will use the new value 1 of `req` only at time 20. This explains the shift in transaction marking in the figure.

Assertions may also be checked in *random simulation* [16, 19, 38] environments. Random simulation can be achieved using testbenches that generate random stimuli using constraints or assumptions. While random simulation can hit a large amount of bugs rather quickly, it is difficult to achieve good coverage of corner cases. Another drawback of random simulation is its speed—resolving imposed constraints can be prohibitively slow.

Simulation provides the most intuitive and user-friendly environment for assertion debugging. Even when assertions are not targeted for simulation, simulation may be used for assertion debugging. It seldom happens that complex assertions are written correctly the first time. Usually failures in new assertions are caused by bugs in the assertions themselves, not by design errors. Before checking assertions in other environments, such as emulation and formal verification, it is highly recommended to debug them in simulation. We discuss assertion debugging in Chap. 19.

1.4.2 Checking Assertions Using Hardware Acceleration

Checking assertions in simulation is intuitive and convenient, but unfortunately, simulation is slow compared to hardware speeds, and as a result, only very short testing sequences may be checked this way. For example, to check a CPU model, an operating system and several typical applications should be run on it, but it would take months or years to simulate a few seconds of the real work.

Solutions to bring the speed of simulation closer to that of the hardware being simulated include hardware acceleration, emulation [39], and rapid prototyping. In these methods, the design model is synthesized into a logic netlist, and this netlist is mapped onto a Field-Programmable Gate Array (FPGA) or an equivalent programmable device. Of course, checking a design in this way is still much slower than running the real device, but it is significantly faster than simulation. Because emulation tests are much longer than the simulation ones, they have a better likelihood of revealing bugs that could not be reached in simulation. To capture these bugs, assertions need also to be synthesized to become part of the emulation model. In theory all SVA constructs are synthesizable, enabling the solution to work in most cases. For some cases, however, this solution falls apart as some complex assertions synthesize into enormous size, consuming a large amount of available gates.

1.4.3 Checking Assertions Using Formal Verification

Formal Verification (FV) [20] is the most powerful method to check design correctness. It conducts exhaustive proof that the design complies with its specification. More precisely, formal verification tools prove assertion correctness under the hypothesis that all assumptions are satisfied. Unlike simulation and emulation, there is no need to provide input stimuli.[8] If a tool can prove the assertion correctness, the assertion is correct for any set of input stimuli under the specified assumptions.

The main limitation of FV methods is the capacity of FV tools. They can handle only relatively small models, even though modern FV tools can efficiently handle designs containing several thousands of state elements, latches and flip-flops. Another important point to keep in mind is that the model for FV should be completely specified, requiring all its input assumptions to be explicitly stated. If some assumptions are missing, spurious assertion failures (so called *false negatives*) may be reported, as discussed in Sect. 20.3.

It follows that in simulation the main verification setup effort is modeling the environment and devising the testbench, while in FV a great deal of effort is spent on specifying assumptions.

[8] Actually many verification tools do require some input information, such as a clock pattern or a reset sequence.

1.4.4 Assertion Efficiency

It is often possible to express the same assertions in multiple ways. A specific style of assertion implementation may have a major effect on simulation or formal verification performance. Therefore, it is important to know how to write assertions efficiently. Unfortunately, in many cases formal verification and simulation impose different requirements on assertions for efficiency considerations, creating situations where efficiency tradeoff between the two methods becomes necessary. Possibilities exist to make a small sacrifice in assertion efficiency in formal verification that can provide a tremendous boost in simulation speed. Many factors are involved in making tradeoffs: complexity of assertions, number of assertions, and algorithms employed by a specific tool. When using a specific simulation or formal verification tool one should follow tool-specific recommendations about assertion efficiency.

1.5 Assertion Reuse

Although SVA is a powerful specification language, writing assertions is not an easy task. Even experienced people rarely write complex assertions correctly for the first time. Debugging assertions is more difficult than debugging RTL because the assertion language is declarative. Fortunately, many assertions are commonly encountered and may be reused by adapting them to different situations. For example, such assertions as "two signals are mutually exclusive", "a request is granted in N cycles", and "an FSM is never stuck" are routine. This presents an opportunity to define them once and then reuse by customizing as a library unit.

SVA provides many features for assertion reuse. Assertion components may be named and parameterized. Several related assertions, together with modeling code may be grouped as a unit for future reuse.

Expression Reuse

Expressions may be named and parameterized using a **`let`** statement, as shown in Fig. 1.10.

In this example, a parameterized expression `$onehot(~sig)` is named `onecold` using a **`let`** statement. `$onehot` is a SVA system function returning *true*[9] when exactly one bit of its argument is set to 1. This **`let`** expression checks for *one cold* encoding which means exactly one bit of `sig` is 0. Notice that an instance of the **`let`** expression is used in assertions `a1`, `a2`, and `a3`. `a1` is an immediate assertion

[9] Strictly speaking, *true* is not defined in SystemVerilog, but we will use it where appropriate as an alias for `1'b1`. Similarly, we will use *false* for `1'b0`.

```
logic a, b, c, d, cond, clk;
let onecold(sig) = $onehot(~sig);
// ...
always_comb begin
  // ...
  a1: assert (onecold({a, b, c}) || d);
end
a2: assert final (onecold({a, b, c}) -> d);
a3: assert property (@(posedge clk) cond |=> onecold({a, b, c}));
```

Fig. 1.10 Expression reuse

```
logic ready, request, grant, clk;
// ...
sequence falling(x);
  (x ##1 !x);
endsequence
a1: assert property (@(posedge clk) falling(ready) |=> ready);
a2: assert property(@(posedge clk)request |=> falling(grant));
```

Fig. 1.11 Sequence reuse

ensuring that exactly one signal among `a`, `b`, and `c` is low when `d` is low. `a2` is a final assertion ensuring that when exactly one signal among `a`, `b`, and `c` is low, `d` must be high. Another variation is the concurrent assertion `a3` which specifies that after condition `cond` is true, exactly one signal among `a`, `b`, and `c` is 0.

We describe the **`let`** statement in Sect. 8.1.

Sequence Reuse

It is possible to assign names to sequences of signal values in time and to reference these sequences by name in assertions, as shown in Fig. 1.11.

`falling` is a sequence name, and `x` is its argument. `(x ##1 !x)` defines a sequence of values of signal `x` in time. Its meaning is that the value `x = 1` is followed by the value `x = 0` in the next clock cycle.[10] Sequence `falling` is reused in concurrent assertions `a1` and `a2`.

Assertion `a1` states that `ready` flag may drop at most for one clock cycle. More precisely, if `ready` gets deasserted after being asserted then at the next clock cycle `ready` should be asserted again. The operator `|=>` means "then at the next clock cycle", and it is called *non-overlapping suffix implication*. Assertion `a2` states that

[10] For simplicity, here and in future examples we ignore the possibility of unknown and high-impedance values `X` and `Z` unless explicitly stated.

```
logic end_reset, operational, enter_deadlock_area, stuck, clk;
// ...
property forever_n(x, n);
  nexttime[n] always x;
endproperty
a1: assert property (@(posedge clk)
      end_reset |-> forever_n(operational, 100));
a2: assert property (@(posedge clk)
      enter_deadlock_area |-> forever_n(stuck, 5));
```

Fig. 1.12 Property reuse

`request` should be granted (`grant == 1`) in the next cycle, and one cycle later `grant` should be deasserted.

We describe sequences in Chaps. 6 and 11.

Property Reuse

Like expressions and sequences, properties may also be assigned a name to be used in concurrent assertions, as shown in Fig. 1.12.

In this example, `forever_n` is a property specifying that after `n` clock cycles (operator **`nexttime`**`[n]`) `x` should be true forever (operator **`always`**). This property is then reused in assertions `a1` and `a2`. Assertion `a1` states that 100 clock cycles after reset phase was completed (`end_reset` asserted) the device should be operational forever (`operational` should always be high). Assertion `a2` states that 5 cycles after entering a deadlock area (`enter_deadlock_area` asserted) signal `stuck` should be asserted forever.

We describe properties in Chaps. 5 and 10.

Assertion Libraries

Although the language features for naming an expression, a sequence or a property are beneficial for reuse in writing individual assertions, they are not sufficient for building a library of assertions. Commonly, an element from an assertion library encapsulates one or more related assertions, and some code to support the expressions used within the assertions, such as an FSM state or a variable value computed from a function.

A more suitable feature than what we have described so far is a **`checker`**. The **`checker`** construct is similar to a module in that it can contain assertions and modeling code, but its instantiation and parameterization accommodate the flexibility and usage that are specific to assertions.

```
1 checker mytrig (sequence trig, property prop, event clk);
2   a1: assert property (@clk trig |-> prop);
3   c1: cover property (@clk trig);
4 endchecker : mytrig
5 module m (input logic done, ready, clock, output logic idle);
6   ...
7   assign idle = done || !ready;
8   mytrig check_consistency(done, idle, posedge clock);
9 endmodule : m
```

Fig. 1.13 A simple **checker**

The example shown in Fig. 1.13 illustrates the concept of **checker**.

mytrig is a **checker** which gets three arguments: trig, prop, and clk. trig should be a sequence, prop should be a property, and clk should be an event. mytrig consists of assertion a1 checking that whenever trig happens prop is true (operator |->, called *overlapping suffix implication*), and a cover statement c1 monitoring whether trig happens.

The checker is instantiated in module m, Line 8, with actual arguments done, idle, and **posedge** clock. Even though done and idle are signals, it is valid to pass them to the checker as actual arguments because Boolean expressions are special cases of sequences and properties.

We describe checkers and their use in Chaps. 9, 23 and 24.

1.6 SVA and PSL

Besides SVA, PSL (Property Specification Language) [6] is another standard assertion specification language that is widely used in the industry. The goal of PSL is to provide a language subset for assertions that could work in conjunction with a variety of languages. To that end, the syntax is designed to be as neutral as possible, customized with a syntactic *flavor* for the individual language hosting the PSL features, such as SystemVerilog flavor and VHDL flavor. The semantics related to the integration of PSL with the host language is left open for the tools to define, to suit the environment of the tool.

Many of the PSL language features are semantically equivalent to those of SVA, but there are some differences of importance. One of the PSL features is the Optional Branching Extension (OBE) which defines operators for temporal properties in terms of branching time. The OBE features are meant only for formal verification and do not fit the simulation paradigm. SVA does not have the notion of branching time [35]; the time used in SVA is always linear (see Chap. 21), but all the linear time operators in SVA can be simulated. The OBE operators in PSL cannot be simulated.

PSL has an important mechanism of *vunits* (verification units) for encapsulating verification code. One vunit may inherit another in order to modify a portion of the verification environment. SVA has the **checker** construct and the **bind** statement. Vunits and checkers implement two different approaches to verification environment design: vunits are based on overriding and name matching, while checkers are based on argument mapping.

In contrast to PSL, SVA provides immediate and deferred assertions. The use and semantics of assertions in procedural code is undefined in PSL. Also, PSL lacks any notion of properties being invoked recursively. Since SVA is an integral part of SystemVerilog, its simulation semantics is well defined and SVA can be used much more widely within the context of SystemVerilog than is possible with PSL. Some examples of the benefits are:

- Sequences can be used outside the context of assertions.
- Integration with functional coverage features is powerful.
- Sampling of variables is precisely defined.
- Type compatibility and conversion is handled smoothly.

Exercises

1.1. What is the main difference between the assertion specification language in SystemVerilog and the language subset used for RTL description?

1.2. Modify the RTL code in Figs. 1.2 and 1.4 to take reset signal `rst` into account: when `rst` is asserted checking of active transactions should be stopped.

1.3. Implement the following assertion

```
assert property(@(posedge clk) req[*2] |=> grant[*2]);
```

in RTL: two consecutive requests should be followed by two consecutive grants.

1.4. What kinds of assertions exist in SVA? What is the difference between them?

1.5. Compare formal specification languages with natural languages. What are the advantages of formal languages?

1.6. What are the main advantages and disadvantages of checking assertions in (conventional) simulation?

1.7. Why is it useful to check assertions in emulation?

1.8. What are main advantages and disadvantages of checking assertions using formal verification?

1.9. Why is assertion reuse important? Which constructs exist in SystemVerilog for assertion reuse?

1.10. What is the intended use of checkers in SystemVerilog?

1.11. What are the main similarities and the main differences between SVA and PSL?

1.12. *Simultaneous reads and writes*

(a) Express a statement forbidding simultaneous reads and writes as an immediate, deferred and concurrent assertion. Reuse the common part in all assertions.
(b) Write a checker forbidding simultaneous reads and writes. Also check that both reads and writes actually happen.

1.13. *Request is always granted*

(a) Write a concurrent assertion stating that each request should be granted at the next cycle.
(b) Is it possible to express the same thing as an immediate assertion?
(c) As a deferred assertion?

1.14. Write the following assertion: When `reset` is deasserted it remains low forever.

Chapter 2
SystemVerilog Language Overview

The limits of my language mean the limits of my world.

— Ludwig Wittgenstein

Our objective in this chapter is to provide an overview of some SystemVerilog features that are important for understanding assertions, needed for writing assertions, or used in conjunction with assertions to support other tasks. The SystemVerilog features common with Verilog are considered known and are not discussed in this chapter.

We begin this chapter with a discussion about compilation and elaboration. Then we provide a brief overview of several SystemVerilog constructs that are not part of Verilog, and therefore may be not commonly known, but are referred later in this book. These constructs include structured procedures, clocking blocks, interfaces, programs and packages. The SVA-related constructs, such as checkers are not considered in this section, because they are studied in detail in next chapters.

The reader familiar with the material described here may safely skip this chapter. A detailed description of SystemVerilog language, of testbench writing, methodology and design is out of the scope of this book and can be found in the books [16, 30, 59, 61, 64]. The SystemVerilog Language Reference Manual [8] (referred in the rest of this book as LRM) is, of course, the most comprehensive reference.

2.1 Compilation and Elaboration

SystemVerilog code, prior to be simulated or formally verified by a tool, needs to be preprocessed and loaded. This initial step is divided into two major phases: compilation and elaboration. The goal of *compilation* is to read one or more source files, to

E. Cerny et al., *SVA: The Power of Assertions in SystemVerilog*,
DOI 10.1007/978-3-319-07139-8_2

perform syntactic and semantic analysis and to check for syntax or semantics errors. The source files can be compiled all at once or divided into multiple sets of files so that each set can be compiled separately into what is known as a *compilation unit*.

Typically, for simulation the compilation is performed by a separate tool, called a *compiler*. This allows compiling the SystemVerilog model only once and checking it multiple times on different tests. Some simulators, however, contain a compiler as their integral part and perform compilation anew for each simulation session.

To build the final simulation model, all compilation units together must go through the *elaboration* phase which binds all components by evaluating parameter values and constant expressions, connecting instances, building hierarchies, and resolving references. For noninterpretive simulators, another step is needed to create object code from the elaborated model to build the final executable simulation model.

It is important to distinguish between the computations performed at compile time, elaboration time and simulation time. For example, compiler directives, such as **`include**, **`define** and **`ifdef** are evaluated at compile time. As stated above, evaluation of constant expressions, including parameter evaluation is done at elaboration time. The same is true for generate constructs. On the other hand, assignment statements and procedural control statements are evaluated at simulation time. We clarify the difference between compile time, elaboration time and simulation time constructs with the following examples.

Example 2.1. Consider the following statements:

```
`define k 2 + 3
let m = 2 + 3;
int n = 2 + 3;
```

All of them seem very similar, and all of them evaluate to 5, but their evaluation time is different, and so is their semantics. Namely, `k` is evaluated at compile time, `m` is evaluated at elaboration time, and `n` is evaluated at simulation time.

The declarations `logic [`k:0] a;` and `logic [m:0] a;` are legal, but `logic [n:0] a;` is not, because the bounds of a vector must be known at the elaboration time.

The statement `` `ifdef k `` is legal, but both `` `ifdef m `` and `` `ifdef n `` are illegal, because **`ifdef** requires a macro identifier.

Note also that the compile time definition is processed syntactically whereas elaboration time and simulation time constructs are evaluated semantically. For example, `logic [2*`k:0] a;` is equivalent to `logic [7:0] a;` because `` 2*`k `` is substituted literally by `2*2 + 3` (therefore, it is recommended to put parentheses in the **`define** statement: `` `define k (2 + 3) ``). But `logic [2*m:0] a;` is equivalent to `logic [10:0] a;`, because the evaluation is done at elaboration time, and the values of `m` is first evaluated to 5, and only then it is multiplied by 2. For a more detailed discussion on the differences between **`define** and **let** statements see Sect. 8.1. □

Example 2.2. Generate **if** and procedural **if**. Consider the following module declaration:

```
1 module m #(bit FLAG = 1) (input logic in1, in2, select,
2                           output logic out1, out2);
3 if (FLAG) assign out1 = in1;
4 else assign out1 = in2;
5 always_comb
6   if (select) out2 = in1;
7   else out2 = in2;
8 endmodule : m
```

In this example the **if** statement on Line 3 is a generate **if**, because an **if** outside a procedural context is recognized as generate. This statement is processed at elaboration time using the value of FLAG, which is a module parameter. An attempt to put a non-constant expression, such as `select`, in this case is illegal.

The **if** statement on Line 6 is a normal procedural **if** statement. It is executed at simulation time, and its argument need not to be an elaboration time constant. The procedural **if** statement is recognized by its procedural context of the **always_comb** procedure. □

2.2 SystemVerilog Procedures

Most SystemVerilog design elements (modules, interfaces, programs and checkers) can contain *structured procedures*, previously called *procedural blocks*. There are the following structured procedures in SystemVerilog: initial, always and final procedures, functions and tasks. Some design elements have limitations on the procedures they may contain. For example, programs cannot contain always procedures; packages can contain no procedures.

Initial and general purpose always procedures are well-known and are not discussed here. We will limit our consideration by outlining specialized always procedures and by final procedures.

2.2.1 Specialized Always Procedures

In SystemVerilog there are several specializations of always procedures:

- **always_comb**
- **always_latch**
- **always_ff**

These procedures follow the same syntax as general always procedure though several limitations apply.

2.2.1.1 Procedure always_comb

The **always_comb** procedure is intended to represent combinational logic. It has an implicit sensitivity list including all variables read within the block or within functions called from this block. The **always_comb** procedure is automatically executed at time 0 to have its outputs consistent with its inputs. Consult the LRM for other rules and restrictions imposed on the **always_comb** procedure.

Example 2.3. The following code illustrates a mux implementation using **always_comb**.

```
logic in1, in2, sel, out;
always_comb
  if (sel) out = in2;
  else out = in1;
```

The signal `out` cannot be assigned anywhere else in the code because it is assigned within an **always_comb** procedure. □

2.2.1.2 Procedure always_latch

The **always_latch** procedure is intended to represent latched logic. From the formal point of view there is no difference between the **always_comb** and **always_latch** procedures, and the appropriate procedure should be chosen to clarify the design intent. However, the tools may impose additional checks. For example, unlike **always_latch**, they may enforce that the variables assigned inside **always_comb** are fully assigned.

Example 2.4. The following code illustrates a latch implementation using **always_latch**.

```
logic data, clk, q;
always_latch
  if (clk) q = data;
```

The signal `q` cannot be assigned anywhere else in the code because it is assigned inside an **always_latch** procedure. □

2.2.1.3 Procedure always_ff

The **always_ff** procedure is intended to represent sequential logic, such as flip-flops and registers. Unlike the general always procedure, the **always_ff** procedure contains exactly one event control and no blocking timing controls (such as delay control #). Similar to **always_comb** and **always_latch**, it is illegal to modify signals assigned in this procedure from any other place. Consult the LRM for other restrictions imposed on the **always_ff** procedure.

Example 2.5. The following code illustrates a flip-flop implementation using **always_ff**.

```
logic data, q, clk, rst;
always_ff @(posedge clk or negedge rst)
  if (!rst) q <= 1'b0;
  else q <= data;
```

The signal q cannot be assigned anywhere else in the code because it is assigned inside an **always_ff** procedure. □

2.2.2 *Final Procedure*

The **final** procedure is the opposite of the **initial** procedure. It executes at the end of the simulation. Often it is used as a clean-up routine and for displaying or storing information such as simulation final results, statistics, and coverage data. The users can declare more than one final procedure, in which case, they are executed sequentially, but in an arbitrary order.[1] Effectively, the final procedures constitute a single process in which the procedures execute sequentially. Because the final procedure executes in zero time, the statements allowed in a final procedure are those allowed in a function. Consult the LRM for an exact description of the final procedure.

Example 2.6. The following code illustrates the use of the final procedure to display the total number of assertion failures.

```
logic clk, rdy, rst;
int fCount = 0;
...
always @(posedge clk)
  rdy_fail: assert (rdy -> !rst) else fCount++;
final
  $display("Number of assertions rdy_fail failed: %d",fCount);
```

Assertion rdy_fail increments fCount each time it fails. At the end of simulation, the final procedure prints the total number of the assertion rdy_fail failures. □

[1]SystemVerilog LRM suggests that the order of execution for final procedures be deterministic for a tool.

2.3 Clocking Blocks

2.3.1 Clocking Block Declaration

The *clocking block* construct is aimed to provide a flexible scheme for synchronizing and sampling of signals with respect to a design clock. Often the input signals are driven from a testbench to design units, while the testbench is modeled by a *program* (see Sect. 2.5). New values to signals for a test are set in the testbench with appropriate delays using a clocking block.

It is possible to enclose property and sequence definitions in a clocking block, using the common event control of the clocking block. This assists in grouping related properties and sequences as well as it provides the convenience of leaving out the explicit specification of a clock for each individual declaration of a property or sequence in the clocking block.

For use of clocking blocks in testbenches, readers are advised to refer to books on SystemVerilog testbench (such as [59]) or to the LRM. Clocking declarations may be included in modules, interfaces, programs and checkers. Along with other information, a clocking block may be declared with

- A name
- An event expression
- Variables sampled and driven by the clocking block
- A list of sequences and properties

There is one restriction on the declarations of properties and sequences: no explicit clock is allowed in the declarations. Consequently, multiply clocked properties or sequences cannot be declared inside clocking blocks. Note also that assertions cannot be written within a clocking block.

Example 2.7. The code below contains a clocking block named `cblk` declared with the clocking event `@(posedge clk)`.

```
logic req, gnt;
  // Evaluation of gnt
  ...
clocking cblk @(posedge clk);
  input  req, gnt;
  property p_req_granted;
        req |=> gnt;
  endproperty
endclocking
a1: assert property (cblk.p_req_granted)
  else $error("Request not granted");
a2: assert property(@(posedge clk) req |=> !req)
  else $error("Request asserted during two consecutive cycles");
```

Variables req and gnt are used as inputs in the clocking block and they get sampled with the clocking event. The property p_req_granted is declared in the clocking block. Because a clocking block creates a scope, the assertion a1 refers to property req_granted using its hierarchical name as cblk.req_granted. The assertion a1 inherits its clock from the property cblk.req_granted (see Sect. 12.2.6), and therefore no explicit clock specification in assertion a1 is needed.

The property req |=> !req is written directly in the body of assertion a2, and not declared in the clocking block. Therefore assertion a2 must specify its clocking event explicitly. □

2.3.2 *Default Clocking*

One clocking block may be specified as default for a given design element. Specifying a default clocking allows to omit clock specification in concurrent assertions belonging to the same design element (e.g., module, interface, etc.). In this case the clocking event of the default clocking becomes an implicit clock of the assertion. When the default clocking is used only for assertion purposes, its body is usually left empty and no name is given to the default clocking block.

Example 2.8. The following code illustrates a typical usage of the default clocking with assertions.

```
module m (input logic req, clk, output logic gnt);
default clocking @(posedge clk); endclocking
// Evaluation of gnt
...
property p_req_granted;
  req |=> gnt;
endproperty
a1: assert property (p_req_granted)
  else $error("Request not granted");
a2: assert property(req |=> !req)
  else $error("Request asserted during two consecutive cycles");
endmodule : m
```

In this example the body of default clocking is empty and this block is left anonymous. We could provide it a name:

```
  default clocking dfltclk @(posedge clk); endclocking
```

but it serves, essentially, no purpose. The concurrent assertions a1 and a2 do not have an explicit clock specified with them, and therefore a clocking event @(posedge clk) from the default clocking block is inferred for them.

Note the difference between this example and Example 2.7. In Example 2.7 the property `p_req_granted` is written in the scope of a clocking block, and to instantiate it assertion `a1` needs to specify the name of the clocking block `cblk` as part of the hierarchical property name. In this example property `p_req_granted` does not belong to the scope of the clocking block, and its name is directly visible by assertion `a1`. In Example 2.7 the clocking event of the clocking block is inferred for property `p_req_granted`, and assertion `a1` inherits this clock from *the property*. Assertion `a2` there has to specify its clock explicitly. In the current example the clocking event of the default clocking is inferred for all assertions in the module, because no assertion here has an explicitly specified clock. □

2.4 Interfaces

This section contains an introduction to *interfaces*—design elements aimed to encapsulate the communication between modules. The reader can find the details in [61] and the LRM. The interface groups signals together, and it may also contain assertions and communication logic. Unlike the modules, interfaces may be passed via ports as a group.

The contents of an interface are not limited to signals. In fact, it may contain most entities allowed in modules such as

- Data types and variables
- Clocking blocks, functions and tasks
- Initial, always, and final procedures to define additional behavior
- Sequences, properties, and assertions

In the simplest case the interface is just a named bundle of signals, as shown in Example 2.9.

Example 2.9. Consider two modules: `producer` and `consumer`. Their communication is implemented via interface `data`.

```
1  interface data (input logic clk);
2  logic [7:0] a, b;
3  logic active_a, active_b;
4  endinterface : data
5
6  module producer (data i);
7    initial begin
8      i.active_a = 1'b1;
9      i.active_b = 1'b0;
10   end
11   always @(posedge i.clk) begin
12     i.active_a <= !i.active_b;
13     i.active_b <= !i.active_a;
14     i.a <= $random;
15     i.b <= $random;
16 end
17 endmodule : producer
18
19 module consumer (data i, output logic [7:0] out);
20   always @(posedge i.clk) begin
21     if (i.active_a) out <= i.a;
22     if (i.active_b) out <= i.b;
23   end
24 endmodule : consumer
25
26 module top;
27   logic clk = 0;
28   logic [7:0] result;
29   initial repeat (100) #5 clk = !clk;
30   data ifc(clk);
31   producer prod(.i(ifc));
32   consumer cons(.i(ifc), .out(result));
33 endmodule : top
```

In this example interface `data` (Lines 1–4) is just a collection of signals. The values of these signals are generated by the module `producer` (Lines 6–17) and then the values of `a` and `b` are routed to the output by the module `consumer` (Lines 19–24). These modules and the interface are instantiated in module `top` (Lines 26–33). Instead of individual signal and port declarations for `a`, `b`, `active_a` and `active_b` in the modules, all these declarations are packed in the interface. Individual signals from the interface are accessed directly from the interface instance. □

Interfaces are a natural place to put assertions or to instantiate checkers to specify the obligations imposed on the modules on one side of the interface and the expectations of the modules on its other side (see Sect. 1.2.2). This allows to avoid duplication from specifying the same assertions at the boundaries of each connected module.

Example 2.10. An assertion checking that both `active_a` and `active_b` are not asserted together can be added to interface `data` from Example 2.9:

```
interface data (input logic clk);
logic [7:0] a, b;
logic active_a, active_b;
a_active: assert final (!(active_a && active_b))
  else $error("a and b cannot be active simultaneously");
endinterface : data
```

Putting assertion `a_active` in interface `data` avoids its duplication if it were put in the modules. In the latter case it should have been written twice: once as an assertion in the module `producer`, and once as an assumption in the module `consumer`. Of course, the same effect could be achieved by putting this assertion in the module `top`. However, putting the assertion in the interface is preferable: this assertion will be checked in the unit-level testing or formal verification of component modules, and not only in the integration testing.

Discussion: Module `consumer` observes, but does not drive, the interface signals `active_a` and `active_b`. Therefore, in formal verification of `consumer` it is desirable to treat `a_active` as an assumption rather than as an obligation. Formal verification tools generally treat **`assert`** statements as obligations and **`assume`** statements as assumptions, but they also typically provide controls for overriding the default interpretation. A tool may also provide an automatic mechanism for characterizing an assertion based on its signal support, e.g. inferring that an **`assert`** referencing only inputs be treated as an assumption rather than as an obligation. □

2.5 Programs

For writing testbenches SystemVerilog provides special design elements, called *programs*. Their primary purpose is to generate and send stimuli to primary inputs of a DUT and to receive its responses to validate the design behavior. A program is similar to a module construct in its declaration of ports and body, but has a number of limitations. For example, a program may contain continuous assignments, initial and final procedures, but not always procedures. Programs may instantiate checkers, but not other design elements (e.g., modules or other programs). The main program feature distinguishing programs from modules is their simulation semantics intended to avoid most races between the design and the testbench (see Chap. 3). A detailed description of programs the reader can find in [59] and in the LRM.

Example 2.11. We assume that the module `router` consumes request packets and issues acknowledgment packets in the next clock cycle upon a request packet reception. The test environment generating 100 request packets with random IDs and random data is implemented by the program `test`.

```
typedef enum logic {REQ = 1'b1, ACK = 1'b0} dirType;

typedef struct packed {
  dirType rq;
  logic [6:0] id;
  logic [23:0] data;
} packetType;

program test (input logic clk, packetType received,
              output packetType sent);
logic [6:0] sent_id;
initial begin
  repeat (100) begin
    @(posedge clk);
    sent_id = $random;
    sent = '{REQ, sent_id, $random};
    @(posedge clk);
    a1: assert final (received.rq == ACK)
      else $error("Corrupted packet");
    a2: assert final (received.id == sent_id)
      else $error("Lost packet");
  end
end
endprogram : test

module router (input packetType inpkt, logic clk,
               output packetType outpkt);
...
endmodule : router
module top;
  logic clk = 1'b0;
  initial repeat (400) #5 clk = !clk;
  packetType inpkt, outpkt;
  test t(.clk(clk), .received(outpkt), .sent(inpkt));
  router r(.*);
endmodule : top
```

Program `test` checks also the correctness of the received packets. Assertion `a1` checks that the received packet is an acknowledgment packet, and assertion `a2` checks that the ID of the acknowledgment packet coincides with the ID of the request packet sent in the previous cycle. □

2.6 Packages

The design element *package* is intended for encapsulation and reuse of common declarations. One can use packages as libraries of useful declarations, such as type definitions, function, task, let declarations, properties, sequences, and checkers. A package introduces its own name space to avoid conflicts with local names where a package element is referenced.

Example 2.12. A package is a natural container for common let declarations, sequences, properties and checkers to be used project-wide, such as the package `common_props` shown below.

```
package common_props;

  let one_cold(sig) = $onehot(~sig);

  property req_granted (sequence req, property gnt);
    req |=> gnt;
  endproperty : req_granted
  ...
endpackage : common_props
```

In this package it is shown a let declaration for one cold encoding of the bits of `sig`, and a property declaration to check that the request `req` is granted (`gnt` is received) in the next clock cycle. In order for the property to be generic, `req` and `gnt` are not limited to signals, but `req` may be an arbitrary sequence, and `gnt` may be an arbitrary property (see Sect. 6.4). □

To reference a declaration from a package one has to provide the package name followed by `::` within the name declared in the package.

Example 2.13. The code below shows how to use declarations from the package `common_props` from Example 2.12.

```
module m (input logic rq, done, clk, ...);
  wire [7:0] data;
  ...
  a1: assert final (common_props::one_cold(data));
  ...
  a2: assert property (@(posedge clk)
                        common_props::req_granted(rq, done));
  ...
endmodule : m
```

In module `m` the let and the property names from the package are prefixed with the package name `common_props::`. □

To make specific names declared in a package visible in the current scope, one can use an *import* statement containing the list of names to be referenced. Alternatively, one can use a wild card import to make all construct names belonging to the package visible in the current scope.

Example 2.14. Instead of explicit specification of the name of package `common_props` defined in Example 2.12 one can use an import statement in module `m` from Example 2.13:

```
module m (input logic rq, done, clk, ...);
  import common_props::one_cold, common_props::req_granted;
  wire [7:0] data;
  ...
  a1: assert final (one_cold(data));
```

```
  ...
  a2: assert property (@(posedge clk) req_granted(rq, done));
  ...
endmodule : m
```

It is also possible to import all the names defined in the package `common_props` into module `m` using the wild card notation:

```
module m (input logic rq, done, clk, ...);
  import common_props::*;
  wire [7:0] data;
  ...
  a1: assert final (one_cold(data));
  ...
  a2: assert property (@(posedge clk) req_granted(rq, done));
  ...
endmodule : m
```

The latter version looks more convenient, but the wild card notation may lead to a clutter with the local name space and with namespaces of other packages when multiple wild card imports are used. □

Exercises

2.1. What are the sizes of wires `x` and `y`?

```
`define a 3 + 5
module m (...);
  let b = 3 + 5;
  wire [`a * 4] x;
  wire [b * 4] y;
  ...
endmodule : m
```

What will happen if the statement `let b = 3 + 5;` has been substituted by `int b = 3 + 5;`?

2.2. What is wrong with the following code?

```
module m (input logic a, b,  c, clk, output logic x, y, z);
  initial begin
    y = 1'b0;
    z = 1'b0;
  end
  always_comb x = a & b;
  always_latch y = a | c;
  always_ff @(posedge clk) begin
    if (y) x <= a;
    @(posedge clk) x <= x & b;
  end
  final @(posedge clk) z <= x | y;
endmodule : m
```

2.3. Correct the following code to make it legal.

```
module m(input logic a, b, c, output logic o);
  if (a) o = b;
  else o = c;
endmodule : m
```

2.4. What is illegal in the following code?

```
module m (input logic a, b, clk);
  clocking mycblk @(posedge clk);
    property p; a |=> b; endproperty
    a1: assert property (p);
  endclocking
  a2: assert property (p);
endmodule : m
```

2.5. What is the purpose of the following SystemVerilog design elements?

- module
- interface
- program
- checker
- package

Chapter 3
SystemVerilog Simulation Semantics

Imitation is at least 50 percent of the creative process.

— Jamie Buckingham

In this chapter we discuss SystemVerilog simulation semantics: how the SystemVerilog model simulation is performed. This is necessary to understand the simulation semantics of assertions described further in this book. We cover simulation semantics of the language features only as needed to provide a frame of reference for completing the discussion on semantics of assertions. The exhaustive description of SystemVerilog simulation semantics may be found in the LRM. See also books [59, 61].

3.1 Event Based Simulation

SystemVerilog constructs do not exist in isolation and the interaction between them in simulation is rather complex. The description of their behavior in simulation is known as the *simulation semantics* of the language. The simulation semantics of SystemVerilog is described in terms of *events* and processes. Examples of processes are structured procedures (initial and all kinds of always procedures, see Sect. 2.2.1), continuous assignments, etc. All processes are scheduled concurrently, i.e., they may be executed in parallel. However, in many cases the result of process evaluation should be such as if a specific order of process evaluation were imposed.

E. Cerny et al., *SVA: The Power of Assertions in SystemVerilog*,
DOI 10.1007/978-3-319-07139-8_3

Example 3.1. The module `procReq` contains a series of assignments:

```
1 module procReq(input logic req, gnt1, gnt2, clk);
2   wire tmp, proceed;
3   logic allow;
4   assign tmp = allow & gnt1;
5   assign proceed = tmp & gnt2;
6   always @(posedge clk) allow <= req;
7   always @(posedge proceed) processData();
8 endmodule : procReq
```

Assuming that input `clk` transitions from 0 to 1 and no other input changes at that time, the result would be as if the following evaluation order were imposed:

1. Assignment to `allow` in Line 6
2. Assignment to `tmp` in Line 4
3. Assignment to `proceed` in Line 5
4. Evaluation of the subroutine (task or function) `processData` if `proceed` becomes true in Line 7

As we will see in this chapter, this order of evaluation is obtained by creating events, scheduling events, and performing the computations directed by the scheduled events, all carried out in the order established by the semantic framework to obtain the intended result. The parallelism between the continuous assignments and always statements in this example is broken down into ordered discrete events. Thus, in this case, the parallelism is unrolled into a sequential order as directed by the occurrence of events. In other cases, true parallelism may exist between statements, allowing indeterminate order of statement execution and values of variables. □

Another important facet of SystemVerilog is Programming Language Interface (PLI) (or its newer version VPI) described in the LRM, which provides an interface from the evaluation of language constructs to the external environment using other programming languages or scripts. The interface is used to inspect values, change values or get callbacks. There are certain points in the semantic structure where specific groups of VPI functions are allowed to take place. We, however, do not delve into the details of that allotment. The rest of the semantics are largely unaffected by its exclusion.

3.2 The Simulation Engine

There are two types of events that help explain the event-driven simulation engine: *update event* and *evaluation event*. This notion of event should not be confused with **`event`** construct in SystemVerilog which is a data type used to name and trigger events. An update event occurs whenever there is a change in the value of a variable. The update event may trigger other activities and events dependent on the change in value.

Example 3.2. In Example 3.1, the nonblocking assignment

```
allow <= req;
```

causes an update event if the value of variable `allow` changes as a result of the assignment. There are many language constructs whose execution is tied to the occurrence of update events. The continuous assignment statement

```
assign tmp = allow && gnt1;
```

in Example 3.1 is executed when and only when either the update event on `allow` or `preGrant` occurs. The always procedure

```
always @(posedge clk) allow <= req;
```

executes when and only when the event **posedge** `clk` occurs. □

The execution of a statement may not materialize immediately, but be scheduled as an evaluation event in a queue corresponding to a region for execution, based on the type of the statement and its context. By scheduling evaluation events in various queues and by executing them later from the queues, the intended order between statements is accomplished. The execution of an evaluation event can result in further update events or evaluation events which are again scheduled. The creation and execution of these events, together with their scheduling in queues is what keeps the simulation engine running. As long as there are scheduling events left to process, the engine keeps executing statements and progresses through time. The simulation ends only when there are no more events left in the queues.

Example 3.3. When the evaluation event for the always procedure

```
always @(posedge clk) allow <= req;
```

is executed, an evaluation event for the nonblocking assignment `allow <= req;` emerges and gets scheduled. When this evaluation event for the nonblocking assignment is executed, an update event for variable `allow` is issued. □

3.3 Bringing Order to Events

Now, we can see the important role of queues in the assembly of discordant events into a predictable simulation execution model. First, we focus on the execution of statements specified in the context of design code, rather than assertions, or programs that represent testbench code. Events issued from the design code are grouped in a *region set* called the *Active* region set. We elaborate upon the notion of a region set, including the Active region set and other region sets, in Sect. 3.5. For now, we limit our discussion to the activities within the regions of the Active region set and to the queues they contain.

The queues represented in Fig. 3.1 belong to the regions from the Active region set. There are three principal regions and queues corresponding to them in this region set: Active region, Inactive region and NBA region.

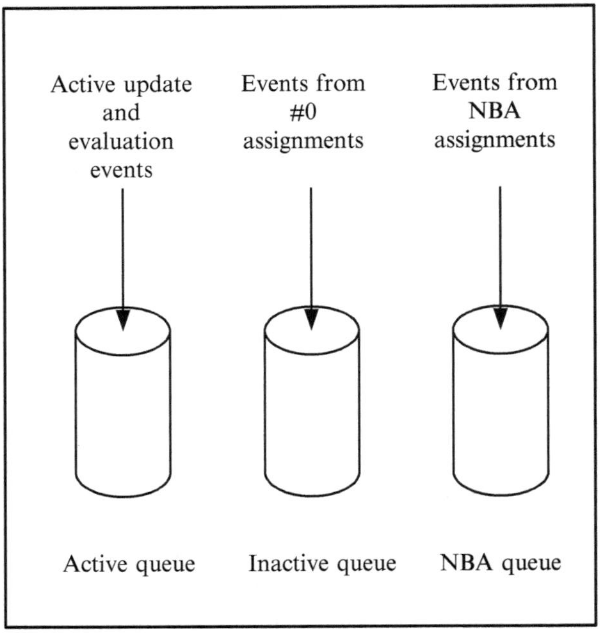

Fig. 3.1 The Active region set

As the name suggests, the Active region queue contains events pending for immediate execution. Events in this queue may be executed in any order, implying parallelism between events. After an event from the queue has been executed, such as updating a variable value, it is removed from the queue. Initially at time 0, the initial processes are scheduled in the Active region queue.

If a statement is encountered with `#0` as the delay control, an evaluation event for the statement is entered into the Inactive region queue. When all events from the Active region queue have been executed with the queue becoming empty, the events from the Inactive queue are transferred to the Active region queue, resulting in an empty Inactive queue. The execution of the events from the Active region queue resumes once again.

Finally, if the execution encounters a nonblocking assignment, the expression on the right-hand side of the assignment is evaluated and an update event for the variable on the left-hand side with the computed value is appended to the NBA region queue. The events from this queue start execution only when the Active and Inactive region queues are empty. Unlike the Active region queue, the events in the NBA region queue are executed in the same order as they are entered in the queue. Therefore, the sequential order of nonblocking assignments in a procedure is replicated in the queue and those statements are executed in order.

Ordinarily, nonblocking assignments model register transfer statements triggered by the clock derived from its enclosing always procedure. As a result, the register stores the new value and propagates it. If there is a combinational logic driven by the register, its value is further propagated by the new update event entered in the Active region queue. Algorithmically, the following steps are taken.

1. Execute events from the Active region queue. New events may be issued and entered in the appropriate queues. Execute until all events are consumed.
2. Transfer events from the Inactive region queue to the Active region queue and return to Step 1. Skip this step if there are no events in the Inactive queue.
3. Execute events from the NBA region queue. New update events issued from this execution are entered in the Active region queue. Return to Step 1.

The above algorithm is iterated until there are no more events left in any queue in the current time.

3.4 Determinism and Nondeterminism

As we saw in the previous section the order of event execution is not always deterministic. The simulators normally will order the events in an optimal way to avoid redundant computations, but one cannot rely on that. RTL should be well-formed: the results of its simulation should be deterministic regardless of the actual execution order of events.

Example 3.4. To illustrate the concept of determinism, consider the following continuous assignments:

```
wire a, b, c, d;
...
assign a = b & c;
assign b = d;
```

Assume that nets `c` and `d` received new values. Then the continuous assignments to `a` and `b` are scheduled for execution. The order of their execution is arbitrary. If the assignment to `a` executes first, then it is scheduled for execution again after execution of the assignment to `b` since its left-hand side depends on `b`. At the end of the simulation step (assuming that the values of `c` and `d` do not change anymore) `a` and `b` receive the same values regardless of the actual order of assignment execution and update event processing. Normally, the simulators will first execute the second assignment and then the first one to avoid redundant computations. □

The determinism does not guarantee the absence of glitches. A *glitch* is a situation when a signal retains the same value at the end of the simulation tick as it has at its beginning, but the value changes somewhere in the middle of the simulation tick.

Example 3.5. The following code may result in a glitch in the signal `a`.

```
wire a, b, c;
...
assign a = b | c;
```

Assume that the value of `b` changes from 1 to 0, and the value of `c` changes from 0 to 1. Since at the beginning of the simulation tick the values of `b` and `c` were 1 and 0, correspondingly, the value of `a` was 1. At the end of the simulation tick (assuming no more value transitions for `b` and `c`), the values of `b` and `c` are 0 and 1, correspondingly, and therefore, the value of `a` remains 1. In the middle, however, the situation may be different.

If both `b` and `c` change prior to evaluation of `a` or if `c` changes first, the value of `a` remains unchanged during the entire simulation tick. If the order of evaluation is `b`, `a`, `c`, `a`, then the first time `a` gets value of 0 and the second time it becomes 1 again. In the latter case we have a glitch: though the initial and the final value of a is 1 in this simulation tick, it is changed to 0 in the middle of the simulation tick. □

If the RTL is not well-formed, the final result may depend on the evaluation order. Such a situation is called a simulation *race*.

Example 3.6. The following code is an example of a simulation race.

```
logic clk, a, b;
...
always @(posedge clk) a <= b;
```

Assume that both `b` and `clk` transition from 0 to 1. If the value of `b` changes *before* the change of `clk`, `a` will receive 1, the *new* value of `b`. If the value of `b` changes *after* the change of `clk`, `a` will retain 0, the *old* value of `b`. Thus the final value of `a` depends on the actual order of the evaluation of events. □

3.5 Region Sets

The regions Active, Inactive and NBA are all confined to the Active region set. The execution iterates over the queues in a region set until all events scheduled in any of its region queues are executed. Certain events, however, may be scheduled in the other regions.

The processing of events is sequentially ordered into distinct regions, where each region manages and executes events that are scheduled in the region. The processing proceeds from one region to the next and can iterate until no further processing is needed in any region for the given time step.

We discuss the role of the following regions and region sets. Other regions are not important for the understanding of assertion semantics.

1. Preponed region
2. Active region set
3. Observed region
4. Reactive region set
5. Postponed region

The processing of regions also takes place in the order as shown above.

One region differs from another because of the kind of events that are handled by it. As we saw in the previous section, the Active region set handles events from the design code. As we will explain below, concurrent assertion evaluation is performed in the Observed region. An update event on a port connection to a program schedules an evaluation event in the Reactive region set. Nevertheless, events scheduled in other regions cannot be executed from a region which is currently being processed. This makes a region safe from execution interference of statements that belong to a semantically different region.

The Preponed region is a precursor to the time slot.[1] No value changes or events occur in this region. In most cases sampling of signal values used in concurrent assertions is performed in the Preponed region as explained in Sect. 4.4.3. On the contrary, the Postponed region is the tail end of the time slot meant for finishing simulation tasks in this slot that do not include value changes or events. In this region the action blocks for final assertions are executed. Both of these regions are entered only once.

There are other important regions in the simulation engine to support System-Verilog features: the Observed region and the regions belonging to the Reactive region set.

The Observed region is meant for the evaluation of sequences, properties and concurrent assertions. Signal values remain constant during the Observed region. The evaluation mechanism and the queues that reside in this region are quite different than in the Active region. Nevertheless, events originated in this region do get scheduled into the Active and Reactive regions.

The Reactive region set executes statements from programs and checkers. Programs are intended for writing testbenches, as external environments for designs, feeding stimuli, observing design evaluation results and building tests to exercise the design (see Sect. 2.5). This region set is a mirror image of the Active region set, with similar events, regions, queues and statement execution to the Active region set. The corresponding regions and queues are called Reactive, Re-Inactive, and Re-NBA. These regions can also schedule events to the Active region.

The simulation engine processes one region at a time, and transitions from one region to the next only after exhausting events and evaluations in a region. The order of the movement between the regions is fixed as follows: regions from the Active region set, the Observed region, and regions from the Reactive region set. The iterative motion between the regions continues until there are no more events or evaluation tasks left in any region. The regions and their order are depicted in Fig. 3.2.

[1] We use terms *time slot*, *time step* and *simulation step* interchangeably.

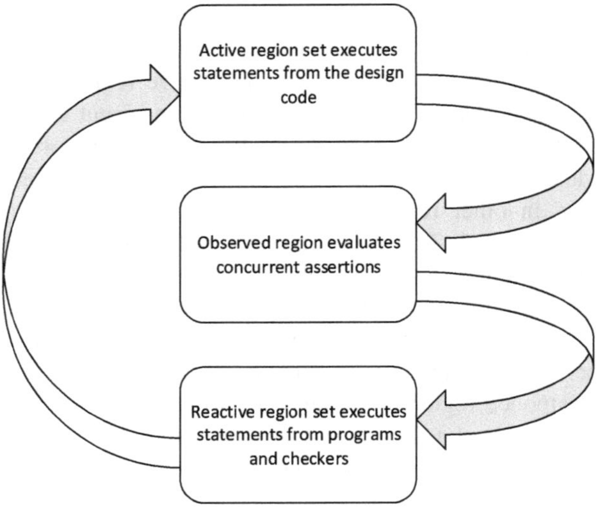

Fig. 3.2 Three main regions and sets

Simulation keeps running until there are no more events to execute.

Example 3.7. To explain how simulation is performed across these regions, consider module `procReq` processing requests and program `test` generating stimuli for it.

```
module procReq  (input logic req, gnt, clk);
  logic allow;
  wire proceed;
  assign proceed = allow && gnt;
  always @(posedge clk) allow <= req;
  always @(posedge proceed) processData();
  a1: assert property(@(posedge clk) req |=> proceed || !gnt);
endmodule : procReq

program test(input logic sync, output logic request, grant);
  logic oldreq = 1'b0;
  assign grant = oldreq;
  initial begin
    request = 1'b0;
    for (int i = 0; i < 50; i++) begin
      @(posedge sync);
      oldreq <= request;
      request <= $random;
    end
  end
endprogram : test

```

```
23 module top();
24   logic r, g;
25   logic clock = 1'b0;
26   always #5 clock <= !clock;
27   procReq dut(r, g, clock);
28   test tb(clock, r, g);
29 endmodule : top
```

Signal value changes ordered by regions until time 5 are shown in Table 3.1.[2] See also Exercise 3.3.

Table 3.1 Signal value changes (Example 3.7)

Initialization
`allow` = X (Line 2), `proceed` = Z (Line 3),
`req` = X, `gnt` = X, `clk` = X (Line 1)
`r` = X, `g` = X (Line 24), `clock` = 0 (Line 25),
`sync` = X, `request` = X, `grant` = X (Line 10), `oldreq` = 0 (Line 11)
Time 0
Active region
`clk` = 0 (Line 1), `proceed` = X (Line 4), `sync` = 0 (Line 10)
Reactive region
`grant` = 0 (Line 12), `request`= 0 (Line 14)
Active region
`gnt` = 0, `req` = 0 (Line 1), `proceed` = 0 (Line 4)
Time 5
NBA region
`clock` = 1 (Line 26),
Active region
`clk` = 1 (Line 1), `sync` = 1 (Line 10)
NBA region
`allow` = 0 (Line 5)
Observed region
Assertion `a1` evaluation. No failure detected (Line 7)
Re-NBA region
`oldreq` = 0 (Line 17),
`request` = 1 (Line 18, it is assumed that `$random` returns an odd number)
Active region
`req` = 1 (Line 1)
Time 10
...

[2]The LRM is ambiguous about net initialization: *The default initialization value for a net shall be the value Z*. Our understanding is that in the beginning of simulation all nets are initialized with Z; hence `proceed` initially assumes the value of Z. However, not all simulators implement the net initialization this way.

□

As we could see, the normal flow between regions: Preponed → Active → Inactive → NBA → Observed → Reactive → Re-Inactive → Re-NBA → Postponed often complicates by iterations between regions. For example, the computations in the Reactive region may schedule a new event in the Active region. It is also possible that the assertion evaluation in the Observed region is triggered by events in the Reactive region. One such scenario is when an assertion clock is modified in the Reactive region set.

Example 3.8. We will modify the code from Example 3.7 to generate the clock by the program `test`, instead of the module `top`.

```
1  module procReq(input logic req, gnt, clk);
2    logic allow;
3    wire proceed;
4    assign proceed = allow && gnt;
5    always @(posedge clk) allow <= req;
6    always @(posedge proceed) processData();
7    a1: assert property(@(posedge clk) req |=> proceed || !gnt);
8  endmodule : procReq
9
10 program test(output logic request, grant, sync);
11   logic oldreq = 1'b0;
12   assign grant = oldreq;
13   initial begin
14     request = 1'b0;
15     sync = 1'b0;
16     for (int i = 0; i < 100; i++) begin
17       #5 sync <= !sync;
18       if (i % 2) begin
19         oldreq <= request;
20         request <= $random;
21       end
22     end
23   end
24 endprogram : test
25
26 module top();
27   logic r, g, c;
28   procReq dut(r, g, c);
29   test tb(r, g, c);
30 endmodule : top
```

Since the clock is generated in the Re-NBA region (Line 17), it schedules the event in the Observed region which will be executed after the completion of the Re-NBA region. Thus, the order of the simulation regions at time 5 in this case will be: Reactive region set → Active region set → Observed region. See also Exercise 3.4.

□

3.6 A Time Slot and the Movement of Time

In Example 3.7, we noted that the events get scheduled at different times, such as at time 0 and time 5. Each event is associated with a simulation time, which is the time maintained by the simulator to account for the delays in the design. Without the delays in the design, the simulation time will not advance and all events will occur at time 0.

The time delays in the system are specified with a scale and a precision. Nonetheless, time is discrete and there exists a global time precision which is the smallest unit of time in the system being simulated. *1step* denotes the smallest time precision.

Example 3.9. In the module `m` below, the smallest unit of time (one step) is `1ns`.

```
module m(...);
  timeunit 1ns;
  ...
endmodule : m
```

□

We have seen how the event queues and the regions establish the order of processing within a time slot. The time within a time slot remains constant, and thus, all events scheduled within a time slot refer to the same time. When all events are processed for a time slot, the simulation control moves to the nearest time slot containing scheduled events.

Example 3.10. Consider the program `test` declaration from Example 3.8.

```
program test(output logic request, grant, sync);
  logic oldreq = 1'b0;
  assign grant = oldreq;
  initial begin
    request = 1'b0;
    sync = 1'b0;
    for (int i = 0; i < 100; i++) begin
      #5 sync <= !sync;
      if (i % 2) begin
        oldreq <= request;
        request <= $random;
      end
    end
  end
endprogram : test
```

When the execution reaches the statement `#5`, an evaluation event is scheduled for a future time slot. For example, if the current time is 10, the next statement is scheduled for time step 15 in the Reactive region set. When the simulation control transitions to time 15, the scheduled event gets executed, etc. □

The time advances only when the events of the current time slot are exhausted.

3.7 Simulation Semantics of Assignments

We briefly review the impact of assignments on scheduling events. We assume that the assignments are executed in the Active region set. This happens, for example, for assignments belonging to a module. The same rules apply to assignments belonging to a program with an only difference that they are executed in the Reactive region set.

- A continuous assignment schedules an update event in the Active region to update the value of its left-hand side, whenever there is a change in the right-hand side expression value.
- A blocking assignment without a delay executes immediately, and issues update events for statements dependent on the new value of the left-hand side. An assignment with 0 intra-assignment delay computes the right-hand side and schedules an evaluation event in the Inactive region to make the assignment, issue other update events if necessary, and continue the sequential execution from that statement. For a greater delay, it schedules like for 0 delay in the Active region for the future time.
- A nonblocking assignment schedules an update event in the NBA region to update the left-hand side based on the current value of the right-hand side. For a delayed statement, it schedules the event for a future time in the NBA region based on the delay.

The event processing rules discussed in this chapter apply to the Active and Reactive regions as explained with the use of their corresponding queues. In the Observed region, only certain special statements are executed. Largely, the evaluation of concurrent assertions is carried in the Observed region. Chapters 4 and 14 discuss assertion simulation semantics in more detail.

Exercises

3.1. Identify all potential cases of glitches and races in the following code. Assume that `clk` changes at most once per time step.

```
module m (input logic in, clk);
  wire w, not_w, w_or_not_w;
  logic next_w, v, not_v, v_or_not_v, next_v;
  logic onev = 1'b1, onew = 1'b1;
  assign w = in;
  assign not_w = !in;
  assign w_or_not_w = w || not_w;

  always @(posedge clk) begin
    next_w <= w;
    onev <= v_or_not_v;
    onew <= w_or_not_w;
  end
  always_comb begin
    v = in;
    not_v = !in;
    v_or_not_v = v || not_v;
  end
  a1: assert property (@(posedge clk) w_or_not_w);
  a2: assert property (@(posedge clk) v_or_not_v);
endmodule : m
```

3.2. Explain why SystemVerilog requires that all assertion clocks tick at most once per simulation cycle (= time step).

3.3. Continue Table 3.1 in Example 3.7 for time steps 10 and 15. Assume that in time step 15 `$random` returns an even number.

3.4. Fill a table analogous to Table 3.1 for the code in Example 3.8.

Part II
Basic Assertions

Chapter 4
Assertion Statements

Language is a mixture of statement and evocation.

— Elizabeth Bowen

In this chapter, we describe SVA assertion statements:[1]

- Assert statements
- Assume statements
- Restrict statements
- Cover statements

The term *assertion* is overloaded in SVA; in a narrower sense it means an **`assert`** statement, and in a broader sense it means any assertion statement listed above. In this chapter, we use the term assertion in its narrow meaning. We indicate the meaning explicitly when it is not clear from the context.

Assertion, assumption, and cover statements may be of the following kinds: immediate and concurrent. Immediate assertions are further subdivided into simple immediate and deferred. The deferred ones are of two kinds—observed and final. They are used when filtering of 0-width glitches is needed in simulation. Restrict statements may only be concurrent.

For convenience, we briefly recapitulate main results of Sect. 1.4, how different kinds of assertion statements are checked in simulation and in formal verification. We then describe basic simulation semantics for different kinds of assertion statements. Their understanding is important to correctly choose the kind of assertions in each particular case. Knowing principles of assertion simulation is also

[1]In SVA, there is also **`expect`** statement used mostly in testbenches, but we do not describe it in this book.

E. Cerny et al., *SVA: The Power of Assertions in SystemVerilog*,
DOI 10.1007/978-3-319-07139-8_4

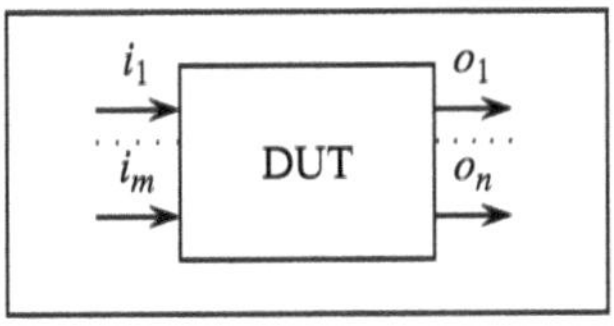

Fig. 4.1 DUT and its environment

important to correctly interpret simulation results on traces and waveform diagrams. The simulation semantics for concurrent assertions outside procedures is discussed in Sect. 4.4.6 and for procedural assertions in Sect. 14.5.

4.1 Assertion Kinds

Conventionally, the code representing the hardware design as the object of verification is called *Device Under Test* (DUT), while the testbench and other supplementary code is called *environment*, as shown in Fig. 4.1. For example, a CPU as a DUT could have an environment consisting of a program generating the stimuli, a chipset model and a memory model.

Assertions specify the desired behavior of DUT for checking. Black-box assertions specify relationships between DUT inputs $i_1, \ldots, i_m$ and DUT outputs $o_1, \ldots, o_m$, while white-box assertions specify relationships between internal signals.

As we mentioned in Sect. 1.3, in SystemVerilog there are two main kinds of assertions:

- Immediate
- Concurrent

The immediate assertions are

- Simple
- Deferred

The deferred assertions can further distinguish whether to filter glitches that arise due to statement execution in the *active region*, the so-called *observed deferred assertions*, or arising also due to execution in the *active* and *reactive* regions, the so-called *final deferred assertions*.

If not specified otherwise, in the rest of the book we refer to *simple immediate* assertions as *immediate*, and to *observed and final deferred* assertions as *deferred*.

Assertions have the following syntax.

assertion ::= name: assert_keyword (assertion_body) action_block

The syntax for different types and kinds of assertions differs only by the keyword. *assert_keyword* is **assert** for simple immediate assertions, **assert** #0 for observed deferred assertions, **assert final** for final deferred assertions, and **assert property** for concurrent assertions. *assertion_body* is a non-temporal expression for immediate and deferred assertions, and a temporal expression for concurrent assertions. An action block contains code to be performed in case of assertion success (*pass action*) and failure (*fail action*). Both pass and fail actions of immediate and concurrent assertions (but not deferred) may be blocks with several statements each. Restrictions apply to the statements that can appear in action blocks of deferred assertions. If no fail action is specified with an assertion, an `$error` system task is called with a default tool specific error message. Although assertion name is optional, it is highly recommended to always specify it. The significance of the name is that it gets reported by simulators, FV and debug tools. If an assertion name is omitted, verification tools assign a tool-specific name to the assertion. In that case it may be impossible to refer to the specific assertion using assertion control tasks.

Always specify assertion names.

4.2 Immediate Assertions

Immediate assertions are the simplest kind of assertions. These assertions are Boolean and unclocked, and they tightly follow the simulation flow. Immediate assertions may be placed only in procedural code.

4.2.1 Immediate Assertion Simulation

Immediate assertions are akin to other procedural statements and behave like procedural **if** statements. The assertion condition is evaluated each time the control flow reaches the assertion. The evaluation is performed immediately with the values taken at that moment for the assertion condition variables. If the assertion condition is true, that is, it has a nonzero known value, the pass action is executed; otherwise, when the condition is false, that is, it has a zero value, or its value is `x` or `z`, the fail action is executed. Since the assertion condition is non-temporal, its execution computes and reports the assertion results at the same time.

The region in which immediate assertions execute depends on where these assertions are placed in the source code. From the definition of an immediate assertion, it follows that normally in modules and in interfaces, the immediate assertions and their action blocks are executed in the Active region. In programs, they are executed

```
always_comb begin
  // ...
  a1: assert (a == b) begin
        a1_success++;
        $info("a1: a and b have value %b", a);
      end
      else begin
        a1_failure++;
        $error("a1 failure: a = %b, b = %b", a, b);
      end
end
```

Fig. 4.2 Immediate assertion

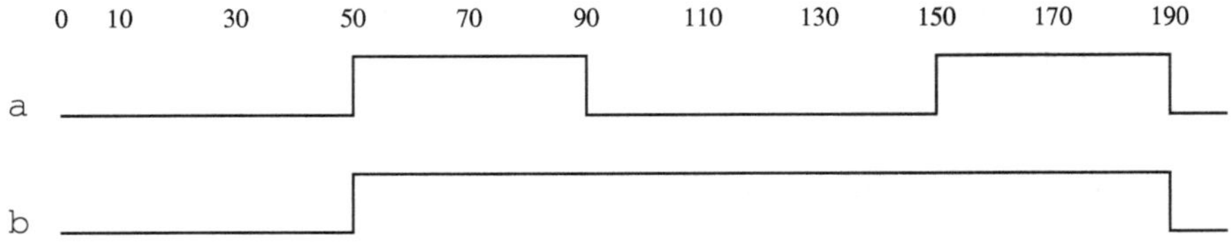

Fig. 4.3 Timing diagram for assertion `a1`

in the Reactive region (Sect. 3.5). Immediate assertions in programs and checkers execute in the Reactive region.[2]

Figure 4.2 contains an example of immediate assertion `a1`, and the corresponding timing diagram for the values of its expression variables is shown in Fig. 4.3. Recall (Sect. 2.2.1.1) that the **`always_comb`** procedure executes unconditionally at time 0, and note that at time 0 both `a` and `b` have the value `1'b0`. Therefore, the pass action is executed at time 0. The pass action increments counter `a1_success` and prints a message `a1: a and b have value 0`. Because **`always_comb`** is sensitive to the arguments of assertion `a1`, each time the value of either `a` or `b` changes, a reevaluation of the assertion condition is prompted, followed by the execution of the appropriate action block. Thus, the pass action is executed at time 50, 150, and 190, while the fail action block is executed at 90.

In practice, the pass action of assertions is seldom used as assertions are expected to succeed. The success information may provide confidence at the beginning of design and test development, but quickly becomes superfluous later on. The most common way is to specify only a fail action to issue an error message:

```
a1: assert (a == b)
  else $error("a1 failure: a = %b, b = %b", a, b);
```

[2]The LRM contains contradictory text that should be resolved in the subsequent revision: In Clause 17.5 it states that initial procedures in checkers can contain immediate assertions, but in Clause 17.2 it does not allow immediate assertions in checker body.

```
1 assign a = ...;
2 assign b = ...;
3 always_comb a1: assert (a == b);
```

Fig. 4.4 Glitch in immediate assertion

4.2.2 Simulation Glitches

The code in Fig. 4.2 is not as straightforward as it looks; the way variables `a` and `b` get their values greatly affects the behavior of assertion `a1`. Let us modify this code by explicitly specifying the assignments of `a` and `b`. The resulting code is shown in Fig. 4.4 where we deleted the assertion action blocks for convenience.

In this case, the continuous assignments and assertion `a1` are executed in three separate processes: each continuous assignment is a separate process by itself and assertion `a1` executes in the scope of **`always_comb`**. We had to place assertion `a1` there, as immediate assertions are allowed only in procedural code. Consider the assertion behavior at time 50. SystemVerilog does not impose any predefined order of process execution for parallel processes such as the ones in this example, so let us assume the following order:

- Line 1 is executed. `a` is assigned the value 1, `b` is still keeping its old value 0.
- Line 3 is executed. Because `a` and `b` at this point have different values, assertion `a1` fails, and an error message is issued.
- Line 2 is executed. `b` is assigned the value 1.
- Line 3 is executed again. Now the assertion passes, and no message is issued.

As a result, the user will think that the assertion failed at time 50, though essentially it passed, and its failure was just a simulation glitch. Had we assumed a different simulation order (Lines 1, 2, and 3), no assertion failure would be reported.

This example clearly shows that, due to their vulnerability to 0-delay simulation glitches, using immediate assertions may be problematic. So, when should immediate assertions be used? From their simulation semantics it follows that they should be used in the following cases:

- Debugging simulation results by traversing the simulation flow and detecting situations such as glitches. However, as we pointed out earlier, immediate assertions may only discover glitches when they manifest in simulation, and not the real glitches in the circuit.
- When delay controls are specified that make the code impervious to glitches.
- In **`program`** testbenches which observe only the stable values of design variables because programs execute in the Reactive region set.

In all other cases, we recommend to use deferred assertions.

Immediate assertions are sensitive to simulation glitches. Use them only when you need to follow exactly the simulation flow, in **program**-based testbenches, or when your code contains delay controls. In all other cases when unclocked Boolean assertions are required use deferred assertions.

4.2.3 *Effect of Short-Circuiting*

Immediate assertions (simple and deferred) can be placed in functions. Functions can be called in expressions involving logic operators. This may lead to some unexpected results because of short-circuiting in the evaluation of the expressions. Consider the following example:

```
function bit check3bits (bit [2:0] expr, value);
  a2: assert (expr > value) else
    $error("a2 failure: expr = %b, value = %b", expr, value);
  return (expr > value);
endfunction : check3bits
assign combined = v && check3bits(x, 3'b1);
```

Suppose that both `v` and `x` become `0` at some simulation time step. The continuous assignment is evaluated, but since the first operand `v` of `&&` is `0`, the simulator determines that `combined` is `0` and by the rule of short-circuiting it does not need to evaluate the function call. Therefore, the assertion is not evaluated either and the error goes undetected. If the order of operands in the assignment were exchanged, as in `assign combined = check3bits(x, 3'b1) && v;` the error would be detected because the first operand is always evaluated. A similar situation can arise when `&&` is replaced by `||` and v takes the value of `1`.

4.3 Deferred Assertions

The official name of deferred assertions is "deferred immediate assertions" since they are a variant of immediate assertions. However, we will call them simply "deferred assertions", reserving the name "immediate assertions" for immediate assertions that are not deferred.

Deferred assertions are unclocked Boolean assertions, and they differ from immediate assertions in the following ways:

- Deferred assertions come in two forms identified by the keywords **assert** `#0` for *observed deferred assertions* and **assert final** for *final deferred assertions*. In the rest of the book we refer to the latter simply as *final assertions*.
- Deferred assertions are not sensitive to simulation glitches.
- Deferred assertions may be placed both inside and outside procedural code.

- Default reporting and the execution of action blocks of deferred assertions are scheduled in the Reactive region in the case of observed deferred assertions and in the Postponed region for the final deferred assertions.
- An action block of a deferred assertion may only be a subroutine call. The contents of the body of the subroutine of final deferred assertions is restricted to passive statements because the subroutine executes in a region that does not allow any further generation of events in the current time step.

4.3.1 Deferred Assertion Simulation

Deferred assertions evaluate like immediate assertions, but the results are tentative, contingent upon possible re-execution of the same assertion in the time slot. This can happen, for example in Fig. 4.4 when the enclosing process of an assertion is sensitive to variables with values that fluctuate before they stabilize in the Active region. The immediate assertions shown in that figure would fire on every such change if the expression `a == b` becomes false.

The difference between observed and final deferred assertions is in the extent of glitch filtering. Observed assertions filter glitches that occur in a single scheduling region set, Active or Reactive. The example in Fig. 4.4 illustrates a glitch occurring in the Active region. In contrast, final assertions filter glitches that are created by the interaction between Active and Reactive regions. Such a glitch can be created when the assignment to `a` in Fig. 4.4 is made in the Active region (i.e., inside a **`module`**) and that of `b` in the Reactive region (i.e., inside a **`program`**) in the same time slot. In that case the glitch spans both regions and would not be filtered by the observed deferred assertion.

The reporting of deferred assertions is delayed (hence their name), and the deferred assertion actions are placed into a *deferred assertion report queue* of the currently executing process. If a *deferred assertion flush point* is reached in the process, the queue is cleared and thus no reports will be generated.

When a process in which a deferred assertion exists retriggers in the same time slot, the results of the assertion, in contrast to immediate assertions, are not reported immediately. Instead, depending on the kind of deferred assertion, reporting is scheduled in the Observed region for observed assertions and in the Postponed region for final assertions, pending further determination. If the process retriggers in the same time slot before the Observed region, respectively Postponed region, is reached, the previous result of the assertion is *flushed* from the deferred assertion report queue. Therefore, the simulation control in the Observed region, respectively Postponed region, sees at most one copy of the result of the assertion, which is then said to *mature*. In the case of observed assertions, the result and if present the action block subroutine of the matured entry are scheduled in the Reactive region and executed in there. For final assertions, once the Postponed region is reached, each pending report that has not been flushed from the deferred final assertion queue matures and is executed.

```
1  module mod1(output b, ...);
2    assign b = ...;
3    ...
4  endmodule
5  module mod2(...);
6    wire a, b;
7    mod1(.*);
8    assign a = ...;
9    a2: assert #0 (a == b);
10 endmodule
```

Fig. 4.5 Observed Deferred assertion

Let us recall that a deferred assertion is evaluated as a result of a process execution control reaching the location of the assertion. This linkage of an assertion with the process is essential when determining whether to flush the assertion from the deferred assertion report queue. When a process is retriggered it causes only those previous entries of the assertion results to be flushed that are linked with the process. To explain this, consider a function containing a deferred assertion. The function can be invoked from two different processes in the same time slot, resulting in a separate entry in the deferred assertion report queue for each process, but for the same assertion. If an entry is later flushed it is determined according to the process that initiated that action.

The following two examples revisit the example in Fig. 4.4 by replacing the immediate assertion first with an observed deferred assertion and then with a final deferred assertion. In the latter case, the scopes of the assertion and the assignments are also indicated.

Example 4.1. Using observed deferred assertion as shown in Fig. 4.5.

Consider what happens at time 50 for the same simulation order as described in Sect. 4.2.2.

Active region:

- Line 8 is executed. `a` is assigned the value 1, `b` is still keeping its old value 0.
- Line 9 is executed. Since `a` and `b` at this point have different values, the fail action entry of the assertion `a2` is placed into the deferred assertion report queue.
- Line 2 is executed and `b` is assigned the value 1.
- Line 9 is executed again. The previous fail action entry of this assertion is flushed from the deferred assertion report queue, and the success action entry is placed there instead. As the success action is void, effectively only the success result is placed into the entry.

Observed region: The observed deferred assertion report queue entry matures.

Reactive region: All actions from the deferred assertion report queue are executed. Since in our case the queue entry is without an action block, no actions are executed. A tool may, however, choose to report a success result for `a2`.

```
1  program prog(output b, ...);
2    assign b = ...;
3    ...
4  endprogram
5  module mod2(...);
6    wire a, b;
7    prog(.*);
8    assign a = ...;
9    a3: assert final (a == b);
10 endmodule
```

Fig. 4.6 Final Deferred assertion

As we can see, in spite of a glitch in simulation, this glitch does not affect the behavior of deferred assertion `a2`. □

Example 4.2. Using final deferred assertion as shown in Fig. 4.6.

Consider what happens at time 50 for the same simulation order as described in Sect. 4.2.2.

Active region:

- Line 8 is executed. `a` is assigned the value 1, `b` is still keeping its old value 0.
- Line 9 is executed. Since `a` and `b` at this point have different values, the fail action entry of the assertion `a3` is placed into the deferred assertion report queue.

Reactive region:

- Line 2 is executed and `b` is assigned the value 1.

Active region:

- Line 9 is executed again due to a change of value in `b`. The previous fail action entry of this assertion is removed from the deferred assertion report queue, and the success action entry is placed there instead. As the success action is void, effectively only the success result is placed into the entry.

Postponed region: The final deferred assertion report queue entry matures. All actions from the deferred assertion report queue are executed. Since in our case the queue entry is without an action block, no actions are executed. A tool may, however, choose to report a success result for assertion `a3`.

As we can see, in spite of a glitch in simulation that spanned the Active-Reactive-Active regions, this glitch does not affect the behavior of deferred assertion `a3`. If the final assertion were replaced by an observed one, the glitch would not be filtered out and the failure caused by a value change of `a` as well as the success caused by the value change of `b` would be reported. □

The above examples illustrate one case of flushing the results of deferred assertions. There are other circumstances where the act of flushing is also needed. In all, flushing should be perceived as unconditional when its enclosing process encounters one of the following situations.

- Process, which is previously suspended due to a **wait** statement or an event control, resumes after its **wait** statement or event control is enabled.
- Process is retriggered due to a value change of a variable in its sensitivity list.
- Process is explicitly disabled using the **disable** statement.

Since the deferred assertions are glitch-free, they are the preferable way to express unclocked Boolean assertions in RTL. However, when the code has event controls, the behavior of deferred assertions becomes unintuitive, and it may result in missed failure reporting.

Do not use deferred assertions when the code has delay controls. Use immediate assertions instead.

Use final deferred assertions instead of observed ones except when the action block is required to execute some action that is not allowed in the Postponed region.

Final deferred assertions are more robust in that their reporting will not change when a portion of a module-based design is replaced by more abstract model in a program. For example, as illustrated by the replacement of module `mod1` in Fig. 4.5 by a program `prog` in Fig. 4.6.

4.3.2 *Deferred Assertion Actions*

The complex bookkeeping of deferred assertions explains the restriction imposed on their actions that they may consist only of a single subroutine call. A subroutine in SystemVerilog can be either a **task** or a **function**.

Example 4.3. The following observed deferred assertions are legal:

```
int err_cnt = 0;
task err(bit a, b, string s);
  err_cnt++;
  $error("%s failure: a = %b, b = %b", a, b);
endtask
da1: assert #0 (a == b)
  else err("da1", a, b);
da2: assert #0 (a == b) $info("a and b have value %b", a);
da3: assert #0 (a == b) $info("a and b have value %b", a);
  else $error("da3 failure: a = %b, b = %b", a, b);
da4: assert #0 (a == b);
```

The following observed deferred assertions are *illegal*, as their fail actions contain either more than one statement, or are not a subroutine call:

```
da5: assert #0 (a == b) else begin
  err_cnt++; $error("p1 failure: a = %b, b = %b", a, b); end
da6: assert #0 (a == b) else ctr++;
da7: assert #0 (a == b) else
  begin
    $error("da7 failure: a = %b, b = %b", a, b);
  end
```
□

If the above observed deferred assertions are replaced by final ones, then assertion `da1` becomes also illegal because the action block modifies the global variable `err_cnt`.

Delayed execution of deferred assertions puts forth a question about which values of the subroutine arguments are used during action execution. The answer is twofold:

- If a subroutine argument is passed by value, the argument value is used at the instant when the deferred assertion expression is evaluated.[3]
- If a subroutine argument is passed by reference, the argument value from the reporting region is used. That is, in the Reactive region for observed assertions and in the Postponed region for the final assertions.

Since system tasks `$display`, `$error`, etc. pass their arguments by reference, it means that when these tasks are used with deferred assertions, the argument values from the Reactive, respectively Postponed, region are printed. Although these values in these region may differ from the values used during the deferred assertion expression evaluation, in practice the values of variables representing actual design signals remain the same, and thus the deferred assertion reporting provides accurate reporting of signal values.

4.3.3 Standalone Deferred Assertions

Unlike immediate assertions, deferred assertions may also be placed outside procedural code. In such cases, a deferred assertion is semantically treated as if the assertion were enclosed within an **`always_comb`** procedure.

This was illustrated in Figs. 4.5 and 4.6. Explicit **`always_comb`** statement used with the immediate assertion in Fig. 4.4 is not necessary. This feature of deferred assertions makes their usage more intuitive and convenient.

At this point is it useful to point out that immediate assertions of any kind may behave differently in an **`always_comb`** procedure and in an **`always`** `@*` procedure.

---

[3]This means that the value of the variable must be stored in the deferred assertion report queue with the assertion identification.

In both cases the sensitivity list is inferred, but the inference is different leading to potentially different behavior of the assertions:

- **always_comb** executes at time 0, while **always** @* will execute only when triggered by a value change of a signal in the sensitivity list. Consequently, an assertion may fire at time 0 in the former case, but not in the latter.
- **always_comb** is sensitive to changes of static variables read inside a function, while **always** @* is not. Therefore, an immediate assertion inside a function may not fire in the latter case unless the variables read in the assertion are among the arguments of the function.
- Variables that are both read and written in an **always_comb** procedure do not appear on the implicit sensitivity list while in **always** @* they do. Therefore, an assertion that refers to such variables in an **always_comb** procedure may not fire.

> In RTL design code use **always_comb** to model combinational logic and if needed provide checking using deferred assertions.

4.3.4 Effect of Short-Circuiting in Deferred Assertions

Short circuiting effects on deferred assertions can be even more non-intuitive than on simple immediate assertions. The following code is similar to that in Sect. 4.2.3 except that an observed deferred assertion is used:

```
function bit check3bits (bit [2:0] expr, value);
  a2: assert #0 (expr > value) else
    $error("a2 failure: expr = %b, value = %b", expr, value);
  return (expr > value);
endfunction : check3bits
assign combined = v && check3bits(x, 3'b1);
```

Suppose that at some simulation time `v` became `1` and `x` became `2`. The continuous assignment is evaluated, the assertion passes and is enqueued in the deferred assertion result queue. Later in the time step, `v` and `x` change to `0`. The continuous assignment is re-evaluated and the result of the previous execution is flushed. Since `v` is `0`, the simulator determines that `combined` is also `0` and by the rule of short-circuiting it does not need to evaluate the function call. Therefore, the assertion is not evaluated either and neither success nor failure is reported in the Reactive region. If the order of operands in the assignment were exchanged, as in `assign combined = check3bits(x, 3'b1) && v;` the error would be detected because the first operand is always evaluated. Alternately, `&&` can be replaced on scalar bits by `&` (a bitwise "and") in which case short-circuiting does not apply.

4.4 Concurrent Assertions

Concurrent assertions have the same format as the immediate simple and deferred ones, however, their action blocks are not limited to subroutine calls as in the case of deferred assertions. The action blocks may contain any statements and their body may have a more complex structure:

> concurrent_assertion_body ::=
>   [clocking_event] [**disable iff** (reset)] property

Here, the square brackets are not part of the syntax, they show that the corresponding constructs are optional.

Example 4.4. The following concurrent assertion

```
a1: assert property (@(posedge clk)
  disable iff (rst) a |-> nexttime[2] b);
```

is controlled by clocking event `posedge clk` and has a disabling condition `rst`. The assertion property is `a |-> nexttime[2] b`. □

Like immediate assertions of any kind, concurrent assertions can be used inside modules, interfaces, programs and checkers. They can be placed:

- in **always** procedures
- in **initial** procedures
- standalone (also called static)—outside any procedure

For an assertion placed in either of the two procedures, it gets evaluated only when the control point reaches the assertion statement. Commonly, an assertion is placed in an **initial** procedure with the intention of evaluating the assertion only once. For example, assertion `a1` checks that `ready` is low at the first tick of the clock:

```
initial a2: assert property (@(posedge clk) !ready);
```

In other cases, the assertion is monitored continuously: at each tick of its clock. For example, assertion `a3` checks that `ok` is high at every tick of the clock:

```
a3: assert property (@(posedge clk) ok);
```

Or, an assertion monitored continuously can be placed in an **always** procedure as:

```
always @(posedge clk)  begin
  d1 <= i1 | i2 ;
  a4: assert property (d1 |=> i3 | i4);
  dout <= f_ecap(d1);
end
```

Procedural assertions embedded in always procedures are discussed in detail in Chap. 14. Their use in checkers is discussed in detail in Chap. 9 and in Chap. 24.

Assertions can be viewed as an observer machine that produces pass/fail output. This would be the case when assertions are included as part of the design in emulation or in formal verification (see Chap. 21). In simulation, however, a different view can be taken that is more suited for understanding the assertion behavior and debugging. In this view, an assertion is considered as a machine that issues an evaluation attempt or transaction at every tick of the leading clock of the assertion (while it is enabled) and ends by its success or failure. Each such transaction thus has a certain duration in time and can be controlled and analyzed separately from other such evaluation attempts. This chapter discusses this attempt-based view of assertion evaluation in simulation and examines efficiency issues connected with this view.

4.4.1 Simulation Evaluation Attempt

To explain the notion of a simulation attempt, let us consider the following assertion:

```
a1: assert property(@(posedge clk)a ##1 b);
```

In this case the assertion has only a single clock that determines the instants at which the assertion evaluation progresses. This is also the so called *leading clock* of the assertion. When later we discuss assertions that refer to multiple clocks, the one clock among them that starts the evaluation attempts (there can be only one such clock in SystemVerilog Assertions) is also referred to as the leading clock.

The clocking event is the rising edge of `clk` which is also the assertion leading clock. In simulation, the assertion will observe the values of signals `a` and `b` starting at every tick of **`posedge`** `clk`. Thus, at clock tick t the evaluation of `a` is performed, and if the result is true, it is followed by the evaluation of `b` at the next tick, $t + 1$. If either `a==0` at t or `b==0` is true at $t + 1$, the evaluation starting at t fails at time t or time $t + 1$, respectively. However, if both `a==1` at t and `b==1` at $t + 1$ are true, then the evaluation that started at t succeeds at $t + 1$. An evaluation starting at clock tick t is independent of the evaluations starting at all other clock ticks. Consequently, the evaluation result of the attempt at t is independent of the results of attempts at all other clock ticks, and is thus called an *evaluation attempt* (starting at tick t).

Each such attempt has a start time corresponding to the simulation time of the leading clock tick, and an end time corresponding to the time of the clock tick at which the evaluation attempt either succeeds or fails. Furthermore, several attempts starting at different times may be under evaluation at the same time in a sort of pipeline fashion. For instance in our simple example, if `a` and `b` were true for several clock ticks, there would be two concurrent attempts. If the sequence definition spanned more than two cycles, more than two attempts could independently be evaluated at the same time.

An evaluation attempt has a start time and an end time. The end time may also be infinite, but in simulation bounded by the extent of the simulation. Evaluation of more than one attempt may be in flight at the same time.

If an assertion is always enabled, then there will be a series of evaluation attempts, starting at every tick of the leading clock. This is the case when the assertion is placed outside any procedure. However, if it is placed in an always procedure, then the start of an attempt is also controlled by the execution of the body of the always procedure reaching the position of the assertion. Furthermore, if an assertion is placed in an initial procedure, there will be only one evaluation attempt starting at the first tick of the leading clock.

We can thus see that a concurrent assertion outside a procedure will execute "always", inside an **initial** procedure it will execute once, and inside an **always** procedure it may execute more than "once". This is to some extent different from the more classical interpretation of assertions that are not part of a design language (e.g., in PSL). There, unless an explicit top-level **always** operator is specified in the assertion, the evaluation only starts at the first clock tick. For users who have been using PSL and plan to use SystemVerilog assertions, care must be taken not to include this top-level **always** operator, because it could lead to some performance penalty in simulation, although the behavior remains correct as illustrated in Example 4.5.

Example 4.5. Operator **always** as a top-level property in an assertion.

```
module m;
  bit clk, a, b;
  default clocking ccc @(posedge clk); endclocking
  a_always: assert property(always (a ##1 b));
  //...
endmodule
```

Assertion `a_always` does not make practical sense for the following reason: it will start an evaluation attempt at every tick of it clock. The top-level operator is **always**, and thus in each such attempt it will redundantly fork off a new version of the always operator to perpetually check for every pair of consecutive values of `a` and `b`. Unless the simulator can detect this situation and do something about it, there may be an ever increasing number of evaluation attempts that never terminate, causing a heavier and heavier burden on the simulator. See also Sect. 6.3.2. □

Example 4.6. The following two assertions provide an interesting illustration of the meaning of an evaluation attempt.

```
module m;
  bit clk, a, b;
  a1: assert property(@(posedge clk) a |=> b);
  initial
    a2: assert property(@(posedge clk) always (a |=> b));
endmodule
```

Assertion `a1` states that if `a` is true then it must be followed by `b` true at the next clock tick. Since the assertion is not in any procedure, it will evaluate attempts starting at every clock tick.

The body of assertion `a2` also states that if `a` is true then it must be followed by `b` true at the next clock tick. However, since `a2` is in an initial procedure, there will be only one attempt started at the first clock tick, but then due to the **`always`** property operator it will evaluate `a |=> b` continuously starting at every clock tick.

Is there a difference between the two assertions? If we are concerned about the first failure that may occur during a simulation, i.e., `a` being true at a clock tick followed by `b` false at the next clock tick, then there is no difference in reporting the failure. However, if we wish to detect any subsequent failures or we wish to know when the failing sequence of `a` and `b` started, then assertion `a1` provides this information while `a2` may not. This is because `a2` runs only one attempt, and if `a |=> b` fails somewhere it is a failure of **`always`** and thus a failure of only that attempt. In the case of `a1`, the failing evaluation attempt of `a |=> b` will be reported with its start and end times. Thereafter, the assertion evaluation continues and may report other failing attempts. □

4.4.2 Clock

Concurrent assertions must be controlled by a *clocking event*, or *clock* and all the assertion evaluation attempts are synchronized with this event. Subexpressions of the property expressions may have their own clocking events. Multiclocked properties are relatively rare, and we postpone their description until Chap. 12. For now, we assume that the subexpressions of the assertion property expression do not use a different clock.

Since concurrent assertions are clocked, the main assertion clock must be present in the assertion. This clock either should be explicitly specified or inferred from the context: from an event control in the surrounding **`always`** or **`initial`** procedure, or from **`default clocking`**. We discuss in detail clock inference rules in Sects. 12.2.2 and 14.2.

All signal values participating in the property expression are sampled at the assertion clock tick only, and their values between the clock ticks are completely ignored. For example, the assertion

```
a1: assert property (@(posedge clk) a);
```

passes for the signal waveforms shown in Fig. 4.7. Although `a` is low between time 55 and 65, this is considered to be a glitch and is ignored, because the values of `a` are sampled on **`posedge`** `clk`, i.e., at times 10, 20, etc.

The role of a clock in concurrent assertions is to convert the continuous time into the discrete one. This is important since SVA temporal operators are defined for discrete time.

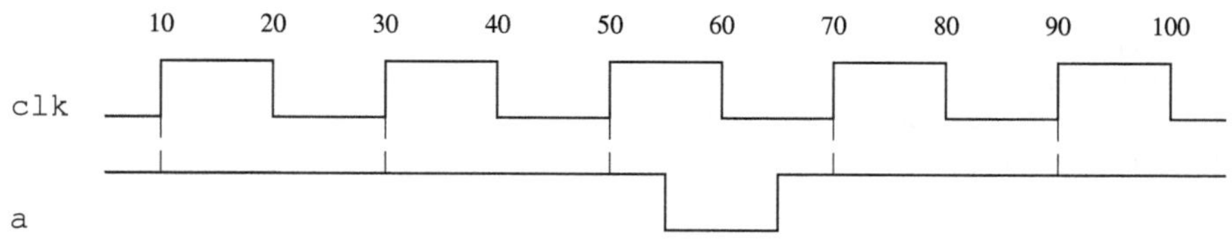

Fig. 4.7 Signal glitches

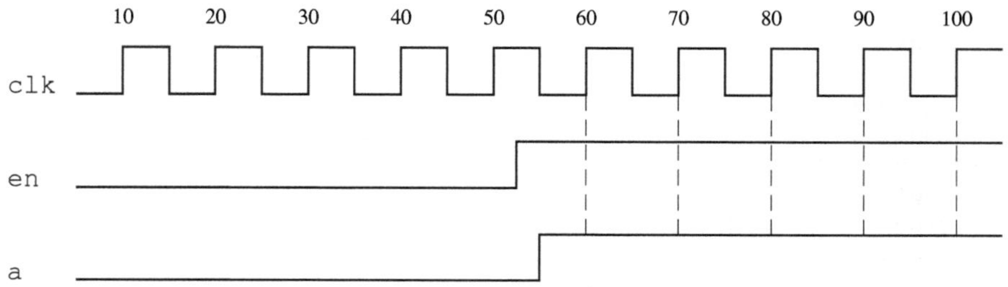

Fig. 4.8 Gated clock

The clocking event follows standard SystemVerilog semantics, and it is based on the clock signal changes: **posedge** clk triggers when clk becomes 1, **negedge** clk triggers when clk becomes 0, clk triggers any time when clk changes value, and **edge** clk triggers when clk changes to 0 or to 1.[4]

For example, the assertion

```
a2: assert property (@clk a);
```

does *not* mean that a is checked each time when clk is high, but rather each time when clk changes.

Gated Clock

An assertion clock may be gated. For example, the following assertion has a gated clock which is active only when en is high:

```
a3: assert property (@(posedge clk iff en) a);
```

When en is low clock ticks are ignored. For example, this assertion passes in the case shown in Fig. 4.8 because clk is disabled while a is low until time 55.

[4] clk and **edge** clk behave the same way when clk is of type **bit**, but they behave differently when the type is **logic**.

```
1  module top;
2    bit sys_clk;
3    global clocking GCLK @(edge sys_clk);
4    endclocking
5    // code to animate sys_clk in simulation
6    ...
7    mod1 mod1_inst();
8    mod2 mod2_inst();
9  endmodule
10 module mod1();
11   logic a, b;
12   ...
13   m_check m_check_inst(a, b);
14 endmodule
15 module mod2();
16   logic x, y, clk;
17   global clocking @(posedge clk);
18   endclocking
19   ...
20   m_check m_check_inst(x, y);
21 module m_check(input logic a, b);
22   a_simple:  assert property (@$global_clock a |=> b);
23 endmodule
```

Fig. 4.9 Global clocking declaration

Global Clocking

SystemVerilog provides the capability to specify a clocking event for the entire or parts of the elaborated model as the *primary system clock*, called also *global clock*. This is done with a *global clocking* declaration, which is a special form of clocking block declaration. Its purpose is to provide relation to the primary system clock used in formal verification and emulation. There can be several global clocking specifications in the design affecting different parts of the design hierarchy, but at any hierarchical level at most one of them is effective and can be referred to using the `$global_clock` system function.

Figure 4.9 illustrates the syntax of a global clocking declaration. One global clocking is specified in Line 3 within the module `top`. This declaration says that the event **`edge`** `sys_clk` is the primary system clocking event for the underlying hierarchy until another global clocking declaration is encountered (if any). Since the model has a global clocking declaration, the system function `$global_clock` may be used to reference the system clocking event. Such a reference appears in Line 22. In this model, module `m_check` is instantiated in `mod1` and in `mod2`. In the instance `mod1_inst` `$global_clock` resolves to the global clocking in the nearest parent module which is `top` and behaves the same way as the hierarchical reference `top.GCLK`. However, in the instance `mod2_inst`, `$global_clock` resolves to the hierarchically nearest specification in Line 17 within the module `mod2`. It behaves as **posedge** `top.mod2_inst.clk`. In this example, the module `m_check` sees a different primary system clock depending on where the checker is instantiated in the design.

Global clocking name is optional, and `GCLK` in Line 3 could be omitted as shown in Line 17.

A common purpose of declaring a global clock is to specify the primary clock for formal verification. The ticks of the primary clock are at the finest granularity of time in a formal model, and the global clock is assumed to tick forever. Global clocking in formal verification is discussed in Chap. 21. For consistency with simulation, it is also recommended that in simulation all events be synchronized with the global clock. Nevertheless, the simulator is not required to check this property of the global clock.

The previous SystemVerilog Language Reference Manual (LRM) from 2009 [5] allowed a single non-hierarchical specification of global clocking. This however turned out to be insufficient in the context of System on Chip designs where different subsystems—Intellectually Property blocks may come from different providers and have their own global clocking declaration. The 2012 LRM [8] corrected this weakness and provided a way to specify a hierarchical global clocking as illustrated in Fig. 4.9. Designs that used the 2009 interpretation of global clocking may have to be slightly modified to conform to the new rules. In particular, global clocking declared in an autonomous top-level module without any underlying hierarchy will not be visible in any part of the design.

In the presence of global clocking, a number of sampled value functions may be used that are synchronized to the global clock. These are discussed in Sect. 7.2.2. The global clocking declaration also defines a specific event to be referenced by `$global_clock` and to govern the global clocking sampled value functions in simulation. Global clock use cases are examined in subsequent chapters.

4.4.3 Sampled Values for Concurrent Assertion

In simulation, all values of variables appearing in a concurrent assertion used sampled values of the variables. In most cases, such as static design variables, the sampled value is the value of the variable in the Preponed region as explained in Sect. 3.5. It means that concurrent assertions use the values of signals at the beginning of the simulation step in which the clocking event occurs. Assertion value sampling makes concurrent assertions insensitive to simulation glitches. The assertion body is evaluated using values collected in the Preponed region. The action blocks are executed in the Reactive region.

One should be aware of this peculiarity of concurrent assertions to correctly analyze their behavior in timing diagrams and simulation traces. Since the assertion action blocks are executed in the Reactive region, the signal values used in the action blocks may be inconsistent with the sampled values of the signals used in the assertion. To remedy this, the sampled value of the variables should be used

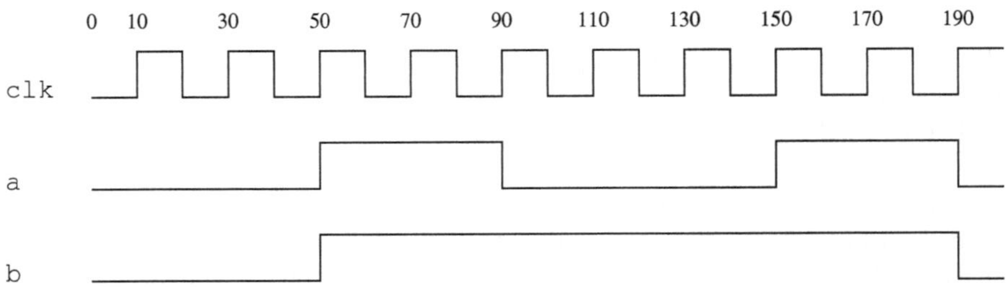

Fig. 4.10 Timing diagram for assertion a2

explicitly by calling system function `$sampled` in the action blocks, as explained in Sect. 7.2.1.1.

Example 4.7. Below, we reuse the example from Sect. 4.2.1 by substituting immediate assertion by a concurrent one.

```
a2: assert property (@(posedge clk) a == b)
  begin
    a2_success++;
    $info("a2: a and b have value %b", a);
  end
  else begin
    a2_failure++;
    $error("a2 failure: a = %b, b = %b", a, b);
  end
```

The timing diagram with the addition of the `clk` waveform is reproduced in Fig. 4.10.

The pass action of assertion `a2` executes at times 10, 30, 50, 70, 90, 170, and 190. Its fail action executes at times 110, 130, and 150. As an example, consider assertion status at time 90. Assertion `a2` samples values of signals `a` and `b` in the Preponed region, that is *before* the value of `a` changes. Both `a` and `b` at time 90 have the sampled value 1, the assertion passes at that time, and its pass action is executed; counter `a2_success` is incremented by one, and the message `a2: a and b have value 0` is printed. Yes, it is 0 that is printed, not 1 because the action blocks of concurrent assertions are executed in the Reactive region, and the value of `a` there is already 0. The message is misleading, since the assertion uses the old value of `a`, while its action block uses the new one! To correct this, we should explicitly specify the sampled values of `a` and `b` in the action blocks: `$info("a2: a and b have value b", $sampled(a))` in the pass action block, and `$error("a2 failure: a = %b, b = %b", $sampled(a), $sampled(b))` in the fail action block. □

There are several exceptions to the general rule of sampling signal values in concurrent assertions, such as the sampled values of automatic variables, local variables and active free checker variables, and in common assertion usage the assertion main reset expression (the argument of **disable iff** statement) which is never sampled as explained in detail in Sect. 13.1.1. Sampling of variables is explained in detail in Appendix A.

The value of the main assertion reset is not sampled in concurrent assertions.

4.4.4 Reset

Usually, we are interested in checking an assertion only when the reset signal is inactive. **disable iff** operator is used to specify the reset expression of the assertion. If the main assertion clock is explicitly specified, **disable iff** should immediately follow it. If the assertion clock is omitted, **disable iff** should be the first operator in the assertion body. There may be at most one **disable iff** operator in the whole assertion.

If the disable condition is true between the start of an evaluation attempt in the Observed region and the end of the evaluation attempt then the overall evaluation of the property results in disabled. The evaluation attempt is discarded (neither success nor failure). Such attempts are called *disabled*. The reset is asynchronous, in the sense that it is monitored at every time step, and not only at the ticks of the assertion clock. Therefore, it also makes it sensitive to 0-width simulation glitches. The sensitivity can be removed by applying `$sampled` function to the disable expression (or its component variables). This makes the disable condition insensitive to such glitches like the body of the concurrent assertion.

Example 4.8. Consider assertion `a1`

```
a1: assert property(@(posedge clk) disable iff (rst)
  req |-> nexttime[2] ack);
```

with the timing diagram shown in Fig. 4.11.

In this example, there are three assertion evaluation attempts beginning at times 20, 40, and 70 that satisfy the antecedent signal `req`.[5] First two of them are disabled, and therefore the assertion does not fail even though there is no `ack` received. The last attempt is normal, and it succeeds because `ack` is received in two cycles after `req` is issued. Refer to Chap. 13 for detailed discussion about assertion resets. □

As we mentioned, the **disable iff** expression acts asynchronously. Its evaluation goes on evaluating regardless of the occurrence of the clock tick of the associated assertion. And, should it evaluate to true, all evaluation attempts of that assertion that are "in flight" are declared disabled.

[5]One could be tempted to say that the attempts begin at times 15, 35, and 65, but recall that the attempts are synchronized with the rising edge of the clock.

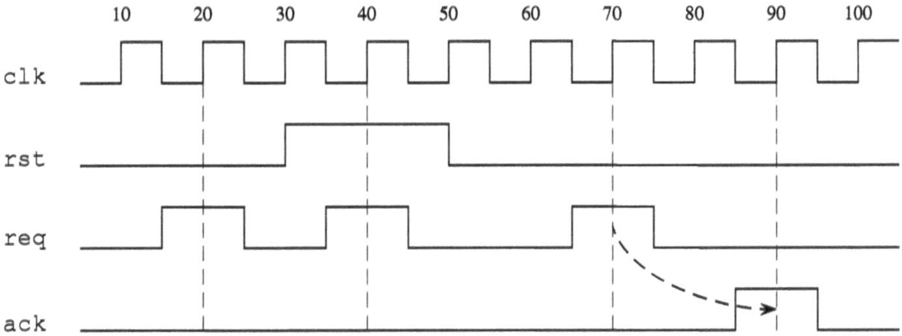

Fig. 4.11 Assertion with reset

Example 4.9. An example of **`disable iff`**.

```
module reqgen(input logic  busy, clk, rst, output logic req);
  wire idle;
  assign idle = !busy;
  always @(posedge clk or posedge rst) begin
    if (rst) req <= '0;
    else req <= !busy;
  end
  req_when_idle: assert property (
    @(posedge clk) disable iff (rst) idle |=> req)
      $display("req_when_idle completed");
endmodule: reqgen
```

In the usual way, assertion `req_when_idle` is scheduled in the Observed region as a result of the occurrence of its clock (**`posedge`** `clock`) in the Active region. None of the signals in the assertion expression, namely, `idle` and `req`, have any effect on the assertion outside the time slots in which the clock tick occurs. The **`disable iff`** expression (`rst`) is an exception to this general rule. `req_when_idle` is disabled in the time slot when signal `rst` becomes true. Its action block is scheduled in the same Reactive region. □

> Unlike other expressions in the body of a concurrent assertion, the **`disable iff`** expression is not sampled and can affect the result of the assertion from any scheduling region in which it becomes true.

4.4.5 Boolean Expressions

Boolean expressions are elementary building blocks for assertions. There are two places in an assertion statement where Boolean assertions are used:

- operands of property and sequence operators,
- argument of **disable iff** as the top-level assertion reset or of **default disable**.

As we saw in the previous sections, Boolean expressions as the operands of property and sequence operators use the sampled values of variables, while as the arguments of **disable iff** or **default disable** they use the current values of the variables.

A Boolean expression is composed of SystemVerilog expressions, but with some restrictions on the operators, variables and their types.

Boolean expressions used in concurrent assertions must satisfy the following constraints:

- The resulting expression type is cast-compatible with an integral type. Integral data types include integer data (**int** and **integer**), packed array, packed structure, packed union, enum variable, or time variable.
- Non-static class properties and methods are not referenced.
- Evaluation of the expression has no side effects. For instance, unary increment and decrement, and binary assignment operators such as `++`, `--`, `+=`, `-=` may not be used.
- **chandle**.[6]
- Functions called in Boolean expression may not have output or **ref** formal arguments (**const ref** is allowed.) Functions must not preserve any state information from call to call.

Arrays are commonly compared in the subexpressions of a Boolean expression, such as

```
logic [15:0] aA [4], aB[4];
a1: assert property (@(posedge clk)(aA == aB));
```

Elements of dynamic arrays, queues, and associative arrays accessed in an assertion must exist within the scope of the assertion until the assertion expression evaluation completes.

A variable in any subexpression of a Boolean expression, in general, must be declared as static. The static variable declared in a task, clocking block, module, program, or interface can be referenced. Limited use of automatic variables is allowed in procedural concurrent assertions as explained in Sect. 14.3.

4.4.6 Event Semantics for Concurrent Assertions

Let us now consider a concurrent assertion which is not embedded in procedural code and decompose the assertion execution into various kinds of essential computational steps in the event semantic framework described in Sect. 3.5 (the `triggered`

[6] **chandle** is a data type used to represent storage for pointers passed to SystemVerilog from C code.

sequence method signals that the end of a sequence has been reached, and it is described in Sect. 11.2.1).

```
sequence ack; @(posedge clk)enable ##[1:10] end_ack;
endsequence
a1: assert property (@(posedge clk)
          req |-> busy until ack.triggered)
    else $error("Assertion a1 fails");
```

We can subdivide the activity as follows:

1. Sample signals `enable`, `req`, `end_ack`, and `busy` which are needed for concurrent assertion evaluations.
2. Detect the occurrence of clock `@(posedge clk)`, since sequence `ack` and assertion `a1` evaluations are activated by the clock.
3. Start a new attempt for sequence `ack`.
4. Determine whether there is a match for sequence `ack`.
5. Start a new attempt for assertion `a1`.
6. Resume evaluation of the previous attempts for this clock tick.
7. For an attempt of `a1` resulting in a failure, schedule the action block execution in the Reactive region.
8. For an attempt of `a1` resulting in a success, report a success.
9. Execute action blocks.

These steps are taken in specific regions. Each region contributes to the execution of concurrent assertions as shown in Fig. 4.12.

Normally, the detection of an assertion clock occurs in the Active region. What sets a concurrent assertion apart from an immediate assertion is that the former is triggered by its clock. Consequently, its evaluation resumes at the arrival of the clock and gets suspended at the end of the time slot in which the clock occurs.

Let us now consider assertions embedded in an always procedure. Beside the clock, the new attempts of procedural concurrent assertions are bound to the procedural context in which they are specified. For instance,

```
function bit f1(bit arg);
//...
endfunction
always @(posedge e1)  begin: B1
  a <= b + c ;
  if (c1_enb)
    a2: assert property (f1 (a) ##2 f1(a));
  dout <= f1(a);
end
```

An evaluation attempt of assertion `a2` can only be initiated if the simulation control reaches the statement. In this case, it means that,

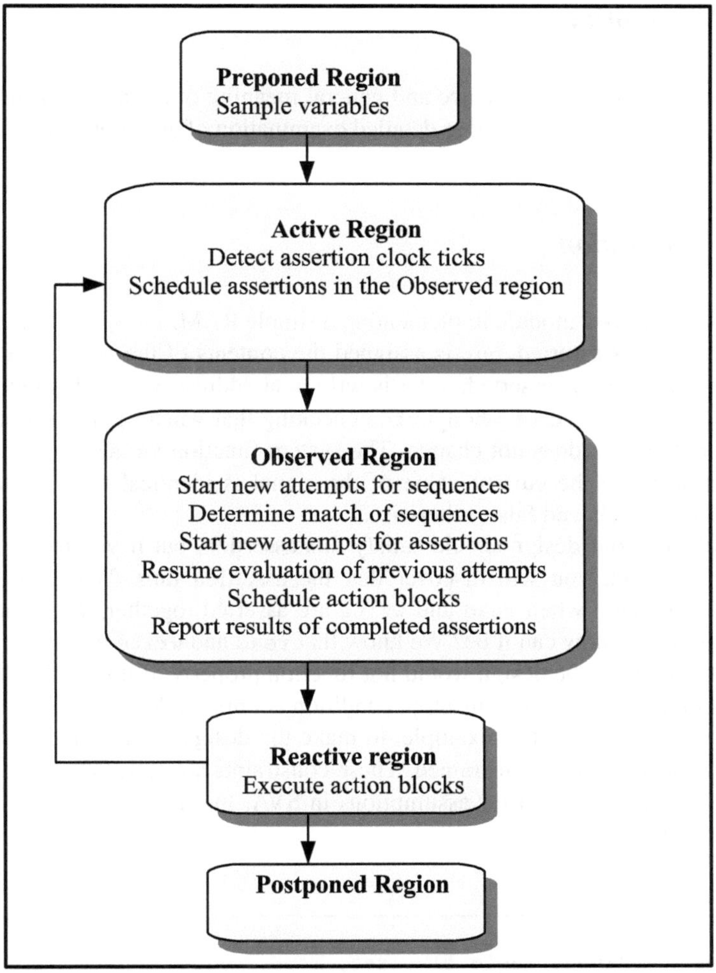

Fig. 4.12 Assertion processing in regions

1. Event (**posedge** e1) occurs, and
2. Signal `ci_enb` is true.

Otherwise, a new attempt of `a2` is not fired off. However, the previous attempts of `a2` in progress are no longer coupled with (2); they are only sensitive to (1), hence their evaluation is resumed should (1) occur.

Additional queues are called upon in the Observed region to schedule new attempts and track other attempts that are in progress. Two queues, procedural assertion queue and matured assertion queue are added in the Observed region to dispatch the evaluation. The event semantics of procedural concurrent assertions is described in greater detail in Chap. 14.

4.5 Assumptions

We have mentioned the existence and general meaning of assumptions in previous chapters, here we provide a more detailed examination of assumptions.

4.5.1 Motivation

Figure 4.13 shows a module implementing a simple RAM.

When `read` is asserted, `out` is assigned the contents of the memory at address `addr`. When `write` is asserted, `data` is written at address `addr`. This module also contains assertion `stable_when_write` checking that when `write` is asserted the module output `out` does not change. The system function `$stable` (Sect. 7.2.1.4) returns true when the current value of the signal is identical to its value at the previous clock tick, and false, otherwise.

It looks like our design should satisfy this assertion, but if you try to formally verify the model, you will discover that the assertion fails. Why? The answer may be surprising: when `read` and `write` are asserted together, the value of `out` may change. But how can it be? We know that `read` and `write` cannot be asserted together, otherwise the design would not function properly. But how should formal verification tools guess that this input condition is impossible?

As we can see from this example, to make the design work properly its inputs should be appropriately constrained. These constraints are called *assumptions*, and there is a special notation for assumptions in SVA. In our example, the following assumption is missing:

```
module ram (input logic clk, rst, read, write,
            logic [7:0] addr, logic [15:0] data,
            output logic [15:0] out);

logic [15:0] mem[255:0];

always @(posedge clk or posedge rst) begin
  if (rst) out <= '0;
  else begin
      if (read) out <= mem[addr];
      if (write) mem[addr] <= data;
  end
  stable_when_write: assert property (disable iff (rst)
    write |=> $stable(out));
end
endmodule : ram
```

Fig. 4.13 Simple RAM

```
mutex: assume final (rst || $onehot0({read, write}))
  else $error("read/write contention");
```

Or alternately as a concurrent assumption

```
mutex: assume property (@(posedge clk) disable iff (rst)
  $onehot0({read, write})) else $error("read/write contention");
```

This assumption reads that at all times signals `read` and `write` are mutually exclusive, that is, they cannot have value 1 simultaneously. To express this, we use the system function `$onehot0` (Sect. 7.1.1) returning true when at most one bit of a vector is 1.

Assumptions are an important part of a design specification. Although we illustrated a design specification in Sect. 1.2.1, Fig. 1.6 only with assertions, a real design specification should have assumptions as well. Assumptions are also important for ABV, as they document the conditions that guarantee proper functioning of the module. It is difficult to underestimate the importance of assumptions in system integration: if each module clearly specifies its interface, most integration errors are discovered early as assumption violations.

4.5.2 Assumption Definition

The goal of assumptions is to *constrain a system behavior*. A system is composed of the DUT (model) and its environment (see Sect. 4.1, Fig. 4.1). The system behavior may be constrained either by constraining the DUT or by constraining its environment. If the DUT is deterministic, as normally happens when it is implemented in RTL,[7] there is not much sense in constraining its behavior. Therefore, assumptions are used to constrain the behavior of the environment. Assertions and assumptions play dual roles—assertions specify the behavior of the DUT and assumptions specify the behavior of its environment.

Syntactically assumptions are similar to assertions, but they use the keyword **`assume`** instead of **`assert`**.

assumption ::= name: assume_keyword (assumption_body) action_block

Similar to assertions, assumptions may be immediate (keyword **`assume`**), deferred (keyword **`assume`** `#0` and **`assume final`**), and concurrent (keyword **`assume property`**).

Immediate Assumptions. Figure 4.14 shows a typical example of an immediate assumption:

[7] In this section, we limit our discussion to deterministic models; study of nondeterministic models is postponed to Chap. 23.

```
function int quotient(int dividend, divisor);
  assume (divisor != 0) else $fatal("Division by zero");
  quotient = dividend / divisor;
endfunction
```

Fig. 4.14 Immediate assumption

Note that in this case we do need an immediate assumption, the deferred version would not work. Indeed, even if there is a simulation glitch, we should make sure that `divisor` is non-zero.

Deferred Assumptions. We have seen an example of a deferred assumption in Sect. 4.5.1, and we repeat it here for completeness:

```
mutex: assume final ($onehot0({read, write}))
  else $error("read/write contention");
```

In this case, the deferred form is preferable as we want this assumption to be standalone.

Concurrent Assumptions. The following is an example of a concurrent assumption constraining `sig` to remain stable for two clock ticks whenever `trig` is 1.

```
stable_input: assume property (@(posedge clk) disable iff (rst)
  trig |=> $stable(sig))
    else $error("sig is not stable");
```

4.5.3 *Checking Assumptions*

Checking assumptions has its own specifics: assumptions play different roles in simulation and in formal verification.

4.5.3.1 Assumptions in Simulation and Emulation

Handling assumptions in simulation and emulation is not different from handling assertions. In simulation, it is checked that the constraints imposed by assumptions hold. In case when these constraints are violated, an error is flagged.

In simulation keywords **`assert`** and **`assume`** are synonymous; their choice emphasizes the verification intention. Assertions are used to check the DUT behavior, while assumptions are used to check the environment correctness, that is, the DUT input values are correct.

4.5.3.2 Assumptions in Formal Verification

Any specific DUT behavior may be described by a corresponding signal trace, a sequence of all DUT signal values in time. This is what we see in simulation if we request the dump of all DUT signals.[8]

In FV, assumptions are used to constrain the set of legal traces of a DUT, that is, the assumptions *are not checked* in FV. The role of assumptions and assertions in FV is absolutely different: *assertion satisfaction is checked provided that all assumptions hold.*

Thus, in the example from Sect. 4.5.1, assumption `mutex` is used to keep only those traces where `read` and `write` are not 1 at the same time. Assertion `stable_when_write` holds for those traces, while it fails without this `mutex` constraint.

4.5.3.3 Assumptions in Random Simulation

In the case where randomization is done only in the environment, for example, using constraint solving SystemVerilog Testbench (SVTB), there is no difference between random and deterministic simulation as far as checking assumptions is concerned. However, when the DUT input stimuli are generated directly, the assumptions in random simulation may act as constraints. Therefore, the role of assumptions in random simulation is similar to their role in FV—to limit the legal traces of DUT. But if in FV we consider all legal traces simultaneously, in random simulation we generate only one legal trace.

Tools may eventually provide means for constraining simulations using assumptions, but it involves several difficult issues to be resolved and may imply some restrictions on the form of the assumptions. See also the discussion on free variables and assumptions in checkers used in random simulation in Sect. 23.3.

A distribution operator **`dist`** in SystemVerilog can be used in assumptions for tuning them for random simulation. It is best to explain its usage on the following example.

The assumption

```
m1: assume property (@clk a dist {1 := 2, 3 := 1, 4 := 5});
```

means that `a` may only get values 1, 3, or 4. If this assumption is used as a constraint in a random testbench, the specified weights, or frequencies are taken into account. In our example, `a` assumes values 1, 3, and 4 with the respective weights of 2, 1, and 5.

If the distribution weight is omitted, 1 is assumed by default.

---

[8]In simulation all traces are, of course, finite. In FV, we can also consider infinite traces. This is discussed in Chap. 21.

We are not going to further elaborate the distribution usage in assumptions as it falls beyond the scope of this book.

In formal verification **dist** acts as **inside** operator, and the weight specifications are ignored. It is also legal, though meaningless, to use distributions in **assert** and **cover** statements, and they are also treated as **inside** operators there.[9]

4.6 Restrictions

Sometimes it is difficult to formally verify a block. It becomes necessary to verify special cases separately and then combine them together. We can take an Arithmetic Logic Unit (ALU) as an example. ALU can perform several commands, such as addition, subtraction, arithmetic shift, etc. It thus makes sense to split the verification process into several cases corresponding to the commands, i.e., we separately verify addition, subtraction, etc.

To specify the case of addition it is natural to use an assumption, such as

```
m1: assume property (@clk opcode == OP_ADD);
```

where `opcode` is an operation code control variable in ALU, and `OP_ADD` means addition.

This will do the job in FV, but in simulation this assumption is likely to fail because the simulation test cases are not guaranteed to limit the ALU commands to addition only. The workaround is to wrap such assumptions in **`ifdef**, which makes the code less readable and dependent on the custom setup.

In SystemVerilog, there is a cleaner solution, called *restriction*. Other than the keyword **restrict**, the restriction syntax is the same as the syntax of assertions and assumption. Unlike assertions and assumptions, restrictions have only concurrent form, and they cannot have action blocks:

restriction ::= name: **restrict property** (property) ;

Restrictions are treated as assumptions in FV, but they are completely ignored in simulation. They are meant for limiting formal proofs to particular cases. Consequently, the above example should be rewritten as:

```
r1: restrict property (@clk opcode == OP_ADD);
```

Since restrictions are ignored in simulation, using actions with them is meaningless, this is why the restriction syntax does not allow actions.

[9]It is conceivable that simulators could also check that the values of the expression in the **dist** operator used in an assertion or cover satisfy the specified distributions at the end of simulation.

4.7 Coverage

The last assertion statement **cover** has also been mentioned in previous chapters. In this section, we provide general information about **cover** statements, while the detailed discussion about coverage can be found in Chap. 18.

4.7.1 Motivation

Assertions and assumptions define how the DUT and its environment *must* behave. It is also highly desirable to document how they *can* behave. When testing the design, it is necessary to make sure that the tests cover important scenarios, and different corner cases. This is achieved by *functional coverage*—a methodology to specify the scenarios to be covered. For example, for the model shown in Fig. 4.13 it is useful to check the following scenarios:

- `read` is asserted.
- `write` is asserted.
- Both `read` and `write` are simultaneously deasserted.

The corresponding coverage statements are shown on Fig. 4.15

4.7.2 Coverage Definition

Cover statement is used to register when a specific scenario happens in the design. It has the following syntax:

cover_statement ::= name: cover_keyword (cover_body) pass_action

Unlike assertions and assumptions, coverage statements have only pass action which is executed when the coverage condition (scenario) is met. The corresponding statistical information is reported by a simulator and is registered in the coverage database. Like assertions and assumptions, coverage statements may be immediate, deferred (observed and final), and concurrent.

```
read_asserted: cover property(@(posedge clk) disable iff (rst)
  read);
write_asserted: cover property(@(posedge clk) disable iff (rst)
  write);
rw_deasserted: cover property(@(posedge clk) disable iff (rst)
  !(read || write));
```

Fig. 4.15 Coverage statements for simple RAM

Immediate Coverage. The immediate coverage statement acts like an `if` statement: when the body expression is true, the pass action is performed. For example, the following statement prints a message each time when the counter reaches its maximal value:

```
c1: cover (ctr_max) $display("Counter reached its maximal
    value");
```

Deferred Coverage. Deferred coverage is similar to immediate coverage, but it is glitch-free (see Sect. 4.3). In the following example, the message is issued each time the bus is active. It uses observed deferred cover, but it could also be a final observed cover if glitches spanning entire time slots should be filtered out. `bus_drivers` is a vector of wires driving the bus:

```
wire [15:0] bus_drivers;
c2: cover #0 (bus_drivers !== 16'bz) $display("Bus is active");
```

If a final deferred cover is used, the recording of hits in a coverage database by the simulator would have to take place in the postponed region. The LRM is not clear whether a change of state of a database is permitted in this region, even though it does not generate or schedule any event.

4.7.2.1 Concurrent Coverage

There are two versions of concurrent coverage: property coverage and sequence coverage. Property coverage has the keyword **`cover property`**, and its body may contain an arbitrary property, as concurrent assertions and assumptions.

property_coverage_body ::=
    [clocking_event] [**`disable iff`** (reset)] property

If an evaluation attempt is successful, the pass action is executed only once for each evaluation attempt.

The sequence coverage has the similar syntax, but it has the keyword **`cover sequence`** instead of **`cover property`**, and its body is limited to a sequence. [10]

sequence_coverage_body ::=
    [clocking_event] [**`disable iff`** (reset)] sequence

Here, the pass action is executed at each sequence match.

The example in Fig. 4.16 illustrates the difference between the two forms.

The body of both statements is identical, the intention is to cover a scenario when `write` followed by several (maybe zero) `busy` cycles is followed by a `read`. But their behavior is different. For the trace shown in Table 4.1 `c3` reports the hit

[10] See Chap. 6.

```
c3: cover property (@(posedge clk) write ##1 busy[*] ##1 read);
c4: cover sequence (@(posedge clk) write ##1 busy[*] ##1 read);
```

Fig. 4.16 Concurrent coverage

Table 4.1 Stimuli for c3 and c4

Clock cycle	write	busy	read
t	1	0	0
$t+1$	0	1	0
$t+2$	0	1	1
$t+3$	0	0	1

only once per attempt—at time $t+2$, while c4 reports the hit twice—at times $t+2$ and $t+3$.

In practice, property coverage is much more useful than sequence coverage. The main use of sequence coverage is to react on each sequence match in the pass action block to trigger some testbench actions.

For functional coverage collection use **cover property** statements.

4.7.3 Checking Coverage

Usually coverage is checked in simulation, but there is an added value to check coverage in FV too.

4.7.3.1 Checking Coverage in Simulation

Checking coverage in simulation is somewhat similar to checking transaction completion for assertions—the simulator reports when a given sequence of signals happens. For example, for the cover statement

```
cover property(@(posedge clk) write ##1 read);
```

the simulator will report each time it detects write signal followed by read.

The simulators usually register successful completion of evaluation attempts in a coverage database and collect the coverage statistics across available tests.

Table 4.2 Checking assertions in simulation and in FV

Assertion	Simulation	Formal verification
`assert`	Check whether an assertion is violated on a given simulation trace	Check whether an assertion can be violated while respecting all specified assumptions. If yes, may report a counterexample
`assume`	Check whether an assumption is violated on a given simulation trace (same as for assertions)	Use as a constraint when checking assertions and coverage points. The assumption correctness is not checked
`restrict`	Ignore	Use as a constraint when checking assertions and coverage points (same as for **`assume`**)
`cover`	Check whether a coverage condition is met on a given simulation trace	Check whether the coverage condition can be met while respecting all specified assumptions. If yes, may report a coverage witness

4.7.3.2 Checking Coverage in Formal Verification

In FV, the coverage condition (called also coverage point) is checked for its feasibility, that is, whether it can be reached when all the assumptions are met. FV tools either report that a coverage point cannot be reached, or provide a reachability witness, as discussed in Sect. 20.1.

4.8 Summary of Checking Assertions

Table 4.2 summarizes checking SVA assertions in simulation,[11] and in FV.

Exercises

4.1. What do assertions check?

4.2. Write four versions of an assertion checking that all bits of some packed bit vector are set to 1: immediate, deferred observed, deferred final, and concurrent. Explain the difference between them.

4.3. What is the purpose of assumptions?

[11] We ignore here the fact that assumptions may be used as constraints in random simulation. This feature of assumptions is seldom implemented in commercial simulation tools.

4.4. Write four versions of an assumption stating that the signal parity checksum is 0: immediate, deferred observed, deferred final, and concurrent. Explain the difference between them.

4.5. What is the difference between assumptions and restrictions?

4.6. Is it possible to specify actions with restrictions? Why?

4.7. What is the purpose of cover statements?

4.8. What is the difference between assertions and cover statements?

4.9. What is the difference between property and sequence coverage?

4.10. Write the following statements:

- Two inputs of the block must be mutually exclusive.
- Two outputs of the block must be mutually exclusive.
- We will conduct our verification session only in case when two inputs of the block are mutually exclusive.
- We want to check that two inputs of the block may be mutually exclusive.
- We want to check that two outputs of the block are not necessarily mutually exclusive.

4.11. What is the difference between the following statements is simulation and in formal verification:

- `p1: assert property(@(posedge clk)a)$display("a is high");`
- `p2: assume property(@(posedge clk)a)$display("a is high");`
- `p3: restrict property(@(posedge clk)a);`
- `p4: cover property(@(posedge clk)a)$display("a is high");`
- `p4: cover sequence(@(posedge clk)a)$display("a is high");`

4.12. Develop a small design that if simulated will illustrate a difference in assertion firing between an observed deferred assertion and a final one having the same Boolean expression.

Chapter 5
Basic Properties

Where there is no property there is no injustice.

— John Locke

Properties play a central role in SVA because they form the bodies of concurrent assertions. A *property* is a temporal formula that can be either true or false on a given trace.[1] As explained in Sect. 4.5.3, by trace we understand a series of all DUT signal values in time. Properties are interpreted over traces. In this chapter, we assume that each point on the trace corresponds to a tick of the clock on which the property is evaluated. This definition will not work for multiply clocked properties, but we currently limit our consideration to singly clocked properties only. The more general case is described in Chaps. 12, 13, and 22. Also, in formal verification (FV) we may assume that all the traces are infinite, which corresponds to the case when the property clock ticks infinitely many times. Of course, simulation traces are always finite. We informally describe the property semantics here, and defer the formal semantics of properties until Chap. 22.

This chapter is one of the central chapters of the book, and a good grasp of the material described here will be instrumental in understanding the rest of the book and in writing basic temporal assertions. Its contents are addressed both to users of simulation and to users of FV. FV issues are always commented so that the reader interested in only assertion simulation can skip them. However, we would like to stress that understanding FV issues may provide a deeper insight into the nature of temporal assertions, even for those who are not planning to use FV.

[1]An exception is *disabled* status of a property when the **`disable iff`** operator is used. See Sect. 13.1.1.

E. Cerny et al., *SVA: The Power of Assertions in SystemVerilog*,
DOI 10.1007/978-3-319-07139-8_5

Table 5.1 Basic property operators

Operator	Associativity
`not`	–
`nexttime`	–
`and`	Left
`or`	Left
`until`	Right
`until_with`	Right
`always`	–
`s_eventually`	–

We describe only the elementary and most commonly used property operators here. The less frequently used and more complex operators are discussed in Chap. 10.

In this chapter, we use the following convention: letters `a`, `b`, `c`, and, `e` denote integral expressions or signals; `p` and `q` denote properties. All letters may be optionally indexed. We enumerate clock ticks with integer numbers starting with 0. In the diagrams, we mark the ticks where a Boolean expression holds with a black dot, and the ticks where a property holds with a black triangle.

The property operators described in this chapter are summarized in Table 5.1. They are grouped by their precedence, from highest to lowest.

Properties can be built from simpler properties in a recursive manner. First we define "primitive" properties, and then we show how to build new properties using property operators from existing ones. The primitive properties are constructed from sequences. However, the notion of a sequence is less intuitive before introducing the notion of a property. Therefore, in this chapter we consider a Boolean, which is the simplest form of sequence, as the primitive form, and postpone the introduction of sequences until Chap. 6. In Chap. 6, we clarify why primitive properties are in fact sequential properties and that Boolean properties are the simplest case of sequential properties.

Example 5.1. To illustrate what we mean by recursive definition of properties, consider property **`nexttime always`** `e`, the meaning of which is explained later. This property is built by applying property operator **`nexttime`** to the simpler property **`always`** `e`. The latter property, in its turn, is built by applying property operator **`always`** to the primitive property `e`. □

5.1 Boolean Property

The simplest property is a *Boolean property*—an integral expression `e`, which is treated as Boolean. A Boolean expression is *true* if it has at least one bit set to 1, and *false* otherwise, i.e., when all its bits have values 0, `x` or `z`. Now, extending the meaning, we define a Boolean property informally for a point on a trace as follows:

Fig. 5.1 Boolean property

Boolean property `e` is *true in clock tick i* iff Boolean expression `e` is true in clock tick i.

Boolean Expressions and Properties. What is confusing is how we can distinguish between Boolean expression and Boolean property if they both have exactly the same syntax? The answer is simple: if a Boolean is used in a context where a property is expected, then it is a Boolean property. For instance, on the one hand, `e` in the property expression **`nexttime always`** `e` from Example 5.1 is a Boolean property since a property operator **`always`** expects a property as its argument. On the other hand, in the expression `~e`, `e` is a Boolean expression, but not a Boolean property, since the bitwise negation operator `~` requires a Boolean[2] expression, and it cannot accept a property as its argument.

Our definition of Boolean property explains the meaning of the sentence "`e` is true in clock tick i". However, in the general definition of property given in the beginning of this chapter we defined the property truth relative to the entire trace, and not in a specific clock tick. So what does it mean that `e` is true on a whole trace, and not just in a specific clock tick of this trace? This question is general and it may be answered for any property, not just for a Boolean one. Consequently, we give a general answer to this question for arbitrary property `p`.

Property `p` is true on a trace iff it is true in clock tick 0 of this trace.

This implies that Boolean property `e` is *true* iff `e` is true in the clock tick 0, as shown in Fig. 5.1.

Example 5.2. If a DUT contains three signals `a`, `b`, and `c`, and in clock tick 0 `a = 1`, `b = 1`, and `c = 0` then the property `a|b` holds since `a|b` is true in clock tick 0. □

Boolean expressions when used as Boolean properties cannot have side effects. For example, `a++ == b` cannot be used as Boolean property.

Although Boolean properties are very common as part of more complex properties, Boolean properties alone can be used for specifying the initial state of the system, as in the following example:

[2]Recall that according to our definition any integral expression is also a Boolean.

Example 5.3. Assume that initially the reset `rst` is active, and check that initially the value of `ready` is low.

Solution:

```
initial begin
  m1: assume property (@(posedge clk) rst);
  a1: assert property (@(posedge clk) !ready);
end
```

□

The behavior manifested by a Boolean property alone when placed in an **initial** procedure is useful because only one property evaluation attempt is executed. Placing such an assertion elsewhere would lead to unconditional evaluation of the Boolean at every assertion clock tick.

5.2 Nexttime Property

Property **nexttime** p, as its name suggests, is *true in clock tick i* iff property p is true in clock tick $i + 1$. For a trace, according to the general definition (Sect. 5.1), property **nexttime** p is *true* iff property p is true in clock tick 1, as shown in Fig. 5.2.

If p is a Boolean expression e, **nexttime** e means that e is true in the clock tick 1.

Multiple nexttime Operators. What happens if we apply the **nexttime** operator twice: **nexttime nexttime** p? According to the definition, it means that **nexttime** p holds in clock tick 1, which is equivalent to the statement that p holds in clock tick 2. Similarly, **nexttime nexttime nexttime** p means that p holds in clock tick 3, and so on. Since having a big chain of **nexttime** operators makes the property unreadable, SVA provides a shortcut **nexttime**[n], where n is an elaboration time constant. For example, **nexttime**[3] p is a shortcut for **nexttime nexttime nexttime** p. It is also legal to specify **nexttime**[0] p, which is roughly equivalent to just p in the case of singly clocked assertions.[3] Its semantics in multiply clocked properties is described in Chap. 12.

nexttime is seldom used on its own. The following example illustrates such usage.

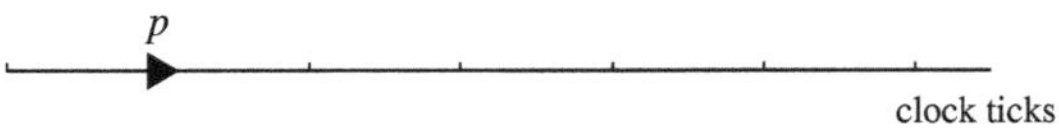

Fig. 5.2 **nexttime** property

[3]This is further discussed in Sect. 10.5.

Example 5.4. Reset `rst` should be low in clock `clk` tick 9.

Solution:

```
initial a1: assert property(@(posedge clk) nexttime[9] !rst);
```

Discussion: This property does not say anything about the behavior of `rst` in clock ticks 0–8. □

Finite and Infinite Traces. The definition of property **`nexttime`** is not as trivial as it may seem. When we require a property to be true in the next clock tick we implicitly assume that the clock ticks at least one more time. This observation holds when clocks tick infinitely many times, and hence traces are infinite. Infinite traces allow us to ignore this important question in the definition of property operators, namely, what happens if the clock stops ticking in the middle of evaluation. Note that this question is important even for Boolean properties: How can we define the truth of a Boolean property on the empty trace? We do consider property behavior on finite traces in Chaps. 10 and 22, but in this chapter we ignore it because we wish to concentrate on the main behavior of property operators.

Efficiency Tip. **`nexttime`** with a big factor is inefficient both in simulation and in FV. Try to keep the factor small, especially in complex assertions. In simulation, it is recommended not to exceed several hundred for simple argument properties, and in FV not to exceed a couple of tens. The common rule is the smaller the better.

5.3 Always Property

Property **`always`** `p` is *true in clock tick i* iff `p` is true in all clock ticks $j \geq i$. It follows (see Sect. 5.1) that property **`always`** `p` is *true* iff property `p` is true in every clock tick, as shown in Fig. 5.3.

Property **`always`** `p` defines a series of "instances" of property `p` starting at clock ticks 0, 1, ... We say that an **`always`** property defines a series of *evaluation attempts* of the underlying property. For the **`always`** property to be true, all of the underlying attempts must evaluate to true.

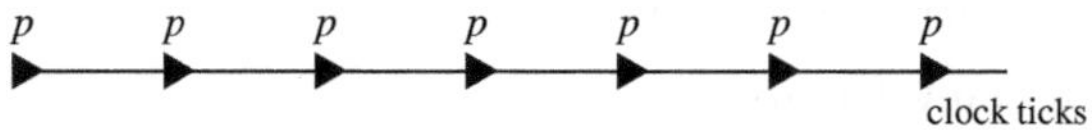

Fig. 5.3 **`always`** property

Example 5.5. For a Boolean expression e, **always** e is true iff e is true in every clock tick. □

Example 5.6. What is the meaning of **nexttime always** p?

Solution: According to the definition of **nexttime** property, **always** p should hold in clock tick 1. Therefore, property p should hold in every clock tick starting from clock tick 1. □

Example 5.7. What is the meaning of **always nexttime** p?

Solution: According to the definition of **always** property, **nexttime** p should hold in every clock tick. Therefore, property p should hold in every clock tick starting from clock tick 1.

Discussion: Properties **always nexttime** p and **nexttime always** p (Example 5.6) are equivalent. □

Example 5.8. What is the meaning of **always always** p?

Solution: According to the definition of **always** property, **always** p should hold in every clock tick. Therefore, property p should hold in every clock tick.

Discussion: It follows that **always always** p is equivalent to **always** p, and that the outer **always** in this case is redundant. □

Efficiency Tip. Simulation performance of **always always** p may be much inferior to that of **always** p (see Sect. 4.4.1) if it is required to maintain information about all evaluation attempts that are in progress (for example, for debugging purposes).

5.3.1 Implicit Always Operator

The **always** operator is useful for specifying system invariants. As we discussed in Sect. 4.4.6, all concurrent assertions placed outside procedural code are continuously monitored. This means that there is an implicit outermost **always** operator which defines a series of evaluation attempts. Consequently, the following assertions a1 and a2 are equivalent in the sense that either both pass or both fail.

```
a1: assert property (@clk p);
```

is equivalent to

```
initial a2: assert property (@clk always p);
```

However, simulation reporting may be different, as explained in Sect. 4.4.1, because a1 has as many evaluation attempts as there are clock ticks, while a2 has only one attempt.

The explicit **always** operator is rarely used. The vast majority of assertions are either written outside procedural code or inside **always** procedures[4] and thus have the implicit outermost **always** operator.

As in the case of the explicit double **always** operators, an **always** property in the body of a continuously monitored assertion may result in degradation of simulation performance.

Efficiency Tip. Do not explicitly specify the outermost **always** in continuously monitored assertions.

Example 5.9. Check that signal `sig` may only have values 0, 1, 2, or 4.

Solution:

```
a1: assert property (@(posedge clk) sig inside {0, 1, 2, 4});
```

assuming that `a1` is a standalone assertion.

Discussion: Following the efficiency tip, this assertion should *not* be written as

```
a2: assert property (@(posedge clk)
                     always sig inside {0, 1, 2, 4});
```

Although assertions `a1` and `a2` are equivalent, simulation performance of `a2` may be worse. □

5.4 S_eventually Property

Property **s_eventually** `p` is *true in clock tick* i iff `p` is true in some clock tick $j \geq i$. It follows that property **s_eventually** `p` is *true* iff property `p` is true in some clock tick, as shown in Fig. 5.4.

The reader may wonder why the keyword **s_eventually** has a prefix **s_** . As explained in Chap. 10, **s_eventually** is a strong operator, and in SVA names of strong operators have a prefix **s_** .

Similar to **always** `p`, property **s_eventually** `p` defines a series of evaluations of `p`. But for **s_eventually** `p` to succeed only requires that at least one of those evaluations succeed.

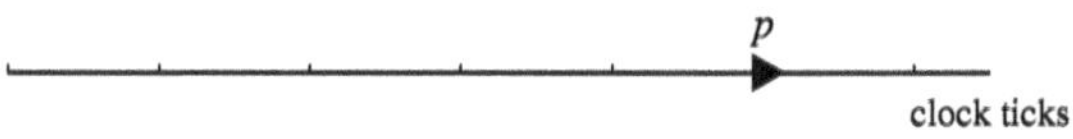

Fig. 5.4 **s_eventually** property

[4]See Chap. 14 for discussion about procedural concurrent assertions.

Example 5.10. Reset `rst` should eventually be deactivated.

Solution:

```
initial a1: assert property (@(posedge clk) s_eventually !rst);
```
□

Checking `s_eventually` Property. The `s_eventually` property differs significantly from all other properties described in this chapter.[5] Consider, for instance, assertion `a1` from Example 5.10. Suppose that we simulated the DUT for 10,000 clock ticks and `rst` remained high all the time. Does it mean that assertion `a1` is wrong? No, it does not, since it might pass if we simulate few more clock cycles. The failure of this assertion can only be observed on an infinite trace in which `rst` is always high. Such assertions are called *liveness* assertions, and they are studied in Chap. 21. Generally speaking, liveness assertions cannot be falsified in simulation, but only in FV. However, we can say something about property `s_eventually` even observing its behavior in simulation: It looks suspicious if the condition of `s_eventually` does not happen during simulation. For instance, when executing any test from the testbench, it is reasonable to expect that in Example 5.10 a zero value of `rst` will be observed. In the case when the condition of `s_eventually` has never been observed, a simulation tool usually issues a warning message at the end of simulation.[6]

Example 5.11. What does `s_eventually always p` mean?

Solution: According to the definition of `s_eventually` this property means that there exists some clock tick where `always p` is true. This is equivalent to the statement that `p` is true from some clock tick on.

Discussion: Strictly speaking, this property can neither fail nor pass in simulation. As mentioned earlier, a simulation tool may issue a warning message (or failure) at the end of simulation if in the last clock cycle a Boolean `p` is false. □

Efficiency Tip. Checking property `s_eventually always p` in simulation may be costly, especially if it is not in the scope of an `initial` procedure.

---

[5] Here, for convenience, we use the terms "property" and "assertion" interchangeably in the context of failure or success.

[6] As we explain in Chap. 10, since `s_eventually` is a strong operator, at the end of simulation when there are no more clock ticks and `rst` was high all the time, the simulator may declare failure of the property.

Example 5.12. Reset `rst` remains low starting from some moment.

Solution:

```
initial
  a2: assert property (@(posedge clk) s_eventually always !rst);
```

Discussion: Note the difference between assertion `a2` and assertion `a1` from Example 5.10. Assertion `a2` checks that `rst` at some moment becomes and *remains* low, whereas assertion `a1` only checks that `rst` becomes low.

Exercise 5.5 discusses the meaning of `a2` when it is a stand-alone assertion. As we mentioned above, the stand-alone version of this assertion may be inefficient in simulation. □

Example 5.13. What does property **`always s_eventually`** `p` mean?

Solution: According to the definition of **`always`** this property means that property **`s_eventually`** `p` holds in every clock tick i. This is equivalent to saying that for every clock tick i there is a clock tick $j \geq i$ where `p` is true. So,

```
always s_eventually p
```

holds if `p` is true infinitely many times, in other words, `p` is true infinitely often.

Discussion: This property can only be verified on infinite traces in formal verification because simulation traces are finite. □

Efficiency Tip. Checking property **`always s_eventually`** `p` may be costly in simulation.

Example 5.14. A pending request `req` should be eventually granted (`gnt` is asserted). This includes the case when the request is granted immediately.

More specifically, it is given that when the request becomes active it remains active until it is granted.

Solution: Consider an arbitrary clock tick i. The assertion should be satisfied in both of the following cases:

1. `req` is never asserted in all clock ticks $\geq i$.
2. `req` is asserted for the first time ($\geq i$) in some clock tick $j \geq i$, and `gnt` is asserted for the first time ($\geq j$) in some clock tick $k \geq j$. In this case, `req` will also be pending until clock tick k.

Both cases imply that the attempt of the property **`s_eventually`** `req -> gnt` starting in the clock tick i is satisfied, and assertion

```
a1: assert property(@(posedge clk) s_eventually req -> gnt);
```

covers the desired behavior of the pending request. Note that `a1` only checks the behavior of the grant in response to `req`. It does not check that a grant is not issued without a request. It also does not check that request, once issued, persists until granted.

Of course, this assumes that we verify the property on an infinite trace, hence in FV. In that case it is not difficult to see that the opposite is also true: Assertion `a1` implies the required behavior. Indeed, if `req` goes high in some clock tick i, it will remain high until granted by definition. The attempt of the property starting

```
s_eventually req -> gnt
```

in clock i guarantees that this `req` is granted.

Discussion: This property is a special case of property **`always s_eventually`** `p` described in Example 5.13. Assertion `a1` is standalone, and therefore it has an implicit **`always`** operator. Its performance in simulation may be poor if `req` is asserted and no `gnt` is asserted for a long time (if ever). Simulation performance efficiency is discussed in detail in Sect. 19.3.

In this example, we assumed that the request remains pending until granted. Example 6.23 describes the case of an arbitrary, not necessarily pending request.

□

Fairness.

Example 5.15. A device must be available infinitely many times. The device availability is indicated by high value of the `ready` signal.

Solution:

```
a1: assert property (@(posedge clk) s_eventually ready);
```

Discussion: This assertion is standalone, hence there is an implicit top-level **`always`** operator. □

The property **`always s_eventually`** `e` is very important for verifying liveness properties in FV. It expresses the notion of *fairness*. Fairness indicates that some resource eventually becomes available, as in Example 5.15. The absence of fairness is called *starvation*, the situation when the requested resource is never available: Imagine a car waiting at an intersection forever on the red light when the traffic lights are broken. We get back to the notions of fairness and starvation in Chap. 21. In simulation, of course, this assertion cannot be verified (it cannot fail) because simulation will end in a finite number of clock ticks.

5.5 Basic Boolean Property Connectives

The following Boolean connectives between properties exist in SVA:

- **`not`**—negation
  `not` `p` is true iff `p` is false.
- **`and`**—conjunction
  `p` **`and`** `q` is true iff both `p` and `q` are true.

- **or**—disjunction
  `p or q` is true iff either `p` or `q` (or both) are true.

The above definitions with obvious modifications apply also to the way property truth is determined in clock tick i. For example, `p and q` is true in clock tick i iff both `p` and `q` are true in clock tick i.

In the special case of Boolean properties, it is possible to rewrite Boolean property connectives in a different way. For example, `e1 and e2` means that both `e1` and `e2` are true, which can be also expressed as `e1 && e2`.[7] The domain of the operators `&&` and **and** is different: `&&` may be used with Boolean expressions only, while **and** requires sequence or property arguments. Since Boolean expressions in this case may also be considered as Boolean properties, both operators may be applied to them. Note also that `&&` has greater precedence than **and**.

Similarly, `e1 or e2` is equivalent to `e2 || e2`, but `not e` is equivalent to `!e` only if the property clock eventually ticks (see Chap. 22).[8]

Example 5.16. The following expression is syntactically illegal: `(a and b) || c`. Although `a and b` is logically equivalent to `a && b`, it is not a Boolean expression. The operator `||`, however expects both its operands to be Boolean expressions. □

Example 5.17. What does `not always p` mean?

Solution: According to the definition of property **not**, this property is true iff `always p` is false, which means that `p` is false at least in one clock tick. This is exactly `s_eventually not p`. Similarly, `not s_eventually p` is `always not p`.

Discussion: For a Boolean expression `e`, `not always e` may be rewritten as `s_eventually !e` and `not s_eventually e` as `always !e`. □

Example 5.18. Reset `rst` must be asserted during the first two cycles.

Solution:

```
initial a1:
  assert property (@(posedge clk) rst and nexttime rst);
```

Discussion: Note that `rst && nexttime rst` is syntactically illegal: `&&` expects both of its operands to be Boolean expressions, while `nexttime rst` is a property, but not a Boolean expression.

A more elegant way to write the same property is described in Sect. 6.5. □

---

[7]Even though `&&` is a short circuit operator (that is, its second operand is not evaluated if the first operand is evaluated to false), and **and** is not, there is no difference between them in this case, since expressions used in assertions cannot have side effects. See Sect. 5.1.

[8]The difference may not be observable in simulation depending on whether the simulator takes into account the strength of properties at the end of simulation. `not e` is a strong property, while `!e` is weak.

Example 5.19. What is the meaning of `(always p) and (always q)`?

Solution: According to the definition of property **and** this property is true iff both `p` and `q` hold in each clock tick, that is, the original property is equivalent to `always (p and q)`.

Discussion: `(always p) or (always q)` is *not* equivalent to `always (p or q)`. Consider a case when `p` holds in all odd clock ticks, and `q` holds in all even ticks. Then `always (p or q)` is true, whereas `(always p) or (always q)` is false. □

5.6 Until Property

Property `p until q` is *true in clock tick i* iff `p` is true in every clock tick $j \geq i$ until, but not including, the first clock tick $k \geq i$ where `q` is true. If there is no such k, p should be true in all clock ticks $j \geq i$.

It follows that property `p until q` is *true* iff the property `p` is true in every clock tick until (but not including) the first clock tick where `q` is true. See Fig. 5.5. If `q` never happens `p` should be true forever.

Note that `p until q` does not mean that `p` cannot be true starting from the clock tick when `q` becomes true. It only means that `p` *does not have* to be true after `q` becomes true for the first time. Also, the operator **until** is *nonoverlapping*: `p` does not have to be true when `q` becomes true for the first time (though, of course, `p` *may* be true at this moment).

There is also an *overlapping* version of **until** called **until_with**. The only difference between **until** and **until_with** is that for **until_with** to be true `p` *must* be true at the moment `q` becomes true for the first time, as stated in the following definition:

Property `p until_with q` is true iff the property `p` is true in every clock tick until (and including) the first clock tick where `q` is true. See Fig. 5.6. If `q` never happens `p` should be true forever. We leave the definition of the truth of property **until_with** in clock tick i as an exercise to the reader.

`p until_with q` is equivalent to `p until (p and q)`. (Why? See Exercise 5.8.)

Example 5.20. Table 5.2 contains an initial trace fragment of signals `a`, `b`, `c`, `d`, and `e`.

- `a until b` is true since `b` is true in clock tick 0.
- `a until c` is true since `a` is true in clock tick 0, and `c` is true in clock tick 1.

Fig. 5.5 **until** property

Fig. 5.6 **until_with** property

Table 5.2 Initial trace fragment for Example 5.20

Clock tick	a	b	c	d	e
0	1	1	0	0	0
1	1	1	1	0	0
2	0	1	0	1	0
3	0	0	1	1	1

- `a` **`until`** `d` is true since `a` is true in clock ticks 0 and 1, and `d` is true in clock tick 2.
- `a` **`until`** `e` is false since `a` is false in clock tick 2, and `e` is false in clock ticks 0, 1, and 2.
- `b` **`until_with`** `a` is true since both `a` and `b` are true in clock tick 0.
- `b` **`until_with`** `c` is true since `b` is true in the clock ticks 0 and 1, and `c` is true in clock tick 1.
- `b` **`until_with`** `e` is false since `b` is false in clock tick 3, and `e` is false in clock ticks 0, 1, and 2.

□

Example 5.21. `ready` should be low until `rst` becomes inactive for the first time.

Solution:

```
initial a1: assert property(@(posedge clk) !ready until !rst);
```

□

Example 5.22. There cannot be a `read` before the first `write`:

Solution:

```
initial
  a2: assert property (@(posedge clk) !read until_with write);
```

□

Exercises

5.1. Write a restriction saying that initially all bus drivers are disconnected (have a high impedance value).

5.2. Write the following assertion: `rdy` should be low while `rst` is active.

5.3. What do the following properties mean?

(a) **`always nexttime always`** `p`
(b) **`nexttime always nexttime`** `p`

5.4. Discuss the usage of the **`always`** operator in assertions.

5.5. What is the meaning of the following assertion (note that it does not belong to an **initial** procedure)?

```
a1: assert property (@(posedge clk) s_eventually always !rst);
```

5.6. What is the meaning of the following properties?

(a) `s_eventually nexttime p`
(b) `nexttime s_eventually p`
(c) `s_eventually nexttime always p`
(d) `always nexttime s_eventually p`

5.7. What do the following properties mean?

(a) `1 until q`
(b) `p until 1`
(c) `0 until q`
(d) `0 until_with q`
(e) `p until 0`
(f) `nexttime(p until q)`
(g) `(nexttime p)until q`
(h) `p until nexttime q`
(i) `p until_with nexttime q`
(j) `p1 until (p2 until p3)`
(k) `p1 until_with (p2 until_with p3)`
(l) `always(p until q)`
(m) `(always p)until q`
(n) `p until always q`
(o) `p1 until (p2 until always p3)`

5.8. Prove that `p until_with q` is equivalent to `p until (p and q)`.

Chapter 6
Basic Sequences

A sequence works in a way a collection never can.

— George Murray

In Chap. 5, we showed how to build complex properties from the elementary building blocks. We considered Boolean properties as the simplest building block. SVA provides sequences as more elaborate building blocks for properties. Since the simplest sequence is a Boolean expression, we could say that properties are built not from Boolean expressions, but from sequences.

A *sequence* is a rule defining a series of values in time. A sequence does not have a truth value, it has one initial point and zero or more *match*, or *tight satisfaction* points. Like properties, sequences are clocked. If the clock is not written explicitly, we assume that the sequence inherits this clock from the property to which it belongs. Starting from a specific initial point in a trace, a sequence defines zero or more *finite* fragments on this trace, each beginning at the initial point and ending at a tight satisfaction point. We will call the length of a trace fragment defined by a sequence match simply the length of the sequence match. In the following sections, we define the sequence match separately for each kind of a sequence. Before we proceed to the accurate definitions, we informally illustrate the notion of a sequence on the following example.

Example 6.1. Sequence `a ##[1:2] b` defines the following scenario: `a` is followed by `b` in one or two clock ticks. Let the initial point of this sequence be clock tick 2. Then this sequence has a match if `a` is true in clock tick 2 and either `b` is true in clock tick 3 or `b` is true in clock tick 4. Thus, the following matching outcomes are possible:

E. Cerny et al., *SVA: The Power of Assertions in SystemVerilog*,
DOI 10.1007/978-3-319-07139-8_6

1. a is false in clock tick 2 or b is false in clock ticks 3 and 4. In this case, the sequence has no match.
2. a is true in clock tick 2, b is true in clock tick 3, and b is false in clock tick 4. In this case, the sequence has a single match at clock tick 3.
3. a is true in clock tick 2, b is false in clock tick 3, and b is true in clock tick 4. In this case, the sequence has a single match at clock tick 4.
4. a is true in clock tick 2 and b is true in clock ticks 3 and 4. In this case, the sequence has two matches, or two tight satisfaction points: 3 and 4.

Sequence `a ##[1:2] b` defines 0, 1, or 2 trace fragments. In case 1 it defines zero fragments, in case 2 it defines one fragment, 2:3, in case 3 it also defines one fragment, 2:4, and in case 4 it defines two fragments, 2:3. and 2:4. □

In this chapter, we use the following conventions: letters a, ..., e denote Boolean expressions, r and s denote sequences, and p and q denote properties. We numerate clock ticks with integer numbers starting with 0. In diagrams, we designate trace fragments defined by sequence matches with ovals.

The sequence and property operators described in this chapter are summarized in Table 6.1. They are grouped by their precedence, from highest to lowest. Additional sequence operators are covered in Chap. 11.

6.1 Boolean Sequence

Boolean expression e defines the simplest sequence—a *Boolean sequence*. This sequence has a match (or a tight satisfaction point) at its initial point if e is true. Otherwise, it does not have any satisfaction points at all. This is illustrated in Fig. 6.1.

Example 6.2. The initial fragment of the trace of signals a and b is shown in Table 6.2.

Table 6.1 Basic sequence and property operators

Operator	Associativity
[*...]	–
[*]	–
[+]	–
##	Left
or	Left
|->	Right
|=>	Right

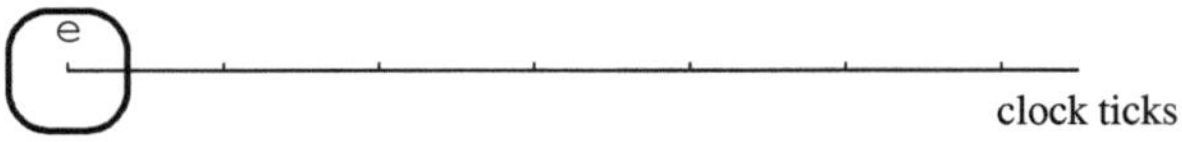

Fig. 6.1 Boolean sequence

Table 6.2 Initial trace fragment from Example 6.2

Clock tick	a	b
0	1	1
1	0	0
2	0	1
3	1	1

Boolean sequence `a && b` matches trace fragment 0:0 if its initial point is 0, and trace fragment 3:3 when its initial point is 3. For initial points 1 and 2 there are no sequence matches. □

6.2 Sequential Property

Although a sequence itself cannot be true or false, it is possible to associate with a sequence a *sequential property* (or *sequence property*) in the following way:[1]

The *sequential property* defined by sequence `s` is true in clock tick i iff there is no finite trace fragment $i:j$ witnessing inability of sequence `s` with the initial point i to have a match. Sequence `s` should not admit an empty match (the notion of empty match is explained below in this section).

This definition is applicable only to sequential properties in the context of assertions or assumptions. In the context of cover statements, the definition of sequential property is different, as explained in Chap. 18. Except for Chap. 18 we will assume that in all examples sequential properties are written in the context of assertions or assumptions.

Although it is not easy to understand the full meaning and the rationale of the definition of a sequential property at this point, because we have not yet described the SVA constructs in which all nuances of this definition come into play, we need this definition now to build properties from sequences. We will explain some of its aspects here, while the other aspects will become clear to the reader only later.

Consider the most important special case of this definition: if a sequence has at least one match, then the corresponding sequential property is true. The entire definition is broader since it allows in some cases sequential properties to be true even if their sequences do not have any match. However, if there is some number $L > 0$ such that all matches of sequence `s` have a length $\leq L$ then if `s` does not have any match then the sequential property `s` is false. Indeed, in this case the fact that sequence `s` does not have any match on a trace fragment of length L witnesses its inability to have any match. We will call such sequences *bounded sequences*. It is easy to see that all Boolean sequences are bounded, as all their matches are one clock cycle long.

[1]Actually, there is more than one way to associate a property with a sequence, as explained in Chaps. 10 and 22. The definition provided in this section relates to the *weak* sequential property.

Example 6.3. Sequence `a ##[1:2] b` informally described in Example 6.1 is bounded.

Solution: This sequence can only have matches of length 2 or 3, and its match upper bound $L = 3$: if this sequence does not have a match on a trace fragment of three clock cycles, it does not have matches at all. □

The definition of sequential properties for bounded sequences may be simplified:

The sequential property corresponding to bounded sequence `s` is true iff sequence `s` has at least one nonempty match.

So, what is the nature of *unbounded* sequences? They should admit arbitrarily long matches. Examples of unbounded sequences are provided in Sect. 6.10.

Another point to be clarified is the notion of *empty match*. The match is empty if the trace fragment it defines is empty. We will provide examples of empty match in Sects. 6.5.1 and 6.8. Note that a Boolean sequence cannot have an empty match. It either has a match of size 1 if its Boolean expression is true, or it does not have matches at all.

Pay attention to the clock tick where the truth value of a sequential property is defined.

The truth value of sequential property `s` corresponds to the *initial* point of sequence `s`, and *not* to the point of its tight satisfaction.

Example 6.4. Let `a` be true in clock tick 2, and false in all other clock ticks, and `b` be true in clock tick 3, and false in all other clock ticks. Then sequential property `a ##[1:2] b` is true in clock tick 2 (not 3!), and false in all other clock ticks. □

Let us apply the definition of a sequential property to a Boolean sequence. Since Boolean sequences are bounded, Boolean sequential property `e` is true iff Boolean sequence `e` has a match, that is, when `e` is true. We come to the conclusion that Boolean sequential properties are exactly Boolean properties we described in Sect. 5.1. Therefore, sequential properties generalize Boolean properties, and it is possible to define all property operators from Chap. 5 independently of the general sequential properties, as we mentioned in the introduction to that chapter.

6.3 Sequence Concatenation

From two sequences `r` and `s`, one can build a new sequence `r ##1 s` by *concatenating* these two sequences: there is a match of sequence `r ##1 s` if there is a match of sequence `r` and there is a match of sequence `s` starting from the clock tick immediately following the match of `r`, as shown in Fig. 6.2.

In other words, a finite trace matches `r ##1 s` iff it can be split into two adjacent fragments, the first one matching `r`, and the second one matching `s`. If both operands of sequence concatenation are bounded sequences, its result is also bounded.

Example 6.5. What is the meaning of `a ##1 b`, where `a` and `b` are Boolean expressions?

Solution: Let the initial point be clock tick i. Sequence `a ##1 b` has a match iff `a` is true in clock tick i and `b` is true in clock tick $i + 1$. Sequence `a ##1 b` may have only one tight satisfaction point, in clock tick $i + 1$. Therefore, this sequence means that `b` immediately follows `a`.

Discussion: Tight satisfaction of sequence `a ##1 b` depends neither on the value of `b` in clock tick i nor on the value of `a` in clock tick $i + 1$. Although this looks obvious, this is a source of confusion for many people who *erroneously* believe that for this sequence match `b` must become true for the first time in clock tick $i + 1$. □

Example 6.6. Write a sequence capturing the following scenario: request `req`, immediately followed by retry `rtry`, immediately followed by acknowledgment `ack`.

Solution: `req ##1 rtry ##1 ack`. □

6.3.1 *Multiple Delays*

We will begin this section with a motivational example.

Example 6.7. Write the following sequence: request `req` followed by acknowledgment `ack` in two cycles.

Solution: This description is equivalent to "request, immediately followed by anything, immediately followed by acknowledgment". "Anything in a given clock tick" means a Boolean sequence that has a match in this clock regardless of the values of `req` and `ack`. To match, the corresponding Boolean expression should be true. Recall (Sect. 5.1) that true may be expressed with any nonzero known value, for example, value 1. So, the desired sequence may be written as `req ##1 1 ##1 ack`. □

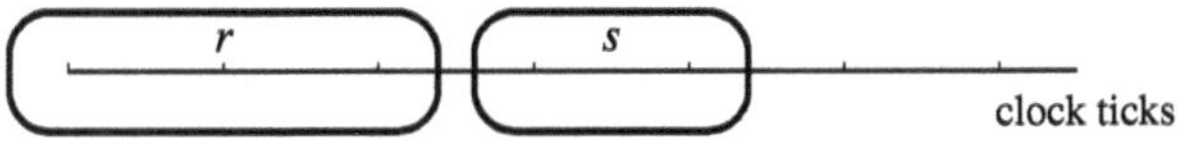

Fig. 6.2 Sequence concatenation

Fig. 6.3 `r ##2 s`

Example 6.7 illustrates a typical situation when two sequences are not adjoining, but there is a constant number of clock ticks between them (Fig. 6.3). There is a special syntax to capture this situation:

```
r ##n s
```

`n` must be a nonnegative elaboration time integral constant. We will call this interval between two sequences in clock cycles *delay* (not to be confused with the delay operator # in SystemVerilog).

Efficiency Tip. Big delay values are inefficient both in simulation and in FV.

Example 6.8. Using this syntax, the sequence from Example 6.7 may be rewritten as `req ##2 ack`. □

6.3.2 Top-Level Sequential Properties

As stated in Sect. 6.2, there is a sequential property associated with each sequence not admitting an empty match. With the exception of Boolean sequential properties, top-level sequential properties are relatively rare in assertions and assumptions, and they are normally used to specify reset sequences.[2]

Example 6.9. Reset `rst` must be initially high and be low in clock tick 20.

Solution:

```
initial a1: assert property (@(posedge clk) rst ##20 !rst);
```

Discussion: We do not claim anything about the reset behavior in all clock ticks other than 0 and 20. □

You should be very careful when using sequential properties in continuously monitored assertions and assumptions since their meaning may be different from your intent, as illustrated by the following examples.

Example 6.10. What does the following assertion mean?

```
a1: assert property (@(posedge clk) a ##1 b);
```

[2]Top-level sequential properties are very common in cover statements, see Chap. 18.

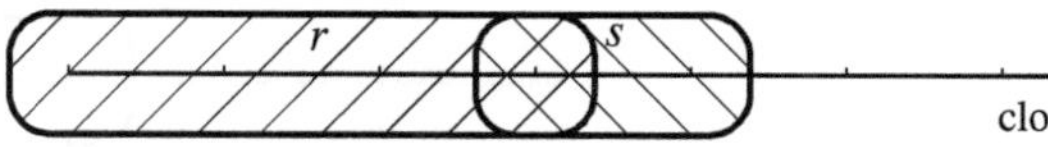

Fig. 6.4 Sequence fusion

Solution: Since this assertion is continuously monitored (has an implicit outermost **always** operator), sequential property `a ##1 b` must be true in each clock tick. Hence, `a` must be true in each clock tick, and `b` must be true starting form clock tick 1.

Discussion: This assertion does *not* mean interleaving of `a` and `b`. □

Example 6.11. We want to state that the value of `sig` toggles every cycle: 0101... or 1010.... The following assertion

```
a1: assert property (@(posedge clk) sig ##1 !sig);
```

does *not* check this condition. Similar to Example 6.10, it means that `sig` is true in each clock tick, and also that `sig` is false starting from clock tick 1. Therefore, this assertion is contradictory: it requires that starting from clock tick 1 `sig` be simultaneously true and false. Example 7.26 explains how to implement this assertion correctly. □

6.3.3 Sequence Fusion

Sequence fusion is an overlapping concatenation. The fusion of sequences `r` and `s`, denoted as `r ##0 s`, is matched iff for some match of sequence `r` there is a match of sequence `s` starting from the clock tick where the match of `r` happened (see Fig. 6.4). Note the difference between sequence concatenation `r ##1 s` and sequence fusion. For sequence concatenation, we start matching sequence `s` from the *next* clock tick after a match of `r` has happened, while for sequence fusion we start matching `s` from the *same* clock tick of the match of `r`.

Example 6.12. What is the meaning of fusion of two Boolean sequences `a` and `b`?

Solution: `a ##0 b` can have a match iff both `a` and `b` are true simultaneously. Therefore, `a ##0 b` is semantically equivalent to `a && b`. Note, however, that `a ##0 b` and `a && b` are not syntactically interchangeable. `a ##0 b` is a sequence, so it cannot be used as an operand in a Boolean expression. For example, `a && b || c` (where `c` is a Boolean expression) is legal, whereas `(a ##0 b) || c` is syntactically illegal.[3] □

Example 6.13. What is the meaning of `(a ##1 b) ##0 (c ##1 d)`, where `a`, `b`, `c`, and `d` are Boolean expressions?

---

[3] Also `&&` is not the same as `##0` when match items are attached to the first sequence, see Chap. 16.

Fig. 6.5 `##3 s`

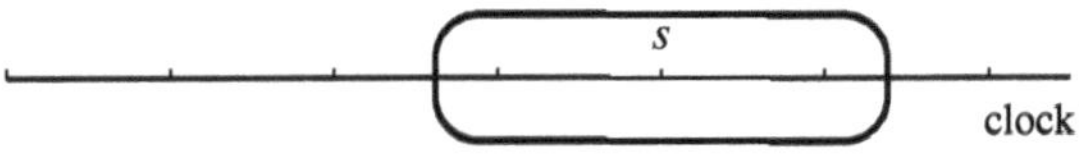

Solution: Let the initial point of the sequence be clock tick i. Sequence `a ##1 b` has a match iff `a` is true in clock tick i and `b` is true in clock tick $i + 1$. The match of sequence `a ##1 b` happens in clock tick $i + 1$. Therefore, for the fusion to have a match, sequence `c ##1 d` should have a match starting from clock tick $i + 1$. It means that `c` should be true in clock tick $i + 1$, and `d` should be true in clock tick $i + 2$. Bringing it all together, we have that sequence `(a ##1 b) ##0 (c ##1 d)` is equivalent to sequence `a ##1 b && c ##1 d`. □

Example 6.14. Write a sequence describing two back-to-back transactions. The transactions are represented with sequences `trans1` and `trans2`.

Solution: Consider two interpretations:

- `trans2` starts in the clock tick when `trans1` finishes.
- `trans2` starts in the clock tick next to the endpoint of `trans1`.

The first scenario may be expressed as `trans1 ##0 trans2`, while the second scenario may be expressed as `trans1 ##1 trans2`. □

We will provide more examples of sequence fusion in the following sections.

6.3.4 *Initial Delay*

We will start this section with a motivational example:

Example 6.15. Skip $n > 0$ cycles before matching sequence `s`.

Solution: `1 ##n s`. Recall that value 1 means true, matching anything at the initial clock tick of the resulting sequence. □

The situation described in Example 6.15 is typical, and for convenience there is a special syntax to specify the number of clock ticks to be skipped before beginning a sequence match (Fig. 6.5), i.e., to specify *initial sequence delay*: `##n s`. The delay is the number of clock ticks and must be a nonnegative elaboration time integral constant. Our definition works for nonzero `n`. The case of `n` $=$ 0 is considered in Sect. 6.9.

Efficiency Tip. Large initial delays are inefficient in FV, and may also negatively affect simulation performance. See Sect. 19.3 for details.

Example 6.16. The value of `a` should be always true starting from clock tick 2.

Solution:

```
a1: assert property (@(posedge clk) ##2 a);
```

Discussion: The same intent may be expressed using property operator **`nexttime`**:

```
a2: assert property (@(posedge clk) nexttime[2] a);
```

□

6.4 Suffix Implication

In Sect. 6.2, it was shown how properties may be built from sequences by promoting sequences to sequential properties. There are additional ways to build properties from sequences, the most important one being the *suffix implication*. A suffix implication is built from a sequence (`s`) and a property (`p`). `s` is called the *antecedent*, and `p` is called the *consequent*. A suffix implication is true when its consequent is true upon completion of its antecedent. Below we provide a more accurate definition.

There are two versions of suffix implication: *overlapping*, denoted as `s |-> p`, and *nonoverlapping*, denoted as `s |=> p`. In the overlapping implication the consequent is checked starting from the moment of *every* nonempty match of the antecedent. In the nonoverlapping implication the consequent is checked starting from the next clock tick after each match of the antecedent.

Nonoverlapping implication `s |-> p` is *true in clock tick* i iff for *every* tight satisfaction point $j \geq i$ of `s` with initial point i, property p is true in clock tick j. For each match of the antecedent, the consequent is separately evaluated. According to the property truth definition from Sect. 5.1, property `s |-> p` is true iff it is true in clock tick 0.

Nonoverlapping implication `s |=> p` is defined as
`(s ##1 @$global_clock 1)|-> p`. The meaning of this definition is explained in Chap. 12. For singly clocked assertions, this definition may be simplified:
`s ##1 1 |-> p`.

We want to stress that both overlapping `s |-> p` and nonoverlapping `s |=> p` implications and their consequent `p` are *properties*, whereas their antecedent is a *sequence*, and it is *not* promoted to a sequential property.

Except for Boolean assertions, suffix implication is the most common way of building assertions. Antecedent `s` represents a *triggering condition*: when this condition holds, consequent `p` is checked. The suffix implication is very often used with stand-alone assertions (having an implicit outermost **`always`** operator)—the antecedent defines "interesting" attempts where we want to check the consequent.

Example 6.17. When `rdy` is asserted `rst` must be low.

Solution: We can use the overlapping implication. When `rdy` is true, Boolean sequence `rdy` has a match. At this point we need to check a Boolean property stating that `rst` is false:

```
a1: assert property (@(posedge clk) rdy |-> !rst);
```

Discussion: When both antecedent and consequent are Boolean, the suffix implication is equivalent to the logical implication:

```
a2:assert property (@(posedge clk) rdy -> !rst);
```

Of course, it is illegal to use a suffix implication in a Boolean expression. Logical implications may be used in all kinds of assertions: immediate, deferred, and concurrent, while suffix implications are allowed in concurrent assertions only. □

Example 6.18. `done` must be asserted in the next clock tick after `sent` has been asserted.

Solution: This assertion means that if in some clock tick `sent` has been asserted then at the next clock tick `done` should be asserted. There are several possibilities to divide this property into an antecedent and a consequent. We can say that the antecedent is `sent`, and the consequent is `done` starting from the next clock tick. The resulting assertion will be:

```
a1: assert property(@(posedge clk) sent |-> nexttime done);
```

The same intent may be expressed using sequential property in the consequent:

```
a2: assert property(@(posedge clk) sent |-> ##1 done);
```

We can move the delay into the antecedent to get the same effect:

```
a3: assert property(@(posedge clk) sent ##1 1 |-> done);
```

Note that assertion

```
a4_illegal: assert property(
  @(posedge clk) sent nexttime 1 |-> done);
```

is *illegal* since the antecedent of a suffix implication is a sequence, not a property, and using property operator **nexttime** is forbidden in sequences.

Of course, the best way to implement the same assertion is using non-overlapping implication as shown below.

```
a5: assert property(@(posedge clk) sent |=> done);
```

Assertion `a5` is equivalent to assertion `a3` by the definition of the nonoverlapping implication. □

Example 6.19. In Example 6.18 we showed that it is possible to move a unit delay from antecedent to consequent and vice versa when the first operand in the consequent (or the last one in the antecedent) is 1. However, in the case of an arbitrary operand this is wrong. For example, property `write ##1 done |=> read` means that if `done` follows `write` then `read` must be asserted in the next clock tick after `done`. If there is no `done` after `write` the property passes. Property `write |=> done ##1 read` means a different thing: each `write` must be followed by a series of `done` and `read`. If there is no `done` after `write` the property fails. □

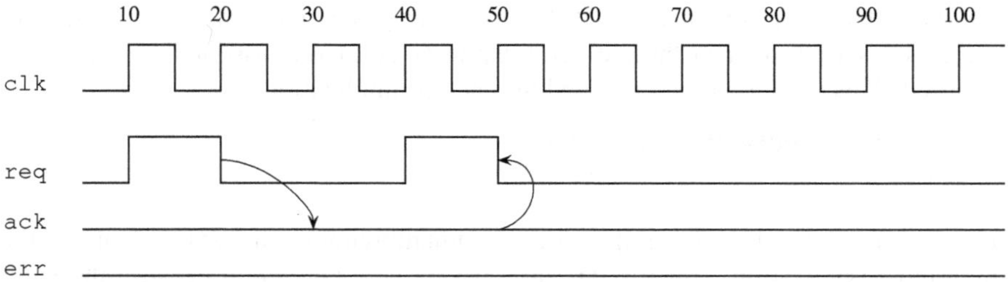

Fig. 6.6 Overlapping implication

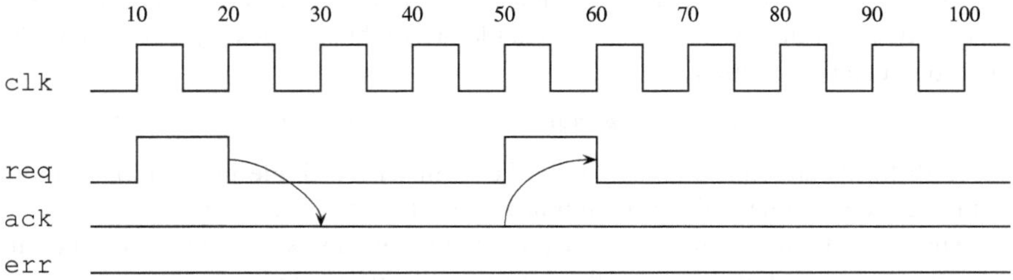

Fig. 6.7 Nonoverlapping implication

Example 6.20. If there is no acknowledgment `ack` within three clock ticks after request `req` was issued, request `req` must be resent unless an error indicator `err` is set.

Solution: From this formulation it is not clear when the request should be resent exactly: in three or in four clock ticks? We will consider both cases (assertions `a1` and `a2`) differing by the type of the implication:

```
a1: assert property (@(posedge clk)
  req ##1 !ack[*3] |-> req || err);
a2: assert property (@(posedge clk)
  req ##1 !ack[*3] |=> req || err);
```

The corresponding timing diagrams are shown in Figs. 6.6 and 6.7. □

6.4.1 *Nested Implication*

Suffix implications can be nested as illustrated in the following example.

Example 6.21. If `start` is asserted two clock ticks before `send`, then acknowledgment `ack` should arrive in three clock ticks after `send` was asserted.

Solution: This assertion means that if `start` is asserted and if two clock ticks later `send` is asserted, then in three clock ticks after `send` was asserted, `ack` must be asserted. This can be directly mapped into nested implications:

```
a1: assert property (@(posedge clk)
  start |-> ##2 send |-> ##3 ack);
```

Discussion: Nested suffix implication is unambiguous: the antecedent of the outermost implication is `start`, and not `start |-> ##2 send` because the latter expression is a property, and not a sequence, whereas the antecedent must be a sequence.

The same assertion may be reformulated in the following way: each time `send` is issued two cycles after `start`, `ack` should arrive three cycles after `send`, and the assertion can be rewritten as

```
a2: assert property (@(posedge clk) start ##2 send |-> ##3 ack);
```

Both forms are equivalent. Usually the form with a single implication is more intuitive than the one with nested implications, and it is easier to debug.

The rule of transforming nested implications works with appropriate modifications for any initial delays in the consequent. For example, `r |=> s |=> p` is equivalent to `r ##1 s |=> p`, and `r|->s|->p` is equivalent to `r ##0 s |-> p`.

□

6.4.2 Examples

In this section, we provide several important examples illustrating use of various sequence and property operators combined with suffix implication.

Example 6.22. Request `req` should be active until `grant` is asserted.

Solution: This assertion may be formulated as "Whenever the request is high, it should remain high until (not including) `grant` is asserted":

```
a1: assert property(@(posedge clk) req |-> req until grant);
```

Discussion: If the request should remain asserted also in the first clock tick when the grant is asserted the assertion should be modified as:

```
a2: assert property(@(posedge clk)
                    req |-> req until_with grant);
```

□

Efficiency Tip. Boolean antecedents in suffix implication are efficient in FV, but they may be not very efficient in simulation when the consequent requires a long time for its completion. See Chap. 7 and Sect. 19.3 for a detailed discussion.

Example 6.23. Request `req` must be granted. This assertion means that each time request `req` is high, grant `gnt` should be high in some clock tick in the future.

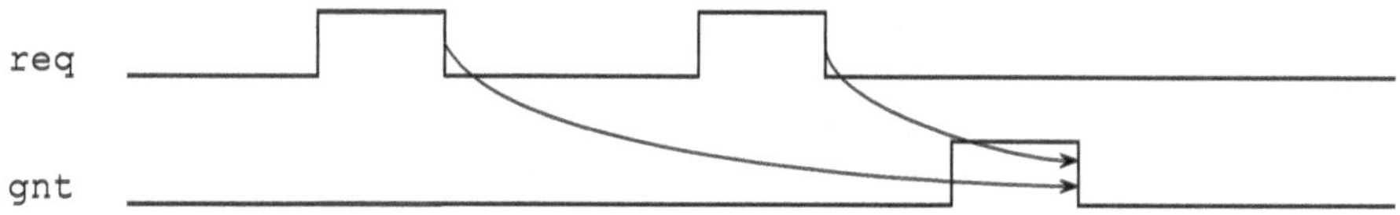

Fig. 6.8 Two requests corresponding to the same grant

Solution:

```
a1: assert property (@(posedge clk) req |=> s_eventually gnt);
```

Discussion: This assertion does not distinguish between grants to different requests. For example, there may be the same grant for several requests, as shown in Fig. 6.8. This example reflects the fact that the normal semantics of assertions is global, and not pipelined: there is no easy way to distinguish between different attempts (transactions) of the same assertion. The pipelined semantics in our case means that each request should have its *own* grant. Chapter 15 explains how to implement pipelined semantics in assertions using local variables. □

Example 6.24. After request `req` has been sent, acknowledgment `ack` should come before data (`data_ready`).

Solution:

```
a1: assert property (@(posedge clk)
  req |-> !data_ready until_with ack);
```

Discussion: In this implementation, the acknowledgment is allowed to be issued in the same clock tick as the request; see Exercise 6.5 for the case when the acknowledgment is expected to come strictly after the request. If the request is never acknowledged – there is neither acknowledgment nor data—the assertion passes.

□

6.4.3 Vacuous Execution

The definition of the suffix implication states that the consequent is checked starting from the moment of *every* nonempty match of the antecedent. It follows that in the case when the antecedent does not have a match, the suffix implication holds. This case is called *vacuous execution* of the implication.

If an assertion is continuously monitored, it is natural that many of its attempts terminate vacuously. For instance, if in Example 6.22 each attempt is nonvacuous then the request is always active, which is not likely to happen in practice. But if all the assertion attempts pass vacuously, it indicates a serious problem in validation. What would you say about a civil engineer who constructed a bridge, and to the question "Will this bridge withstand if a heavy truck crosses it?" he will answer

"Of course, it will. There was no truck crossing it until now, and the bridge hasn't collapsed yet"?

There exist different definitions of vacuity [13, 36], the LRM provides a minimal set of rules that the tools are expected to check. We are not going to provide the entire list of vacuous scenarios in this book. The main source of the vacuous execution for assertions is the case we described: when the antecedent is false. Chapter 10 provides further details on vacuity.

Example 6.25. Property `ok |-> !err` passes vacuously iff `ok` is false. But the equivalent property `err |-> !ok` passes vacuously iff `err` is false. Therefore, the assertion vacuity depends on the exact style in which it is written. □

6.5 Consecutive Repetition

We will begin this section with a motivational example.

Example 6.26. Write a sequence stating that a transmission phase lasting three consecutive cycles is followed by a receiving phase lasting two consecutive cycles. The transmission phase is represented by signal `trn`, and the receiving phase is represented by signal `rcv`

Solution: `trn ##1 trn ##1 trn ##1 rcv ##1 rcv`.

Discussion: This sequence defines a trace fragment of five clock cycles such that in the first three cycles `trn` is true, and in the last two cycles `rcv` is true. For example, `trn` may also be true in the fourth or the fifth cycle of the fragment, and also in any clock cycle outside this trace fragment. □

The sequence from Example 6.26 looks verbose, but the situation it describes is typical. In SVA, there is a special operator to denote a *consecutive repetition* of a sequence `s` `n` times, where `n` is a nonnegative elaboration time integral constant: `s[*n]` (see Fig. 6.9).

Example 6.27. Using the shortcut notation for the consecutive repetition, the sequence from Example 6.26 may be rewritten as `trn[*3] ##1 rcv[*2]`. □

Example 6.28. The assertion from Example 5.18 "Reset `rst` must be asserted during first two cycles" may be more elegantly expressed using sequences:

```
initial a1: assert property (@(posedge clk) rst[*2]);
```
□

Fig. 6.9 `r[*2]`

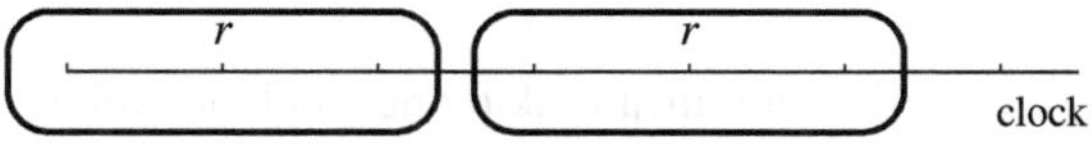

Example 6.29. Signal `sig` remains high during five cycles.

Solution:

```
a1: assert property (@(posedge clk) !sig ##1 sig |=> sig[*4]);
```

Discussion: This assertion states that once the signal goes high it remains high during four additional clock ticks. It does not forbid the signal to stay high for longer time. See Exercise 6.6 for an alternative interpretation. □

6.5.1 Zero Repetition

It is possible to define *zero repetition* of sequences: sequence `s[*0]` is a sequence admitting only an empty match. In other words, sequence `s[*0]` matches on any trace, but the trace fragment it defines is empty—it does not contain any clock tick. Because of this characteristic of zero repetition, it is also called *empty sequence*. Empty sequence is a strange creature, its behavior significantly differs from the behavior of "normal" sequences. It is rarely written explicitly, but its implicit use in delay and repetition ranges is rather common, therefore it is extremely important to understand its behavior. The meaning of empty sequence concatenation and fusion is not obvious, and it is clarified below.

The empty sequence cannot be promoted to a sequential property, as the definition of sequential property in Sect. 6.2 excludes sequences admitting an empty match. Thus, assertion **`assert property`** `(@(`**`posedge`** `clk)s[*0]);` is *illegal*.

6.5.1.1 Concatenation with Empty Sequence

In this section, we will clarify the meaning of sequence `r[*0] ##1 s`. Let the initial point of the resulting sequence be clock tick i. According to the definition of sequence concatenation (Sect. 6.3), `r[*0] ##1 s` has a match in clock tick $j \geq i$ iff the interval $i : j$ may be split into two consecutive parts, the interval where sequence `r[*0]` has a match and the interval where sequence `s` has a match. Since `r[*0]` matches an empty trace fragment, the match of sequence `r[*0] ##1 s` coincides with the match of sequence `s`. In other words, `r[*0] ##1 s` is equivalent to `s`. Similarly, `r ##1 s[*0]` is equivalent to `r`.

Concatenation with an empty sequence clarifies the semantics of sequence concatenation operator `##1`.

`r ##1 s` does *not* mean "skip one clock tick after match of `r` and then match `s`", but "start matching `s` after match of `r`".

6.5.1.2 Fusion with Empty Sequence

In this section, we will clarify the meaning of sequence `r[*0] ##0 s`. According to the definition of sequence fusion (Sect. 6.3.3), the match of sequence `r[*0] ##0 s` requires the clock tick of the match of `r[*0]` be the first clock tick of sequence `s`. Since `r[*0]` does not match any positive number of clock ticks, a match of sequence `r[*0] ##0 s` is impossible. Similarly, sequence `r ##0 s[*0]` cannot be matched, either. Therefore, a fusion with an empty sequence does not have a match.

This result reveals a very important fact:

Sequence fusion never admits an empty match.

6.5.1.3 Empty Sequence in Antecedent

What happens when the antecedent of a suffix implication is an empty sequence? In case of the overlapping implication `s[*0] |-> p`, its antecedent does not have nonempty matches, and according to the definition of the overlapping implication, `s[*0] |-> p` trivially holds (is a tautology).

The situation with the nonoverlapping implication is completely different. According to its definition, `s[*0] |=> p` is equivalent to `s[*0] ##1 1 |-> p`, which is, in its turn, equivalent to `1 |-> p` (see Sect. 6.5.1.1). The latter is equivalent to `p`.

Although it is rarely used in its pure form, this seemingly pathological behavior of the empty sequence in antecedents is important to understand the semantics of more complex antecedents admitting empty matches, described in Sect. 6.8.1.

6.6 Sequence Disjunction

Sequence disjunction `r or s` is a sequence which has a match whenever either `r` or `s` (or both) have a match.

Boolean Disjunction. For Boolean expressions `a` and `b` sequence `a or b` has a match iff `a || b` is true. Therefore, in case of Boolean values, a sequence disjunction behaves as a logical disjunction.

Sequence Disjunction versus Property Disjunction. Sequence disjunction and property disjunction have exactly the same syntax. The following rule shows how they are distinguished: if both `r` and `s` are sequences then `r` **`or`** `s` is a sequence disjunction, otherwise, it is a property disjunction. For example, the formula `(a ##1 b)`**`or`** `##1 c` is a sequence disjunction, and `(a ##1 b)`**`or nexttime`** `c` is a property disjunction. In the context where both sequences and properties may appear, the exact decision is not important since in this case both definitions yield equivalent results. Why?

Example 6.30. There are two types of transactions: the read transaction, in which read request `read` is followed by `data_ready` in three cycles, and the write transaction, in which write request `write` is followed by `done`. Write a sequence representing a generic transaction.

Solution: The sequence representing the read transaction is `read ##3 data_ready`; the sequence representing the write transaction is `write ##1 done`. The generic transaction is their disjunction, `read ##3 data_ready` **`or`** `write ##1 done`. □

Multiple Matches. All sequences we considered until now could have at most one match. Sequence disjunction introduces sequences that can have multiple matches.

Example 6.31. Sequence `a[*2]` **`or`** `b[*3]` may have 0, 1, or 2 matches. □

6.7 Consecutive Repetition Revisited

In Sect. 6.5, we introduced consecutive repetition with factors `n` $\geq$ 0. The rules of building consecutive repetition may be summarized as follows:

- `s[*0]` is an empty sequence.
- `s[*n]`, where `n` $>$ 0 is defined recursively:

```
s[*n] ≡ s[*n-1] ##1 s.
```

In this section, we will define repetition ranges.

6.7.1 *Repetition Range*

Instead of a fixed number of repetitions one can specify a repetition range: finite `s[*m:n]` and infinite `s[*n:$]`. `m` and `n` should be elaboration time integral constants, `m` $\leq$ `n`, and `$` stands for an "infinite number".

6.7.1.1 Finite Repetition Range

Consider an example of a finite repetition range first. What is the meaning of sequence `s[*2:4]`? Intuitively, it means that sequence `s` is repeated from 2 to 4 times. More formally, `s[*2:4]` has a match iff either `s[*2]` has a match, or `s[*3]`, or `s[*4]` has a match. That is, `s[*2:4]` is equivalent to `s[*2] or s[*3] or s[*4]`. This leads us to the following recursive definition:

`s[*n:n]` $\equiv$ `s[*n]`.
`s[*m:n]` $\equiv$ `s[*m:n-1] or s[*n]`, `m` $<$ `n`.

Efficiency Tip. Big repetition factors, and ranges with big finite upper bounds, are inefficient both in simulation and in formal verification.

6.7.1.2 Infinite Repetition Range

Consider now an example of an infinite repetition range: intuitively `s[*1:$]` means that `s` happens one or more times. Following the definition of finite repetition range we would like to define `s[*1:$]` as `s or s[*2] or s[*3]`.... Unfortunately, such a definition does not work as it produces an infinite formula. Therefore, we need to define `s[*1:$]` directly.

Let the initial point of sequence `s[*1:$]` be clock tick i. Sequence `s[*1:$]` has a tight satisfaction point (match) $j \geq i$ iff there is some number `n` > 0 such that j is the tight satisfaction point of sequence `s[*n]`. In other words, sequence `s[*1:$]` is tightly satisfied on trace fragment $i : j$ if it is possible to divide this trace fragment into one or more consecutive fragments so that each such fragment tightly satisfies `s`.

After we have defined infinite repetition range `[*1:$]` we can define any infinite repetition range as follows:

`s[*0:$]` $\equiv$ `s[*0] or s[*1:$]`.
`s[*n:$]` $\equiv$ `s[*n-1] ##1 s[*1:$]`, `n` > 1.

> `s[*n:$]` does *not* mean that sequence `s` is repeated infinitely many times, but that it is repeated `n` or more (finite) number of times.

There are shortcuts `s[*]` provided for `s[*0:$]` (zero or more times), and `s[+]` for `s[*1:$]` (one or more times) that we will widely use.

Example 6.32. Describe a transaction as a sequence. A transaction starts with Beginning Of Transaction signal (`bot`) and ends with End Of Transaction signal (`eot`). Each transaction is at least two cycles long, and transactions cannot overlap.

Table 6.3 Initial trace fragment in Example 6.34

Clock tick	a	b	c
0	1	0	0
1	1	1	0
2	0	0	1

Solution: Since the transaction is at least two cycles long, `bot` and `eot` of the same transaction cannot be asserted in the same clock cycle. The fact that transactions cannot overlap means that between `bot` and the next `eot` there are no other occurrences of `bot`. We allow `bot` of the next transaction to occur in the same clock cycle with `eot` of the previous transactions. Transactions should also be well formed; therefore no two `eot` should occur without a `bot` occurrence between them. Therefore, the sequence describing the transaction is

```
bot ##1 (!bot && !eot)[*] ##1 eot.
```
□

Example 6.33. Write the following assertion: `rdy` becomes asserted the first time when the reset sequence is over. The reset sequence is represented by the high value of `rst`.

Solution:

```
initial a1: assert property (@(posedge clk)
  rst && !rdy [*] ##1 !rst && rdy);
```

Discussion: The same assertion may be more intuitively rewritten using property operator **`until`**:

```
initial a2: assert property (@(posedge clk)
  rst && !rdy until !rst && rdy);
```
□

Example 6.34. We revisit Example 5.21: `ready` should be low until `rst` becomes inactive for the first time.

Solution: This time we are going to implement this assertion using sequences:

```
initial assert property(@(posedge clk) !ready[*] ##1 !rst);
```

Discussion: In this case, the property operator **`until`** may be implemented as a sequential property (see also Example 6.33). If `a` is a Boolean and `s` is a sequence, `a` **`until`** `s` is equivalent to sequential property `a[*] ##1 s` in the assertion or assumption context. However, for a general sequence `r`, `r` **`until`** `s` and the sequential property `r[*] ##1 s` are *not equivalent*. To understand this, compare `(a ##1 b)[*] ##1 c` and `a ##1 b` **`until`** `c` for values of `a`, `b`, and `c` shown in Table 6.3. Sequence `(a ##1 b)[*]` has a match in clock tick 1, therefore sequence `(a ##1 b)[*] ##1 c` has a match in clock tick 2, and the corresponding sequential property passes. Sequential property `a ##1 b` is false in clock tick 1 (recall that the truth value of a sequential property relates to the initial point of its sequence), while `c` is true the first time in clock tick 2. Therefore, property `a ##1 b until c` fails. □

Example 6.35. When the first operand of **`until_with`** is an integral expression, and the second one is a sequence, `a until_with s` may be implemented with sequential property `a[*] ##0 s`. □

Efficiency Tip. Infinite repetition ranges may be inefficient in simulation. Their efficiency is explained in detail in Sect. 19.3. Infinite repetition ranges are efficient in FV if their lower bound is small.

Previous examples show that sequence and property operators are in some cases interchangeable in assertions. In such cases, it is a matter of style or of simulator efficiency, which solution to choose. The sequence operators are often more concise, but in many cases property operators are more readable. Sequence operators are more flexible as they can appear in both sequences and properties, whereas property operators may appear only within another property or assertion. On the other hand, property operators are more generic, as they can have both sequences and properties as their operands, while sequences may only have sequence (or Boolean) arguments. This fact makes property-based implementation more suitable for assertion libraries.

We will end this section with an example illustrating some semantic subtlety of the use of infinite repetition range in the antecedent.

Example 6.36. What does the following assertion mean?

```
a1: assert property (@(posedge clk) a[+] |-> p);
```

Solution: For each evaluation attempt `a1` checks that when there is a continuous series of `a`, property `p` holds. For example, if there is a series `aa` at the beginning of some attempt, `p` should hold in the first two clock ticks of this attempt. But checking `p` in the second clock tick of this attempt is redundant: it will be checked in the first clock tick of the next attempt. Therefore, the original assertion is equivalent to

```
a2: assert property (@(posedge clk) a |-> p);
```

Discussion: There should be no difference between the efficiency of `a1` and `a2` in FV, but `a2` is extremely inefficient in simulation. □

Efficiency Tip. Never implement the property `s |-> p` as `s[+] |-> p`.

6.8 Sequences Admitting Empty Match

Sequences admitting an empty match introduce many subtle points that should be well understood. Unlike other sequences, *sequences that admit an empty match cannot be promoted to properties* according to the definition of sequential property (Sect. 6.2). One such sequence is an empty sequence described in Sect. 6.5.1. The empty sequence is rarely used explicitly, but it is often used implicitly as part of other sequences, for example, in repetition ranges of the form `[*]` or `[*0:n]`. In this section, we provide several examples illustrating a peculiar behavior of sequences admitting an empty match.

Example 6.37. The following assertion is *illegal*:

```
assert property (@(posedge clk) a[*]);
```

The reason is that in this assertion the sequence `a[*]` is promoted to a property while it admits an empty match. □

Example 6.38. Sequence `a ##1 b[*0:2] ##1 c` matches traces `ac`, `abc`, and `abbc`. Note that in the trace `ac` there is no gap between `a` and `c`: `b[*0]` does not have any duration in time! □

Example 6.39. Sequence `a[*] ##0 b[*]` is equivalent to `a[+] ##0 b[+]`, since the sequence fusion does not admit an empty match (Sect. 6.5.1).

Discussion: `a[*] ##1 b[*]` is *not* equivalent to `a[+] ##1 b[+]`. □

6.8.1 Antecedents Admitting Empty Match

Sequences admitting an empty match have many subtle points when they are used as antecedents. We will illustrate their behavior on sequence `a[*]`. Other sequences admitting an empty match exhibit a similar behavior. Consider overlapping implication `a[*] |-> p` first. According to the definition of zero-or-more repetition (Sect. 6.7.1.2), it is equivalent to `a[*0] or a[+] |-> p`. All nonempty matches of the antecedent are matches of `a[+]`, and therefore `a[*] |-> p` is equivalent to `a[+] |-> p`. Therefore, it is possible to limit antecedents of overlapping implications with sequences that do not admit an empty match. The sequences admitting empty matches add nothing new in this case.[4]

Now consider nonoverlapping implication `a[*] |=> p`. It can be rewritten as `(a[*0] or a[+]) ##1 1 |-> p`, which is equivalent to `(a[*0] ##1 1) or (a[+] ##1 1) |-> p` (see Exercise 6.2). The first disjunct of the antecedent is equivalent to 1, and the whole property can be reduced to `p` (see Sect. 6.5.1.1). Therefore, in the nonoverlapping implication the antecedents admitting an empty match are *completely redundant* when the assertion is continuously monitored. To summarize, antecedents admitting an empty match in suffix implications usually indicate a problem in the assertion.

[4]Of course, with the exception of the empty sequence, which has only the empty match.

Never use antecedents admitting an empty match in suffix implication.

Note also that in case of antecedents admitting an empty match, the transformation described in Example 6.18 is not valid. For example, `s[*0] ##1 1 |-> p` (i.e., `1 |=> p`) is `p`, but `s[*0] |-> nexttime p` is a tautology.

6.9 Sequence Concatenation and Delay Revisited

In Sect. 6.3, we introduced sequence concatenation operators with factors $n \geq 0$. The rules of building sequence concatenation may be summarized as follows:

- `r ##0 s` is a sequence fusion.
- `r ##1 s` is a sequence concatenation.
- `r ##n s`, where $n > 1$ is defined in terms of repetition:

`r ##n s` ≡ `r ##1 1[*n-1] ##1 s`.

We also defined there initial delay operators with factors $n > 0$. Now we can provide a definition for any factor $n \geq 0$:

`##n s` ≡ `1[*n] ##1 s`.

According to this definition `##0 s` is equivalent to `s` (why? see Sect. 6.5.1.1), which is intuitive, as `##0 s` means "wait 0 clock ticks before the sequence beginning".

In a fashion similar to consecutive repetition ranges (Sect. 6.7.1), it is possible to specify *delay ranges*, finite or infinite, between two sequences, as follows:

`r ##[0:0] s` ≡ `r ##0 s`.
`r ##[m:n] s` ≡ `(r ##1 1[*m-1:n-1] ##1 s)`, where $n \geq m > 0$.
`r ##[0:n] s` ≡ `(r ##0 s) or (r ##[1:n] s)`, where $n > 0$.
`r ##[m:$] s` ≡ `(r ##1 1[*m-1:$] ##1 s)`, where $m > 0$.
`r ##[0:$] s` ≡ `(r ##0 s) or (r ##[1:$] s)`, where $n > 0$.

Informally speaking `r ##[m:n] s` means that there are `m` to `n` clock ticks from the tight satisfaction point of `r` to the initial point of `s`. `r ##[m:$] s` means that there are `m` or more clock ticks from the tight satisfaction point of `r` to the initial point of `s`. As in the case with consecutive repetition there is a shortcut `r ##[*] s` for `r ##[0:$] s` ("zero or more clock ticks"), and `r ##[+] s` for `r ##[1:$] s` ("one or more clock ticks").

It is also possible to define *initial delay ranges*:

`##[m:n] s` ≡ `1 ##[m:n] s`, where $n \geq m \geq 0$.
`##[m:$] s` ≡ `1 ##[m:$] s`, where $m \geq 0$.

Informally speaking `##[m:n] s` means skipping from `m` to `n` clock ticks before the initial point of `s`, and `##[m:$] s` means skipping `m` or more clock ticks before the initial point of `s`. As usual, there is a shortcut `##[*] s` for `##[0:$] s` ("zero or more clock ticks"), and `##[+] s` for `##[1:$] s` ("one or more clock ticks").

Efficiency Tip. Big delay factors, and ranges with big finite upper bound are inefficient both in simulation and in FV. Infinite delay ranges may also be inefficient in simulation. Their efficiency is explained in detail in Sect. 19.3. Infinite delay ranges are efficient in FV if their lower bound is small.

Example 6.40. Write a sequence describing the scenario when `ready` is asserted at the end of the transaction (signal `etrans` asserted) or in the next clock tick after it.

Solution: `trans ##[0:1] ready` □

Example 6.41. Write the following sequence: grant `gnt` asserted from two to four clock ticks after request `req` was asserted.

Solution: `req ##[2:4] gnt`

Discussion: If the initial point is 0, this sequence means: `req` is true in clock tick 0, and `gnt` is true either in clock tick 2, 3, or 4. It does not say anything about `req` behavior after clock tick 0: `req` does not have to be deasserted there, though it can be. Neither does this sequence claim that `gnt` is false in clock ticks 0, 1, 5, ... □

Example 6.42. Request `req` must be granted (grant `gnt` should be asserted) within five clock ticks.

Solution:

```
a1: assert property (@(posedge clk) req |-> ##[1:5] gnt);
```

Discussion: In this example, it is also possible to have one `gnt` issued for several requests, as shown in Fig. 6.8. □

Example 6.43. The device should become ready (`ready` asserted) from 10 to 12 cycles after power-on.

Solution:

```
initial a1: assert property (@(posedge clk) ##[10:12] ready);
```

Discussion: This assertion states that `ready` is asserted either in clock tick 10, 11 or 12. It does not state that `ready` must be continuously asserted in clock ticks 10, 11 and 12. Neither does it state that ready cannot be asserted before clock tick 10.

□

Example 6.44. Write a sequence describing a scenario when request `req` is granted (`gnt` is received).

Solution: This means that `gnt` is asserted in one or more clock ticks after `req`: `req ##[+] gnt`. □

Example 6.45. Write a sequence describing a scenario when request `req` is granted (`gnt`) in two or more cycles.

Solution: `req ##[2:$] gnt`.

Discussion: This sequence does not specify that there be no grant in clock tick 1, it just requires at least one grant to happen in two or more clock ticks. □

Example 6.46. What is the meaning of the following assertion?

```
initial a1: assert property (@(posedge clk) ##[*] s |-> p);
```

Solution: According to the definition of the overlapping implication, at each tight satisfaction point of sequence `##[*] s` property `p` should be true. This is equivalent to the requirement that `p` is true starting at all tight satisfaction points of sequence `s` with initial points 0, 1, ... Therefore, assertion `a1` is equivalent to assertions `a2` below:

```
a2: assert property (@(posedge clk) s |-> p);
```

Discussion: Assertions `a1` and `a2` are also equivalent to assertion `a3`.

```
a3: assert property (@(posedge clk) ##[*] s |-> p);
```

Although assertions `a1`, `a2`, and `a3` are logically equivalent, their simulation performance is likely to be very different. Assertion `a2` is the most efficient, assertion `a1` is much less efficient, but using assertion `a3` may lead to a tremendous performance degradation (see Sect. 19.3). FV performance for these three assertions is generally the same. □

Efficiency Tip. Never use infinite initial delay in antecedents.

6.10 Unbounded Sequences

In Example 6.42, we discussed the situation when a request had to be granted within five clock ticks. The suggested assertion was

```
a1: assert property (@(posedge clk) req |-> ##[1:5] gnt);
```

We want now to modify the problem and require that the request `req` be granted some time in the future, without specifying any upper bound. It may seem that the only change required is to replace the range upper bound 5 with `$`:

```
a2_wrong: assert property (@(posedge clk) req |-> ##[1:$] gnt);
```

or, equivalently,

```
a3_wrong: assert property (@(posedge clk) req |=> ##[*] gnt);
```

However, both assertions `a2_wrong` and `a3_wrong` are *wrong*. To understand why, recall the definition of sequential property from Sect. 6.2 that we reproduce here for convenience.

The sequential property defined by sequence s is true in clock tick i iff there is no finite trace fragment $i : j$ witnessing inability of sequence s with the initial point i to have a match. Sequence s should not admit an empty match.

When we introduced this definition, we were familiar only with bounded sequences—sequences all of whose matches happen within a finite time interval. For bounded sequences, the definition of sequential property can be simplified: the sequential property defined by a bounded sequence is true iff the sequence has at least one match (Sect. 6.2).[5] Using the sequence operators described in this chapter, sequences built on top of infinite delay or repetition ranges are *unbounded*, i.e., they can have arbitrarily long matches. For such sequences, the definition of sequential property cannot be simplified any more.

In our example, sequence `##[*] gnt` is unbounded. Any finite trace fragment does not witness the inability of this sequence to match. Indeed, the fact that `gnt` did not assume high value during first 1,000 clock ticks starting from the sequence initial point, does not prevent `gnt` to assume high value in the future. Therefore, sequential property `##[*] gnt` is true, regardless of the actual values of `gnt`, and assertions `a2_wrong` and `a3_wrong` pass even if `req` is never granted. The correct solution was provided in Example 6.23:

```
a4: assert property (@(posedge clk) req |=> s_eventually gnt);
```

An alternative way to encode this assertion is discussed in Chap. 11.

Although a sequential property of the form `##[*] s` is meaningless, since it is a tautology, the sequence `##[*] s` itself is not. We saw in Example 6.46 that operator `##[*]` modified the meaning of the implication. Namely, assertion

```
initial
  a_always: assert property (@(posedge clk) ##[*] s |-> p);
```

checks implication `s |-> p` in every clock cycle, whereas assertion

```
initial
  a_once: assert property (@(posedge clk) s |-> p);
```

checks this implication only once.

The definition of sequential property should be clear at this point for both bounded and unbounded sequences. However, its rationale has not been explained yet. Even worse, the behavior or assertions `a2_wrong` and `a3_wrong` is nonintuitive according to this definition. Nevertheless, this definition has many advantages that will be explained only in Chaps. 10 and 22. In this chapter, we will only provide one additional example, further clarifying the definition of sequential property.

We mentioned in Example 6.34 that a sequential property of the form `a[*] ##1 b` is equivalent to property `a until b`. This equivalence is intuitive, and it is conditioned by the definition of the sequential property. Indeed, property `a[*] ##1 b` fails iff there is a finite fragment of the trace witnessing that sequence `a[*] ##1 b` cannot have a match. This only happens if before the first occurrence of

[5] As follows from the definition, all sequence matches should be non-empty.

b there is a clock tick with a low value of a, i.e., exactly when property a **until** b fails. If b never happens and a always happens then any finite fragment of the trace cannot witness the inability of sequence a[*] ##1 b to match, since potentially b could happen immediately after the end of this fragment. This is, again, consistent with the definition of **until**, which requires a always to happen if b never happens.

Exercises

6.1. Write a sequence implementing the scenario "Asserted request is deasserted in the next clock tick".

6.2. Show that if r, s, and t are sequences then

(a) (r **or** s) ##n t is equivalent to (r ##n t) **or** (s ##n t)
(b) r ##n (s **or** t) is equivalent to (r ##n s) **or** (r ##n t).
Is (r ##n s) **or** t equivalent to (r **or** t) ##n (s **or** t)?

6.3. Modify the transaction definition from Example 6.32 so that the beginning of the next transaction cannot happen in the same cycle as the end of the previous one.

6.4. During transaction execution the ready flag must be low. The transaction is delimited by bot and eot signals. What happens if transactions can overlap?

6.5. Modify the assertion from Example 6.24 to account only for the acknowledgment coming strictly after the request.

6.6. Modify the assertion from Example 6.29 to require the signal to be active during exactly five cycles.

Chapter 7
Assertion System Functions and Tasks

It's a question of whether we're going to go forward into the future, or past to the back.

— Dan Quayle

This chapter describes system functions and tasks designed to be used in assertions. System tasks control the execution of assertions and their action blocks. Assertion system functions are divided into two groups:

- Bit vector functions
- Sampled value functions

7.1 Bit Vector Functions

Table 7.1 contains a list of available *bit vector functions* along with their description. All bit vector functions, with the exception of `$countbits`, have a bit vector as their single argument. Function `$countbits` requires an additional list of arguments to specify the control values to match.

It is possible to write user functions that accomplish the same thing, however, writing them can be rather cumbersome and certainly less efficient in simulation. This is why the functions are included as part of the standard. The EDA tool providers can thus implement them in an as efficient way as possible within the tools.

E. Cerny et al., *SVA: The Power of Assertions in SystemVerilog*,
DOI 10.1007/978-3-319-07139-8_7

Table 7.1 Bit vector functions

Name	Description
`$countbits`	Count number of bits in a vector which match the argument values
`$onehot0`	Check that at most one bit in a vector is high
`$onehot`	Check that exactly one bit in a vector is high
`$countones`	Count number of bits in a vector with value high
`$isunknown`	Check whether a vector has a bit with value x or z

7.1.1 Count Bits with Specific Values

`$countbits` returns the number of bits of its argument bit vector having the value of one of the control bits. Bit vector function
`$countbits(e, list_of_control_bits)` has the following arguments:

- `e` — a bit vector, i.e., a packed or unpacked integral expression.[1]
- `list_of_control_bits` — a comma separated list of control bit arguments carrying the values `1'b1`, `1'b0`, `1'bx` or `1'bz`.

At least one control bit must be specified. A repetition of a control bit is ignored. Also, when the width of a control bit argument is bigger than one, its LSB is taken as the control bit.

Example 7.1. No bus driver should be in high impedance state when signal `en` is true.

Solution:

```
dr_t: assert final (en -> ($countbits(bus_in, 1'bz) == 0));
```
□

`$countbits` is a generalized bit vector function of all other bit vector functions described in the following sections. The equivalent forms of the other bit vector functions are shown below. Assume that `e` is a bit vector expression.

- `$countones(e)` is same as `$countbits(e, 1'b1)`
- `$onehot(e)` is same as `$countbits(e, 1'b1)== 1`
- `$onehot0(e)` is same as `$countbits(e, 1'b1)<= 1`
- `$isunknown(e)` is same as `$countbits(e, 1'bx, 1'bz)!= 0`

Note that using functions `$countones`, `$onehot` and `$onehot0` is not always safe in the presence of unknown values `x` or `z`. In this case using `$countbits` may be a better choice.

[1]It is treated as a vector of equal size assigned from `>>{e}`, where `>>{}` is a streaming operator. See [8].

7.1.2 Check for Mutual Exclusion

System function `$onehot0` checks that all bits of its single argument are mutually exclusive. More precisely, it returns `1'b1` if at most one bit of its argument is set to 1. Otherwise, it returns `1'b0`. Bits carrying the values `x` or `z` are treated as 0.

Example 7.2. All bus drivers, that is, the bits set to `1'b1`, of the vector `bus_in` must be mutually exclusive.

Solution:

```
a_mutex: assert final ($onehot0(bus_in));
```

□

Example 7.3. `read` and `write` requests cannot appear together.

Solution:

```
a_norw: assert property (@(posedge clk)
                         $onehot0({read, write}));
```

We use a concatenation operator to build a bit vector from two signals.

Discussion: If one of the signals, say `write`, has value `x` then assertion `a_norw` passes, which is problematic, because `x` may correspond to both 0 and 1, and `$onehot0` does not guarantee the mutual exclusion. Therefore, in order to consider `x` and `z` as possible 1, it is safer to rewrite our assertion using `$countbits`:

```
a_norw_x: assert property (@(posedge clk)
    $countbits({read, write}, 1'b1, 1'bx, 1'bz) <= 1);
```

□

7.1.3 One-Hot Encoding

System function `$onehot` checks that exactly one bit of its argument is set to 1. If this condition is met, it returns `1'b1`, otherwise, it returns `1'b0`. In a similar way, `$onehot0` checks that at most one bit of its argument is set to 1. If this condition is met, it returns `1'b1`, otherwise, it returns `1'b0`.

Example 7.4. Check that a control `state` of an FSM has a one-hot encoding when `rst` is low.

Solution:

```
a_onehot:  assert property (@(posedge clk) disable iff (rst)
  $onehot(state));
```

Discussion: To check one-cold encoding use `$onehot(~state)`. In the presence of unknown values, `$countbits` function should be used to check one-hot and one-cold encodings (see Exercise 7.2). □

7.1.4 *Number of 1-Bits*

System function `$countones` returns a value equal to the number of bits of its argument set to `1`.

Example 7.5. The system stores the maximal number of simultaneously active transmitters in register `trmax`. The transmitter activity is encoded by vector `transmitters` in which each bit represents activity of the corresponding transmitter. Check that the total number of simultaneously active transmitters does not exceed the number stored in `trmax`. Assume that all bits of `transmitters` have defined values.

Solution:

```
a_tract: assert property (@(posedge clk)
  $countones(transmitters) <= trmax);
```
□

`$onehot0(sig)` is equivalent to `$countones(sig)<= 1`, `$onehot(sig)` is equivalent to `$countones(sig)== 1`. We recommend to use a specific system function, and not `$countones` whenever possible since it makes the user intent clearer and may be handled more efficiently by tools. When there is a danger of incorrect handling of undefined values `x` and `z` as 0, `$countones` function should be used.

7.1.5 *Unknown Bits*

System function `$isunknown` returns `1'b1` if *any* bit of its argument has the value `x` or `z`.

Example 7.6. All bits of `data` should have known values (i.e., 0 or 1) when `read` is active.

Solution:

```
a_valid_data: assert property (@(posedge clk)
  read |-> !$isunknown(data));
```
□

7.2 Sampled Value Functions

This section describes *Sampled Value Functions* (SVF)—system functions accessing present, past, and future sampled values of an integral expression. One can divide sampled value functions into two groups: general sampled value functions, and global clocking sampled value functions. Although sampled value functions have a temporal nature, they may be used wherever integral expressions are legal with some exceptions explained below and in later chapters.

7.2.1 *General Sampled Value Functions*

Table 7.2 contains the list of available general sampled value functions. Below we describe each function in more detail.

Table 7.2 General sampled value functions

Name	Description
`$sampled`	Return sampled value of expression
`$past`	Return past value of expression
`$rose`	Check whether expression value rose
`$fell`	Check whether expression value fell
`$changed`	Check whether expression value changed
`$stable`	Check whether expression value remained stable

7.2.1.1 Present Sampled Values

System function `$sampled` takes an integral expression as its argument, and returns its sampled value. In most cases expression sampling is done in the Preponed region. See Appendix A for a detailed definition of expression sampling. Using system function `$sampled` in concurrent assertions and in checker **`always_ff`** procedures is redundant since the expressions used there are already sampled as explained in Sects. 4.4 and 9.2.2.2.

Example 7.7. The following deferred cover statement prints a message when signal `sig` value changes:

```
c_changed: cover final (sig != $sampled(sig))
  $info("\%t: sig value changed", $time);
```

In this cover statement, the value of `sig` from the Observed region (which is normally the final value of `sig` in the current simulation step) is compared against `$sampled(sig)`, the value of `sig` at the beginning of this simulation tick. If these values are different, a message is issued. This statement does not work in FV since in FV all signal values are conceptually sampled (see Chap. 21). There are more conventional ways to detect signal changes in a clock-based design, which work both in simulation and in FV as explained later in this chapter. □

Example 7.8. As mentioned above, the use of `$sampled` system function in concurrent assertions is redundant. Assertion

```
a1: assert property (@(posedge clk) $sampled(a));
```

is exactly the same thing as

```
a2: assert property (@(posedge clk) a);
```

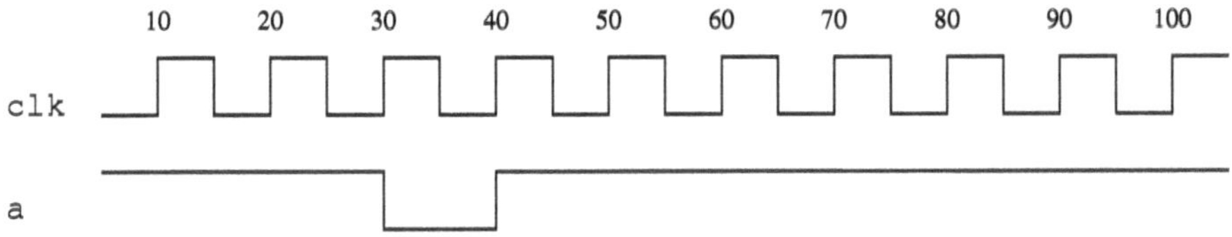

Fig. 7.1 Assertion violation

The situation with resets is different. In assertion

```
a3: assert property (@(posedge clk) disable iff (rst) a);
```

the value of `rst` is *not* sampled. To make it sampled, it should be specified explicitly:

```
a4: assert property (@(posedge clk)
  disable iff ($sampled(rst)) a);
```

For further discussion about **`disable iff`**, see Chap. 13 □

The main use of system function `$sampled` is in an action block of concurrent assertions, as illustrated in the following example.

Example 7.9. Assertion

```
a1: assert property (@(posedge clk) a)
  else $error("Error: a = \%b.", a);
```

is violated at time 40 for the waveform shown in Fig. 7.1 (recall that in the assertion body the sampled value of `a` is used, and it is 0 at time 40). But the issued error message will be `Error: a = 1.` The reason is that the action block is executed in the Reactive region (Sect. 4.4) when the value of `a` is already 1. To make the reporting consistent, the function `$sampled` has to be explicitly invoked in the action block:

```
a2: assert property (@(posedge clk) a)
  else $error(''Error: a = \%b.'', $sampled(a));
```

Note that this rule is applicable to action blocks of concurrent assertions only. Specifying sampled values in action blocks of immediate or deferred assertions will produce inconsistent error messages because these assertions do not use sampled values. □

7.2.1.2 Past Sampled Values

Sampled value function `$past(e, n, en, @clk)` has the following arguments:

- `e` — an integral expression.
- `n` ≥ 1 — a constant expression specifying the number of clock ticks (delay).
- `en` — a gating expression for the clocking event.
- `clk` — a clocking event.

Table 7.3 Sampled and past values of a

Time	`$sampled(a)`	`$past(a,,,@(posedge clk))`
30	0	0
40	1	0
42	0	1
50	0	1
60	1	0
70	1	1
80	0	1
90	0	0

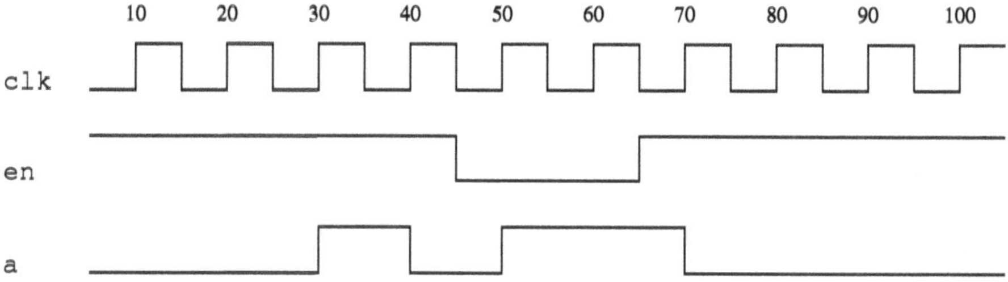

Fig. 7.2 Timing diagram for Examples 7.10, 7.15, 7.19, and 7.23

All arguments but the first one are optional and have default values:

- If the clocking event is omitted, it is inferred from the context as described in Sect. 7.2.1.5. For example, if `$past` is invoked in a singly clocked assertion then the clock of this assertion is assumed, both in the assertion body and in the assertion action blocks.
- The gating condition defaults to `1'b1` — no clock gating.
- The number of clock ticks defaults to 1.

The last optional arguments may be skipped, like in `$past(a)`. If the intermediate ones are omitted, a comma should be placed for each omitted argument:

```
$past(a,,,@(posedge clk))
```

`$past` returns the sampled value of `e` that was `n` strictly prior time steps ago in which event `@(clk iff en)` occurred, i.e., the value is taken `n` ticks of `clk` ago, but counting only those clock ticks in which `en` was high.

Example 7.10. Table 7.3 contains sampled and past values of `a` for some time steps for the waveforms shown in Fig. 7.2. For example, to find `$past(a,,,@(posedge clk))` at time 40 we must take the sampled value of `a` at time 30, which is 0. To find `$past(a,,,@(posedge clk))` at time 42 we must take the sampled value of `a` at the time of the last strictly preceding clock tick, which is 40, and this value is 1. □

Values Before Initial Clock Tick. The definition of `$past` given above is incomplete: what happens if for a given time step there are not enough previous clock ticks? In this case, `$past(e)` returns the initial value of `e`. The initial value of a static variable is that as computed using the initial values stated in the declaration of the variables involved in `e`. If a static variable has no explicit initialization, the default value of the corresponding type is used, even if the variable is assigned a value in an initial procedure.

> FV tools may ignore variable initialization everywhere, except in **`checker`** constructs.[2] Also, many FV tools consider all variables to be of two-state value type, and therefore they assume that `$past(e)` is 0 at clock tick 0 for any e.

Example 7.11. For the following declaration:

```
logic a = 1'b1, b = 1'b0;
logic c;
bit d;
wire w = a;
initial c = 1'b1;
```

the initial value of

- `a` is `1'b1`.
- `b` is `1'b0`.
- `a|b` is `1'b1`.
- `c` is `1'bx`, even though `c` is assigned a value in the initial procedure.
- `d` is `1'b0`.
- `w` is `1'bz`. `w` is a net, and **`wire`** `w = a;` is an implicit continuous assignments, and not initialization.

Beyond clock tick 0 the past values of these signals are their initial values. Note that we number clock ticks starting from 0. For example,

```
$past(a,,,@(posedge clk)) = 1'b1
```

and

```
$past(c,,,@(posedge clk)) = 1'bx
```

Beyond clock tick 1 (including)

```
$past(a, 2,,@(posedge clk)) = 1'b1
```

[2]FV tools usually work with the synthesis model of DUT, and variable initialization is non-synthesizable. However, FV algorithms can deal with initial states and the tools can infer initial states by analyzing the variable initializations.

and

```
$past(c, 2,,@(posedge clk)) = 1'bx
```

etc. As mentioned in the note preceding this example, some FV tools will assume the past value of all these expressions to be `1'b0`. □

Example 7.12. Consider the following code fragment.

```
logic a;
a1: assert property(@(posedge clk) $past(a));
```

Assertion `a1` fails at clock tick 0, since `$past(a)` returns `x`—the initial value of `a`.

This example shows that one should be careful when using `$past` in assertions because it may lead to nonintuitive assertion behavior in the initial clock ticks. □

Example 7.13. Each grant `gnt` should be immediately preceded by a request `req`.

Solution:

```
a1: assert property (@(posedge clk) gnt |-> $past(req));
```

Discussion: What happens if `gnt` is active at the initial clock tick? If `req` has not been explicitly initialized, `$past(req)` is `x` or 0 depending on `req` type, and the assertion fails. In this case, this is a desired behavior. □

Example 7.14. Gray Encoding: Check that signal `sig` has a *Gray* encoding, i.e., its two consecutive values differ in only one bit.

Solution: Here is the first solution:

```
a1_wrong: assert property (@(posedge clk)
  $onehot(sig ^ $past(sig)));
```

The result of the exclusive OR ^ operator between the current and the past values of `sig` contains ones in the bits that changed. We want to make sure that exactly one bit of the result is 1. But what happens at the initial clock tick? `$past(sig)` will return `x`, the result of the exclusive OR will also be `x`, and assertion `a1_wrong` will fail in the initial clock tick! To handle the initial situation correctly, we need to delay the first check of the assertion:

```
a2: assert property (@(posedge clk)
  nexttime $onehot(sig ^ $past(sig)));
```

Here `sig` may have any value at the initial clock tick.

Discussion: One can argue that a real-life implementation should take into account the reset sequence during which no check is performed and therefore no initial delay is necessary:

```
a3_problematic: assert property (
  @(posedge clk) disable iff (rst) $onehot(sig ^ $past(sig)));
```

but this is also problematic. For example, if during reset and immediately after it `sig` was equal to 0, assertion `a3_problematic` will fail immediately after reset.

Table 7.4 Past values of a relative to `@(posedge clk)`

Time	`$past(a,1,en)`	`$past(a,2,en)`
30	0	0
40	0	0
50	1	0
60	1	0
62	1	0
70	1	0
80	1	1
90	0	1

Therefore, even in the presence of a reset sequence, the initial delay may be necessary:

```
a4: assert property (@(posedge clk) disable iff (rst)
  nexttime $onehot(sig ^ $past(sig)));
```

A similar situation occurs when the `$assertcontrol` system task is used to disable the assertion execution during the reset sequence (Sect. 7.3.1) □

> Be careful with handling initial clock ticks when using `$past`. In many cases, it requires introducing an initial delay.

Gated Clock. The third argument of `$past` specifies the gating condition. To compute a past value relative to a gated clock, the third argument should be explicitly specified. To compute `$past(a, n, en, @clk)`, it is necessary to take the sampled value of `a` at `n` clock cycles strictly prior to the current simulation tick, where only "enabled" clock ticks in which `en` is true are counted.

Example 7.15. Table 7.4 contains values of `$past(a, 1, en, @(posedge clk))` and of `$past(a, 2, en, @(posedge clk))` for the waveforms from the timing diagram shown in Fig. 7.2. As an example consider simulation time steps 62 and 80. At time 62, the previous clock tick when `en` was high is at time 40, therefore `$past(a, 1, en, @(posedge clk))` is 1—the sampled value of `a` at time 40. `$past(a, 2, en, @(posedge clk))` at time 62 is the same as `$past(a, 1, en, @(posedge clk))` at time 40, i.e., 0.

`$past(a, 1, en, @(posedge clk))` at time 80 is 1—the sampled value of `a` at time 70. `$past(a, 2, en, @(posedge clk))` at time 80 is the same as `$past(a, 1, en, @(posedge clk))` at time 70. The last enabled clock tick prior to 70 is at time 40, therefore the result is 1. □

Example 7.16. Verify the following implementation of a flip-flop:

```
always @(posedge clk)
    if (en) q <= d;
```

Solution:

```
a_ff:  assert property (@(posedge clk)
  nexttime q == $past(d,, en));
```

□

Efficiency Tip. `$past(..., n, ...)` is not efficient for big delays `n`. It is recommended to minimize the number of calls of `$past`, and to avoid calling it whenever it is not essential. The width of the expression `e` in `$past(e, ...)` is not critical in simulation, but may significantly affect FV performance: every additional bit in `e` introduces performance penalty.

Example 7.17. Given the definition:

```
logic check;
logic [31:0] a, b, c;
```

assertion

```
a1: assert property (@(posedge clk) ##1 check |-> $past(c) ==
    $past(a) + $past(b));
```

is better to rewrite as:

```
a2: assert property (@(posedge clk) ##1 check |-> $past(c == a +
    b));
```

In fact, assertion `a1` has three invocations of `$past`, while `a2` has only one. Assertion `a1` has $32 * 3 = 96$ application of `$past` to individual bits, while `a2` is applied to a single bit—the result of comparison. This is essential for FV and helps in simulation. □

Sampled Value Functions Outside Concurrent Assertions. `$past` and other sampled value functions, except for the global clocking future value functions described in Sect. 7.2.2, are not limited to concurrent assertions. However, it should be understood that the values returned are based on the sampled values of the argument, as discussed in the following example.

Example 7.18. The following code

```
logic a, b;
always @(posedge clk)
  a <= $past(b);
```

has the same effect as

```
logic a, b, temp;
always @(posedge clk) begin
  temp <= $sampled(b); a <= temp;
end
```

provided that `a` is not assigned elsewhere.

To understand this, we need to apply the definition of `$past(b)`—the sampled value of `b` on the prior clock rise. The value of `b` immediately after the prior clock rise is modeled by the variable `temp`, and `a` is equal to the sampled value of `temp` at that time step, i.e., to the value of `temp` immediately before the clock rise. □

7.2.1.3 Rose and Fell

Sampled value function `$rose(e, @clk)` returns *true* iff the Least Significant Bit (LSB) of `e` has changed to 1, and false, otherwise. The sampled value function `$fell(e, @clk)` returns true if the LSB of `e` has changed to 0, and false, otherwise. More precisely

```
$rose(e, @clk) ≡
  $past(LSB(e),,,@clk)!== 1 && $sampled(LSB(e))=== 1.
$fell(e, @clk) ≡
  $past(LSB(e),,,@clk)!== 0 && $sampled(LSB(e))=== 0.
```

System functions `$rose` and `$fell` compare the past value with the current *sampled* value of the expression. The clocking event argument is optional and if omitted, its value is inferred from the context as explained in Sect. 7.2.1.5.

Example 7.19. For the timing diagram in Fig. 7.2, `$rose(a, @(posedge clk))` returns *true* for time steps $30 < t \leq 40$ and $50 < t \leq 60$. For all other time steps, it returns *false*.

`$fell(a, @(posedge clk))` returns *true* for time steps $40 < t \leq 50$ and $70 < t \leq 80$. For all other time steps, it returns *false*. □

Example 7.20. If an expression changes its value from `3'b100` to `3'b001`, `$rose` returns *true*, and `fell` returns *false*, since only the LSB counts. If an expression changes its value from `x` to 1, `$rose` returns *true*, and if it changes its value from `x` to 0, `$fell` returns *true*. If an expression becomes `x`, both `$rose` and `$fell` return *false*. □

Example 7.21. Assume that `sig` is of type **`bit`**. The assertion from Example 6.29 "signal remains high during 5 cycles" may be rewritten as follows using function `$rose`:

```
a1: assert property (@(posedge clk) ##1 $rose(sig) |=> sig[*4]);
```

What happens if we omit the initial delay in assertion `a1`?

```
a2: assert property (@(posedge clk) $rose(sig) |=> sig[*4]);
```

Solution: The behavior of these assertions differs in case `sig` is high at clock tick 0. At clock tick 0, the antecedent of `a2` is true because the past value of `sig` prior to the initial clock tick is 0. Therefore, assertion `a2` requires `sig` to remain true for four additional clock ticks. Assertion `a1` does not have a nonvacuous attempt starting at clock tick 0, and it skips the comparison with the "prehistoric" value of `sig`.

Discussion: It is difficult to argue which solution is more natural, but consider the dual case when we want to check that the signal remains low during 5 cycles:

```
b1: assert property(@(posedge clk) ##1 $fell(sig) |=> !sig[*4]);
b2: assert property (@(posedge clk) $fell(sig) |=> !sig[*4]);
```

The behavior of assertions `a1` and `b1` is similar, but the behavior of assertions `a2` and `b2` is different! If at clock tick 0 the value of `sig` is low, `$fell(sig)` will return false, so that assertions `b1` and `b2` are equivalent. One could argue that if `sig` were of type **`logic`**, the behavior of assertions `a2` and `b2` would be consistent, as `$fell` returns *true* in case of transition from `x` to 0. This is true in simulation, but usually not in FV, because most FV tools treat all variables as two-valued as mentioned above.

This example shows again that one should be very careful with the behavior of sampled value functions in the initial clock tick, and that it is better to introduce an initial delay. Specifying a reset with assertions `a2` and `b2` makes their behavior unpredictable at the moment when the reset goes low, as explained in Example 7.14.

To conclude this example, we mention that the assertion suggested in Example 6.29 is more efficient than assertion `a1`. □

Example 7.22. Assertion triggers in Examples 6.23, 6.42, and 6.24 are level-sensitive events: "When request *is* high ...". In many cases, it is desirable to have edge-sensitive triggers: "When request *becomes* high ...". For instance, Example 6.42 may be reformulated as: "When request becomes high it should be granted within 5 cycles." The corresponding assertion is (note the **`nexttime`** operator!):

```
a1: assert property (@(posedge clk)
  nexttime ($rose(req) |-> ##[1:5] gnt));
```

or

```
a2: assert property (@(posedge clk)
  ##1 $rose(req) |-> ##[1:5] gnt);
```

The same assertion is probably better written as[3]:

```
a3: assert property (@(posedge clk)
  !req ##1 req |-> ##[1:5] gnt);
```

□

[3]Assertions `a1` and `a2` are equivalent, but assertions `a1` and `a3` are not completely equivalent: If at clock tick 0 `req` has value x, and in clock tick 1—value 1, the consequent at clock tick 1 is checked in `a1`, but not in `a3`. If the variable values are 2-valued, both assertions are equivalent. The situation when assertions behave differently in presence of unknown values is quite common, we will not always explicitly comment on it.

7.2.1.4 Changed and Stable

Sampled value function `$changed(e, @clk)` returns *true* iff the past value of `e` is different from its current value, and *false* otherwise. The sampled value function `$stable(e, @clk)` returns *true* if the past value of `e` is identical to its current value, and *false*, otherwise. More precisely,

```
$changed(e, @clk) ≡ $past(e,,,@clk)!== $sampled(e).
$stable(e, @clk)  ≡ $past(e,,,@clk)=== $sampled(e).
```

`$stable(e, @clk)` is the same as `!$changed(e, @clk)`. `$changed` and `$stable` compare the past value with the current *sampled* value of the expression.

As in other clocked sampled value functions, the clock event argument is optional and if omitted, its value is inferred from the context (Sect. 7.2.1.5).

Example 7.23. For the timing diagram shown in Fig. 7.2, system function `$changed(a, @(posedge clk))` returns *true* for time steps $30 < t \leq 60$ and $70 < t \leq 80$. For all other time steps, it returns *false*. The value of `$stable` is the inverse. □

Example 7.24. If `e` remains `x`, `$stable(e)` is true. If `e` changes from 0 or 1 to either `x` or `z`, `$stable(e)` is false. □

Example 7.25. Signal `sig` should be always stable.

Solution: One can attempt to write this assertion as:

```
a1_wrong: assert property (@(posedge clk) $stable(sig));
```

but this implementation is *wrong*. The problem is, as usual, with clock tick 0. At clock tick 0 `$stable` returns *true* only if the new value of `sig` coincides with its initial value. Therefore, assertion `a1_wrong` checks that `sig` always preserves its initial value. As an illustration, consider the case when `sig` is of type **`logic`**, and when it is not explicitly initialized—the most common case in practice. In this case, assertion `a1_wrong` succeeds iff `sig` is always `x`, apparently not what was intended to check. As always in such cases, the solution is to delay the assertion execution:

```
a2: assert property (@(posedge clk) nexttime $stable(sig));
```

Discussion: Assertion `a2` is still problematic: it checks signal stability only on rising `clk`. In many cases in practice it is desirable to ensure absolute signal stability, for example, in verification of clock domain crossing (Sect. 1.2.3). We postpone this discussion until Sect. 7.2.2. In this section, we consider signal stability only relative to some specific clock. □

Example 7.26. We are now ready to implement the assertion from Example 6.11 with the help of a sampled value function: "`sig` toggles every cycle: 0101... or 1010...".

Solution:

```
a_toggle: assert property (@(posedge clk) ##1 $changed(sig));
```
□

Example 7.27. Each time signal `sig` changes, it should remain stable for four additional cycles.

Solution:

```
a1: assert property(@(posedge clk)
                    ##1 $changed(sig) |=> $stable (sig)[*4]);
```

Discussion: It is possible to specify signal stability at clock phases instead of clock cycles.

```
a2: assert property(@(edge clk)
                    ##1 $changed(sig) |=> $stable(sig)[*4]);
```

□

Example 7.28. Signal `sig` is stable between the start and the end events. The events are represented by the high value of signals `start_ev` and `end_ev`.

Solution:

```
a_stable: assert property (@(posedge clk)
  start_ev |=> $stable(sig) until_with end_ev);
```

Discussion: The limitation that both `start_ev` and `end_ev` are signals may be relaxed. `start_event` may be an arbitrary sequence, and `end_ev` may be an arbitrary property. In such a case, the stability is checked only after sequence `start_ev` matches, and until the starting clock tick of a successful evaluation of property `end_ev`. □

7.2.1.5 Clock Inference

All sampled value functions described in Sect. 7.2.1 but `$sampled` have a clocking event as their argument. This argument may be omitted, in which case the clocking event is inferred from the context in a similar way to the inference of the clocking event in properties and sequences (Chaps. 12 and 14):

1. If the function is called from a concurrent assertion, the appropriate clocking event from the assertion is used.
2. If the function is called from an action block of an assertion statement, the leading clock of the assertion is used.
3. If the function is called from an **`always`** or **`initial`** procedure, the procedure clock is inferred. The rules of clock inference in procedures are described in Chap. 14.
4. Otherwise, the event from default clocking is used as described in Sect. 12.2.2.

Table 7.5 Past global clocking sampled value functions

Name	Description
`$past_gclk`	Return expression value in previous tick of global clock
`$rose_gclk`	Check whether signal value rose from last tick of global clock
`$fell_gclk`	Check whether signal value fell from last tick of global clock
`$changed_gclk`	Check whether signal value changed from last tick of global clock
`$stable_gclk`	Check whether signal value remained stable relative to last tick of global clock

7.2.2 Global Clocking Sampled Value Functions

Global clocking sampled value functions are sampled value functions controlled by the global clock. The global clock is the primary system clock, as explained in Sect. 4.4.2. There are two groups of global clocking sampled value functions: past and future.

> Global clocking sampled value functions may be used only if **`global clocking`** has been defined.

7.2.2.1 Past Global Clocking Sampled Value Functions

Table 7.5 summarizes past global clocking functions. Past global clocking SVF are simple shortcuts for corresponding general sampled value functions described earlier.

```
$past_gclk(e)     ≡ $past(e, 1, 1, @$global_clock)
$rose_gclk(e)     ≡ $rose(e, @$global_clock)
$fell_gclk(e)     ≡ $fell(e, @$global_clock)
$changed_gclk(e)  ≡ $changed(e, @$global_clock)
$stable_gclk(e)   ≡ $stable(e, @$global_clock)
```

7.2.2.2 Future Global Clocking Sampled Value Functions

Table 7.6 summarizes future global clocking sampled value functions.

The future global clocking functions provide information about the future behavior of an expression. Unlike the past global clocking functions, there are no expressions using the general sampled value functions corresponding to future global clocking functions. Function `$future_gclk(e)` returns the sampled value

Table 7.6 Future global clocking sampled value functions

Name	Description
`$future_gclk`	Return expression value in next tick of global clock
`$rising_gclk`	Check whether signal value is rising
`$falling_gclk`	Check whether signal value is falling
`$changing_gclk`	Check whether signal value is changing
`$steady_gclk`	Check whether signal value is not changing

of `e` at the next tick of the global clock.[4] Other future global clocking SVF can be defined through function `$future_gclk`:

```
$rising_gclk(e) ≡
   $sampled(LSB(e))!== 1 && $future_gclk(LSB(e))=== 1
$falling_gclk(e) ≡
   $sampled(LSB(e))!== 0 && $future_gclk(LSB(e))=== 0
$changing_gclk(e) ≡
   $sampled(e)!== $future_gclk(e)
$steady_gclk(e) ≡
   $sampled(e)=== $future_gclk(e)
```

There are several restrictions imposed on future value functions: they cannot be used in reset conditions, outside concurrent assertions,[5] and they cannot be nested.

Example 7.29. The following use of future functions is *illegal*:

```
always @(posedge_clk) a <= $future_gclk(b) && c;
a1_illegal: assert final (a -> $future_gclk(b));
a2_illegal: assert property (@(posedge clk)
  disable iff (rst || $rising_gclk(interrupt)) req |=> gnt);
a3_illegal: assert property (@(posedge clk) req |-> $future_gclk
    (ack && $rising_gclk(gnt));
```

In the first statement `$future_gclk(b)` is used in an assignment, `a1_illegal` is not a concurrent assertion, in `a2_illegal` a future SFV is used in a reset condition, and in `a3_illegal` two future SVF are nested. □

Efficiency Tip. Global clock future value functions are usually more efficient in FV than past sampled value functions. In simulation, the picture is the opposite.

Stability Assertions. The following examples illustrate how a future sampled value function may be used to write assertions checking signal stability.

Example 7.30. `sig` value does not change.

[4]Formal interpretation of `$future_gclk(e)` is provided in Chap. 21.

[5]Future global clocking sampled value functions are legal in let-statements used in concurrent assertions, and in definitions of sequences and properties.

Solution: We have already considered this assertion in Example 7.25, but the implementation there had two drawbacks: we had to delay the assertion by one clock tick to handle the past value correctly at clock tick 0, and we could specify stability only relative to a clocking event. Controlling the assertion by the global clock and using a future value function overcomes these drawbacks:

```
a_stable: assert property (@$global_clock $steady_gclk(sig));
```

In this implementation, there is no need to have an initial delay since function `$steady_gclk(sig)` checks that the `sig` value is identical to the one in the next clock tick, hence there is no reference to the values of `sig` prior to clock tick 0, as in the case with `$stable`.

As `$global_clock` indicates the primary system clock, all signal changes that we are willing to consider are synchronized with it as explained in Sect. 4.4.2 and Chap. 21, and if there is a signal change between the ticks of the global clock, it is considered to be a glitch and is ignored at the RTL level. Recall, however, that it is the user responsibility to define **global clocking**. □

Example 7.31. Signal `sig` should be stable between two consecutive clock ticks. In other words, the signal value can change only when the clock is rising.

Solution:

```
a1: assert property (@$global_clock disable iff(rst)
  $changing_gclk(sig) |-> $rising_gclk(clk));
```

Discussion: In this assertion, we consider the clock to be a regular signal, and not an event. The same assertion may be expressed using past sampled value function, but this is less natural and usually less efficient in FV:

```
a2: assert property (@$global_clock disable iff(rst)
  ##1 $changed(sig) |-> $rose(clk));
```

Equivalently, we could use past global clocking function (`$changed_gclk(sig)`, etc.), but this is more verbose. □

Clock as a Sampled Signal.

Example 7.32. Enable `en` should be low when `sig` is going to change.

Solution: The easiest way to write this assertion is to use `sig` as a clock:

```
a1: assert property(@sig !en);
```

Discussion: This solution works always in FV, but it does not always work in simulation. The reason is that in simulation it is required that the clock value changes at most once in any simulation tick, otherwise the simulation of the assertion may not work correctly.

To overcome the limitation, the assertion can be rewritten in the following way:

```
a2: assert property(@$global_clock $changing_gclk(sig) -> !en);
```

This example shows that the global clock may be used as a "carrier" for other signals that cannot be used as a clock directly. □

Clock Fairness.

Example 7.33. Clock is always ticking.

Solution: It is required to check the clock fairness, i.e., that in any given moment there is some future moment when the clock ticks (see also Example 5.15):

```
a_fclk: assert property(@($global_clock)
  s_eventually $rising_gclk(clk));
```

Discussion: The assertion clock is `$global_clock` relative to which the ticking of `clk` is checked. This assertion works only if `$global_clock` ticks on both edges of `clk` or faster. If `$global_clock` is the same as **posedge** `clk`, ticking of the `clk` cannot be detected: when **posedge** `clk` happens the sampled value of `clk` is always 0.

Efficiency Tip. Assertion `a_fclk` is efficient in FV.[6] In simulation, this assertion is not optimal, as it may accumulate several overlapping attempts, as explained in Sect. 19.3.

However, checking this assertion in simulation is not useful. The simulation run is finite, and thus by analyzing the run it is not possible to conclude that the clock is always ticking. It is only possible to check that the clock ticks at least some predefined number of times during the simulation run (see Exercise 7.8). For practical needs, it may be sufficient to examine the simulation trace. □

One could wonder why general sampled value functions contain only the past version. Why is there no `$future(e, @clk)` function in the language? The answer is simple: we do not know whether an arbitrary clock is fair, i.e., never stops ticking. If `clk` stops ticking at some moment, the value `$future(e, @clk)` is undefined. As for the global clock—the primary system clock—is concerned, it is required to be fair by definition, and `$future_gclk(e)` always makes sense. This consideration is important in FV, where infinite traces are handled. See Sect. 21.2.3.1 for an additional explanation.[7]

---

[6]This assertion is still rather expensive to check in FV. What we mean that among various versions of the same assertion, this assertion is one of the most efficient.

[7]In simulation, the assertion behavior is usually ignored at the last tick of the global clock.

7.3 Tasks for Controlling Assertions and Runtime Violations

During the course of development, assertions often fail because the code is incomplete or is in the process of getting fixed. Such failures are impediments to development, particularly when the developer is well aware of the causes of the failures. This section describes the system control tasks to control assertion executions, assertion action block executions, **expect** statement execution and run time violations for `unique`, `unique0` and `priority`. Using these tasks, the developer can selectively turn off assertion or other monitoring checks until the code is brought to a state where the tests can actually be exercised.

Section 7.3.1 describes tasks for controlling assertion evaluation attempts. Section 7.3.2 describes tasks for controlling action blocks. Section 7.3.3 describes a more generalized task `$assertcontrol` to control both assertion execution as well as action block execution.

7.3.1 Tasks for Controlling Evaluation Attempts

It is often necessary to stop the evaluation and reporting of unwanted assertion failures during special initialization, power down, or reset operation phases of the design under verification. If there is some Boolean expression that characterizes such a phase, then the expression can be used in a **disable iff** clause in subsequent assertions to disable assertion failures during those phases. For other situations when a testbench initializes the design, the assertions can be disabled explicitly by using *assertion control tasks*:

- `$asserton`
- `$assertoff`
- `$assertkill`

By default all assertions are enabled from time 0. To stop the start of any attempts from time t, it is sufficient to call `$assertoff` before the Observed region of the simulator at time t. This, however, does not stop any evaluation attempt that has been started earlier. To restart the evaluation, at some later time t', the task `$asserton` is called.

Therefore, to avoid executing all assertions during the initialization phase, `$assertoff` should be called from the testbench at time 0, followed by `$asserton` when exiting the initialization phase. This is illustrated in the following example.

Example 7.34.

```
module m;
  bit clk;
  default clocking ck @(posedge clk); endclocking
  a: assert property( ... );
  initial begin
    $assertoff();
```

```
    //... await reset activity completed ...
    @ck; //synchronize and start assertion from this tick
    $asserton();
  end
  //... other code ..
endmodule
```

In this case, assertion `a` will start evaluation attempts starting from the moment statement `@ck` unblocks.

What if the call to `$asserton()` were made in a **program** that executes in the Reactive region as in this code snippet?

```
module m;
  bit clk;
  prg my_program();
  default clocking ck @(posedge clk); endclocking
  a: assert property( ... some_property ...);
  //... other code ..
endmodule
program prg;
  initial begin
    $assertoff();
    //... await reset activity completed ...
    @m.ck; //synchronize and start assertion from this tick
    $asserton();
  end
  //... other code...
endprogram
```

Since the call is made from the Reactive region, the assertion is not yet enabled when the evaluation of `a` is to start in the Observed region, the first evaluation of `a` will only happen one clock tick later. Also, the `$assertoff` does not execute until the Reactive region of time step 0, hence the assertion evaluation attempt at time 0 is enabled and will execute if there is a clock tick. □

If the initialization phase is to be executed again later in the simulation, we should again stop the assertions, but in that case we should consider using `$assertkill`. It stops the subsequent evaluation attempts, and also any evaluation attempts that have started earlier (that would still be continuing evaluation during the initialization phase). Calling this task thus avoids failures of evaluation attempts that do not matter anymore, even though they started during the"normal" design phase.

The above approach will work well if we need to stop all assertions, but what if some assertions should run during the special phase? The tasks accept a list of arguments that define the modules or scopes and the depth down to individual assertions to which the task call applies. The argument list is of the same form as for the well-known `$dumpvars` task:

assert_control_task ::= name[`(`levels[`,` list_of_scopes_or_assertions]`)`] `;`
name ::=

```
      $asserton
    | $assertoff
    | $assertkill
```
list_of_scopes_or_assertions ::= scope_or_assertion , {scope_or_assertion}
scope_or_assertion ::=
  scope_identifier
  | assertion_identifier
  | hierarchical_identifier

The *levels* argument is a nonnegative integer constant. When set to 0, it applies to all the items on the *list_of_scopes_or_assertions* and all the scopes below. When set to 1, it applies only to the items in the current scope but not to the hierarchy under it. When set to some `n > 1`, it applies to the items and to `n-1` levels of hierarchy below. Of course, if the item is a full hierarchical path to an assertion, the task applies only to that assertion since there are no levels below.

7.3.2 Tasks for Controlling Action Blocks

Assertions have optional *action blocks* (Chap. 4), that is, blocks of procedural code that are executed when the assertion attempt succeeds (pass action block) or fails (fail action block). In the case of covers, only the pass action block is available because cover failure is not interesting. The following example shows a concurrent assertion `a` and a concurrent cover `c`, each with action blocks.

Example 7.35.
```
a: assert property(p1)
    begin // pass action block
        process_pass();
        $info("assertion PASSED");
    end
    else begin // fail action block
        process_failure();
        $error("assertion FAILED");
    end
c: cover property(p2)
    begin // pass action block
        process_cover();
        $info("COVERED");
    end
```
□

By default, the action blocks execute on every success and failure of the assert and cover statements. This behavior may not be desirable, for example, when the property passes vacuously or when running massive regression tests where only failures should be reported.

SystemVerilog provides several tasks for controlling the execution of action blocks:

```
$assertpassoff
$assertpasson
$assertfailoff
$assertfailon
$assertvacuousoff
$assertnonvacuouson
```

The arguments to the system tasks are the same as for the assertion control system tasks described in Sect. 7.3.1.

As it can be seen, the assertion action control tasks come in pairs, one for disabling and one for reenabling the action.

`$assertpassoff` stops the execution of pass action blocks in the scope specified by the arguments to the task call until `$assertpasson` is called affecting the same scope.

`$assertfailoff` stops the execution of fail action blocks in the scope specified by the arguments to the task call until `$assertfailon` is called affecting the same scope.

`$assertvacuousoff` stops the execution of pass action blocks for vacuous evaluation attempts in the scope specified by the arguments to the task call until `$assertpasson` is called affecting the same scope.

All the disabling tasks do not affect evaluation attempts that are already in progress as well as action blocks that are currently executing.

The following are typical cases of using these tasks:

- In regression tests, disable pass action blocks on **`assert`** and **`assume`** statements unless their execution has some functional impact on the testbench behavior. In covers, it could remain enabled if they provide information for some user-specific coverage analysis tools.
- Even when pass action blocks are enabled, we recommend to disable their execution on vacuous evaluation attempts because they may lead to an incorrect interpretation of the verification results.
- If default failure messages issued by the verification tool are sufficient, and fail and pass action block execution has no functional impact on the testbench, it may be preferable to disable the execution of fail action blocks. Verification tools often have means to control how many failures of assertions are reported by default reporting mechanisms. This is useful for catching the first few failures without cluttering the log with redundant repetitions of the same message.

7.3.3 General Assertion Control Task

The preceding sections discussed control tasks that mostly perform a singular function. For example, `$assertoff` turns off assertions and `$assertpassoff` disables the execution of action blocks associated with assertion successes. Moreover, the selection of statements by these tasks is limited and not adequate in situations when only certain categories of statements are needed to be controlled. It is not possible to

Table 7.7 Control type values

Value	Control	Description
1	Lock	Prohibit any control changes
2	Unlock	Allow control changes
3	On	Enable
4	Off	Disable
5	Kill	Kill assertions
6	PassOn	Enable execution of success action blocks
7	PassOff	Disable execution of success action blocks
8	FailOn	Enable execution of fail action blocks
9	FailOff	Disable execution of fail action blocks
10	NonvacuousOn	Enable execution of action blocks on non-vacuous success
11	VacuousOff	Disable execution of action blocks on vacuous success

turn off assert statements while leaving cover statements on. Similarly, concurrent assertions cannot be set apart from immediate assertions when different control is required for each group. Such fine-grained control is often needed in debugging during development phases of a design project.

We now discuss `$assertcontrol`[8], a control task that assimilates the functionality of the tasks described in the preceding sections as well as provides additional fine-grained control and flexibility. It has the following syntax:

```
$assercontrol(control_type[,[assertion_type][,[directive_type]
                [,[levels][,list_of_scopes_or_assertions]]]]);
```

The first argument *control_type* is an integer which specifies the type of control that is applied on the selected assertions. The control functions and their corresponding argument values are enumerated in Table 7.7. The control functions include all the features of the tasks in the preceding sections.

A model under development in a diverse environment of collaborating engineers is often entangled with control tasks that are suitable to individual portions of the design, but may be intruding on other portions when integrated to execute together. In such cases, the designer often needs to have overriding control to make progress. A locking feature is provided to assist in overriding control. When *lock* is enabled on an assertion, subsequent change in control is prohibited until *lock* is disabled. For instance, if an assertion was turned off before *lock* was enabled, it remains turned off regardless of any further invocation of `$asserton` or `$assertcontrol` to turn it on. *lock* can be disabled at a later time to continue controlling the assertion execution. Similar actions of *lock* are applicable on controlling action blocks as well. Applying *lock* to assertions contained in a power domain while controlling other assertions in the system is a common application.

[8] `$assertcontrol` was introduced in 2012 LRM [8].

Table 7.8 Assertion type values

Value	Assertion type
`1`	Concurrent
`2`	Simple Immediate
`4`	Observed Deferred Immediate
`8`	Final Deferred Immediate
`16`	Expect
`32`	Unique
`64`	Unique0
`128`	Priority

Table 7.9 Directive type values

Value	Directive type
`1`	Assert
`2`	Cover
`4`	Assume

The second argument *assertion_type* selects the type of statements on which the control function is applied. Table 7.8 specifies the statement type options and the associated argument values. The available statements for selection include assertion types such as concurrent and immediate, unique, unique0 and priority for runtime violations of case and if statements, and expect statements. Each argument value for the selected statement type is specified by a single designated bit in the argument integer, such that more than one type of statement can be selected by computing the union of the corresponding values. For example, concurrent assertion statements are selected by value 1, final deferred immediate type is selected by 8, so value 9 selects both of them. The argument itself is optional, which defaults to value 255 when the argument is omitted. Value 255 includes all types.

The third argument *directive_type* selects the assertion directive types and is useful only when the selection via argument *assertion_type* is of assertion statement types. The directive type options are enumerated in Table 7.9. Like the *assertion_type* argument, multiple directive types can be selected by computing the union of the corresponding values. For example, value 3 will select assert and cover directives. The argument is optional, which defaults to value 7, when the argument is omitted. Value 7 selects all directive types.

The fourth argument *levels* specifies levels as described in Sect. 7.3.1. This argument is also optional and defaults to level 0 when omitted, selecting full hierarchies of scopes specified by argument *list_of_scopes_or_assertions*.

The fifth argument *list_of_scopes_or_assertions* specifies a list of items on which the selection of previous arguments is applied as described in Sect. 7.3.1. Each item is either an assertion or a scope to control. If the argument is omitted, then the selections are applied on all top-level scopes.

Although `$assertcontrol` requires more arguments to specify, the task offers comprehensive functionality. It simplifies many situations where a manual grouping of assertions either by name or other means would be necessary for controlling

Table 7.10 Control task equivalents

Control task	Equivalent representation with `$assertcontrol`
`$asserton` (level, list)	`$assercontrol(3,15,7,`level, list)
`$assertoff` (level, list)	`$assercontrol(4,15,7,`level, list)
`$assertkill` (level, list)	`$assercontrol(5,15,7,`level, list)
`$assertpasson` (level, list)	`$assercontrol(6,31,7,`level, list)
`$assertpassoff` (level, list)	`$assercontrol(7,31,7,`level, list)
`$assertfailon` (level, list)	`$assercontrol(8,31,7,`level, list)
`$assertfailoff` (level, list)	`$assercontrol(9,31,7,`level, list)
`$assertnonvacuouson` (level, list)	`$assercontrol(10,31,7,`level, list)
`$assertvacuousoff` (level, list)	`$assercontrol(11,31,7,`level, list)

them. If a designer is interested in just turning off final deferred immediate, `$assertcontrol` can be invoked with statement type argument of value 8. In the example below, all final deferred immediate statements will be turned off under scope `top.inst1.s1`.

```
$assertcontrol(4,8,1,0,top.inst1.s1);
```

`$assertcontrol` subsumes functionality provided by all other control tasks. Table 7.10 shows the equivalent assertion control task representation to `$assertcontrol`. In that table *level* represents the level of hierarchy and *list* represents the list of assertions or scopes.

Exercises

7.1. Write an assertion checking that exactly `n` bits of a signal are high (low).

7.2. Write an assertion checking one-cold encoding for four-state value variables.

7.3. It was mentioned in Example 7.9 that invoking system function `$sampled` in action blocks of immediate or deferred assertions for reporting purposes will result in inconsistent messages. Explain why.

7.4. Modify the assertions in Example 7.14 to allow `sig` either remain unchanged or change in one bit in consecutive clock cycles. Write two versions of this assertion: one which does not impose any constraints on the initial value of `sig`, and the other requiring `sig` to be 0 upon the termination of the reset sequence.

7.5. Rewrite assertion

```
b1: assert property(@(posedge clk) ##1 $fell(sig) |=> !sig[*4]);
```

from Example 7.21 without using sampled value functions.

7.6. What is the assertion behavior in Example 7.26 if the initial delay `##1` is dropped? Assume that `sig` is of type **`bit`**.

7.7. What does the following assertion mean?

```
assert property(@($global_clock)
  s_eventually $changing_gclk(clk));
```

What is the difference between this assertion and the assertion from Example 7.33

7.8. Write an assertion stating that the clock ticks at least `n` times, where `n` is an elaboration time constant. See Example 7.33.

7.9. There are two clocks: fast (`fclk`) and slow (`sclk`), and `sclk` is 8 times slower than `fclk`. Write the following assertion: signal `sig` may only change on the third rising edge of `sclk` after its last change.

Part III
Metalanguage Constructs

Chapter 8
Let, Sequence and Property Declarations; Inference

The beginning of wisdom is to call things by their right names.

— Chinese proverb

In SystemVerilog, modules, programs, interfaces, checkers, functions, and tasks provide means for reuse, and for abstracting and hiding details. SystemVerilog assertions provide such means too. This is achieved using parameterized **`let`**, **`sequence`**, and **`property`** declarations. Their argument lists as well as instantiation semantics are quite different from the other reuse features. In addition, certain kinds of actual arguments can be inferred from the instantiation context. Similar to sequences and properties, **`let`** declarations allow to abstract expressions, making code more readable and reusable. **`let`** can be used anywhere, not only in assertions. One of their intended uses is for defining reusable parameterizable expressions for immediate and deferred assertions.

8.1 Let Declarations

`let` declarations are a way to define parameterizable templates for forming expressions. **`let`** can be declared in any declarative scope, wherever variables can be declared, and can be instantiated wherever expressions can be used. **`let`** declarations are similar to text macros, but they are better adapted for use in SystemVerilog expressions because they are part of the core language. They follow normal scoping rules. The formal arguments of a **`let`** definition may be typed and can have default actual arguments. However, there are some practical restrictions to be placed both on the form of the expression definition and on the arguments, even though they are not stated explicitly in the SystemVerilog LRM. These will be discussed later in the section.

E. Cerny et al., *SVA: The Power of Assertions in SystemVerilog*,
DOI 10.1007/978-3-319-07139-8_8

The following example shows some simple **let** usage and illustrates the effect of scoping rules applied to these declarations. They are contrasted with similar macro declarations.

Example 8.1. Effect of scoping rules.

```
module m;
  logic clk, a, b, c, d;
  let let_exp = a && b;

  always @(posedge clk) begin: B1
    let let_exp = c || d;
    a <= let_exp;
    b <= a;
  end
  assign c = let_exp;
  assign d = b;
endmodule
```

There are two **let** declarations of the same name `let_exp`, but each having a different expression associated with the name. The first one is defined in the module scope, and the second one in the **always** procedure scope. The variables used in the expression on the right-hand side of the let declaration must be visible at the point of declaration. Wherever a **let** is instantiated the nearest visible **let** declaration is used. The expression on the right-hand side of the declaration is substituted in the place of the instance. Module `m` definition is thus equivalent to the following code:

```
module m;
  logic clk, a, b, c, d;
  always @(posedge clk) begin: B1
    a <= c || d;
    b <= a;
  end
  assign c = a && b;
  assign d = b;
endmodule
```

Suppose now that we replace the **let** declarations by macro definitions. The original module definition becomes

```
'define macro_exp a && b
module m;
  logic clk, a, b, c, d;
  always @(posedge clk) begin: B1
'define macro_exp c || d
    a <= 'macro_exp;
    b <= a;
  end
  assign c = 'macro_exp;
  assign d = b;
endmodule
```

After macro substitution, module m is quite different from the one after **let** substitution

```
module m;
  logic clk, a, b, c, d;
  always @(posedge clk) begin: B1
    a <= c || d;
    b <= a;
  end
  assign c = c || d;
  assign d = b;
endmodule
```

The difference is in the assign statement where in the case of **let**, it is the definition from the module scope that is used, whereas in the case of a macro, it is the latest definition in the lexical order.[1] To imitate the scoping of **let** using macro definition, the symbol would have to be explicitly *undefined* when exiting the scope and again *defined* to the original expression, as shown next. Clearly, this is much more tedious and error prone.

```
`define macro_exp a && b
module m;
  logic clk, a, b, c, d;
  always @(posedge clk) begin
`define macro_exp c || d
    a <= `macro_exp;
    b <= a;
`undef macro_exp
`define macro_exp a && b
  end
  assign c = `macro_exp;
  assign d = b;
endmodule
```

□

Unlike macro definitions, **let** declarations follow normal scoping rules. The right-hand side expression from the definition is substituted in place of the instance.

The instantiation of **let** is quite different from, e.g., calling a function because the **let** body is substituted in the place of the instance. Unlike functions, **let** instantiations cannot be recursive.

let declarations can have formal arguments. Again, unlike in functions, the actual arguments are substituted for every occurrence of the corresponding formal argument when the instance is replaced by the **let** body. The actual argument variables must be visible in the instance scope. The formal argument types are

[1]The compiler may issue a warning for the second macro definition, saying that the symbol has been redefined.

restricted to the event type and the integral types allowed in assertions, and may be typed or untyped. If untyped, the actual argument expression is enclosed in parentheses before substitution in the place of the formal argument. The extra pair of parentheses is added to preserve the precedence of evaluation as indicated by the parent expression containing the **let** instance. If the formal argument is typed, then the self-determined type of the result of the evaluation of the actual argument must be type cast compatible. Provided that the types are compatible, the actual expression is cast to the type of the formal argument before being substituted in place of the occurrences of the formal arguments in the **let** body.

The following example shows the use of an untyped formal argument.

Example 8.2.

```
let orReduct(x) = |x;
module m;
  logic [2:0] sel;
  logic [7:0] [1:0] data;
  logic a, b, v, w;
  assign v = a && orReduct(sel);
  assign w = orReduct(data);
  //...
endmodule
```

After substitution, the code takes the following form:

```
module m;
  logic [2:0] sel;
  logic [7:0] [1:0] data;
  logic a, b, v, w;
  assign v = a && (|sel);
  assign w = (|data);
  //...
endmodule
```

□

In the above example, notice that, because the formal is untyped, there is more latitude in using actual argument expression of different types in the **let** instance; no type conversion is performed and the validity of the substituted expression is completely determined by the expression within which it is substituted.

The formal arguments can also have a default actual argument. When no actual argument is provided in a **let** instance, the default one is used. The following example is similar to the preceding one except that the **let** is now defined inside the module and the formal argument x has a default actual expression `sel`. The first instance does not provide an actual argument, hence the default `sel` is taken. The result is the same as in the preceding example.

Example 8.3.

```
module m;
  logic [2:0] sel;
  logic [7:0] [1:0] data;
  logic a, b, v, w;
  let orReduct(x = sel) = |x;
  assign v = a && orReduct();
```

```
  assign w = b || orReduct(data);
  //...
endmodule
```

□

When typed formal arguments are used, type checking is performed to ensure that the actual argument is type cast compatible with the formal argument. If that is so, then type casting takes place. This is illustrated in the next example.

Example 8.4.

```
typedef bit [1:0] my_t;
let orReduct(my_t x) = |x;
module m;
  logic [2:0] sel;
  logic [7:0] [1:0] data;
  logic a, b, v, w;
  assign v = a && orReduct(sel);
  assign w = orReduct(data);
  //...
endmodule
```

The resulting code after substitution is as follows:

```
typedef bit [1:0] my_t;
module m;
  logic [2:0] sel;
  logic [7:0] [1:0] data;
  logic a, b, v, w;
  assign v = a && (|my_t'(sel));
  assign w = (|my_t'(data));
  //...
endmodule
```

Because of the type cast, only the low-order 2 bits of either argument are converted to **bit** and used in the reduction **or** operator. □

> Typed arguments enforce type compatibility and casting, but limit flexibility as compared to untyped arguments.

Let us recapitulate the way the variables in the default argument expression and the variables in the **let** body that are not formal arguments are resolved. The rules are similar to those for function and task declarations. Namely, the default arguments and the variables that are not formal arguments are resolved in the declarative context, while the actual arguments are resolved in the instantiation context.[2]

[2]The SystemVerilog LRM provides many examples illustrating the use of **let** in various scoping contexts, with and without typed arguments and type casting, as well as the use of sampled value functions in **let** definitions.

Example 8.5.

```
module m;
  bit clk;
  logic a, b;
  let x = $past(a && b);
  let y = $past(a && b, , ,@(posedge clk));
  always_comb begin
    a1: assert #0 (x);
    a2: assert #0 (y);
  end
  a3: assert property(@(posedge clk) a |-> x);
endmodule
```

In this example, `let` instances are used in three assertions, `a1`, `a2`, and `a3`. While the form of the instance is legal in `a2` and `a3`, it is illegal in `a1`. Why?

The sampled value function requires a clocking event for updating the previous value of its argument. There is no clocking event available and it cannot be inferred when `x` is expanded in `a1`. This results in an illegal use of `$past`. In `a2`, the clocking event is explicitly specified in the sampled value function in the definition of `y`. Therefore, after substitution of `y` into `a2`, a legal form of the deferred assertion `a2` is obtained. Finally, in `a3` there is no explicit clocking event in the sampled value function, after substituting the body of `x` into `a3`, the assertion contains the following body:

```
a3: assert property(@(posedge clk) a |-> ($past(a && b)));
```

The property in the assertion that resulted from the substitution is perfectly legal because the clocking event for the sampled value function is inferred from the assertion. □

We now examine some problematic cases. They are not explicitly identified as illegal in the SystemVerilog LRM, but their use may lead to some unexpected results and should be considered as illegal. The first case involves explicit or implicit form of *variable_lvalue* assignment, that is, the variable or expression on the left-hand side of an assignment.

Example 8.6.

```
let inc1(int x) =  x++;
let inc2(bit [7:0] y) = y+=2;
let combinetwo(integer v, w) = (v = v + w);
module m;
  bit clk;
  integer a, b;
  a1: assert property (@(clk) inc1(a));
  a2: assert property (@(clk) inc2(a) == 1);
  property p;
    @(clk) combinetwo(a, b);
  endproperty
  a3: assert property (p);
endmodule
```

After substitution of **let**, we obtain the following code:

```
module m;
  bit clk;
  integer a, b;
  a1: assert property (@(clk) (int'(a)++));
  a2: assert property (@(clk) (((type bit [7:0])'a+=2) == 1));
  property p;
    @(clk) ((a = a + b));
  endproperty
  a3: assert property (p);
  endmodule
```

Assertion `a1` contains an illegal form—`int'(a)` is an expression, not a variable, hence the increment operator cannot be applied to it.

Assertion `a2` results in an illegal form because type cast is applied to the whole implicit assignment, and it is not clear how the type cast is to be interpreted.

Assertion `a3` is seemingly legal, but there is a side effect of replacing the current value of `a` by its sampled value summed with `b`. The SystemVerilog LRM does forbid assignment expressions in Boolean expressions in sequences and properties, but not **let** statements. Even if supported by a simulator, the outcome may be rather unexpected and difficult to understand. Which value is compared with 1 in assertion `a2`? The old one or the new one? It would appear that it should be the old sampled value, but is it? Is the sampled expression just `$sampled(a)` or is it `$sampled(a + b)` in assertion `a3`?

If the **let** definitions were written by the same person who writes the assertions, perhaps the expected outcome is clear to that person. However, if the **let** declarations are part of a package and the user knows little about its contents, then usage of such forms with side effects becomes problematic and should be illegal. □

A similar recommendation applies to passing such expressions as actual arguments to otherwise simple innocuous **let** definitions. This is illustrated on the next example.

Example 8.7.

```
let inc(int x) =  x;
module m;
  bit clk;
  integer a, b;
  a1: assert property (@(clk) inc(a++));
  a2: assert property (@(clk) inc(a+=1) == 1);
  property p;
    @(clk) inc((a = a + b));
  endproperty
  a3: assert property (p);
endmodule
```

After substitution, the code is as follows

```
module m;
  bit clk;
  integer a, b;
```

```
  a1: assert property (@(clk) int'(a++));
  a2: assert property (@(clk) (int'(a+=1) == 1));
  property p;
    @(clk) (int'(a = a + b));
  endproperty
  a3: assert property (p);
endmodule
```

The outcome of the **let** substitution is equally illegal or confusing as in the preceding example. □

> Do not use expressions with side effects such as increment/decrement and operator assignment expressions in **let** actual arguments and in **let** definitions.

Another issue that is not sufficiently discussed in the SystemVerilog LRM is the application of bit and part selects over **let** instances, part and bit selects over formal arguments in **let** defining expressions, and in passing bit or part select as actual arguments to **let** instances. The problem is that unless the user is aware of the **let** definition details, such use may create illegal expressions once the **let** body is substituted in place of the instance. We illustrate some of the problematic situations on the following example.

Example 8.8.

```
typedef bit[1:0] bt_t;
module m;
  logic [7:0] a;
  logic [7:0] [2:0] b;
  logic c, d, e, f;
  let lt1(bt_t x) = x;
  let lt2(x) = x[1:0];
  let lt3(x) = x;
  assign c = lt2(a)[0];
  assign d = lt1(a[4:0])[0];
  assign e = lt2(a[4:0]);
  assign f = lt3(a)[0];
```

After substitution of let arguments and bodies into the assignments, we get the following equivalent code:

```
typedef bit[1:0] bt_t;
module m;
  logic [7:0] a;
  logic [7:0] [2:0] b;
  logic c, d, e, f;
  assign c = (a[1:0])[0];
  assign d = (bt_t'(a[4:0]))[0];
  assign e = ((a[4:0])[1]);
  assign f = (a)[0];
```

□

The assignments to c, d, and e result in illegal expression forms. This is because a bit or part select is taken from a parenthesized expression which is a part select itself. Assignment to f is equally illegal even though the argument is just a variable identifier in which case it would perhaps make sense to allow substitution without the enclosing parentheses.

> Do not apply any select operators on **let** instances.

8.1.1 Syntax of Let

let declaration and its actual arguments have the following form:

let identifier[(let_port_item{, let_port_item}])] = expression;

let_port_item allows port types that are restricted to integral types and **event**. Also, untyped formal arguments may be used to indicate any type. There is no use of port direction, hence port direction specification is not allowed.

Untyped ports can be used at the beginning of the formal port list, but once a type is specified on a port, then all subsequent ports must have a type or an explicit **untyped**. This is illustrated in the following example:

Example 8.9. Typed arguments.

```
let my_let(x, bit y, z) = y ? x : z;
```

The formal argument x is untyped. The arguments y and z are of type **bit**. This means that their actual arguments must be type compatible to type **bit** and is first cast to **bit** before substituting in the expression `y ? x : z` in the **let** instance. □

The formal arguments can also have default actual argument expressions assigned to the formal arguments.

The right-hand side of the definition and the actual arguments (default or otherwise) are general *expression* forms. As we discussed earlier, the expression must be void of implicit and explicit variable assignments.

A **let** *instance* can be used wherever an expression can be used. The instance consists of the **let** identifier followed in parentheses by a list of actual arguments. An actual argument can be specified in one of several ways, like in functions and tasks:

- Missing—when an actual argument is not provided. The default expression for the corresponding formal argument is used.
- Positional binding—the list of actual expressions is separated by commas. They are associated with the formal arguments in the order of appearance.
- Named binding `.formal_identifier(actual_expression)`—the actual argument is explicitly associated with a named formal argument. The order in which

they are written is immaterial. If actual_expression is missing, then again the default actual expression is used.

If a mix of positional and named binding is used, then the positional form must precede any named form.

8.1.2 Uses of Let

let is useful for abstracting expressions and for configuring them as templates in a checker library. A strong case for this abstraction form is when such a template is picked up from a checker library, and instantiated in immediate and deferred assertions. The following example illustrates this point, but for detailed discussion of this topic, see Chap. 24.

Example 8.10.

```
let onehot0(exp, bit reset = 1'b0) = reset || $onehot0(exp);
let noUnknown(exp, bit reset = 1'b0) = reset || $isunknown(exp);

module check(input
             logic rst,
             logic [15:0] decoded,
             logic [3:0] sel
             );
  a1: assert #0 (noUnknown({decoded, sel}));
  a2: assert #0 (onehot0(sel, rst));
endmodule : check
```

Module check is a combinational checker that verifies in assertion a1 that the ports decoded and sel never have an x or z value in any bit position, and in assertion a2 that the port sel has at most one bit asserted 1 . Assertion a1 is checking the expression regardless of whether rst is asserted or not because the default value of 0 is used as the actual argument for reset. Assertion a2 is disabled when rst is asserted 1. □

Notice that we could have used functions $onehot and $isunknown directly in the assertions, but then the reset argument would have to be always added in the expression of the assertion, which can be error prone. The **let** template in a library allows to hide this detail and makes the use of the functions more user friendly. Similarly, we could define a template for $past that provides different default values for its arguments than those provided by the function definition. The let declaration can be placed in a package and then imported wherever needed.

let declarations are suitable for libraries of property-like templates in packages to be used in deferred and immediate assertions.

To conclude this section, we wish to mention that unlike wires, **let** definitions can be used to define expressions that are temporarily used and not meant to be synthesized. For instance,

```
always_comb w = a & b;
assert #0 (w == 1);
```

If w is used only in the **assert** statement, the synthesis tool may or may not synthesize this signal, even though it is not required in the physical implementation of the design. Using **let** w = a & b; instead avoids this problem. The conventional solution is to use **`ifdef** ... **`endif** to enclose nonsynthesizable code but it is less elegant.

It will be interesting to see what other uses will be devised for **let**, and what kind of support various software tools will provide for debugging code that contains **let** statements. Since **let** is part of the language unlike macros, it may be possible to trace its evaluations and even collect coverage.

8.2 Sequence and Property Declarations

Sequence and property declarations allow users to compose temporal formulas into units that can be instantiated in other such units as well as in assertions. This mechanism thus provides means for abstracting temporal behavior to building more complex temporal formulas. The declarations of a **sequence** and **property** are named, and available with optional arguments. The formal argument list definition in sequence declarations is similar to that in property declarations, except that the latter also allows properties as arguments. The declaration interface and substitution of an instance by the body of its definition are similar to **let**, but more complex due to the presence of clocks and disabling conditions. Furthermore, formal argument list and the association of the formal with actual arguments differs considerably from the port list defined for modules, programs, interfaces, functions, and tasks.

We illustrate briefly some of these constructs in the following example. More details are provided in the subsequent section.

Example 8.11.

```
sequence sf_after_a(
  event clk = $inferred_clk, logic a, sequence sf, int n = 1);
  @(posedge clk) a ##n sf;
endsequence

property seq_impl_prop(logic rst = $inferred_disable,
  event clk = $inferred_clk, sequence sf, property pf);
  disable iff(rst) @clk sf |-> pf;
endproperty
```

The sequence sf_after_a has as its formal argument a clocking event named clk. It has a default actual argument which is, in this case, the system function $inferred_clock. In the absence of an actual argument, it infers the clocking event from the instantiation context of the sequence. The sequence has three additional formal arguments. The first one is **logic** a. The type is explicitly specified to ensure that the actual argument be type compatible with **logic**. The second one is **sequence** sf restricting the actual argument to be an expression or a sequence. An expression is allowed because it is also a simple (Boolean) sequence. The final argument is used as a constant in the delay operator and is thus restricted to an integral type. The body of sf_after_a states that n cycles after a the sequence sf should start evaluating at the next clock tick. When sf matches, the sequence sf_after_a matches as well.

Property seq_impl_prop has two formal arguments clk and sf that are of the same type as in the case of the preceding sequence declaration. It has two additional arguments, however. The first one of them, rst, is used as the **disable iff** expression of the property. It has a default actual argument that is used in the absence of an actual argument in the property instance. It can infer the disabling expression from a **default disable iff** *expr* declaration. The second one is an argument of type **property** and thus the actual argument can be an expression, a sequence or a property, but it cannot be a clocking event. □

Suppose now that there are the following declarations in a module (or program or interface or checker):

Example 8.12.

```
logic reset, clock, v, w, x, y;
default disable iff !reset;
default clocking ck @(posedge clock);
endclocking
sequence s;
  v[*3];
endsequence
a_until: assert property(
  seq_impl_prop(, , sf_after_a(, w, s, 1), (x until y));
c_seq: cover property(disable iff (!reset)
  sf_after_a(, w, s, 1));
```
□

How are the assertion and coverage statements to be interpreted? First, some basic rules are applied in the order as shown below:

1. Substitute actual arguments for all occurrences of the formal arguments in the body specifications.
2. Substitute sequence or property instances by their body specifications.
3. Apply inference rules.
4. Apply clock flow rules to determine sampling clocks for all expressions that need a clocking event (see Chap. 12).

In general, the rules cannot be applied just once as stated, rather it has to be a recursive application of the rules starting from the top property expression in the assertion until all the substitutions are completed, resulting in a property specification containing no unclocked expressions and no instance of another property or sequence.

Let us apply these rules to the example. First consider assertion `a_until`. For substituting `seq_impl_prop` instance body into the assertion requires to determine its actual arguments. The first two actual arguments are missing in the instance specification; therefore, default values are used. In this case, they are inferred. For `rst` the argument `$inferred_disable` is inferred from **`default disable iff`** and thus it is `!reset`. For `clk`, `$inferred_clock` is inferred from **`default clocking`** and thus it is **`posedge`** `clock`.

The third argument is an instance of sequence `sf_after_a` which itself needs its actual arguments. The first argument is again inferred from **`default clocking`** as **`posedge`** `clock`. The second one is variable `w` and the third one is an instance of sequence `s`. The last argument is for `pf`, and it is simply `x` **`until`** `y` where `x` and `y` are variables declared in the module. The assertion with the actual arguments substituted has the following form:

```
a: assert property(seq_impl_prop(!reset, (posedge clock),
  sf_after_a((posedge clock), w, s, 1), (x until y));
```

We can now proceed with the substitution of the arguments into the body of `seq_impl_prop`. It becomes

```
disable iff (!reset) @(posedge clock)
  (sf_after_a((posedge clock), w, s, 1)) |-> (x until y)
```

The next step is to substitute the body of the sequence `sf_after_a`. Its sequence expression after substitution of the actual arguments becomes

```
@(posedge clock) w ##1 (v[*3])
```

where `(v[*3])` is the body of sequence `s`. The final form of the body of assertion `a_until` before clock flow is applied is as follows:

```
disable iff(!reset) @(posedge clock) (@(posedge clock)
  (w ##1 (v[*3])) |-> (x until y)
```

The top-level clock is pushed into the antecedent sequence of the implication and then flows into the consequent property as explained in detail in Chap. 12. The result in this case is a single-clocked property, running on **`posedge`** `clk`.

```
disable iff(!reset) @(posedge clock)
  (w ##1 v[*3]) |-> (x until y)
```

What is the type of expressions `x` and `y`? Since `x` and `y` are variables (i.e., Boolean expressions), these expressions are in fact simple Boolean sequences. They are used in a property context as the operands of the **`until`** operator, hence they are *promoted* to properties.

Regarding the **`cover property`** statement `c_seq`, the sequence instance is identical to one used in the assertion; therefore, the final form of the sequence expression used in the property expression is as follows:

```
disable iff (!reset) (@(posedge clock) w ##1 (v[*3]))
```

In the preceding example, the inference was quite simple, both the clock and the disabling expression were obtained directly from the default declarations. What happens if an instance of a property or sequence is used directly inside the body of another property or sequence definition? How is the clock or disabling expression inferred in such cases? This will be discussed after we examine the syntax for defining sequences and properties.

8.2.1 Syntax of Sequence–Endsequence

The syntax of a sequence declaration is as follows, where items in [] brackets are optional:

sequence identifier (sequence_port_list)
  {local_variable_declarations}
  sequence_expression ;
endsequence [: identifier]

The *sequence_port_list* consists of a possibly empty comma-separated list of individual port declarations. Each such port can have the following components:

[**local** [port_direction]] type
  identifier {dimension} [= default_argument]

where

- *port direction* is only allowed when the keyword **local** is used indicating that it is a local variable port (see Chap. 16). It can be **input**, **output**, or **inout**.
- *type* is a type specification. It is obligatory when the port is a local variable port, otherwise it is optional. In addition to any integral type, the type can also be an **event**, **sequence**, or the keyword **untyped**. **untyped** indicates that the port(s) following the keyword do not have any formal type and is used to explicitly denote ports as untyped, especially when typed ports precede an untyped port.
- *identifier {dimension}* is a usual port name which can have an optional specification of dimensions provided that it is a local variable port or a typed integral port.
- *default argument* is an optional default actual argument. It has to be type compatible with the port type in the case the type is specified.

The declaration of a sequence is an extension of the interface of **let**. The main differences are the addition of types **event** and **sequence**, and the possibility to indicate local variable ports with their direction and type. Keyword **untyped** can be used to indicate ports that have no specified type. Let us concentrate on ports other than local variable ports, the latter are described in detail in Chap. 16. The type of a port may be specified, but it can also be left without a type like in **let** declarations. A port without a type specification or an explicit **untyped** can be specified only before any typed ports is stated on the port list. How is the type determined, used, and verified in that case? For example, consider the following sequence interface:

```
sequence s_def(a, event b, int c, d, untyped s);
```

The port a is untyped and its type correctness depends on where it is used in the sequence. Port b must be a clocking event expression. Ports c and d are of type two-valued integer. The actual arguments bound to these ports must be type compatible with **int**. Finally, since s is preceded by **untyped**, the actual argument bound to this port can be anything compatible with its use in the sequence expression. Suppose that the sequence expression is as follows:

```
@b a && (c == d) ##1 s
```

In this case, a is restricted to be an integral expression, and c and d can be any integral type that can be cast to **int**. The actual argument bound to the last port, s, can be an expression or a sequence expression including a sequence instance. This is because it is used alone as one of the operands of the sequence operator ##1. If it were used as a in an expression, its type would have been restricted the same way as for a.

Consider now a different sequence expression:

```
@b (a + c == d) ##1 s
```

Let v be a variable of type **bit**, and the actual arguments of s_def be 2'b11 + v, (**posedge** clk), 1, 0, v. When the actual arguments are substituted to the sequence expression, we obtain

```
@(posedge clk) ((2'b11 + v) + 1 == 0) ##1 v
```

Suppose that v is 1'b1 over two clock cycles. Will the sequence match or not? Before substituting the expression 2'b11 + v for the formal argument, it is enclosed in parentheses and cast to its self-determined type. Therefore, the result is an unsigned two-bit expression. When the value of v is 1'b1, it is 0 extended to 2'b01 before being added to 2'b11. The addition yields the result 2'b00. This result is then sign extended to **int** which yields 0 + 1 == 0 as the final Boolean expression. The sequence thus does not match at the first clock tick. If v were 1'b0 in the first clock cycle and 1'b1 in the second clock cycle, then the result of the addition is 2'b11, which after sign extension to **int** yields -1. Therefore, the result of the addition with 1 is 0. Consequently, the first expression evaluates to true, and the sequence will match at the second clock tick since v is 1'b1.

> Actual arguments to sequences are enclosed in parentheses and cast to their self-determined type before being substituted for occurrences of formal arguments in the sequence body. The same applies to actual arguments to **property** instances discussed in the next section.

The actual argument for `s` could also be a sequence expression. For example, suppose that the actual argument is `x ##1 y or v ##2 w`. This is syntactically correct sequence expression. The variables `x, y, v, w` must exist in the instantiating context otherwise it is an error. Let the list of actual arguments be

```
2'b11 + v, (posedge clk), 1, 0, x ##1 y or v ##2 w
```

After substitution for the formal arguments, the resulting expression for sequence `s_def` becomes

```
@(posedge clk) (2'b11 + v + 1 == 0) ##1 (x ##1 y or v ##2 w)
```

Notice the parentheses around the sequence expression that was substituted for the occurrence of the formal argument. It is only when the untyped argument is substituted into the sequence expression that its syntactic and semantic validity can be ascertained. In this case, all is well. If the same sequence expression were provided for the untyped formal argument `a`, it would result in an error, because after substitution, the actual argument sequence expression becomes an operand of an addition.

The question is, should untyped arguments be used at all? The advantage of untyped arguments is that they do not restrict the actual argument to any particular type. This can be very useful in the case of integral types where the dimensions need not be specified. The usual type checking is performed after the actual argument has been substituted wherever the formal argument appeared in the sequence definition. Therefore, the resulting error message may point to the body of the sequence and thus may not be easily comprehensible to the user. For instance, the interface of `s_def` could be written as

```
sequence s_def(bit [1:0] a, event b, int c, d, sequence s);
```

This still leaves much freedom as to the actual argument for `s`, because it can be any integral expression or a sequence expression. It would seem that if the sequence definition `s_def` above were to be reused, e.g., as part of a library package, then it is preferable to specify the expected types of the arguments except in the case where any integral type is allowed and type conversion to **`int`** is not desired (see Chap. 9 for further discussion about untyped vs. typed arguments).

The formal argument can also have default actual assignment specified in the header, as shown in the definition of `sequence_port_item`. The default argument specification is resolved in the scope of the sequence declaration. This is different from the usual actual arguments, which are resolved in the scope of the sequence instance. For example, let us consider the sequence header for `s_def` but add default actual argument specifications:

```
sequence s_def(
  a, event b = $inferred_clock, int c, d, untyped s = x);
```

In this case, in the absence of actual arguments for the formals `b` and `s`, like in the instance specification `s_def(2'b11+v, , 1, 0)` or equivalently using named argument association `s_def(.a(2'b11+v), .c(1), .d(0))`, we can see that neither formal `b` nor `s` has an actual argument. Therefore, the clocking expression of the sequence instance will be inferred from the instantiation context (see Chap. 14), and the formal argument `s` will be replaced by `x`. This variable or sequence definition `x` must have been declared in the scope of the declaration of `s_def`.

Next, we discuss **property** declaration, in particular in its differences from a sequence declaration.

8.2.2 Syntax of Property–Endproperty

The syntax of a **property** declaration is quite similar to that of a **sequence** declaration. The differences are as follows:

1. The encapsulation keywords are **property–endproperty**.
2. The formal and actual arguments can also be *property expressions*. The type of a formal argument may thus be **property**.
3. The body of a **property** declaration may contain *property operators* and refer to other *property instances*.
4. Local variable argument can only be **local input** (see Chap. 16).
5. The property may contain **disable iff** `(condition)` definition. The default actual argument to a formal of some integral type (or untyped) can be `$inferred_disable` (see Sect. 8.3).

The first one is obvious. The second and third differences are a natural extension from sequences to properties in that a property can receive a property expression as its argument and can operate on properties. Note also that under the property expression, sequence expressions and integral expressions are type compatible actual arguments. The fourth difference is because local variables do not flow out of properties (see Chap. 16). In properties, local variables have nowhere to flow out since no concatenation of properties exists. The fifth difference provides means to pass a default disable expression to a top-level property instance in an assertion. Some of these points are illustrated in the next example:

Example 8.13.

```
module m;
  bit clock, reset, a, b, c, d;
  default clocking @(posedge clock); endclocking
  ...
  property p1(bit rst, event clk = $inferred_clock, untyped x,
      property p);
    disable iff (rst) @clk !x ##1 x |-> p;
  endproperty

  property p2;
    a |-> (b until c);
  endproperty

  a_imply: assert property(p1(.rst(reset), .x(d), .p(p2)));
  ...
endmodule
```

After clock inference, substitutions, and clock resolution are completed, this combination of properties and their instances results in the following assertion:

```
a_imply: assert property(
  disable iff (!reset)
  @(posedge clock) !d ##1 d |-> a |-> b until c);
```
□

The lack of specified actual argument for `clk` in the instance of `p1` causes to infer the actual from **`default clocking`** declarations. Since `p1` is the top-level property in the assertion, the **`disable iff`** specification is legal.

If, on the other hand, **`disable iff`** were included in `p2` as shown in the next example, the resulting property expression in the assertion would become illegal. This is because **`disable iff`** would not apply to the top-level property.

Example 8.14.

```
module m;
  bit clock, reset, a, b, c, d;
  default clocking @(posedge clock); endclocking
  ...
  property p1(event clk = $inferred_clock, bit x, property p);
    @clk !x ##1 x |-> p;
  endproperty
  property p2;
    disable iff (reset) a |-> (b until c);
  endproperty
  a_illegal: assert property(p1(.x(d), .p(p2)));
  ...
endmodule
```

After substitutions are carried out, the result is the following illegal assertion because **`disable iff`** is not applied to the top-level property expression (see Sect. 4.4.4):

```
a_illegal: assert property( @(posedge clk) !d ##1 d |->
  disable iff (!reset) (a |-> (b until c)) );
```
□

> Placing **`disable iff`** specifications into property definitions must be planned carefully, otherwise it can lead to unexpected compilation errors.

Like **`let`** definition, **`sequence`** and **`property`** definitions may be placed in packages for later reuse. In that case, particularly the use of **`disable iff`** into such reusable properties must be done with even greater care.

In the next sections, we examine the inference of **`disable iff`** expression and of clocking event.

8.3 Disable Expression and Clock Inference

The **`default disable iff`** statement provides a default disable condition for assertions where the disable condition is not explicitly specified. Similarly, **`default clocking`** provides a clocking expression that can be inferred when not explicitly specified. Both default statements can be specified only once in a **`module`**, **`interface`**, **`program`**, and **`checker`**, and their effect extends over the full scope of the object. It does not extend, however, to any instances of such objects. For a detailed discussion about clock and disable expression inference see Chaps. 12–14.

Exercises

8.1. Write down a parameterized **`let`** definition that can be included in a package for the following situation. Include means to make the **`let`** expression take some useful default value when the argument `rst` is `1'b1`. Illustrate its use in some assertion and assignment statement.

(a) Evaluate to `1'b1` only when the vector argument `sig` has at most one `1` or one `0` and the rest is either `X` or `Z`. Otherwise, return `1'b0`. Should the argument `sig` be typed or untyped? Why?
(b) Is it possible in a single **`let`** definition to restrict the above definition to return `1'b1` only if there is at most one `1` or one `0`, and the rest are `z` (i.e., exclude `x`)?

8.2. Write down a simple module that has the variables `clk`, `reset`, `a`, and `b` of type **`logic`**, and add a concurrent assertion that fails when `a` is `1`, and at the next clock tick `b` is `0`, `1`, or `x`. Let the clock and reset be inferred from the module context.

8.3. Provide a solution to Exercise 8.2 such that the property used in the assertion is first defined outside the module, e.g., in a package.

8.4. What kind of actual arguments can be legally passed to the property that you defined in Exercise 8.3. Can you generalize the property to accept a wider range of arguments? How should the specification of the property change?

Chapter 9
Checkers

Contradictions do not exist. Whenever you think you are facing a contradiction, check your premises. You will find that one of them is wrong.

— Ayn Rand

In this chapter, we introduce checkers, units for packaging assertion-based verification code. On the one hand, checkers are similar to modules and interfaces: they define their own hierarchical scope and they may contain most constructs allowed in modules and interfaces. On the other hand, checkers generalize properties and sequences: they are instantiated "in place", their arguments may be sequences, properties, and edge-sensitive events. One can also say that checkers generalize assertions, since they behave as one complex assertion. In the subsequent sections, we elaborate the checker definition, instantiation and simulation semantics.

Using checkers makes the RTL cleaner: most instrumentation code in modules may be moved to checkers, which improves the code modularity, makes module code more readable and less error prone. Synthesis tools are normally supposed to ignore checkers, thus there is no need in conditional compilation statements to isolate the instrumentation code for the synthesis tools.

Although synthesis tools ignore checkers, most of the checker code is synthesizable. Checkers may be synthesized for hardware emulation and even in silicon, if so desired. Since the formal verification tools work on the synthesis model, checkers are formal verification friendly.

E. Cerny et al., *SVA: The Power of Assertions in SystemVerilog*,
DOI 10.1007/978-3-319-07139-8_9

9.1 An Apology for Checkers: Sequential Protocol

9.1.1 Sequential Protocol Specification as Module

We will start with an example of a sequential protocol. This protocol may be informally defined as follows:

- A transaction beginning is defined by `start` condition, and its end is defined by `complete` condition.
- The transactions do not overlap: `start` can occur again only after the next strictly subsequent cycle in which `complete` occurs. `complete` can occur only after the next strictly subsequent cycle in which `start` occurs. The last condition also implies that there can be no occurrences of `complete` before the next cycle of the first occurrence of `start`.
- Neither `start` nor `complete` can occur during two consecutive cycles.
- There are data signals of the same size: `dataIn` and `dataOut`.
- Whenever `start` occurs, `dataIn` is valid. Whenever `complete` occurs, `dataOut` is valid.
- The value of `dataOut` at the cycle in which `complete` occurs must be equal to the value of `dataIn` sampled in the strictly preceding cycle where `start` occurred.

Our goal now is to write a specification checking the protocol correctness. This specification consists of a control part which ensures the sequential pairing of occurrences of `start` and `complete`, and a data part which checks the data correspondence between `dataIn` and `dataOut` for each such pair. The control part may be implemented as shown in Fig. 9.1. We do not specify the clock and the reset in the assertions (except for using reset condition explicitly in `a_initial_no_complete`). We will fix this later by adding a corresponding **`default clocking`** and **`default disable iff`** statements.

Property `match` states that between two occurrences of signal `x` must occur `y`. The fact that the transactions do not overlap is captured by assertions `a_no_start` and `a_no_complete`. These assertions are symmetric and are written as direct instantiations of property `match`.

```
property match(x, y);
  x |=> !x until_with y;
endproperty : match
a_no_start: assert property (match(start, complete));
a_no_complete: assert property (match(complete, start));
a_initial_no_complete:
  assert property ($fell(reset) |-> !complete until_with start);
```

Fig. 9.1 Encoding of control part of sequential protocol

```
1 var type(dataIn) last_dataIn;
2 always @(posedge clk)
3   if ($sampled(start))
4     last_dataIn <= $sampled(dataIn);
5 a_seq_data_check:  assert property (
6     complete |-> dataOut == last_dataIn);
```

Fig. 9.2 Encoding of sequential protocol data check

The fact that there can be no `complete` before the first occurrence of `start` is captured by assertion `a_initial_no_complete`. Would there be no reset, this assertion should have been placed within an **`initial`** procedure. However, if the reset ever becomes active, it will kill the only attempt of the assertion, and the assertion might be never checked. Also, it is natural to interpret the specification in a way that the entire checking starts anew after each reset. Therefore in our implementation we check that there may be no `complete` before start each time the `reset` goes low. Unfortunately, our solution has two problems:

1. It does not cover the case when `reset` is initially inactive.
2. It does not take into account the fact that the disable condition is not sampled in SVA.

Exercises 9.1–9.3 discuss the solution for these problems.

Now let us move to the data part of the specification. The value of `dataIn` at an occurrence of `start` needs to be stored somewhere, since otherwise it is lost and the comparison with `dataOut` cannot be made. Since the protocol is sequential, one auxiliary storage variable can be used to hold the value of `dataIn` from the nearest preceding occurrence of `start`. The bookkeeping of `dataIn` and `dataOut` is done by auxiliary code. Such auxiliary code required for assertions is called *assertion modeling* code. For now, we implement the assertion modeling code in a module. Its implementation is shown in Fig. 9.2 (Lines 1–4). Line 1 contains declaration of the auxiliary storage variable `last_dataIn` whose type should be the type of `dataIn`.

Since this code is not part of the design RTL, we have no idea what changes first: `clk` or `start`. To follow the concurrent assertion semantics we must capture the sampled value of `start`. This is reflected by the explicit call to the sampled value function `$sampled` in Line 3. For the same reason we explicitly sample `dataIn` when capturing its value in Line 4. This situation is typical when the assertion modeling is done in modules or interfaces. Assertion `a_seq_data_check` in Line 5 checks that at the occurrence of `complete` the value of `dataOut` coincides with the captured valued of `dataIn`.

It remains to define the module declaration. There are two questions to be addressed:

- What is the type of `dataIn` and `dataOut`?
- How to define a clock?

Because we want to create a generic specification and need to admit any type for dataIn, we should pass the type of dataIn as a parameter to a module. The clock may be passed as a signal of **logic** data type:

```
module m_seqprotocol #(parameter type dataType = logic)
                      (input logic start, complete, clk, rst,
                       dataType dataIn, dataOut);
```

Consider now the module instantiation. Let the corresponding actual arguments be go, done, clock, din, dout declared as follows:

```
logic go, done, clock, reset;
logic [31:0] din, dout;
```

The instantiation of module m_seqprotocol is as follows:

```
m_seqprotocol #(.dataType(logic [31:0])) inst1(
  go, done, clock, reset, din, dout);
```

If we want the specification be clocked by **negedge** clock instead of **posedge** clock, we simply need to pass the negation of clock:

```
m_seqprotocol #(.dataType(logic [31:0])) inst2(
  go, done, !clock, reset, din, dout);
```

But what should we do if we want the specification be checked on each clock edge? One solution is to duplicate the specification implementation for this case. Fortunately, we can solve this problem by introducing one more parameter telling whether we wish to check the specification on each edge of the clock. The resulting specification is shown in Fig. 9.3.

We use here an additional parameter bothedges to distinguish the case when the assertions are checked on both edges of the clock, and the case when the assertions are checked on the rising edge of the clock. Depending on the value of this parameter we instantiate the corresponding **default clocking** statement (Lines 6 and 9). Note that the **if** statement on Line 5 is a generate **if**. The assertions use the default clocking, but the event control on the Line 19 must be explicitly specified. We use the name dclk of the default clocking event for this purpose.

The protocol instantiation specification for checking data on both edges of the clock will have the following form:

```
m_seqprotocol #(.dataType(logic [31:0]), .bothedges(1)) inst3(
  go, done, clock, din, dout);
```

We see that using modules as containers of the verification code is cumbersome:

- Module instantiation is redundantly verbose because it requires explicitly passing types of actual arguments if types of the corresponding formal arguments have been parameterized. It also requires a special indication whether the modeling code and assertions are clocked on one edge of the clock or on both. For a general-purpose verification code this information should be explicitly provided as parameters.
- In assertion modeling code the signal values must be explicitly sampled.

```
module m_seqprotocol #(parameter int bothedges = 0,
                       parameter type dataType = logic)
                      (input logic start, complete, clk, reset,
                       dataType dataIn, dataOut);
if (bothedges) begin : L
  default clocking dclk @clk; endclocking
end
else begin : L
  default clocking dclk @(posedge clk); endclocking
end
default disable iff reset;

property match(x, y);
  x |=> !x until_with y;
endproperty : match

var type(dataIn) last_dataIn;

always @L.dclk
  if ($sampled(start))
    last_dataIn <= $sampled(dataIn);

a_initial_no_complete:
  assert property ($fell(reset) |-> !complete until_with start);

a_seq_data_check:  assert property (
   complete |-> dataOut == last_dataIn);
a_no_start: assert property (match(start, complete));
a_no_complete: assert property (match(complete, start));
endmodule : m_seqprotocol
```

Fig. 9.3 Sequential protocol specification as a module

- The module should explicitly take care of distinguishing between the single and double edge clocking modes.

There are more drawbacks besides those illustrated in the sequential protocol example:

- Modules cannot accept sequences and properties as their arguments (ports). This ability is important for verification code, for example, when some external event is defined as a sequence.
- Modules cannot be instantiated within procedural code, and therefore cannot be used for checking correctness of the data within procedural flow.
- Modules cannot infer the clocking event and reset from their instantiation context. This makes their instantiation more verbose than necessary.

```
1  checker c_seqprotocol(start, complete,
2                        dataIn, dataOut,
3                        event clk = $inferred_clock,
4                        untyped rst = $inferred_disable);
5  default clocking @clk; endclocking
6  default disable iff rst;
7
8  property match(x, y);
9    x |=> !x until_with y;
10 endproperty : match
11
12 var type(dataIn) last_dataIn;
13
14 always_ff @clk
15   if (start)
16     last_dataIn <= dataIn;
17
18 a_initial_no_complete:
19   assert property ($fell(rst) |-> !complete until_with start);
20
21 a_seq_data_check: assert property (
22    complete |-> dataOut == last_dataIn );
23 a_no_start: assert property (match(start, complete));
24 a_no_complete: assert property (match(complete, start));
25 endchecker : c_seqprotocol
```

Fig. 9.4 Sequential protocol specification as a checker

9.1.2 Sequential Protocol as Checker

To resolve the problems of assertion packaging and modeling, SystemVerilog has a special construct called **`checker`**. We introduce checkers informally in this section on the example of the sequential protocol specification, while providing detailed description in the section that follows. Figure 9.4 shows the checker-based implementation of the sequential protocol.

The checker syntax is similar to that of the module, but it has several important differences: instead of **`module`**—**`endmodule`** the keywords **`checker`**—**`endchecker`** are used (Lines 1–25). By default all checker ports are input, and therefore we do not explicitly specify their direction (Line 1). The checker instantiation follows the substitution semantics similar to that of sequences and properties, therefore the explicit type specification of the checker ports is not necessary. For this reason we omit the type specification of the ports `start`, `complete`, `dataIn` and `dataOut`, and their type is inferred from the checker instantiation. The semantics of checker input formal arguments is similar to the semantics of property arguments, and almost all formal argument types allowed in properties are also allowed in checkers, including **property** and **sequence**.[1] In this example, we have the argument `clk` of type **event**, and also untyped arguments. As in properties, it is legal to use `$inferred_clock` and `$inferred_disable` system functions for checker argument initialization (Lines 3 and 4; see Sect. 8.3): these functions return the inferred values of the clock and reset from the checker *instantiation* context. By default `clk` contains the inferred clock from the checker instantiation context, and `rst` contains the inferred reset.

The **default** statements on Lines 5 and 6 define `clk` and `rst` as default clock and reset for checker assertions. As a result, the values of the clock and reset inferred at the checker instantiation become known inside the checker. Without these statements, no clock and reset inference can be achieved inside the checker. This checker behavior is identical to modules.

The clock `clk` is passed to the checker as an event[2] (Line 3), and not as a signal. As in the case of properties or sequences we can specify the default value of the clock to be inferred from the checker instantiation context using system function `$inferred_clock` (see Sect. 8.2).

The type of reset `rst` is also inferred from the type of the corresponding actual argument, but since the previous port is typed, we need to use explicit **untyped** declaration on Line 4. The default value of reset is inferred from the checker instantiation context using system function `$inferred_disable`. This is, again, similar to the reset passing to properties as described in Sect. 8.2.

As we can see, we do not need to specify parameters for checkers, and this is why the checkers do not have parameters. We do not need to invent tricks to distinguish between single and double edge clocking model.

Consider now the assertion modeling code (Lines 14–16). Notice that we use **always_ff** (see Sect. 2.2.1.3) instead of **always** in the module. This is because the plain **always** is forbidden in checkers.[3] The formal port `clk` designates already the correct event, and we can use it in the event control directly. We do not need explicitly sample signals on Lines 15–16, because the checker does all the required sampling correctly.

Let us proceed now to the checker instantiations, shown in Fig. 9.5.

The checker instantiation syntax (Lines 6–8) is similar to module instantiation. `c_seqprotocol` is the name of the checker, `inst1`, `inst2` and `inst3` are the names of the checker instances. Checker instantiation semantics is similar to property instantiation in an assertion. All argument types that can be used with a property, can also be used with a checker. For example, it is possible to pass sequences to a checker (see Sect. 9.2.1.4).

The checker instantiation `inst1` (Line 6) is equivalent to the following:

```
c_seqprotocol inst1(go, done, din, dout, posedge clock, reset);
```

[1]Except arguments with **local** qualifier, see Sect. 16.2.

[2]Not to be confused with passing events to modules. In case of modules it is possible to pass a variable of the type **event**. Here we pass the entire event expression, such as **posedge** `clk` by substitution.

[3]More exactly, its usage is deprecated in checkers.

```
1 module m(input logic go, done, clock, reset,
2   logic [31:0] din, dout);
3   default clocking @(posedge clock); endclocking
4   default disable iff reset;
5   ...
6   c_seqprotocol inst1(go, done, din, dout);
7   c_seqprotocol inst2(go, done, din, dout, negedge clock);
8   c_seqprotocol inst3(go, done, din, dout, edge clock);
9 endmodule : m
```

Fig. 9.5 Invocations of checker `c_seqprotocol`

However, explicit specification of the clock and reset is redundant because they can be inferred from the instantiation context (Lines 3 and 4). In instantiations `inst2` and `inst3` (Lines 7–8) the clocking event must be specified explicitly because it is different from the default one.

This example clearly demonstrates the advantages of the checker-based implementation:

- Checkers are instantiated "in place", similarly to property instantiations in assertions. They can have untyped formal arguments. Therefore, there is no need in passing redundant information about data type and size to them.
- Checkers can infer clock and reset from the instantiation context
- Checker instantiation syntax is uniform. All information is passed to a checker through ports. There is no need to distinguish between ports and parameters.
- There is no special handling of the double edge clocks. All clocks are passed as event expressions.
- Checker instantiation is concise, as a result of conventions described in the previous bullets.
- There is no need for explicit sampling of expressions in assertion modeling code. The checker does the required sampling automatically.

Checkers also have the following advantages as containers of the verification code:

- Checkers may accept sequences and properties as their arguments.
- Checkers may be instantiated in procedural code (see Sect. 14.8).

9.2 Checker Declaration

The **checker** declaration has the following syntax:

```
checker checker_name (checker_formal_arguments);
  ...
endchecker
```

If a checker has no arguments, the parentheses may be omitted. **endchecker** may be qualified with a label of the checker name, similar to other compound constructs in SystemVerilog. Specifying the checker name with **endchecker** is a good idea as it makes the code clearer, and allows the compiler to check that the beginning and end of the checker match.

Example 9.1. The following checker consists of a single assertion verifying that request `req` is granted at the following clock tick.

```
checker request_granted(req, gnt, clk, rst);
  a1: assert property(@clk disable iff (rst) req |=> gnt);
endchecker : request_granted
```

In this example, all checker formal arguments are untyped. □

9.2.1 Checker Formal Arguments

9.2.1.1 Argument Direction

Checkers may have input and output arguments. The main use of a checker is to be an observer to follow the DUT behavior and to verify its correctness. Therefore, usually a checker has only input arguments, and if no direction qualifier is specified explicitly, all checker arguments are input by default. However, checkers may have also output arguments. We discuss them in Sects. 9.5 and 23.5.

Example 9.2. In the checker declaration in Example 9.1 no argument direction has been specified. Therefore, all its arguments are input. □

The formal rule for the inference of checker argument direction when it has not been explicitly specified, is as follows:

- If the direction of the first checker argument is omitted, it is assumed to be input.
- If the direction of another checker argument is omitted, it is inferred from the direction of the previous argument.

Example 9.3. Consider the following checker argument declaration:

```
checker check(a, b, output logic c, d,
              input event clk, untyped rst);
```

Here, `a` is an input argument because it is the first checker argument, and no direction has been specified. The second argument `b` is also input, because it has no explicit direction, and the previous argument `a` is input. Argument `c` is output because its direction has been explicitly specified. The direction of `d` is output, as inferred from the direction of `c`. The direction of arguments `clk` and `rst` is input.

□

9.2.1.2 Default Arguments

Checker input formal arguments may have default values, similarly to formal arguments of properties and sequences (Sect. 8.2).

Example 9.4. Write the following checker: request `req` is granted (`gnt` asserted) in `n` clock ticks. By default `n = 1`.

Solution:

```
checker request_granted(req, gnt, n = 1, clk, rst);
  a1: assert property(@clk disable iff (rst)
    req |-> nexttime[n] gnt);
endchecker : request_granted
```

Discussion: Even though `n` is untyped, the actual argument must be an elaboration time constant because it is used in the property operator **`nexttime`**`[n]`. □

9.2.1.3 Context Inference

As in the case of properties, there are two special default argument values: `$inferred_clock` and `$inferred_disable`. These system functions return the clocking event and the reset inferred from the checker *instantiation* context. Section 8.3 explains the context inference rules. We get back to these context inference functions in Sect. 9.3 where we discuss checker instantiation (see also the checker motivation example in Sect. 9.1).

Example 9.5. Same example as Example 9.4, with default values for clock and reset.

```
checker request_granted(req, gnt, n = 1,
  clk = $inferred_clock, rst = $inferred_disable);
  a1: assert property(@clk disable iff (rst)
    req |-> nexttime[n] gnt);
endchecker : request_granted
```

Note that `$inferred_clock` and `$inferred_disable` provide default actual arguments to the formal arguments `clk` and `rst` only. There is no inference of clock and reset from these functions in the checker assertions. □

If clock and reset inference is desired in a checker, **`default clocking`** and **`default disable iff`** statements must be specified, as in the case of modules and interfaces.

Example 9.6. Clock and reset inference in checkers.

```
checker request_granted(req, gnt, n = 1,
  clk = $inferred_clock, rst = $inferred_disable);
  default clocking @clk; endclocking
  default disable iff rst;
  a1: assert property(req |-> nexttime[n] gnt);
endchecker : request_granted
```

In this example, clock and the reset are inferred for assertion `a1` from the default declarations; consequently, the assertion is equivalent to

```
assert property(@clk disable iff (rst) req |-> nexttime[n] gnt);
```

□

The clock and reset inference rules in concurrent assertions inside checkers are the same as the inference rules in modules. They are resolved in the scope of the checker definition, not in the scope of its instantiation.

9.2.1.4 Checker Argument Types

Checker formal input arguments may be of the same types as property arguments (Chap. 8), but they cannot have the **local** qualifier. The output checker arguments cannot be **untyped**, **sequence** or **property**. Unlike module ports, checker formal arguments may not be interfaces.

If a checker argument has an explicit direction qualifier (**input** or **output**), its type must be explicitly specified. Otherwise, the type of a checker argument may be omitted, and it is inferred according the following rules:

- The first checker argument is assumed to be untyped. Since it does not have a direction qualifier it is an input argument as explained in Sect. 9.2.1.1.
- The types of other checker arguments are inferred from the type of the previous arguments, similarly to sequences and properties, as described in Sects. 8.2.1 and 8.2.2.

In Example 9.6, the type of the checker arguments was not specified, that is, all arguments were untyped. Untyped arguments are flexible, but can lead to poor error messages. Example 9.7 illustrates using both typed and untyped formal arguments in a checker.

Example 9.7. We modify the checker from Example 9.5 by ascribing types to its formal arguments, and by passing it an error message.

```
checker request_granted(sequence req, property gnt,
  untyped n = 1, string msg = "",
  event clk = $inferred_clock, untyped rst = $inferred_disable);
  default clocking @clk; endclocking
  default disable iff rst;
  a1: assert property(req |-> nexttime[n] gnt) else $error(msg);
endchecker : request_granted
```

What would happen if we pass a property expression **nexttime** `go` to the formal argument `req`? It is likely that we would get an error message similar to the following one:

The formal argument req of checker request_granted is of type sequence, whereas the actual argument nexttime go is an expression of type property.

Consider the message we might get if the `req` argument were untyped, as in Example 9.5:

Type mismatch in
a1: assert property(req |-> nexttime[n] gnt) else $error(msg);
req is of property type, whereas sequence type is expected.

The user understands the first error message, since it explains the problem in terms of the interface of the checker. However, the second error message is confusing because it mentions the checker *internals*, which are normally hidden from the user.

Unfortunately, it is not always possible or desirable to provide a specific type for a checker argument. Although it seems natural to declare `n` as **`int`**, doing so would exclude the possibility of passing the checker an infinite upper bound `$`, since only untyped formal arguments may receive `$`. In our example, we have to explicitly declare `n` as **`untyped`**, otherwise its type would be taken from the previous argument `gnt` that is of type **`property`**. We also have to leave `rst` untyped because providing a specific type, say, **`logic`**, would require type conversion when the actual argument is of a different type. This may affect checker correctness or slow down the simulation. □

Always specify an explicit type for checker formal arguments that are used as sequences, properties, events, or strings. Specifying a type for formal arguments used as Booleans or integral expressions usually limits checker generality.

9.2.2 Checker Contents

Checker contents are similar, but more restricted than those of a module. For example, checkers cannot contain declarations of modules, interfaces, programs, and packages. Modules, interfaces, and programs cannot be instantiated inside checkers.[4] Also general purpose always procedures are deprecated in checkers and should not be used.[5] Consult the LRM for the exhaustive definition of the constructs allowed in checkers.

[4]The LRM is not clear about placing immediate assertions in checkers. In any case it should be safe to place immediate assertions in action blocks of concurrent assertions.

[5]In the SystemVerilog 2009 standard only general purpose always procedures have been allowed in checkers. But the SystemVerilog 2012 standard introduced specialized always procedures in checkers and the general purpose always procedure has been deprecated.

For example, let, sequence and property declarations may be freely used in checkers. As a rule, it is recommended to avoid sequence and property declarations in modules. If sequences and properties used in modules are general enough, they should be defined in some global place, for example, in a package, to be also accessible to different modules. If a module uses several specific sequences and properties, it is better to encapsulate them in a checker to keep the module code clean. Unlike modules, checkers are natural containers for sequence and property encapsulation. Another construct natural in checkers are covergroups (see Sect. 18.3). For checker modeling, see Sect. 9.4.

As modules, checkers may access elements from their enclosing scope through their hierarchical names, except the following:

- Automatic and dynamic variables.
- Elements of **fork**...**join** blocks (including **join_any** and **join_none**).

Also, hierarchical references into checkers are forbidden.

Example 9.8. As mentioned above, checkers may reference signals defined in modules by their hierarchical names. Thus, we can rewrite the checker from Example 9.1 to address the signal names directly.

```
module m(...);
logic req, gnt;
...
endmodule : m

module top;
logic clock, reset;
...
m m1(...);
request_granted c1(clock, reset);
endmodule : top

checker request_granted(clk, rst);
  a1: assert property(@clk disable iff (rst) m1.req |=> m1.gnt);
endchecker : request_granted
```

In this example the checker references signals `req` and `gnt` declared in module `m1` by their hierarchical names, and not through checker ports. This style limits checker generality and is better to be avoided whenever possible. Its usage is acceptable for complex checkers dedicated to the verification of large design blocks. □

9.2.2.1 Generate Constructs

Generate constructs in SystemVerilog are used to perform elaboration time actions to enable flexible tuning of the design at the elaboration time, as described in Sect. 2.1. In checkers, the same generate constructs may be used, and the same rules as everywhere else apply.

Example 9.9. What happens if we pass the value of -1 to `n` in checker `request_granted` (Example 9.5)? Apparently, we should get an error message about assertion `a1` saying that `##-1` is wrong syntax. Since the user is not necessarily familiar with the checker internals, it may be confusing. It would be much clearer if we report the error message relative to the checker interface. We can do that using a generate **`if`** statement as follows:

```
checker request_granted(
  sequence req, property gnt, untyped n = 1, ...);
  if (!$isunbounded(n) && n < 0)
    $error("Delay value n cannot be negative", n);
  ...
  a1: assert property(req |-> nexttime[n] gnt) else $error(msg);
endchecker : request_granted
```

`$isunbounded(n)` is a system function returning *true* if its argument is $. We need this function to isolate the situation when the unbounded value is provided for `n` because we cannot compare $ with 0. When the checker is instantiated with the negative value for `n` the error message is issued at elaboration time. □

Use generate statements and elaboration system tasks to perform custom elaboration-time checks and to issue custom elaboration error messages.

The ability to use generate statements is an important advantage of checkers over bare assertions: checkers are more flexible and tunable. Therefore, it often makes sense to use checkers even when they consist of a single assertion.

Example 9.10. We explain in Sect. 11.1.4 that simulation performance may degrade when the antecedents have distant matches, or remain unfinished. If in the checker `request_granted`, `req` is a long sequence or if there are incomplete pending requests, the simulation time may be negatively affected. It is possible to truncate the antecedent to k clock ticks by intersecting it with `1[*1:k]`, as explained in Example 11.16. This truncation is very inefficient in FV, and it should be avoided there. We can add an optional argument `truncate` to the checker `request_granted`: when a nonzero value, say, 10, is passed to it, the antecedent `req` will be truncated after 10 clock ticks, and when $ is passed, no truncation is performed. The resulting checker is shown in Fig. 9.6.

If `truncate` has a value $, the conditional generate **`if`** in Line 7 will yield the sequence `ante` (antecedent) from Lines 8 to 10 in the elaboration model. If `truncate` is positive, it will yield the sequence `ante` from Lines 14 to 16. If `truncate` has a nonpositive value, an error message is issued.

The reader may wonder why we suggested this solution instead of always defining the antecedent of assertion `a1` as `req` **`intersect`** `1[*1:truncate]`. This is, of course, correct, but both formal and simulation tools may implement `req` **`intersect`** `1[*1:$]` less efficiently than just `req`. □

```
checker request_granted(sequence req, property gnt,
  untyped n = 1, string msg = "",
  event clk = $inferred_clock,
  untyped rst = $inferred_disable, truncate = $);
  default clocking @clk; endclocking
  default disable iff rst;
  if ($isunbounded(truncate))
    sequence ante;
      req;
    endsequence : ante
  else if (truncate < 1)
    $error("truncate value should be positive");
  else
    sequence ante;
      req intersect 1[*1:truncate];
    endsequence : ante
  a1:assert property(ante |-> nexttime[n] gnt)else $error(msg);
endchecker : request_granted
```

Fig. 9.6 Optional antecedent truncation

9.2.2.2 Checker Procedures

A checker may contain the following structural procedures: **`initial`**, **`always_comb`**, **`always_latch`**, **`always_ff`** and **`final`**.

Initial Procedure. In checkers, **`initial`** procedures may contain assertions, let declarations and a procedural event control statement `@`. All other statements are forbidden there. The only purpose of initial procedures in checkers is to enable assertions that execute a single evaluation attempt starting at the first tick of their leading clock.

Example 9.11. Write a checker verifying that reset `rst` is initially high, then eventually goes low and remains low forever.

Solution:

```
checker simple_reset(rst);
  initial
    a1: assert property (@$global_clock rst[+] #=# always !rst);
endchecker : simple_reset
```

Discussion: We need to check only the first evaluation attempt of `a1`, hence it should be placed in the scope of an **`initial`** procedure. We do not wish to associate the behavior of `rst` with any specific clock; therefore, we chose the global clock, the fastest clock of the system, to control this assertion. See Exercise 9.5 for further discussion. □

Always Procedures. Always procedures in checkers are similar to those in modules. Note the following restriction: in **always_ff** procedures only nonblocking assignments are allowed; blocking assignments are forbidden. We postpone examples of their usage until Sect. 9.4.

Final Procedure. **final** procedures in checkers are not different from final procedures in modules (see Sect. 2.2.2). They are executed at the end of simulation. Their main purpose is to print statistical information and to check the final state of the simulation.

final procedures may contain everything that functions may contain. Therefore, checker **final** procedures may only *immediate* assertions (including simple, deferred observed and deferred final).

Example 9.12. We can add a final procedure to the checker `request_granted` defined in Example 9.7 to check that at the end of simulation there is no outstanding request. For simplicity, we assume that both `req` and `gnt` are Boolean, and that the request remains asserted until granted.

```
checker request_granted(req, gnt, n = 1,
                        event clk = $inferred_clock,
                        untyped rst = $inferred_disable);
  default clocking @clk; endclocking
  default disable iff rst;

  a1: assert property (req |-> nexttime[n] gnt);

  final begin
    a2: assert (!rst -> gnt || !req) else
      $warning("Outstanding request at the end of simulation");
  end
endchecker : request_granted
```

□

It is possible to write an entire checker consisting only of the **final** procedure. Its purpose would be to check the quiescent state at the end of simulation to verify that there are no outstanding transactions, and that some important scenarios were observed at least once.

9.2.3 Scoping Rules

Checkers may be declared at the top-level, i.e., in the scope of a compilation unit. All the preceding checker examples in this chapter were top-level checkers. Checkers may also be declared in other scopes: in modules, interfaces, programs, generate blocks, packages, and in other checkers. The reason for declaring checkers in smaller scopes is to make them local to these scopes, and to make the objects

```
1  checker check1(a, b, c, d,
2    event clk = $inferred_clock,untyped rst = $inferred_disable);
3    default clocking @clk; endclocking
4    default disable iff rst;
5    property p1(x, y);
6      x[*2] |-> y;
7    endproperty : p1
8    property p2(x, y);
9      x |-> y[*2];
10   endproperty : p2
11   ...
12   checker check2();
13     property p1(x, y);
14       x[*2] |-> y[*2];
15     endproperty : p1
16     a1: assert property (p1(a, b));
17     a2: assert property (p2(c, d));
18   endchecker : check2
19   check2 check_cd;
20 endchecker : check1
```

Fig. 9.7 Nested checkers

declared in these scopes and in the higher-level scopes visible inside the checker. Another reason is to hide the checker from other parts of the design.

Example 9.13. Figure 9.7 shows nested checkers: checker `check2` is declared inside checker `check1`.

Even though checker `check2` does not have arguments, the arguments and the other objects of checker `check1` are visible in its scope. For example, arguments `a`, `b`, `c`, and `d` of `check1`, and property `p2` declared in `check1` are used in `check2`. Property `p1` used in assertion `a1` (Line 16) is a local property of `check2` (Lines 13–15), and not the property `p1` declared in checker `check1`(Lines 5–7). Property `p1` of `check1` cannot be directly referenced in `check2` because property `p1` of `check2` masks the visibility of property `p1` of `check1`. If we needed to instantiate property `p1` of `check1` in `check2`, we have to reference it by its hierarchical name `check1.p1`.

Checker `check2` also inherits the default clocking and default reset definitions from checker `check1` (Lines 3–4), hence clocking event `@clk` and reset `rst` are inferred in assertions `a1` and `a2`. □

All objects referenced in a checker are resolved in the scope of the checker definition, and not in the scope of its instantiation.

Example 9.14. The checker mycheck will not compile, even though this checker is instantiated in module m:

```
checker mycheck(event clk = $inferred_clock);
  a1: assert property (@clk a);
endchecker : mycheck
module m(input logic clk, ...);
  logic a = ...;
  mycheck check(posedge clk);
  ...
endmodule : m
```

This is because a referred to in assertion a1 is resolved in the scope of the checker definition where no a is declared, and not in the scope of the checker instantiation where a is visible.

If the hierarchical name of a specific instance of module m is top.unit1.block2 then the checker mycheck could be rewritten as follows to reference a by its hierarchical name:

```
checker mycheck(event clk = $inferred_clock);
  a1: assert property (@clk top.unit1.block2.a);
endchecker : mycheck
```

Of course, a cleaner solution would be to pass a as an argument to the checker. This should be done for small checkers like mycheck. Referencing design signals in a checker by their hierarchical names is useful in big checkers verifying behavior of large pieces of a design, however, it limits their reusability in other designs. □

9.2.3.1 Checkers in Packages

Checkers are natural candidates for units of standard or project-wide verification libraries. The question is how to package several checkers in a reusable unit. Another problem is a possible name collision: the name of a library checker may be the same as a name of another checker or module.

SystemVerilog **package** construct is well suited for both tasks: it can contain several checkers (also properties, sequences, let, constants, etc., see Sect. 2.6), and it also introduces its own name space.

Example 9.15. We can place checker request_granted from Example 9.7 in a package named check_lib as follows:

```
package check_lib;
 checker request_granted(sequence req, property gnt,
   untyped n = 1,
   event clk = $inferred_clock, untyped rst = $inferred_disable);
   a1: assert property(@clk disable iff (rst)
     req |-> nexttime[n] gnt);
 endchecker : request_granted
 // Other checkers ...
endpackage : check_lib
```

To instantiate the checker in a module it is necessary to import the package contents first:

```
module m(logic clk, rst, send, ack, ...);
  import check_lib::*;
  ...
  request_granted ack_received(send, ack, 1, posedge clk, rst);
endmodule : m
```

The statement `import check_lib::*` makes the entire contents of the package visible in `m`. Instead of importing the entire contents, we could import only this specific checker using `import check_lib::request_granted`. It is also possible to use the fully qualified name of the checker when instantiating it:

```
check_lib::request_granted ack_received(
  send, ack, 1, posedge clk, rst);
```

without importing the package contents to avoid name collision. □

Checkers in packages cannot refer to the data that do not belong to the scope of the package. Thus, (the corrected version of) checker `mycheck` in Example 9.14 cannot be placed in a package, since it refers to data `top.unit1.block2.a` in a module. This is forbidden in packages.

9.3 Checker Instantiation

A checker may be instantiated in any place where a concurrent assertion may appear, except for **fork**...**join** blocks. This means that checkers may be instantiated both outside and *inside* procedural code. This is one of the important differences between checkers and modules, as modules may be instantiated only outside procedural code. Checker instantiation in procedural code is called a *procedural* checker instance, while the checker instantiation outside procedural code is called a *static* checker instance.

9.3.1 Connecting Checker Arguments

The association of checker actual arguments with its formal arguments has the same syntax as module port association. The argument association may be positional or named, and the name association may be explicit, implicit, and may use wildcards, the same way as modules, properties, and other similar constructs in SystemVerilog. Different argument association forms may be mixed. We recapitulate these rules common for SystemVerilog design elements on the examples of checkers and illustrate different ways of checker argument associations on the instantiation of checker `request_granted` from Example 9.7.

9.3.1.1 Positional Association

In module m1 below, the actual arguments of the checker are passed according to the order of the corresponding formal arguments. The default values passed to n and clk are identified by commas, one comma for each unspecified actual argument. Were these default arguments the last arguments in the list, there would be no need for these commas. Positional argument association is really convenient only when the number of checker arguments is small.

```
module m1(input logic clk, rst, send, ack, ...);
  default clocking @(posedge clk); endclocking
  ...
  request_granted check(send, ack,,, rst);
endmodule : m1
```

9.3.1.2 Explicit Named Association

In module m2 below, the actual arguments of the checker are passed by an explicit indication of the formal argument names. In the case of the named association, the order of the actual arguments is not important. Omitting default values does not require any additional notation. A named association is convenient when the number of checker arguments is large, or when the default values are passed to the arguments in the middle of the argument list.

```
module m2(input logic clk, rst, send, ack, ...);
  default clocking @(posedge clk); endclocking
  ...
  request_granted check(.rst(rst), .req(send), .gnt(ack));
endmodule : m2
```

9.3.1.3 Implicit Named Association

When the names of actual and formal arguments coincide, there is no need to repeat the argument names. In module m3, argument rst is passed by implicit named association.

```
module m3(input logic clk, rst, send, ack, ...);
  default clocking @(posedge clk); endclocking
  ...
  request_granted check(.rst, .req(send), .gnt(ack));
endmodule : m3
```

9.3.1.4 Wildcard Named Association

When the names of several actual and formal arguments coincide, as in module `m4` below, it is convenient to use a wildcard association making all actual and formal arguments to be connected implicitly by name. Note that in our case it is incorrect to use only the wildcard association because the clocking event passed to the checker `request_granted check` would be `clk`, and not **`posedge`** `clk` as intended. We need to explicitly leave the formal argument `clk` unconnected to allow using the default value `$inferred_clock`. We could also explicitly specify `.clk(`**`posedge`** `clk)` instead.

```
module m4(input logic clk, rst, req, gnt, ...);
  default clocking @(posedge clk); endclocking
  ...
  request_granted check(.*, .clk());
endmodule : m4
```

9.3.2 Instantiation Semantics

Roughly speaking, checkers are "inlined" at their instantiation point—the checker contents are inserted in the place of the checker instance. This is similar to sequence and property instantiation, but different from a module instantiation and from a task call.

Actually, checker instantiation is more complicated than straightforward inlining, hence we need to describe all its subtleties. In this section, we concentrate on static checker instances, while the peculiarities of procedural checker instances are discussed in Sect. 14.8.

A static checker is instantiated by substitution "in place" with reservations concerning checker object naming, clock and reset inference, name resolution, argument sampling, and checker variable semantics.

9.3.2.1 Object Naming

Example 9.16. Consider the following code:

```
checker check(a, b, event clk);
  a1: assert property (@clk a |=> b);
endchecker : check
module m1(input logic clock, req, ack);
  ...
  check mycheck(req, ack, posedge clock);
endmodule : m1
```

Instance `mycheck` of checker `check` in module `m1` is not exactly equivalent to

```
module m1(logic clock, req, ack);
  ...
  a1: assert property (@(posedge clock) req |=> ack);
endmodule : m1
```

This is because the true name of the assertion upon the `check` instantiation in `m1` is `mycheck.a1`, and not just `a1`. The checker introduces its own scope. Since assigning a hierarchical name, such as `mycheck.a1` to an assertion directly in the module is illegal, the checker instantiation "in place" is rather conceptual than real. □

9.3.2.2 Context Inference and Name Resolution

As mentioned in Sects. 9.2.1 and 9.2.3, name resolution and context inference are done at the point of the checker declaration and not of its instantiation.

Example 9.17. Consider instantiation of checker `check` from Example 9.16 in module `m2`:

```
module m2(input logic clock, rst, req, ack);
  default disable iff rst;
  ...
  check mycheck(req, ack, posedge clock);
endmodule : m2
```

There is no inference possible, hence implicitly the disable condition is 0, and not `rst`, because reset is resolved at the declaration point. Thus, assertion `a1` is equivalent to

```
a1: assert property (@(posedge clock) disable iff (0)
  req |=> ack);
```

If the checker were *defined* in module `m2` (see Sect. 9.2.3), then `rst` would be inferred from the **`default disable iff`** statement in `m2`, which is visible in the *declaration* of `check`.

```
module m2(input logic clock, rst, req, ack);
  default disable iff rst;

  checker check(a, b, event clk);
    a1: assert property (@clk a |=> b);
  endchecker : check
  //...
  check mycheck(req, ack, posedge clock);
endmodule : m2
```

The assertion would be equivalent to

```
assert property (@(posedge clock) disable iff (rst)
                  req |=> ack);
```

□

9.3.3 Checker Binding

Sometimes it is desirable to keep verification code separate from the design code. For example, a validator may want to write several checkers verifying DUT behavior without modifying RTL (see Sect. 1.2.2). SystemVerilog allows external checker binding to modules or interfaces using the **bind** directive. It is forbidden to bind anything to a checker, not even another checker. Essentially, SystemVerilog allows to bind other design elements, such as modules and interfaces as well in a similar way, but we will limit our description to checker binding.

It is possible either to bind a checker to all instances of a module or interface, or to choose only specific instances where the checker is to be bound.

The following syntax is used to bind a checker to all instances of a module:

```
bind module_name checker_name checker_inst(args);
```

The following syntax is used to bind a checker to a specific instance of a module:

```
bind module_inst checker_name checker_inst(args);
```

To bind a checker to several instances of a module, the module instances should be separated by commas:

```
bind module_name: module_inst_1, ..., module_inst_n
  checker_name checker_inst(args);
```

Binding a checker has the same effect as if it were instantiated at the very end of the target module or interface. If several checkers are bound to the same module (interface), then the order of instantiation is arbitrary.

Example 9.18. Consider a module `trans` instantiated three times in the top-level module `top`:

```
module top;
  logic clock, snda, sndb, sndc, rcva, rcvb, rcvc;
  ...
  trans ta(clock, snda, rcva);
  trans tb(clock, sndb, rcvb);
  trans #(2) tc(clock, sndc, rcvc);
endmodule : top
module trans #(DEL=1) (input logic clock, in,
                       output logic out);
  if (DEL == 1) begin : b
    always @(posedge clock)
      out <= in;
  end
  else begin : b
    logic [DEL - 2: 0] tmp;
    always @(posedge clock) begin
      tmp[0] <= in;
      for (int i = 1; i < DEL - 1; i++)
        tmp[i] <= tmp[i - 1];
```

```
      out <= tmp[DEL - 2];
    end
  end
endmodule : trans
```

This module generates signal `out` from signal `in` by delaying it by several clock cycles specified by the module parameter `DEL`. Note that the **if** statement in module `trans` is a generate **if** (see Example 2.2).

The following checker `eventually_granted` verifies that each request is eventually granted:

```
checker eventually_granted(sequence req, property gnt,
  event clk = $inferred_clock);
  assert property (@clk req |=> s_eventually gnt);
endchecker : eventually_granted
```

We can bind this checker to module `trans` to verify that the high value of signal `in` of module `trans` is eventually transmitted to its output `out`:

```
bind trans eventually_granted
  check_in2out(in, out, posedge clock);
```

The **bind** statement specifies that checker `eventually_granted` is bound to each instance of module `m`. It is equivalent to instantiating the checker at the end of module `trans`:

```
module trans #(DEL=1) (input logic clock, in,
                       output logic out);
  ...
  eventually_granted check_in2out(in, out, posedge clock);
endmodule : trans
```

Checker `eventually_granted` is too general, as it does not check the exact timing. To make more specific checks, it is possible to use checker `request_granted`:

```
checker request_granted(sequence req, property gnt, int n = 1,
  event clk = $inferred_clock, untyped rst = $inferred_disable);
  a1: assert property(@clk disable iff (rst)
    req |-> nexttime[n] gnt);
endchecker : request_granted
```

This time we cannot bind the checker to all instances of module `trans`, since the specific delay values are different in each instance: we need to bind `request_granted` with the delay value of 1 to module instances `ta` and `tb`, and with the delay value of 2 to module instance `tc`:

```
bind trans: ta, tb request_granted
  delay1(in, out,, posedge clock);
bind trans: tc request_granted
  delay2(in, out, 2, posedge clock);
```

□

9.4 Checker Modeling

9.4.1 Checker Variables

Assertion modeling code in checkers is similar to that of hardware modeling in modules and interfaces: one can declare variables in checkers, assign values to them using continuous, blocking and nonblocking assignments, etc. However, some limitations apply. On the other hand, checkers have a mechanism of nondeterministic modeling which is missing in modules and interfaces. This mechanism is described in Chap. 23.

The variables declared in the checker are called *checker variables*. Declaring nets in the checker is illegal. Checker variables may be initialized at their declaration, but assigning them a value in initial procedures is illegal (see Sect. 9.2.2.2). Checker variables may be assigned using continuous assignments or in always procedures. As mentioned in Sect. 9.2.2.2 the following kinds of always procedures may be used in checkers (these procedures are discussed in Sects. 2.2.1.1–2.2.1.3):

- **`always_comb`**
- **`always_latch`**
- **`always_ff`**

Checker always procedures may contain the following statements:

- assignments (blocking and nonblocking)
- **`if`** and **`case`** statement
- loops
- function and task calls
- let declarations

Blocking assignments are allowed only in **`always_comb`** and **`always_latch`** procedures. Of course, **`always_ff`** should contain an event control.

As in programs, all checker variable assignments are executed in the Reactive region set: continuous and blocking assignments are executed in the Reactive region, and nonblocking assignments are executed in the Re-NBA region.

Checker variables provide means for carrying auxiliary computations in support of assertions, called *assertion modeling code*. Having modeling capability in checkers is important; it allows separating instrumentation code for assertions from the RTL code, thus keeping RTL clean and maintainable. This separation is also convenient for synthesis tools as it provides an easily identifiable distinction between RTL and the instrumentation code without the need of conditional compilation.

Example 9.19. Checker `stable_for_two_ticks` verifies that `sig` may change only in clock ticks 2, 4, ... after the reset `rst` becomes low, as shown in Fig. 9.8.

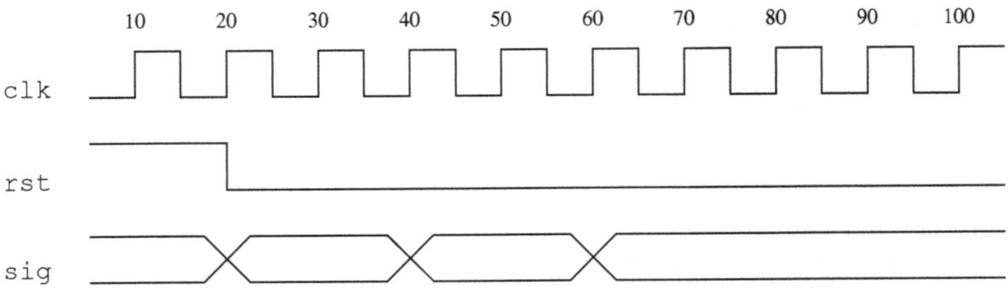

Fig. 9.8 Signal stable for two clock ticks

```
1  checker stable_for_two_ticks(sig,
2    event clk = $inferred_clock, untyped rst = $inferred_disable);
3    default clocking @clk; endclocking
4    default disable iff rst;
5
6    bit toggle = 1'b0;
7    always_ff @clk
8      toggle <= rst ? 1'b0 : !toggle;
9    a1: assert property (!toggle |-> $stable(sig));
10 endchecker : stable_for_two_ticks
```

This checker works as follows. Variable `toggle` is initially low, and it becomes low each time `rst` is deasserted. Otherwise, at each clock tick the value of `toggle` is complemented. Assertion `a1` allows `sig` to change only when `toggle` value is high.

For the waveform shown in Fig. 9.8 the value of `rst` on Line 8 becomes low at time 30 (recall that the value of `rst` is sampled in the checker), and in the next clock tick (time 40) the sampled value of `toggle` is low. Therefore, at time 40, the sampled value of `sig` must be stable: the sampled value of `sig` at time 30 is the same as its sampled value at time `40`. At time `50`, the sampled value of toggle becomes 0, and `sig` may change in this time step, and so on. □

Example 9.20. In the checker from Example 9.19 it is *illegal* to replace

```
bit toggle = 1'b0;
```

with

```
bit toggle;
initial toggle = 1'b0;
```

Initial procedures in checkers may not contain checker variable assignments (see Sect. 9.2.2.2). □

There is an additional restriction imposed on checker variable assignments: it is illegal to reference checker variables in assignments by their hierarchical names.

Example 9.21. The following code is illegal as the checker variable `a` is referenced by its hierarchical name `mycheck.a` in the assignment of `b`.

```
checker check(...);
  bit a;
  ...
endchecker : check

module m_illegal(...);
  ...
  check mycheck(...);
  ...
  wire b;
  assign b = mycheck.a;
endmodule : m
```

In other words, it is forbidden to assign or use checker variables from the outside of the checker (see Sect. 9.2.2). □

Example 9.22. We can generalize the checker `stable_for_two_ticks` from Example 9.19 to check the signal stability for `n` clock ticks. For this purpose, we introduce a counter `ctr`, and we allow `sig` to change only when `ctr` is 0. The system function `$clog2` returns the number of bits necessary to store the value of `n`.

```
checker stable_for_n_ticks(sig, n,
  event clk = $inferred_clock,
  untyped rst = $inferred_disable);

  default clocking @clk; endclocking
  default disable iff rst;
  bit [$clog2(n)-1:0] ctr = '0;

  always_ff @clk begin
    if (rst) ctr <= 1;
    else if (ctr == n - 1) ctr <= 0;
    else ctr <= ctr + 1;
  end

  a1: assert property ($changed(sig) |-> ctr == 0);
endchecker : stable_for_n_ticks
```

Discussion: We could declare `ctr` as **`int unsigned`**, and this solution would work perfectly well in simulation. For efficiency of formal verification it is important to declare all variables with the smallest possible size. This is why we use the system function `$clog2` to designate the smallest variable size. □

> To make checkers efficient in FV, checker variables should be of the smallest size sufficient to store the desired values.

9.4.1.1 Functions in Checkers

The following restrictions are imposed on functions used on the right-hand side of checker variable assignments:

- The functions should not contain **output** or **ref** arguments (**const ref** is allowed).
- The functions should be automatic and not preserve any state information in static variables. The functions must not have side effects.

These restrictions are the same as imposed on function calls in concurrent assertions.

9.4.2 Sampling in Checkers

As we saw in Sect. 9.1.2 the expressions in the right-hand side of nonblocking assignments in checkers are sampled. This is necessary to make assertion modeling in checkers deterministic and consistent with the behavior of concurrent assertions.

Example 9.23. In the following piece of code there would be a race without variable sampling: if `b` changes before `clk`, variable `a` would contain the past value of `b`, otherwise it would contain the current value of `b`.

```
always_ff @(posedge clk) a <= b;
```

If the value of `b` is sampled, `a` would contain the past value of `b` regardless of the race between `b` and `clk`.

□

The exact rule of expression sampling in **always_ff** procedures reads as follows:

> All expressions in an **always_ff** procedure are sampled, except for variables used in the event control.

From this rule it follows that the expressions in non-temporal assertions (immediate, deferred or final) instantiated in this procedure are also sampled. Expressions in other always procedures **always_comb** and **always_latch** are *not* implicitly sampled.

Example 9.24. The code below is a slightly modified example borrowed from the LRM [8]. In the comments, *CV* and *SV* stand for "current value" and "sampled value" respectively.

```
1  checker check(logic a, b, c, clk, rst);
2    logic x, y, z, v, t;
3    assign x = a;                                  // CV of a
4    always_ff @(posedge clk or negedge rst)  // CV of clk and rst
5    begin
6      a1: assert final (b);                    // SV of b
7      if (rst)                                 // CV of rst
8        z <= b;                                // SV of b
9      else z <= !c;                            // SV of c
10   end
11   always_comb begin
12     a2: assert final (b);                    // CV of b
13     if (a)                                   // CV of a
14       v = b;                                 // CV of b
15     else v = !b;                             // CV of b
16   end
17   always_latch begin
18     a3: assert final (b);                    // CV of b
19     if (clk)                                 // CV of clk
20       t <= b;                                // CV of b
21   end
22   ...
23 endchecker : check
```

In the values used in the continuous assignment on Line 3, in the statements belonging to the **always_comb** procedure (Lines 12–15) and in the statements belonging to the **always_latch** procedure (Lines 18–20) are not sampled, according to the sampling definition in checkers. This behavior is the same as in modules.

The behavior in the statements of **always_ff** procedure (Lines 6–9) is different. The values of `clk` and `rst` on Line 7 are not sampled because these variables belong to the event control of the **always_ff** procedure. The value of `b` on Line 6 is sampled, even though it is used in a final assertion because this assertion belongs to **always_ff** procedure. The value of `rst` on Line 7 is not sampled because it is used in the event control (Line 4). The values of `b` and `c` in Lines 8 and 9 are again sampled, because these statements belong to **always_ff** procedure. □

9.4.3 Checker Variables in Final Procedures

In this section we provide an example showing usage of checker variables in a final procedure.

Example 9.25. In Example 9.12 to check for pending requests at the end of simulation, we assumed that `req` remains asserted until the reception of `gnt`. The checker may be modified to work properly even without this assumption. For this purpose, we introduce a checker variable `intrans` which is set to 1 between the assertions of `req` and its `gnt`, or more precisely, from the clock tick after `req` until the clock tick after `gnt`.

For simplicity, we assume again that both `req` and `gnt` are Boolean.

```
checker request_granted(
  req, gnt, n = 1,
  event clk = $inferred_clock,
  untyped rst = $inferred_disable
);
  default clocking @clk; endclocking
  default disable iff rst;
  bit intrans = 1'b0;

  always_ff @clk begin
    if (rst || gnt) intrans <= 1'b0;
    else if (req) intrans <= 1'b1;
  end

  a1: assert property (req |-> nexttime[n] gnt);

  final begin
    a2: assert (!rst -> gnt || !(req || intrans))
      else $error("Outstanding request at the end of simulation");
  end
endchecker : request_granted
```

□

9.5 Checkers with Output Arguments

As we have already mentioned in Sect. 9.2.1, checkers may have output arguments. There are three main use cases where this checker capability is beneficial:

- Returning assertion status from checkers
- Writing modular checkers
- Using checkers as a testbench

In this section we describe the first two use cases, postponing the discussion of the third one to Sect. 23.5. Before we proceed to the use cases we need to describe the rules imposed on checker output arguments. To keep things clear, we illustrate these rules on the example of a trivial checker containing a single assignment and no assertions.

9.5.1 Checker Output Arguments

9.5.1.1 Checker Output Argument Typing

Checker output arguments must be typed, and their type cannot be **`sequence`** or **`property`**. The last two limitations are obvious, the first one is caused by the difficulty to automatically infer a type of the output argument. This limitation is annoying because it prevents checkers with output arguments to be generic.

Example 9.26. The following checker is illegal because its output argument `out` has been left untyped:

```
checker out_illegal (in, output out, input event clk);
  always_ff @clk
    out <= in;
endchecker : out_illegal
```

If we know that the type of `b` is **logic** `[7]`, we can obviously rewrite this checker like this:

```
checker out_concrete (in,
                      output logic [7] out,
                      input event clk);
  always_ff @clk
    out <= in;
endchecker : out_concrete
```

The drawback of this implementation is that the generality has been lost. One workaround could be to declare the argument `out` of a size large enough to suit are needs, for example, 32 bits: **logic** `[32] out`. A better workaround would be using the **type** construct within `out` declaration:

```
checker out_inferred (in,
                      output var type(in) out,
                      input event clk);
  always_ff @clk
        out <= in;
endchecker : out_inferred
```

Here it is explicitly stated that the type of `out` should be the same as the type of `in`, and the concrete type of `in` is inferred from the checker instantiation context.

□

9.5.1.2 Checker Output Argument Initialization

Since assignments in initial procedures in checkers are illegal, the only place of assigning an initial value to a checker output argument, like to any other checker variable, is at its declaration. The syntax of a checker output argument initialization is the same as for specifying a default value for an input checker argument.

Example 9.27. The output argument `out` in the checker `out_inferred` from Example 9.26 has not been explicitly initialized. According to the default initialization rules in SystemVerilog, the initial value of `out` will be `'0` if its actual type is a two-state integral, such as **bit**, and 'x if its actual type is a four-state integral, such as **logic**. If we want it always be initialize it with '0, we should do it explicitly:

```
checker out_inferred (in,
                      output var type(in) out = '0,
                      input event clk);
  always_ff @clk
        out <= in;
```

□

9.5.1.3 Semantics of Checker Output Arguments

Actual checker output arguments, of course, must be L-values.

Example 9.28. The instance good of checker out_inferred defined in Example 9.26 is legal, whereas the instance bad is not:

```
logic x, y, clock;
...
out_inferred good(x, y, posedge clock);
out_inferred bad(x, !y, posedge clock);
```

This is because y is an L-value, and !y is not. □

Checker output argument simulation semantics is similar to that of programs (see Sect. 3.5). The checker instantiation is treated as if there were continuous assignments of the checker output formal arguments to their corresponding actual arguments. These implicit continuous assignments are executed in the Reactive region.

Example 9.29. Consider the following instantiations of checker out_inferred defined in Example 9.26:

```
wire dest;
logic src1, src2, clock;
...
out_inferred inst1(src1, dest, posedge clock);
out_inferred inst2(src2, dest, posedge clock);
```

Here the net dest is driven by outputs of two instantiations of checker out_inferred. These outputs are of type **logic** according to the type of the checker instantiation inputs. The value conflict is resolved according to the standard SystemVerilog resolution rules for nets.

The situation described in this example is rare, and in most cases in practice a checker output is a single driver of the corresponding variable or net. □

9.5.2 *Returning Assertion Status from Checkers*

It may be useful to synthesize assertion in the chip, for example, to detect their violations during post-silicon verification. In this case the corresponding checkers need to be synthesized on the chip like modules. This may be done by introducing an output checker argument to capture the assertion status and to return it to the module where this checker has been instantiated.

Example 9.30. This example, borrowed from the LRM [8] with minor modifications, describes instantiation of a checker verifying the mutual exclusiveness condition in a module and passing the assertion result to a scan latch.

```
1  checker mutex (sig,
2                 event clk = $inferred_clock,
3                 untyped rst = $inferred_disable,
4                 output bit failure = 1'b0);
5    default clocking @clk; endclocking
6    default disable iff rst;
7    a1: assert property ($onehot0(sig))
8      failure = 1'b0; else failure = 1'b1;
9  endchecker : mutex
10
11 module m(input wire bus, logic clock, reset);
12   logic res, scan;
13   ...
14   mutex check_bus(bus, posedge clock, reset, res);
15   always @(posedge clock) scan <= res;
16 endmodule : m
```

Checker `mutex` contains a single assertion `a1` (Line 7) verifying that in its argument `sig` at most one bit is set to 1. Normally assertions have only the fail action, but this assertion has both pass and fail actions to capture assertion success and failure (Line 8). Checker output argument `failure` captures the assertion status.

This checker is instantiated in module `m` as `check_bus` (Line 14) and it checks for mutual exclusion of the bits of `bus` on each riding edge of `clock`. The result is assigned to `res` in the Reactive region. Then it is fed to `scan` (presumably a scan latch) by the nonblocking assignment (Line 15). On each rising edge of `clock` the bits of bus are checked for mutual exclusion and the result is assigned to `res` in the Reactive region. Note that if `clock` is changed in the Active region, `scan` will capture the value of `res` generated on the previous rising edge of `clock`. □

9.5.3 *Writing Modular Checkers*

A module representing a complex design block may be split into several simpler modules representing the component blocks. In this case the higher level module connects instances of the lower level modules. The information is provided to lower level modules through their input ports and is obtained from them through their output ports. Therefore output ports are a key feature in enabling design modularity. The same is true for the verification IP: splitting a checker into smaller checkers is made possible using checker output arguments.

Complex checkers are typical in formal verification when they contain a sophisticated modeling and numerous assertions. To make such checker manageable, it should be split into smaller parts, each part containing either modeling of some block or assertions and assumptions imposed to this block.

Example 9.31. Suppose that we want to verify an FSM implemented in RTL. This FSM may be rather complex, have many states, sophisticated logic required to make this FSM efficient, etc. The complexity of this FSM may prevent its direct formal verification. For the purpose of verification we may not need to account for all its complexity. We can build a model of a more abstract FSM that may be easily verified. For example, the concrete FSM may have two different idle states: `IDLE1` and `IDLE2`. This fact may be irrelevant for formal verification, hence our abstract FSM has a single idle state `AIDLE`. Transitions of the abstract FSM may also be much simpler than those of the concrete FSM, because irrelevant implementation details may be ignored.

The checker verifying the concrete FSM may be split into the following parts:

- Abstract FSM model
- Assertions verifying that the behavior of the abstract FSM is aligned to the behavior of the concrete one
- Assertions verifying the behavior of the abstract FSM (and therefore, the relevant behavior of the concrete one)
- Glue that combines all the above parts together

Let enumeration types `state_t` and `astate_t` contain definitions of the concrete and abstract states, respectively:

```
typedef enum {IDLE1 = ..., IDLE2 = ..., ...} state_t;
typedef enum {AIDLE, ..., AERR} astate_t;
```

The checker implementing the abstract FSM will look like that:

```
checker abstract_fsm(...,
                     event clk = $inferred_clock,
                     untyped rst = $inferred_disable,
                     output astate_t astate); // abstract state
  default clocking @clk; endclocking
  always_ff @clk begin
    if (rst) astate <= AIDLE;
    case (astate)
      AIDLE: astate <= ...;
      ...
      default: astate <= AERR;
    endcase
  end
endchecker : abstract_fsm
```

The checker verify_consistency verifies that the abstract FSM is an abstraction of the concrete one:

```
checker verify_consistency(state, astate, ...,
                           event clk = $inferred_clock,
                           untyped rst = $inferred_disable);
  default clocking @clk; endclocking
  default disable iff rst;

  aidle: assert property (astate == AIDLE <-> state inside {IDLE1
      , IDLE2})
    else $error("Idle states inconsistent");
  ...
endchecker : verify_consistency
```

The checker verify_abstract_fsm verifies the behavior correctness of the abstract FSM:

```
checker verify_abstract_fsm(astate, ...,
                            event clk = $inferred_clock,
                            untyped rst = $inferred_disable);
  default clocking @clk; endclocking
  default disable iff rst;

  never_stuck: assert property (s_eventually astate != AIDLE)
    else $error("FSM stuck");
  unexpected: assert property (astate != AERR)
    else $error("Unexpected behavior detected");
  ...
endchecker : verify_abstract_fsm
```

Finally the top-level checker is as follows:

```
checker verify_fsm(state, ....,
                   event clk = $inferred_clock,
                   untyped rst = $inferred_disable);
  default clocking @clk; endclocking
  default disable iff rst;

  astate_t astate;

  abstract_fsm afsm(..., astate);
  verify_consistency vcons(state, astate, ...);
  verify_abstract_fsm vafsm(astate, ...);
endchecker : verify_fsm
```

Here state is the concrete state generated by the RTL. The ellipses (...) designate other arguments that should be filled in, and are not part of the syntax. Note, that we do not have to explicitly pass the clock and reset to the subcheckers because they may be inferred by default via $inferred_clock and $inferred_disable system functions. □

Exercises

9.1. Assertion a_initial_no_complete from Fig. 9.4 does not behave correctly in the general case when rst is initially low. For what actual arguments corresponding to rst the assertion behaves always correctly? Consider the use cases of simulation and formal verification. Hint: Consider cases when the corresponding actual argument is of type **logic** and **bit**, and when it is explicitly initialized or not.

9.2. Fix assertion a_initial_no_complete from Fig. 9.4 to handle the case when the reset is initially low. Hint: consult Sect. 11.2.1.1.

9.3. In assertion a_initial_no_complete from Fig. 9.4 rst is used both in the implicit **disable iff** clause and in the body of the assertion. Which problem does this cause? How to solve it? Hint: Consider variable sampling in concurrent assertions.

9.4. Write a checker verifying that each request is followed by a grant, and that the request happens at least once.

9.5. Example 9.11 introduces a checker simple_reset to verify that the reset signal is initially high then it eventually goes low and remains low forever.

(a) Modify this checker to make it generic. The new checker should accept the assertion clock as an optional argument. This argument should default to $global_clock.
(b) If the reset remains always high, the checker assertion does not fail in simulation. How should the checker be modified to ensure that a more meaningful scenario is exercised in simulation?
(c) What would the checker verify if followed-by (suffix conjunction) #=# is replaced with suffix implication |=>?

9.6. Implement the checker check_window borrowed from OVL [10] (with minor modifications).

This checker ensures that the condition cond is true in a specified window between a start and complete.

The checker should have the following arguments:

- cond—expression that should be true in the event window
- start—sequence whose completion opens the event window
- complete—sequence whose completion closes the event window
- clk—clock event for the checker. Should be inferred from the context by default
- rst—checker reset signal. Should be inferred from the context by default
- msg—message to be issued in case of the check failure. Default is "Violation"
- collect_cov—elaboration time constant indicating that the coverage about window should be collected. Default is yes. The following coverage information is collected.

   1. The window has been opened
   2. The window has been closes
   3. The number of clock cycles in the window

- `to_assume`—elaboration time constant indicating that the property about the window condition should be assumed. By default the property must be asserted.

  Notes.

1. Consult Sect. 11.2.1 to handle the case when `complete` is an arbitrary sequence. Alternatively, you can implement the simplified version of this checker assuming that `complete` is a signal.
2. Coverage constructs are described in Chaps. 4 and 18.

Part IV
Advanced Assertions

Chapter 10
Advanced Properties

No man acquires property without acquiring with it a little arithmetic also.

— Ralph Waldo Emerson

This chapter briefly recapitulates the basic properties discussed in Chap. 5 and discusses more complex property operators. First, we examine the property equivalents of Boolean operators, namely, **`not`**, **`or`**, **`and`**, **`implies`**, **`iff`**, **`if else`**, and **`case`**. Then we provide a description of temporal operators inspired by Linear Temporal Logic. The operators are described informally; the reader interested in formal semantics should consult Chap. 22. Recursive properties are described in Chap. 17, and the abort operators **`accept_on`**, **`reject_on`**, **`sync_accept_on`**, **`sync_reject_on`** are discussed in Chap. 13. In all examples, we assume that the properties and assertions are in the scope of a default clocking declaration, hence no explicit clocks are specified.

Most of the temporal operators come in two forms, weak and strong. The strong forms are identified by a prefix **`s_`** as in **`s_nexttime`** `p`. In simple terms, the strength determines the property evaluation result when there are not enough clock ticks to complete the evaluation of the operator, e.g., at the end of simulation or when the source stops emitting them. More precise explanation is given in Chap. 21.

Table 10.1 lists all the property operators available in the language. The order of appearance is in decreasing precedence, beginning with the highest on the top. Operators appearing in one block of the table have the same precedence. Blocks are separated by horizontal lines. For example, **`not`**, **`nexttime`**, and **`s_nexttime`** have the same precedence. Additionally, the precedence of any SystemVerilog expression operator is higher than any property or sequence operator.

E. Cerny et al., *SVA: The Power of Assertions in SystemVerilog*,
DOI 10.1007/978-3-319-07139-8_10

10.1 Sequential Property

In addition to the property operators listed in Table 10.1, sequences are promoted to properties when used in a property context. We call them sequential properties, or sequence properties (see also Sect. 6.2). This happens when a sequence is used as the only expression in an assertion or as an operand to a property operator that requires the operand to be a property. Sequential properties can be *weak* or *strong*. A sequence becomes a strong property when it is the argument to a `strong(...)` qualifier, e.g., `strong(a ##1 b)`. A strong sequential property holds if and only if the underlying sequence has a match. Without this `strong` qualifier, a sequential property is weak (see Sect. 6.2, the definition given there corresponds to the week sequential property) in the assertion or assumption context, and strong in the cover context. As we shall see, in situations where a sequence is strong by default, it can be made weak by using the qualifier `weak(...)`. The distinction between strong and weak sequential properties is reflected in their truth value when there are not enough clock ticks to complete the evaluation of the sequence: If there are not enough clock ticks the weak sequence property succeeds, while the strong one fails.

Example 10.1. Consider the following assertions:

```
initial a1: assert property(a ##[+] b);
initial a2: assert property(strong(a ##[+] b));

a3: assert property(!a ##1 a |-> b[*] ##1 c);
a4: assert property(!a ##1 a |-> strong(b[*] ##1 c));

c1: cover property(a ##[+] b);
c2: cover property(weak(a ##[+] b));
```

Table 10.1 Property operators

Operator	Associativity
not	–
nexttime, **s_nexttime**	–
and	Left
or	Left
iff	Right
until, **s_until**	Right
until_with, **s_until_with**	Right
implies	Right
|->, |=>	Right
#-#, #=#	Right
always, **s_always**	–
eventually, **s_eventually**	–
if else, **case**	–
accept_on, **sync_accept_on**	–
reject_on, **sync_reject_on**	–

Note: Recall that `b[*]` is a shortcut for `b[*0:$]` and `##[+]` is a shortcut for `##[1:$]`. The sequence in assertion `a1` is weak because it is used as the property of an assert statement, and there the default is weak. Therefore, when an attempt is triggered by `a` true and `b` does not become true, that evaluation attempt of the assertion succeeds. In this situation, however, `a2` fails because the sequential property is explicitly qualified as strong.

Assertions `a3` and `a4` are a little more complex. Each contains two sequences: `!a ##1 a` and `b[*] ##1 c`. The former sequence is used as the antecedent (or precondition) of the suffix implication `|->`. If a thread of evaluation of that sequence does not complete due to lack of clock ticks or because `!a` is false or is not followed by `a` in the next clock tick, that thread fails, and in the context of the antecedent it contributes no match. In those cases the assertion attempt has a vacuous success (Sect. 10.6). The consequent sequence `b[*] ##1 c` is a weak property in `a3` and a strong property in `a4`. The interpretation is similar to that of `a1` and `a2`. That is, in the absence of a sufficient number of clock ticks or if `b` remains true forever while `c` is never true, the consequent of `a3` succeeds, while that of `a4` fails.

In the case of covers `c1` and `c2`, the default strength of a sequence used as the coverage property is strong. This is done so that a coverage hit is not registered when the sequence does not complete evaluation, e.g., due to lack of clock ticks. To override the default behavior, the qualifier **`weak`** should be used, as in `c2`.[1] □

Let us now examine the various property operators from Table 10.1.

10.2 Boolean Property Operators

The following property operators express Boolean connectives between properties:

- **`not`** `p`—negation
- `p` **`or`** `q`—disjunction
- `p` **`and`** `q`—conjunction
- `p` **`implies`** `q`—implication
- `p` **`iff`** `q`—equivalence
- **`if`**`(b)p` **`else`** `q`—if conditional
- **`case`** `(b)...`—case conditional

Here is their informal description; their formal semantics can be found in Chap. 22.

[1]Though cover `c2` is rather meaningless.

Negation

The property `not p` is true iff the property `p` is false.

Example 10.2. What is `not e` where `e` is a Boolean expression? According to the definition, `not e` is true iff `e` is false as a property. If the clock ticks, then `e` is false as a property iff `!e` is true as a Boolean expression at the first clock tick. If the clock does not tick, then `e` is false as a property iff it is a strong sequential property at the first position of the trace. If the property clock is the global clock, then `not e` is equivalent to `!e` (see Chap. 22, Example 21.29). □

Disjunction

Property `p or q` is true iff either property `p` or property `q` is true (Sect. 5.5).

The syntax is the same as for sequence disjunction discussed in Sect. 6.6. When the sequence disjunction is used as a property it can be replaced by property disjunction with the same constraint on the strength of the sequence disjunction as on both `p` and `q`.

Conjunction

Property `p and q` is true iff both properties `p` and `q` are true (Sect. 5.5).

Note that the syntax of property **and** is similar to that of the sequence **and** operator described as `s1 and s2` in Sect. 11.1.5. However, they have similar meaning only when the sequence **and** is used as a property, and the strength of both `p` and `q` is the same as the strength of the sequence `s1 and s2`.

For further discussion, refer to Sects. 11.1.5 and 5.5.

Implication

The property `p implies q` is true iff either property `p` is false or `q` is true (see Chap. 22).

When `p` and `q` are Boolean expressions `e1` and `e2`, respectively, then `e1 implies e2` is equivalent to `e1 -> e2` provided the clock ticks. This equivalence holds, e.g., if the sampling clock is the global clock, which is guaranteed not to stop. For further discussion on the semantics of **implies**, see Chap. 22.

Previously (Sect. 6.4) we discussed suffix implication `|->`, which also involves a property, but only in the consequent. This has to be contrasted with property implication **implies**:

In `s |-> q`, where `s` must be a sequence and `q` some property, the evaluation of `q` starts at the time when any evaluation thread of `s` has a match. In `p implies q`, `p` and `q` are properties, hence there is no notion of an endpoint and a match. Both `p` and `q` start evaluating at the same time and the truth results are computed using the logical operator **implies**. For example, consider

```
a ##1 b |-> c ##1 d
```

vs.

```
a ##1 b implies c ##1 d
```

In the case of `|->`, the consequent `c ##1 d` will start evaluating when `a ##1 b` matches. In the case of **implies**, both `a ##1 b` and `c ##1 d` start evaluation at the same clock tick. Finally, if `p` is a sequence `s`, writing `strong(s) implies q` is equivalent to `s |-> q` *only when* `s` is a Boolean expression. The Boolean implication `b1 -> b2` is a short-hand for `!bit'(b1) || b2`. Therefore, unlike property and suffix implications, such Boolean expressions have no notion of vacuity of evaluation. That is, the evaluation of `b1 -> b2` as a sequence property is always nonvacuous (see Sect. 10.6).

An interesting example that uses **implies** is as follows.

Example 10.3. The signal `sig` should be high from `m >= 0` clock ticks before event `ev` happens until `n >= 0` clock ticks after it. That is, if `ev` happens at time 10, `m = 2`, and `n = 3`, then `sig` should be high at times 8, 9, ..., 13.

Solution:

```
a1: assert property (
  strong(##m ev) implies sig[*(m + n + 1)]);
```

We have used a strong sequence `strong(##m ev)` in the antecedent of **implies** so as to require that there are enough clock ticks for `ev` to become true.

To express the same using `|->` the assertion becomes more complex:

```
a2: assert property (
  (!sig |-> !ev[*(m+1)]) and (ev |-> sig[*(n+1)]));
```
□

Equivalence

The property `p iff q` is true iff either properties `p` and `q` are both true or they are both false (Chap. 22).

Like the case of **implies**, when `p` and `q` are Boolean expressions `e1` and `e2`, respectively, then `e1 iff e2` is equivalent to `e1 <-> e2` provided the clock ticks, e.g., if the clock is the global clock.

When can **iff** be used? A very useful application is in verifying that two property definitions have the same meaning: The same property may be implemented in different ways because one implementation may be much more efficient in simulation and the other in formal verification. Often the two properties look and feel similar, yet they are not equivalent. If we check their equivalence in formal verification a counterexample explains the difference. Only the assertion comparing the properties and wire or module input declarations of the variables used in the assertion are required in this case; neither a model nor assumptions are needed.

Example 10.4. Check that the properties **always nexttime** e and **nexttime always** e are equivalent.

Solution:

```
wire e;
initial a: assert property (
  (always nexttime e) iff (nexttime always e));
```
□

The following example is perhaps less evident, yet the two properties are in fact equivalent:

Example 10.5.

```
property p1;
  not (a[*] ##1 b);
endproperty
property p2;
  strong(!b[+] ##0 !a);
endproperty
a1: assert property(p1 iff p2);
```
□

If [Else]

The property **if** (b)p is true if Boolean b is false or p is true. The property **if** (b)p **else** q is true if Boolean b is true and p is true, or b is false and q is true.

Note that is possible to express the same using suffix implication, as b |-> p, and (b |-> p) **and** (!(**bit**'(b))|-> q), respectively.

Clearly, the **if-else** form is easier to understand than its equivalent using suffix implication, as illustrated in the following example.

Example 10.6. In assertions a1 and a2, if b is true, a should be false or b should be false one clock tick later, else if b is false then a must be true, followed by b true one clock tick later. Clearly, assertion a1 is easier to understand. Assertion a3 uses an **if** property without the **else** clause. In that case, the equivalent formulation shown in assertion a4 is as easily understood, hence there is no preference between them.

```
a1: assert property(if (b) not strong(a ##1 b)
                      else a ##1 b);
a2: assert property(
      (b |-> not strong(a ##1 b)) and
      (!bit'(b) |-> a ##1 b));
a3: assert property(if (b) a ##1 b);
a4: assert property(b |-> a ##1 b);
```
□

Case

Property **case** is a generalization of **if-else** for a multiple-valued condition `b`.

```
case (b)
  b1: p1;
  ...
  bN: pN;
  default: p;
endcase
```

The property **case** `(b)...` is true iff either `pi` evaluates to true for the first `i` such that the value of `bi` matches the value of `b`, or no `bi` matches the value of `b` and if the *optional* **default** item property `p` is specified then it evaluates to true. If the **default** case item is not specified and no `bi` matches the value of `b`, then the property **case** is vacuously true.

A **case** operator can be used, for example, to define a property in which a sequence delay varies based on some register value. Of course, for practical purposes a small range of delays is assumed:

Example 10.7.
```
property p(bit [2:0] delay);
  case (delay)
    0: a;
    1: nexttime[1] a;
    2: nexttime[2] a;
    3: nexttime[3] a;
    4: nexttime[4] a;
    default: 1'b0; // delay too large
  endcase
endproperty : p
```
□

The same property could be written using a chain of **if-else** property operators, but the meaning of such nested operators is less obvious than when using the **case** operator. It is left to the reader as an exercise at the end of the chapter to rewrite the property in Example 10.7 using **if-else**.

Many of the following operators have been briefly described in Chap. 5, and their formal semantics can be found in Chap. 22. Here, we provide an intuitive recapitulation of the operators, further clarifying the distinction between the strong and the weak forms and between the bounded and unbounded forms. The bounded variants evaluate the operand property over finite, bounded numbers of clock ticks, while the unbounded ones evaluate over indefinite but finite numbers of clock ticks.

10.3 Suffix Operators: Implication and Followed-By

The following are suffix operators:

- Suffix implications `|->` and `|=>`.
- *Followed-by* operators (also named suffix conjunctions) `#-#` and `#=#`.

Table 10.2 Comparison of suffix implication and followed-by

Operator	Antecedent match	Antecedent no match
`|->` `|=>`	Each must yield consequent true	Vacuous success
`#-#` `#=#`	At least one must yield consequent true	Failure

Suffix Implication

The suffix implications have been discussed in Chap. 6. We summarize them here because we need them to describe the *followed-by* operators.

A *suffix implication* operator takes a sequence as the left-hand operand and a property as the right-hand operand: `s |-> p` and `s |=> p`. Whenever `s` matches, property `p` must hold. When `s` has no match then a suffix implication is vacuously true.

The difference between the two forms is that in the case of `|->` the evaluation of property `p` starts at the clock tick that occurs *at* or *after* the tick when `s` matches. In the case of `|=>` the evaluation of `p` starts at the clock tick that occurs *strictly after* the clock tick when `s` matches. When the ending clock of `s` (meaning the clock of the latest evaluated expression of `s`) is the same as the leading clock of `p`, then in the case of `s |-> p` property `p` starts at the same clock tick when `s` matches, while in the case of `s |=> p`, property `p` starts at the clock tick following the match of `s`. This is why `|->` is called an overlapping suffix implication, and `|=>` is a nonoverlapping one.

Suffix Conjunction (Followed-By)

The *followed-by* operators `#-#` and `#=#` also have a sequence as the left-hand operand and a property as the right-hand operand: `s #-# p` and `s #=# p`. If `s` has no match, then *followed-by* evaluates to false. If `s` has one or more matches for a given evaluation attempt, then for the property to evaluate to true, at least one match of `s` must result in `p` evaluating to true (see Table 10.2). In this sense, the behavior is similar to sequence concatenation with `##0` and `##1` cycle delays. The difference is that sequence concatenation requires a sequence as the right-hand side operand, while *followed-by* accepts a property there. This is also the reason that *followed-by* is sometimes called a *suffix conjunction* or *suffix concatenation*. If property `p` is in fact a strong sequential property (e.g., if `s1 #-# s2` appears in a cover), then `s1 #-# `**`strong`**`(s2)` is the same as **`strong`**`(s1 ##0 s2)`. Similarly, `s1 #=# `**`strong`**`(s2)` is the same as **`strong`**`(s1 ##1 s2)`. As to when `p` starts its evaluation relative to the match of `s`, the difference between `#-#` and `#=#` is the same as between the two forms of the suffix implication. The former is overlapping and the latter is nonoverlapping.

A followed-by operator is a dual operator of suffix implication. The following equivalences hold:

`s #-# p` ≡ **not** `(s |-> not p)`, and `s #=# p` ≡ **not** `(s |=> not p)`

This means that even without the availability of the followed-by operators, the same behaviors could be obtained using the right-hand sides of the above equivalences. The intent is, however, more clearly conveyed by the shorter notation when a followed-by operator is used.

The question is where the use of followed-by is appropriate. Its principal usage is in **cover property** statements when the right-hand side argument cannot be restricted to a sequence. This often occurs in properties that are used in checker libraries, where the arguments of the checker are not restricted to be Boolean expressions or sequences only.

Example 10.8. Consider the coverage property: When a "pattern" `x` is detected it is followed by a "pattern" `y`. We may have to restrict `x` to be a sequence, but `y` could be nonrestricted and be any property.

```
property p(sequence x, untyped y);
    x #-# y;
endproperty
```

Its usage could be

```
cov: cover property(p((req[*2]), (s_eventually ack)));
```

□

Example 10.9. Ascertain that a reset condition is true for some `m` initial clock ticks and then it remains false forever. Such a property is often used as an assumption on reset in formal verification.

Solution:

```
initial a: assume property(reset[*m] #=# always !reset);
```

□

We now examine the temporal operators inspired by Linear Temporal Logic.

10.4 Unbounded Linear Temporal Operators

Linear temporal logic (LTL) is a *modal* temporal logic with modalities referring to time, which in SVA means as measured by the occurrence of clock ticks. In LTL, it is possible to write formulae about the future of behaviors following a linear progression of time, such as that a property will eventually be true, that a property will be true until another property becomes true, and so on. The operators can be bounded with some specific ranges or unbounded. The following are linear temporal unbounded property operators available in SVA:

- Weak **until** and **until_with**, and their strong forms **s_until** and **s_until_with**.
- Unbounded weak **always**.
- Unbounded strong **s_eventually**.

Until

There are only the unbounded forms of these operators:

- Unbounded weak: `p1 until p2`
- Unbounded strong: `p1 s_until p2`

The operator comes in two forms, weak **until** and strong **s_until**. The formal semantics is covered in Chap. 22.

The weak **until** property holds true provided that either p2 is true at the first clock tick or p1 holds true at all clock ticks as long as p2 is false. If there is no clock tick at which p2 is ever true, the property evaluates true. The strong form **s_until** is similar except when there is no clock tick at which p2 holds true—the strong form is false in that case.

Example 10.10. Suppose that condition c must hold true between the occurrences of conditions e1 and e2 but not necessarily including these clock ticks.

```
a1: assert property(e1 |=> c until e2);
a2: assert property(e1 |=> c s_until e2);
```

Discussion: Assertion a1 will succeed even if there are not enough clock ticks for detecting e2 true (provided that c holds till then), while a2 will declare failure in that situation due to the use of a strong until operator. □

Always and S_eventually

The following are the unbounded forms of these operators:

- Unbounded weak: `always p`
- Unbounded strong: `s_eventually p`

Property `always p` is true if p holds true at every clock tick. The operator **always** is *weak*, hence when there are no more clock ticks, the property evaluates to true. It is equivalent to `p until 1'b0`. Recall that `1'b0` is Boolean *false* in SystemVerilog.

Property `s_eventually p` is true if there is a sufficient number of clock ticks to find one at which p is true.

What if we negate an **always** property as `not always p`? According to the definition this property is true iff `always p` is false, which means that p is false at least at one clock tick. This is exactly the property `s_eventually not p`. It is a strong eventuality because there must be a clock tick where p is false (otherwise `always p` would be true).

It follows that `s_eventually p` is equivalent to `not always not p`. And also `s_eventually p` is equivalent to `1'b1 s_until p`.

Notice how the negation changes the strength of the resulting property. Negating a weak **always** we obtain a strong **s_eventually**. This is because to falsify

always we require that the operand property p be false somewhere in the future. It must happen, hence the eventuality is strong. **always** and **s_eventually** are *dual* properties.

In case of a Boolean property e, **not always** e may be rewritten as **s_eventually** !e.

Until_with

There are only the unbounded forms of these operators:

- Unbounded weak: p1 **until_with** p2
- Unbounded strong: p1 **s_until_with** p2

The weak **until_with** property holds true provided that either p1 **and** p2 is true at the first clock tick or p1 holds true at all clock ticks until a clock tick when both p1 and p2 hold true. If there is no clock tick at which p1 **and** p2 is true, the weak property evaluates true. The strong form **s_until_with** is similar except when there is no clock tick at which p1 **and** p2 holds true—the strong form is false in that case.

As in the case of the dual operators **always** and **s_eventually**, **until** and **s_until_with** are dual operators, as are **s_until** and **until_with**. The following equivalences hold:

```
p s_until q ≡ not ((not q) until_with (not p))
p until q   ≡ not ((not q) s_until_with (not p))
```

Example 10.11. Write an assertion that verifies the following situation: When Boolean trig is true, property p2 must hold at some clock tick strictly before a clock tick at which property p1 holds.

Solution:

```
a1: assert property (
  if (trig) (not p1) until_with p2);
```

The specification is missing one important point, namely, must p2 ever occur? If not, then the above assertion is correct. If yes, then we should require p2 to be true at some clock tick by using the strong form

```
a2: assert property (
  if (trig) (not p1) s_until_with p2);
```
□

Example 10.12. When req becomes true it must hold until and including gnt. In addition, gnt must happen.

Solution:

```
a1: assert property(
  !req ##1 req |-> req s_until_with gnt);
```

Discussion: If a new request can start immediately at the clock tick following gnt, the assertion would have to be modified as follows because there is no rising edge on req in that situation.

```
a2: assert property(
  (!req || gnt) ##1 req |-> req s_until_with gnt);
```

The sequence in the antecedent matches when either !req ##1 req happens (i.e., rising transition of req) or gnt ##1 req happens which is the case of a continuing request. Note that another assertion should verify that gnt does not occur without a req.

How should we modify the assertions if it is not required that gnt ever happens after asserting req? The answer is similar to the preceding example, Example 10.11, namely, replace the strong form **s_until_with** by the weak one **until_with**. Then, even if gnt never happens after being requested the property will evaluate true. □

10.5 Bounded Linear Temporal Operators

Bounded operators are useful when the property to be verified must be satisfied within some specified range of clock ticks. There is the fixed delay property operator **nexttime**, which is similar to ##m in sequences. The operators **always** and **eventually** are provided with ranges in both strong and weak forms.

The behavior of the bounded operators is as follows.

Nexttime

- **nexttime** p
- **s_nexttime** p
- **nexttime** [m] p
- **s_nexttime** [m] p

The semantics are split according to the value of argument m, which must be an elaboration constant, and according to the strength of the operator:

1. Weak form: **nexttime** p is true at tick t if p is true at tick $t + 1$ or if there is no tick $t+1$ or if there are no clock ticks at all. **nexttime** [0] p has no delay, p has to be true at tick t or there is no tick t. In that sense it is equivalent to 1 |-> p. Property **nexttime**[m] p for some m > 0 is true if p is true at tick t + m or if there are not enough ticks.
2. Strong form: **s_nexttime**[m] p, m > 0 is similar to the weak form except that it does require having a sufficient number of clock ticks, i.e., at least m. It is thus equivalent to **not nexttime**[m] **not** p.

For singly clocked properties `nexttime [0] p` means "either the clock does not tick anymore, or p", and `s_nexttime [0] p` means "the clock ticks at least once, and p".

For example,

```
initial a: assert property(
    nexttime[0] s_eventually e);
```

means that either the clock does not tick at all, or e holds in some finite number of clock ticks. However,

```
initial a: assert property(
    s_nexttime[0] always e);
```

means that the clock ticks at least once and e happens at each clock tick.

When m > 0 the weak form can be defined recursively. This results in a repetitive application m times of **nexttime** on p:

`nexttime nexttime ... nexttime nexttime p`

As mentioned above, the strong form is defined by double negation: Therefore, for `m == 1`, `nexttime not p` says that p does not hold at the next clock tick and it is a weak form. By negating it, a strong property is obtained saying that "it is not true that p does not hold at the next clock tick". That is, it must hold and there must be at least one clock tick.

Example 10.13. When initial reset rst is deasserted, property p must eventually hold after 2 clock ticks, and there must be enough clock ticks:

Solution:

```
initial a: assert property (
  rst ##1 !rst |=> s_nexttime [2] s_eventually p);
```

□

Efficiency Tip. In general, it is recommended to use **s_nexttime** with strong operators, and **nexttime** with weak ones. For instance, `nexttime always p` and `s_nexttime s_eventually p`. It is important to note that p is a property. This situation is common and important for assertion libraries, where there are restrictions on the argument type.

Example 10.14. There cannot be two consecutive requests where req is a Boolean expression. One could be tempted to write the following assertion:

```
a1: assert property (req and nexttime !req);
```

but this is wrong. Assertion a1 is contradictory: `req and nexttime !req` must be true at every clock tick, i.e., req must be true at each clock tick and !req must be true starting from the second clock tick. Already at the second clock tick it thus requires that both req and !req are true!

To write this assertion correctly, we can formulate it as "if there is a request, there should be no request at the next clock tick":

```
a2: assert property (req implies nexttime !req);
```

Or, since req is a Boolean expression (or a signal), the assertion can be written more simply as

```
a3: assert property (req |=> !req);
```

or as

```
a4: assert property (not strong(req[*2]));
```

□

Example 10.15. If there is no acknowledgment ack after req, then in two clock ticks rtry should be asserted.

Solution: This can be reformulated as "if there is a request then either at the next tick there should be an acknowledgment, or in two cycles rtry should be asserted":

```
a1: assert property (
  req implies (nexttime ack) or nexttime [2] rtry);
```

The same assertion may be rewritten as:

```
a2: assert property (
  req implies (nexttime (ack or nexttime rtry));
```

If req, ack, and rtry are Boolean expressions then the assertion can be simplified as follows:

```
a: assert property (req |=> (ack or ##1 rtry));
```

□

Bounded Eventually and Always Operators

- **eventually** [m:n] p
- **s_eventually** [m:n] p
- **s_eventually** [m:$] p
- **always** [m:n] p
- **s_always** [m:n] p
- **always** [m:$] p

The bounded forms of the operators **eventually** and **always**, both weak and strong, are derived operators that are defined using weak and strong forms of **nexttime**, respectively.

1. Let m $\geq$ 0. **eventually**[m:m] p is the same as **nexttime**[m] p. This is a simple equivalence between weak forms. Similarly, (**s_eventually**[m:m] p) is equivalent to **s_nexttime**[m] p.
2. Let m $\geq$ 0, n $>$ m. **eventually**[m:n] p is defined recursively as **eventually**[m:n-1] p **or** **nexttime**[n] p.
   This recursive definition can be expanded into a disjunction of n-m+1 **nexttime** properties. For example, **eventually**[2:4] p is equivalent to **nexttime**[2] p **or** **nexttime**[3] p **or** **nexttime**[4] p.

To define the strong form **s_eventually**[2:4] p, use **s_nexttime**.

3. Let m ≥ 0. **always**[m:m] pis the same as writing **nexttime**[m] p. Since the extent where **always** should hold is only one clock tick, **always**[m:m] is the same as **eventually**[m:m] p as well. The strong forms are defined using **s_nexttime**.
4. Let n > m. **always**[m:n] p is the same as **always**[m:n-1] p **and nexttime** [n] p. The expansion is similar to the expansion of (**eventually**[m:n] p), but the series of **nexttime** properties is formed using conjunctions. For example, **always**[2:4] p is equivalent to **nexttime**[2] p **and nexttime**[3] p **and nexttime**[4] p. The strong form is again defined using **s_nexttime**.
5. Let m ≥ 0. **always**[m:$] p is equivalent to **nexttime**[m] **always** p. The lower bound m just shifts the check for property p to hold forever after m clock ticks. There is no strong form in this case due to the open-ended upper bound, which requires **always** p.
6. Let m ≥ 0. **s_eventually**[m:$] p is equivalent to **s_nexttime**[m] s_eventually p. As in the preceding case, the check for the property p to eventually hold is just shifted by m clock ticks. As a dual property to **always** [m:$] p, there is no weak form.

Like the unbounded case, the form **s_eventually** [m:n] p can also be defined using duality with **always** as **not always**[m:n] **not** p. The strong eventuality requires that p is true within m to n clock ticks and that there are enough clock ticks to cover this range. In the formulation using **always**, **always**[m:n] **not** p will succeed when p fails at each clock tick in the range (there need not be enough clock ticks to cover the range). Again, due to the top-level negation in **not always**[m:n] **not** p, that property will succeed provided p is true at some clock tick in the range. It is thus equivalent to the strong eventuality (**s_eventually**[m:n] p).

Similarly, we can see that (**s_always**[m:n] p) is equivalent to (**not** eventually[m:n] **not** p). Again, the negation of a weak property becomes a strong property and vice versa.

One may ask the question why there is the weak **eventually** [m:n] p with a fixed range, while the open-ended range is only strong **s_eventually** [m:$] p. It works out that if the latter were weak, the property would be in some sense meaningless because in an assertion it cannot ever fail. If there are not enough clock ticks before p is true, then the weak form succeeds. Only the strong form can declare failure in that case. This distinction between open and bounded ranges carries over by duality to the **always** operator. Namely,

```
not eventually [m:n] not p    ≡ s_always [m:n] p
not s_eventually [m:$] not p ≡ always [m:$] p.
```

In the final section of this chapter, we explain how vacuous successes of a property are determined depending on the type of the operator.

10.6 Vacuous Evaluation

When an assertion fails, it provides good information about the cause of the failure, often near its source in the design code. However, what happens if no assertion fails? Does that mean that the design is correct relative to the set of assertions used? In the chapter on coverage (Chap. 18), we can see that assertion success is not a guarantee; we still must make sure that the tests cover as much of the functionality of the design as possible.

Many typical assertions involve one of the following conditional operators:

- suffix implications `|->`, `|=>`
- Boolean implication `->`
- **`if`** - **`else`** operator, or
- implication **`implies`**

Consider the simple Boolean property `x -> y` where `x` and `y` are some Boolean expressions. The interpretation of this property is *if x is true then y must be true*. If `x` is false it does not impose any claim on `y`: the property is true regardless the truth of `y`. This success is called a *vacuous success*. That is, a success that carries no weight as far as the verification of the design is concerned. In practice, a more desirable test is the one in which variable `x` evaluates to true. Therefore, if an assertion has only vacuous successes in a test, it means that either the test never stimulates the design in such a way as to make `x` true, or that the design may never have a value assignment that makes the expression true. In the former case, we must make sure that other tests do trigger the assertion so that it does evaluate nonvacuously. In the latter case, we must examine `x` and the design to see whether the problem is a bad formulation of the expression, or something wrong in the design that prevents the variables from achieving the expected values. Finally, a nonvacuous success of an assertion evaluation attempt requires that the following two conditions be met:

- The assertion property attempt evaluates to true, and
- The evaluation attempt is nonvacuous.

The evaluation attempt of any sequential property, weak or strong (Sect. 10.1), is always nonvacuous since no explicit condition is stated in a sequence. It is essentially a pattern match based on *regular expressions*, and the match is unconditional.

Now we will consider conditional expressions where vacuity is of practical significance. For properties of the form

- `sequence_expr |-> p`
- `sequence_expr |=> p`

the evaluation is nonvacuous if `sequence_expr` has a match and the evaluation thread of `p` that results from the match evaluates nonvacuously.

For property **`if`** `(expr)p1` **`else`** `p2`, an evaluation attempt is nonvacuous when `expr` is true and `p1` evaluates nonvacuously, or when `expr` is false and `p2` evaluates nonvacuously. It follows that vacuity is determined by evaluation of the properties p1 and p2. If expr is true and p1 is a vacuous success, the overall property is a vacuous success; similarly, when expr is false, p2 determines the result. If the **else** clause is missing then the evaluation is nonvacuous if and only if expr is true and the evaluation of p1 is nonvacuous. When expr is false or when p1 is a vacuous success, the evaluation of the overall property is a vacuous success.

Most simulation and formal verification tools report vacuous successes when the top-level property operator is one of the cases listed above. Other more complex forms where nested property operators are involved become computationally costly and more importantly, seldom provide results of practical importance. Nonetheless, the subject of vacuity has been a topic of research, so the interested reader may wish to consult [13, 24, 47], for instance.

An alternative way to define nonvacuous execution of an assertion that is suitable for simulation is to define it recursively for each operator. This is how the SystemVerilog LRM defines nonvacuity. The rules first define nonvacuous attempt evaluation for weak and strong sequences, and then for all property operators in terms of the evaluation of their operands.

Exercises

10.1. Rewrite Example 10.7 using a chain of **if-else** property operators. How could you verify that the two forms are equivalent?

10.2. Which properties are valid (i.e., always true)? Explain.

(a) `((s_eventually p1) and (s_eventually p2)) iff`
`s_eventually (p1 and p2)`
(b) `((s_eventually p1) or (s_eventually p2)) iff`
`s_eventually (p1 or p2)`
(c) `((always p1) implies (always p2)) iff`
`always (p1 implies p2)`
(d) `1'b1 until 1'b0`
(e) `1'b1 s_until 1'b0`

10.3. Write a property never_p that states that its argument property p is never true.

10.4. Write a property `next_ev_p(logic b, property p)` that evaluates to true if and only if its argument p holds at the next occurrence of the Boolean expression b. If there are not enough clock ticks for b to evaluate to true, then the property should fail.

10.5. Write a property `next_ev_a_p(logic b, int m, n, property p)` that generalizes the property from Exercise 10.4 in such a way that p must hold in the range [m:n] of occurrences of b after the start of evaluation of next_ev_a_p. The occurrences of b need not be consecutive. For example, if the range is [2:4], then p must hold from the second to the fourth occurrence of b after the start of evaluation of next_ev_a_p. If there are not enough clock ticks to cover the range of occurrences of b, the property should evaluate false.

10.6. Modify the property from Exercise 10.5 to form property next_ev_e_p(**logic** b, **int** m, n, **property** p) such that p is required to hold at least once during the range [m:n] of occurrences of b.

Chapter 11
Advanced Sequences

Poe's saying that a long poem is a sequence of short ones is perfectly just.

— John Drinkwater

In Chap. 6, we covered basic sequence operators, such as delays, consecutive repetition and disjunction. In this chapter, we learn about the remaining sequence operators. Although these remaining operators do not add any additional expressive power to the language, they are very convenient to use, and make assertions more readable and concise. We also consider sequence methods and discuss using sequences as events.

In examples throughout this chapter, we assume that a **`default clocking`** is defined, and thus omit the clock in assertions unless there is a need to emphasize a specific clock usage. We assume that `e` is a Boolean, `r` and `s` are sequences, and `p` is a property.

11.1 Sequence Operators

The available sequence operators grouped by their precedence from highest to lowest are listed in Table 11.1. For convenience, we also list here the operators covered in Chap. 6.

11.1.1 Throughout

Sometimes it is necessary to make sure that a Boolean condition holds throughout the whole sequence. For this purpose, it is possible to use the sequence operator

E. Cerny et al., *SVA: The Power of Assertions in SystemVerilog*,
DOI 10.1007/978-3-319-07139-8_11

Table 11.1 Sequence operators

Operator	Associativity
`[*...]`	–
`[*]`	–
`[+]`	–
`[->...]`	–
`[=...]`	–
`##`	Left
`throughout`	Right
`within`	Left
`intersect`	Left
`and`	Left
`or`	Left

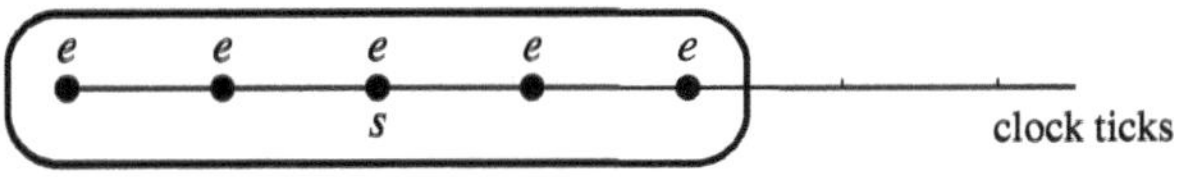

Fig. 11.1
`e` **`throughout`** `s`

`throughout`. The sequence `e` **`throughout`** `s`, where `e` is a Boolean expression and `s` is a sequence, has a match in clock tick t iff `s` has a match at t, and in each clock tick from the start of the evaluation of `s` until and including the match of `s`, the condition `e` is *true* (see Fig. 11.1).

Example 11.1. Write a sequence describing the following scenario: "Three consecutive enabled occurrences of `read` followed by four enabled occurrences of `write`". The occurrences of `read` and `write` are enabled if `en` is asserted.

Solution: The sequence may be implemented as

```
(read && en)[*3] ##1 (write && en)[*4]
```

Using the **`throughout`** operator, the same sequence may be rewritten in a more expressive way:

```
en throughout read[*3] ##1 write[*4]
```

□

Example 11.1 illustrates the fact[1] that **`throughout`** is a convenience operator which does not introduce any additional expressive power to the language. Yet, it greatly improves assertion readability and makes the intent clear.

Efficiency Tip. The **`throughout`** operator is efficient both in simulation and in FV.

[1] It can be proven formally [23].

Fig. 11.2 Sequence `e[->2]`

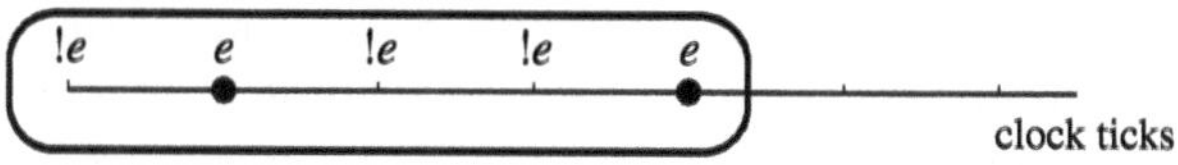

11.1.2 Goto Repetition

Motivation Example.

Example 11.2. After request `req` is serviced by `done` asserted, signal `ready` should be asserted.

Solution: We need to check `ready` in the clock tick following the clock tick when `done` became high *for the first time* after `req` is asserted:

```
a1: assert property (req ##1 !done[*] ##1 done |=> ready);
```
□

The situation when something should happen for the first time described in Example 11.2 is very common. There is a special sequence operator, a *goto repetition*, stating that the condition `e` must happen for the first time: `e[->1]`. For an arbitrary integer constant $n \geq 0$, `e[->n]`, where `e` is a Boolean, is a shortcut for `(!e[*] ## e)[*n]`. This sequence has a match when e happens for the nth time, as shown in Fig. 11.2.

Unlike the consecutive repetition described in Sect. 6.5 which can be applied to arbitrary sequences, the goto repetition may be applied only to Boolean values.

Example 11.3. Using goto repetition, the assertion from Example 11.2 may be rewritten as

```
a2: assert property (req ##1 done[->1] |=> ready);
```
□

It is also possible to specify ranges in goto repetition: `e[->m:n]`, $0 \leq m \leq n$, has a match when e happens for the mth, $m+1$st, ..., and the nth time. The upper bound of the range may also be open-ended (`$`). The formal definitions are as follows (we assume that `m` $\leq$ `n`):

```
b[->m:n] ≡ (!b[*] ##1 b)[*m:n].
b[->m:$] ≡ (!b[*] ##1 b)[*m:$].
b[->m]   ≡ (!b[*] ##1 b)[*m].
```

Example 11.4. After `start` is asserted, at each occurrence of request `req`, starting from the second and ending with the fifth one, enable `en` must be asserted.

Solution:

```
a1: assert property (start ##1 req[->2:5] |-> en);
```
□

Example 11.5. After `start` is asserted, at each occurrence of request `req`, starting from the second one, enable `en` must be asserted.

```
a1: assert property (start ##1 req[->2:$] |-> en);
```
□

Eventuality. It is possible to express eventuality using goto repetition. For example, the property `s_eventually e` is equivalent to `strong(e[->1])`. The operator `strong` is essential here, without it the property is meaningless in the assertion or assumption context as explained in Sect. 6.10.

Example 11.6. After `start_ev`, signal `next` should be asserted at least twice.

Solution: This example is similar to Example 11.5 and one can be tempted to implement this assertion as

```
a1_redundant: assert property (start_ev |=> strong(next[->2:$]));
```

However, assertion `a1_redundant` is an overkill. To check that `next` appears *at least twice*, it is sufficient to check that it appears twice:

```
a2: assert property (start_ev |=> strong(next[->2]));
```
□

Example 11.7. Event `e` must happen at least twice in the entire trace.

Solution: As explained in Example 11.6, the assertion may be written as

```
initial a1: assert property (strong(e[->2]));
```

Discussion: If we need to express that `e` should happen *exactly* twice, we should use followed-by (suffix conjunction):

```
initial a2: assert property (e[->2] #=# always !e);
```

If we need to express that `e` should happen *at most* twice, we should use suffix implication instead of suffix conjunction:

```
initial a3: assert property (e[->2] |=> always !e);
```
□

Next Occurrence.

Example 11.8. When `en` is high, property `p` must be true in the nearest clock tick when signal `e` is true.

Solution:

```
a1: assert property (en ##0 e[->1] |-> p);
```

Discussion: A possible satisfying trace is shown in Fig. 11.3.

The situation where at the next occurrence of `e` property `p` must hold is a common situation. We may define a reusable property named `next_occurrence` as follows[2]:

```
property next_occurrence(e, property p);
  e[->1] |-> p;
endproperty
```

We can now use this definition to rewrite assertion `a1` as[3]

```
a2: assert property (en |-> next_occurrence(e, p));
```

---

[2]In PSL [6], there is a property operator called *next_event* with a similar behavior.

[3]See Sect. 6.4.1 for a discussion about nested implications.

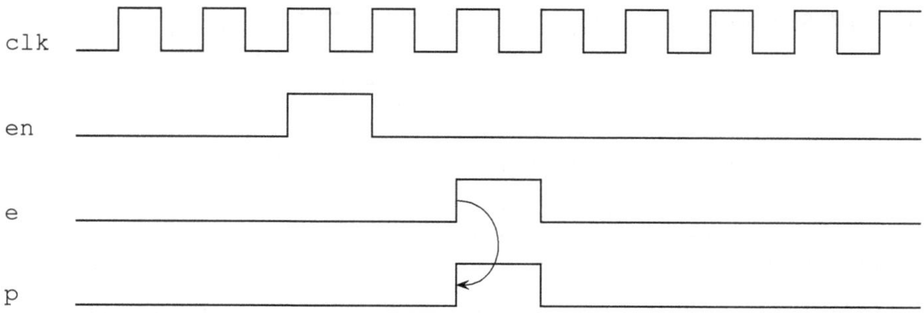

Fig. 11.3 Next occurrence

Such language extensions using property definitions can be placed in packages for reuse. □

Example 11.9. Example 11.8 shows a weak form of the `next_occurrence` property. In the strong version of this property, `e[->1]` must happen, and property `p` should hold when `e` happens. This means that we must replace the suffix implication operator used in Example 11.8 with the followed-by (suffix conjunction) operator:

```
property strong_next_occurrence(e, property p);
  e[->1] #-# p;
endproperty
```
□

Efficiency Tip. Using big factors and ranges in goto repetition is inefficient both in simulation and in formal verification. In simulation, goto repetition may be expensive if it causes long or never-ending attempts, and especially for overlapping attempts, as explained in Sect. 19.3. For example,

```
assert property (a ##1 b[->1] |=> c);
```

is efficient if `b` happens every few clock ticks; it can be extremely inefficient if `a` often happens, and `b` never happens or if its occurrences are rare. The reason is that many property evaluation attempts may be simultaneously accumulated.

11.1.3 Nonconsecutive Repetition

Motivation Example.

Example 11.10. Between the occurrences of the transmission start `start_t` and the transmission end `end_t`, exactly four packets must be sent. Each time a packet is sent, `sent` is asserted. The value of `sent` is not to be checked when `start_t` or `end_t` is asserted.

Fig. 11.4 Sequence e[=2]

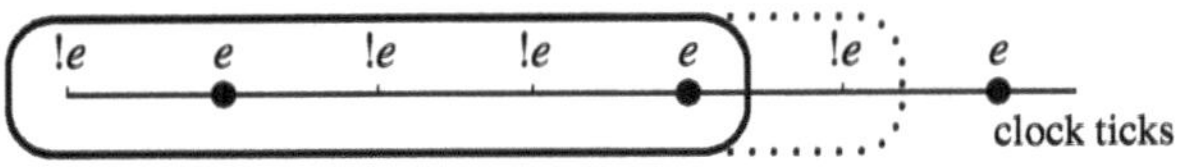

Solution: Using consecutive repetition this assertion may be written as:

```
a1: assert property(start_t |=>
  (!end_t throughout (!sent[*] ##1 sent)[*4]
  ##1 !sent[*]) ##1 end_t);
```

Using goto repetition, this assertion may be rewritten in a more compact way:

```
a2: assert property (start_t |=>
  (!end_t throughout sent[->4] ##1 !sent[*]) ##1 end_t);
```
□

The situation discussed in Example 11.10 where some Boolean must be true a predefined number of times between the match of one sequence and the beginning of another is rather common. There is a sequence operator `[=...]`, called *nonconsecutive repetition*, that designates it. More precisely, sequence `e[=n]` has a match in some clock tick if before this clock tick `e` occurs exactly `n` times, as shown in Fig. 11.4.

Like goto repetition, the nonconsecutive repetition may be applied only to Boolean (integral) values.

Example 11.11. Using nonconsecutive repetition, the assertion from Example 11.10 may be rewritten as

```
a3: assert property (
  start_t |=> (!end_t throughout sent[=4]) ##1 end_t);
```
□

It is possible to specify ranges in nonconsecutive repetition: `e[=m:n]`, $0 \le m \le n$, has a match when e is true for the mth, $m + 1$st, ..., and the nth time. The upper bound of the range may also be infinite (`$`). The formal definitions are as follows (we assume that $\mathtt{m} \le \mathtt{n}$):

```
b[=m:n] ≡ b[->m:n] ##1 !b[*].
b[=m:$] ≡ b[->m:$] ##1 !b[*].
b[=m]   ≡ b[->m] ##1 !b[*].
```

Example 11.12. During one transaction delimited by `start_t` and `end_t`, packets ranging from 2 to 4 should be sent (`sent` asserted). `sent` is not to be checked when `start_t` or `end_t` is asserted.

Solution:

```
a1 assert property (
  start_t |=> (!end_t throughout sent[=2:4]) ##1 end_t);
```
□

Example 11.13. If during one transaction less than two packets are sent (`sent` asserted), the `shortt` bit should be asserted when `end_t` is asserted. `sent` is not to be checked when `start_t` or `end_t` is asserted.

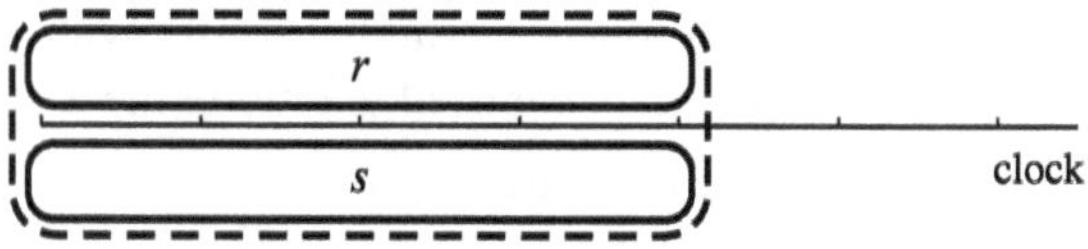

Fig. 11.5 Sequence intersection

Solution:

```
a1: assert property (
  start_t ##1 (!end_t throughout sent[=0:1]) ##1 end_t
    |-> shortt);
```
□

Efficiency Tip. Big factors and ranges in nonconsecutive repetition are inefficient both in simulation and in formal verification. In simulation, a nonconsecutive repetition may be expensive if it causes long or never-ending and overlapping attempts, as explained in Sect. 19.3.

11.1.4 Intersection

Intersection of two sequences `r` and `s` is a sequence `r` **`intersect`** `s`, which has a match when both sequences have a match simultaneously (see Fig. 11.5).

Example 11.14. A command consists of two in-order read actions and one write action. After the command is issued (`command` is asserted), the completion of the write action (`write_complete`), and the completion of the second read action (`read_complete`) should happen simultaneously.

Solution: To express simultaneous completion, we use the **`intersect`** operator:

```
a1: assert property (
  command |->
    write_complete[->1] intersect read_complete[->2]);
```
□

Example 11.15. Each transaction delimited by `start_t` and `end_t` should contain two `read` requests and three `write` requests.

Solution: In this case, we need to spot a clock tick t with the following characteristics:

- There should be exactly two read requests issued before t.
- There should be exactly three write requests issued before t.
- At clock tick t, the first occurrence of `end_t` should happen.

We can express this using the following assertion:

```
a1: assert property (start_t |->
  read[=2] intersect write[=3] intersect end_t[->1]);
```

Assertion `a1` allows the last `read` or `write` request to happen simultaneously with the end of the transaction `end_t` and also with `start_t`. If we require that no `read` or `write` request happens simultaneously with `end_t` and `start_t`, and the transactions do not overlap, we need to rewrite the assertion as

```
a2: assert property (start_t |=>
  !start_t throughout (read[=2] ##1 !read) intersect
  (write[=3] ##1 !write) intersect end_t[->1]);
```

Exercise 11.8 introduces yet another interpretation of this assertion: `read` and `write` request may happen at any time, but when they happen at the end of the transaction they are not counted as part of the current transaction. □

`intersect` operator does not really add more expressive power to the language, but it makes formulas exponentially more concise [23]. One can get a feeling why it is so by attempting to rewrite the assertion in Example 11.14 without **`intersect`** (see Exercise 11.7). Moreover, Example 11.15 is even more convincing. To rewrite assertions `a1` and `a2` without **`intersect`**, it is necessary to explicitly list all possible combinations of `read` and `write`:

```
read[=2] ##[0:1] write[=3] or read[=1] ##[0:1] write[=2] ##[0:1]
  read[=2] or ...
```

Efficiency Tip. Operator **`intersect`** may be expensive in FV. The reason is that most FV engines in their internal representation eliminate the **`intersect`** operator and generate all possible combinations of events. The greater the number of these combinations, the more expensive it is for FV. However, it should be understood that **`intersect`** does not introduce inefficiency by itself (though some FV engines may process **`intersect`** less efficiently than the equivalent explicit representation), but it allows concise coding of complex sequences. Therefore, if **`intersect`** is really needed, there is no choice but to use it. However, if it is possible to write a more specific assertion instead, eliminating the need for **`intersect`**, this should be the first choice. For instance, if it is known in Example 11.15 that all write requests precede the read requests, `read[=2]` **`intersect`** `write[=3]` should be replaced by `write[=3] ##1 read[=2]`.

In simulation, the overhead of **`intersect`** is acceptable except when sequence is compiled into an automaton in which case memory blow-up may occur during the automaton construction (see Sect. 19.3). This blow-up occurs in FV too.

Limiting Sequence Size. Sometimes it is desirable to keep only those sequence matches that occur during some number of first clock ticks. This can be done using the idiom `s` **`intersect`** `1[*1:n]`, where `s` is a sequence, and `n` is an integer constant. Sequence `1[*1:n]` has matches in clock ticks $1,\dots,$ `n`; therefore, only matches of `s` that happen during the first `n` clock ticks are retained, while all others are ignored.

This method may be used to truncate the antecedent sequences to boost simulation performance, as shown in Example 11.16. Note, however, that it is inefficient in FV.

Example 11.16. If acknowledgment `ack` is received after `req`, `ready` should be asserted simultaneously with the acknowledgment receipt.

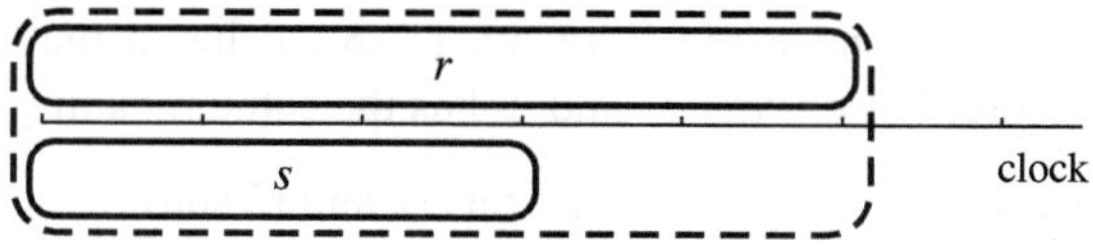

Fig. 11.6 Sequence conjunction

Solution:

```
a1: assert property (req ##1 ack[->1] |-> ready);
```

Discussion: As we discussed in Sect. 11.1.2, this assertion may not be efficient in simulation when `req` persists until `ack`, and `ack` is sent long time after `req` is asserted or if `ack` is not sent at all. It is possible to modify assertion `a1` to limit the time of waiting for `ack` to some predefined number of clock ticks, for example, 10:

```
a2: assert property (
  (req ##1 ack[->1]) intersect 1[*1:10] |-> ready);
```

If `ack` is asserted within 10 clock ticks from `req` issue, assertion `a2` behaves like assertion `a1`; otherwise, the assertion evaluation attempt is ignored. If it is known that `ack` always arrives within 10 clock ticks, assertion `a2` is more efficient in simulation than assertion `a1` for most industrial simulators.

Note that `1[*1:10]` could be replaced by `1[*2:10]` since the antecedent takes at least two clock ticks. □

Efficiency Tip. Antecedent truncation is not efficient for formal verification. As mentioned in Sect. 20.6, the efficiency requirements for assertion checking in emulation are usually aligned with the requirements for FV rather than with simulation. Therefore, in emulation it is also better to avoid antecedent truncation, though it is less critical than in FV.

Throughout. The sequence operator **`throughout`** introduced in Sect. 11.1.1 is a special case of **`intersect`**: `e` **`throughout`** `s` is equivalent to `e[*]` **`intersect`** `s`. Since **`throughout`** does not introduce new event combinations, it is efficient both in simulation and in FV.

11.1.5 Sequence Conjunction

Conjunction of two sequences `r` and `s` is a sequence `r` **`and`** `s`. It has a match in clock tick t iff one of the sequences r and s has a match in that clock tick, and the other sequence has a match in some clock tick $t_1 \leq t$, as illustrated in Fig. 11.6

Sequence conjunction belongs to the **`intersect`** family, and `r` **`and`** `s` is a shortcut for `r ##1 1[*]` **`intersect`** `s` **`or`** `r` **`intersect`** `s ##1 1[*]`.

If both a and b are Boolean, a **and** b has a match iff both a and b are true. Therefore, in this case a **and** b has the same meaning as a && b.[4]

Example 11.17. Two transactions t1 and t2 start at the same time when start_t is asserted. When both transactions complete, ready should be asserted. Transaction completion is signaled by end_t1 and end_t2, respectively.

Solution:

```
a1: assert property (start_t ##1 (end_t1[->1] and end_t2[->1])
     |-> ready);
```

The antecedent matches when the longer of the two transactions completes. □

Sequence Conjunction Versus Property Conjunction. As the sequence and property conjunctions have exactly the same syntax, how to distinguish between a conjunction of two sequential properties and a sequence conjunction promoted to a property? For example, in property en |-> r **and** s, where r and s are sequences, should r and s be interpreted as sequences with **and** as a sequence conjunction, or should they be interpreted as properties with **and** as a property conjunction? The answer is the same as in the case of disjunction (see Sect. 6.6): if the conjunction arguments are sequences, it is a sequence conjunction. Note, however, that essentially the result may be interpreted either way, both definitions agree (this is also the case with disjunction), provided that the resulting sequence is promoted a property.

Efficiency Tip. Sequence conjunction has a reasonable overhead in simulation, but in FV it may be expensive when it defines many different combinations of events (this is similar to the situation with **intersect**, see Sect. 11.1.4). However, top-level conjunction in a sequence promoted to property is not expensive.

Example 11.18. In property en |-> r **and** s, the sequence conjunction in the consequent is not expensive in FV, since r **and** s is promoted to a property, and **and** is its top-level conjunction. This is because in that case sequence **and** can be converted to a property **and** with equivalent behavior.

In property en |-> (r **and** s)##1 a, **and** is not a top-level conjunction promoted to property, hence the conjunction may be expensive.

In property (r **and** s)|-> p, the conjunction is in the antecedent of a suffix implication, the antecedent is never promoted to a property, hence the conjunction may be expensive. □

[4]Except when a or b has a match item, see Chap. 16.

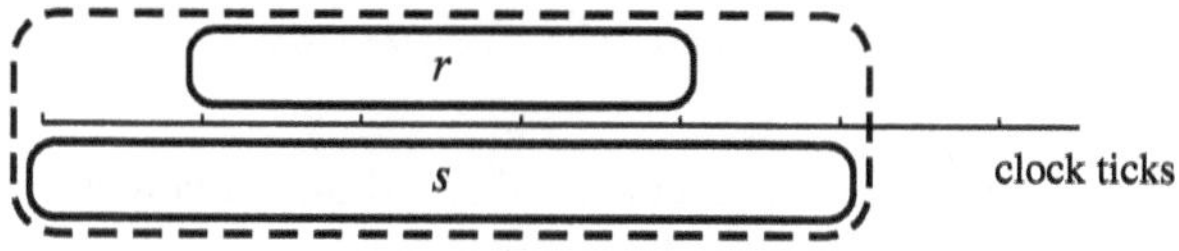

Fig. 11.7 Sequence containment

11.1.6 Sequence Containment

The operator `r` **`within`** `s` checks that sequence `r` is contained within sequence `s`. More precisely, `r` **`within`** `s` has a match in clock tick t iff `s` begins in clock tick t_0 and has a match in clock tick t, and sequence `r` beginning in clock tick t_2 has a match in clock tick t_3, such that $t_0 \leq t_1 \leq t_2 \leq t$, as shown in Fig. 11.7.

The sequence containment operator belongs to the **`intersect`** family, and `r` **`within`** `s` is a shortcut for `1[*] ##1 r ##1 1[*]` **`intersect`** `s`. Notice that `r` may have more than one match while `s` is evaluated.

Example 11.19. There should be at least one `read` request between two `write` requests.

Solution:

```
a1: assert property (write |=> (read ##1 1) within write[->1]);
```

Discussion: Specifying `read ##1 1` and not just `read` is important when the case of `read` appearing together with the second `write` should be excluded. □

Example 11.20. Two consecutive `write` requests cannot appear within a transaction delimited by `start_t` and `end_t` (including transaction delimiters).

Solution:

```
assert property (
  start_t |-> not strong(write[*2] within end_t[->1]));
```

Discussion: We need to specify the **`strong`** qualifier here to keep the property weak because the negation of a weak operator is strong (Chap. 10). Without it, this assertion would check among other things that each transaction eventually completes. This is usually not part of the assertion intent. In addition, checking the eventuality would impose heavy burden on FV tools. □

Efficiency Tip. **`within`** operator has similar overhead as **`intersect`** (Sect. 11.1.4).

Example 11.21. The sequence `(a ##1 1)` **`within`** `b[->1]` is relatively efficient in FV since it is equivalent to

```
!b[*] ##1 a ##0 !b[+] ##1 b
```

In contrast, sequence `a[->2]` **`within`** `b[->2]` is more expensive since it introduces many combinations of `a` and `b`: first `a` then `b` then `a` then `b`; first two `a` and then two `b`; `a` and `b` happening simultaneously, etc. □

11.1.7 First Match of a Sequence

It is sometimes convenient to discard all sequence matches but the first one. This can be achieved using the operator **first_match**. Sequence **first_match**(s) has a match in clock tick t iff sequence s has a match in clock tick t, and it has no match in any clock tick $\tau < t$.

Example 11.22. If a and b have values 1 in clock ticks 0–4 then sequence a[*1:2] ##1 b[*2:3] has matches in clock ticks 2, 3, and 4. In contrast, sequence **first_match**(a[*1:2] ##1 b[*2:3]) has only one match in clock tick 2. □

Example 11.23. When request req is issued and thereafter the first data chunk is received as identified by data bit asserted, acknowledgment ack should be sent.

Solution:

```
a1: assert property(first_match(req ##[+] data) |-> ack);
```

Discussion: Here, the **first_match** operator guarantees that the acknowledgment is only sent when data is asserted for the first time. The same assertion rewritten using the goto repetition should be more efficient:

```
a2: assert property(req ##1 data[->1] |-> ack);
```

□

Example 11.24. Let us modify the requirement of Example 11.23: Acknowledgment ack should be sent in response to a request, when two data chunks are received in consecutive clock ticks for the first time. The first solution can be easily adapted as follows:

```
a3: assert property(first_match(req ##[+] data[*2]) |-> ack);
```

The second solution can also be modified in the following way:

```
a4: assert property(req ##1 data[->1] ##1 data |-> ack);
```

Again, assertion a4 is likely to be more efficient than a3. □

Trailing first_match in Sequential Properties. Trailing **first_match** in sequential properties, both weak and strong is redundant and can be omitted. For simplicity we explain this statement for strong sequential properties. Property **strong**(r ##1 s) is true iff there exists a match of sequence r ##1 s. This match exists if there exists a match of r followed by a match of s, but this is equivalent to the statement that there exists a match of r followed by the *first* match of s. Therefore, **strong**(r ##1 s) is equivalent to **strong**(r ##1 **first_match**(s)).

Similarly, it can be shown that a trailing **first_match** in the outermost **and** or **or** branch in a sequential property is redundant. For example, the sequential property r1 ##1 **first_match**(s) **or** r2 is equivalent to r1 ##1 s **or** r2.

Example 11.25. The property a |-> b ##1 **first_match**(c[*] ##1 d) is equivalent to a |-> b ##1 c[*] ##1 d. The following properties are not equivalent (see Exercise 11.13):

1. `a |-> b ##1 first_match(c[*] ##1 d)##1 e` and
   `a |-> b ##1 c[*] ##1 d ##1 e.`
2. `a ##1 first_match(b[*] ##1 c)|-> d` and
   `a ##1 b[*] ##1 c |-> d.`

□

Efficiency Tip. In general, other than as top-level operator in a sequential property, **`first_match`** is expensive both in simulation and in FV, and should be avoided whenever possible.

11.2 Sequence Methods

There are two methods that may be applied to sequences: `triggered`[5] and `matched`. The difference between sequence operators and sequence methods is that the operators build a new sequence from its operands, whereas the sequence methods return a Boolean value. The syntax of sequence methods is also different: it has the form *sequence_instance.method_name*.

Even though the sequence method `matched` is targeted for multiclock sequences (see Chap. 12), we mention it here to explain its behavior in the simple case of a single clock.

Since sequence methods return a Boolean value (`1'b1` or `1'b0`), they may be used where Boolean expression are used. Nevertheless, several limitations apply. For example, sequence methods cannot be used in sampled value functions (Sect. 7.2).

Sequence methods are evaluated in the Observed region and their sampled value Appendix A is defined to be the current value.

11.2.1 Triggered: Detecting End Point of a Sequence

Given a sequence `s`, the method `s.triggered` returns *true* in clock tick t if there exists a clock tick $t_1 \leq t$ such that when `s` starts evaluating in clock tick t_1, it has a match in clock tick t. `s` must be an instance of a named sequence. The last clocking event of `s` must be the same as the clocking event of the context where `s.triggered` is used. A reference to a formal argument may be used instead of the named sequence, but after actual argument substitution a legal reference to a sequence must result.

[5] In SystemVerilog Standard 2005 [3], there was also the sequence method `ended`, but according to SystemVerilog Standard 2009 [5] `ended` is deprecated, and `triggered` should be used instead.

Example 11.26. The following code is legal:

```
logic a, b, c, d;
// ...
sequence s;
  @(posedge clk) a ##[1:3] b;
endsequence : s
sequence t(x);
  @(posedge clk) x[*5];
endsequence : t
property p(sequence r, untyped y);
  a |-> r(y).triggered;
endproperty : p
a1: assert property (@(posedge clk) c |-> s.triggered);
a2: assert property (@(posedge clk) c |-> t(d).triggered);
a3: assert property (@(posedge clk) p(t, d));
```

`s` and `t(d)` are instances of the named sequences. In `r(y)`, `r` is a formal argument and so is `y`. □

Example 11.27. The following code is illegal:

```
logic a, b, c;
// ...
a1_illegal: assert property (@(posedge clk)
  c |-> (a ##[1:3] b).triggered);
```

`triggered` method is applied to sequence expression `a ##[1:3] b` which is neither a named sequence instance nor a formal argument. □

Example 11.28. Consider the following code:

```
logic a, b, c, d;
sequence s1;
  @(posedge clk) a ##1 b;
endsequence : s1
sequence s2;
  @(posedge clk1) a ##1 b;
endsequence : s2
sequence s3;
  @(posedge clk1) a ##1 @(posedge clk1) b ##1 @(posedge clk) c;
endsequence : s3
a1: assert property (@(posedge clk) d |-> s1.triggered);
a2: assert property (@(posedge clk) d |-> s2.triggered);
a3: assert property (@(posedge clk) d |-> s3.triggered);
```

The clock for the sequences is governed by the clock flow rules described in Sect. 12.2.4.1. Each sequence has its clock explicitly specified, so the clock from the assertion where they are used does not flow in to the sequence. □

Example 11.29. Table 11.2 contains a trace of `a` and `b`, and the values of `s.triggered`, where `s` is defined as

```
sequence s;
  @(posedge clk) a[*1:2] ##1 b[*1:2];
endsequence : s
```

Table 11.2 Sequence end points

Clock tick	0	1	2	3	4	5	6	7	8	9	10	11	12
`a`	1	1	0	1	1	1	0	1	0	0	0	1	0
`b`	0	0	1	1	0	0	0	1	1	1	0	0	1
`s.triggrered`	0	0	1	1	0	0	0	0	1	1	0	0	1

Note that `s.triggered` returns the same value as `r.triggered`, where `r` is defined as

```
sequence r;
  @(posedge clk) a ##1 b[*1:2];
endsequence : r
```

Why? □

Example 11.30. Between request `req` and acknowledgment `ack` (inclusive), `busy` should be asserted. When both `req`, `ack`, and `busy` are Boolean, the desired assertion is

```
a1: assert property (req |-> busy until_with ack);
```

How should we modify this assertion to allow `req` and `ack` be arbitrary sequences? For instance, these sequences could be defined as follows:

```
sequence req;
  start_req ##1 end_req;
endsequence : req
sequence ack;
  enable ##[1:10] end_ack;
endsequence
```

To make the assertion work, in this case we need to assure that `busy` is asserted starting from the *last* clock tick of `req` until the *last* clock tick of `ack`. There is no need to make any changes in `a1` related to `req` handling, as the overlapping implication checks the consequent from the last clock tick of its antecedent. However, `ack` handling requires a modification because otherwise the assertion will check that `busy` is asserted only until the *first* clock tick of `ack`. The required modification is simple: we need to replace `ack` with `ack.triggered`:

```
a2: assert property (req |-> busy until_with ack.triggered);
```

Now suppose that we wish to take reset `rst` into account:

```
a3: assert property (disable iff (rst)
      req |-> busy until_with ack.triggered);
```

It should be noted that **`disable iff`** does not affect the behavior of the `triggered` method, and if sequence `ack` started before `req` was asserted and before `rst` was deactivated, it will not be aborted. However, in this case it would be natural to ignore `ack`. The easiest way to do this is to modify sequence `ack` to take `rst` into account as follows:

Table 11.3 Past temporal operators

Operator	Description
sofar p	Holds in clock tick t iff p holds in all clock ticks $t_1 \leq t$ (Fig. 11.8).
once p	Holds in clock tick t iff p holds in some clock tick $t_1 \leq t$ (Fig. 11.9).
p since q	Holds in clock tick t iff q holds in some clock tick $t_1 \leq t$, and p holds in all clock ticks t_2, $t_1 < t_2 \leq t$ (Fig. 11.10).
p backto q	A weak version of *p since q*: if q has not happened yet, p should hold in all clock ticks $t_1 \leq t$.
previously p	Holds in clock tick t iff $t \neq 0$, and p holds in clock tick $t-1$ (Fig. 11.11).
before p	Holds in clock tick t iff either $t = 0$ or p holds in clock tick $t - 1$. Thus, the only difference between *previously p* and *before p* is that *previously p* is false in clock tick 0, whereas *before p* is true.

```
sequence ack;
  !rst throughout enable ##[1:10] end_ack;
endsequence
```

One problem remains: the reset specified by the **`disable iff`** operator is asynchronous (see Chap. 13), while the behavior of the **`throughout`** operator is synchronous: it is checked only at clock ticks. Usually, this difference is not important. When it is, the sequence to which the method `triggered` is applied as well as the assertion should be controlled by the global clock if it exists. □

To make the sequence to which the method `triggered` is applied sensitive to the assertion disabling condition `reset`, include `!reset` **`throughout`** as the top operator in the sequence. The sequence is disabled synchronously with the clock ticks of the sequence, however.

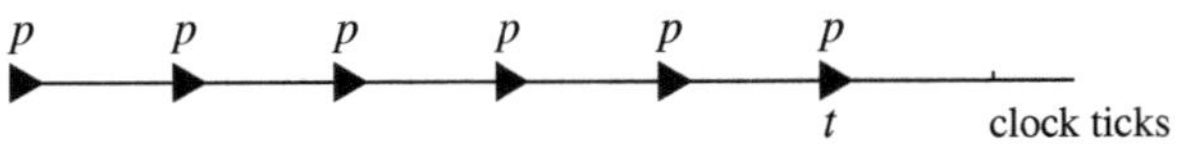

Fig. 11.8 *sofar* property

Fig. 11.9 *once* property

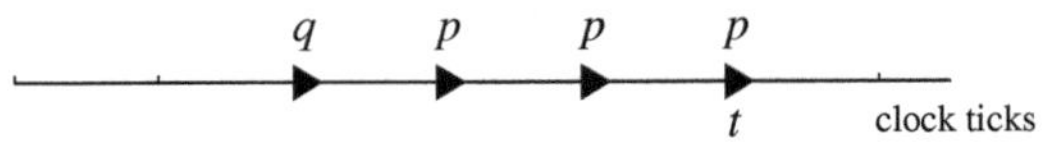

Fig. 11.10 *since* property

Fig. 11.11 *previously* property

11.2.1.1 Past Temporal Operators

In SVA, all property operators are directed to the future. For example, **always** p means that from the current clock tick on the property p holds, **s_eventually** p means that p happens in the current or in a future clock tick, p **until** q means that p holds from now on until q happens, **nexttime** p means, that p holds in the next clock tick, etc.

There exists also past temporal logic [34] in which the operators are directed or operate on past values of signals. We illustrate such past operators in Table 11.3.

It may be shown that past temporal operators do not add any additional expressive power to the language. Everything that may be expressed with future and past temporal operators may be expressed with future temporal operators only. Using past temporal operators just makes the formulas more succinct.

The past temporal operators are *not* part of SVA; nevertheless, it may be of interest to find an appropriate work-around. In the special case when the operands are Boolean, and not arbitrary property expressions, it is natural to use sequence method triggered as shown in Fig. 11.12. The figure shows property definitions that implement several past temporal operators applied to Boolean values.[6] Besides the operators listed in Table 11.3, the figure contains the operator *first* that is not part of a past temporal logic, but is closely related to it. This operator does not have arguments; it returns *true* in clock tick 0, and *false* in all other clock ticks.

Figure 11.12 illustrates several points that we mentioned earlier:

- Sequence methods can be applied to sequence instances only; therefore, we had to create auxiliary sequences, such as seq_previously, seq_once, etc.
- Sequence methods return Boolean values *true* (1Šb1) and *false* (1Šb0). It is correct to use Boolean negation !seq_once(!e).triggered in the implementation of sofar instead of property negation **not** seq_once(!e).triggered. This also applies to the implementation of first.
- It is legal to apply sequence methods to a sequence with arguments, such as seq_previously(e).triggered.

Figure 11.12 does not contain implementations of operators *before* and *backto*. Their implementation is left to the reader as exercise (Exercises 11.15 and 11.16)

[6] For some tools, it may be more efficient to implement seq_not_first using modeling code to set a flag after clock tick 0.

```
sequence seq_previously(e);
  e ##1 1;
endsequence : seq_previously
property previously(e);
  seq_previously(e).triggered;
endproperty : previously

sequence seq_not_first;
  ##1 1;
endsequence : seq_not_first
property first;
  !seq_not_first.triggered;
endproperty : first

sequence seq_once(e);
  e ##[*] 1;
endsequence : seq_once
property once(e);
  seq_once(e).triggered;
endproperty : once

property sofar(e);
  !seq_once(!e).triggered;
endproperty : sofar

sequence seq_since(e1, e2);
  e2 ##1 e1[*];
endsequence : seq_since
property since(e1, e2);
  seq_since(e1, e2).triggered;
endproperty : since
```

Fig. 11.12 Past temporal properties

11.2.1.2 Triggered Outside Assertions in RTL

Using `triggered` method is not limited to assertions, it is also legal in procedural code. For example, if `a` and `b` are wires, and `s` is a sequence, the following statement is legal:

```
assign a = s.triggered || b;
```

Even though this code looks innocent, it is dangerous in modules and interfaces to use `a` in a concurrent assertion. Consider the following assertion:

```
a1: assert property (@(posedge clk) a);
```

The semantics of the assign statement evaluation with respect to a change in the value of `s.triggered` is not well defined. We know that the value of `s.triggered` persists only until the end of the simulation tick, and at the beginning of the next simulation tick the new value of `s.triggered` is reset to 0. However, it is not

```
1  module m(input logic clk, a, b, ...);
2  logic c;
3  sequence ab;
4    @(posedge clk) a ##1 b;
5  endsequence
6  always @(posedge clk) begin
7    c <= ab.triggered;
8    //...;
9  end
10 //...
11 endmodule : m
```

Fig. 11.13 `triggered` method in procedural code

clear whether these changes in value cause events to trigger the assign statement evaluations.

If the intent of the assignment statement is to provide a name for the expression `s.triggered || b` to be reused in other contexts, for example, in concurrent assertions, then a better solution is to use **`let`** which does not perform any assignment, but only associates a name with an expression:

```
let a = s.triggered || b;
```

Then assertion `a1` is internally expanded into

```
a1: assert property (@(posedge clk) s.triggered || b);
```

which yields the desired result.

Using the `triggered` method in procedural code in modules or interfaces is often meaningless, as shown in the following example.

Example 11.31. In the code in Fig. 11.13, the value assigned to `c` in Line 7 is always 0 because the value of `ab.triggered` is evaluated in the Observed region, while the nonblocking assignment is evaluated in the NBA region, before the `ab.triggered` has been evaluated. Since the sequence match event is not in the sensitivity list of the event control in Line 6, there will be no reevaluation of the **`always`** procedure in the same clock tick. □

`triggered` method may be safely used in procedural code in programs and checkers because the procedures are executed after the Observed region. The value of `triggered` has the correct value at that moment.

Do not use the `triggered` sequence method in procedural code in modules and interfaces outside concurrent assertions. `triggered` method may be safely used in procedural code in programs and checkers.

Using the `triggered` sequence method in **`let`** definition with subsequent **`let`** instantiation in concurrent assertions is safe.

11.2.2 The `triggered` Method in Checkers

In Sect. 11.2.1.2 we stated that using `triggered` sequence method on the right-hand side of a nonblocking assignment in modules is meaningless because the nonblocking assignment is performed in the NBA region, while the `triggered` method is evaluated only later in the Observed region.

This is not true for checkers: the `triggered` sequence method may be safely used in checker variable assignments, because checker variables are assigned in the Re-NBA queue of the Reactive region, after the evaluation of `triggered`.

> It is safe to use the `triggered` sequence method in checker variable assignments.

Example 11.32. Disable checking assertions in a checker between the match points of sequences `stop_check` and `start_check`.

Solution:

```
checker toggle_check(sequence stop_check, start_check,
  event clk = $inferred_clock);
  bit rst = 1;
  default clocking @clk; endclocking
  default disable iff rst;

  always @clk begin
    if (stop_check.triggered) rst <= 1'b1;
    else if (start_check.triggered) rst <= 1'b0;
  end

  a1: assert property(...);
  // More assertions here ...
endchecker : toggle_check
```

Discussion: We defined a checker variable `rst` which is 1 between the match point of sequence `stop_check` until the match point of sequence `start_check`. For example, if the stop sequence consists of two consecutive `stop` signals, and the start sequence consists of two consecutive `start` signals, then the actual checker arguments would be `stop[*2]` and `start[*2]`. □

11.2.3 Matched

The method `matched` returns the status of sequence termination but in the strictly subsequent clock tick. Therefore, in the case of single clock, `s.matched` is equivalent to `s1.triggered`, where `s1` is defined as follows:

```
sequence s1;
  s ##1 1;
endsequence
```

Internally, `matched` stores the result of its source sequence match until the arrival of the *next* destination clock tick after the match. There are no limitations imposed on the last clocking event of sequence `s`.

There is not much sense in using `matched` in singly clocked assertions, but it is helpful in multiply clocked assertions where `triggered` might not be directly used. We explore this topic in depth in Chap. 12.

Unlike `s.triggered`, `s.matched` cannot be used outside sequences.

11.3 Sequence as Events

Sequences may be used as events, both edge sensitive and level sensitive.

11.3.1 Sequence Event Control

The syntax of the sequence event control is `@sequence_instance`; where `@sequence_instance` is an instance of a named sequence. Arbitrary sequence expressions cannot be specified with the event control. For example, `@(@(posedge clk)a ##1 b)` is illegal, as `@(posedge clk)a ##1 b` is a sequence expression, but not an instance.

Sequence event control is especially convenient in programs (see Sect. 2.5) to specify starting points of testbench execution. For example, we may wish to start some testbench activity when the initialization of the subsystem is complete (`ready` is asserted), and when the system has entered the normal power mode (`pmode == normal` when the last power mode switch `pswitch` occurred). The resulting code is shown in Fig. 11.14.

When program `test` starts executing, it immediately blocks until sequence `operational` matches. At that point, the program execution is resumed, task `run_test` is called, followed by `Test started` message display.

Note the following:

- Since sequence `operational` is stand-alone, and not part of a property or of an assertion, it cannot infer its clocking event from an enclosing property or assertion. Therefore, the sequence clocking event `@(posedge clk)` must be explicitly specified unless the sequence belongs to the scope of a clocking block, or default clocking is specified.
- The sequence event occurs if *any* attempt of the sequence has a match. This is why the sequence matching starts only from moments when `pswitch && pmode == normal` is true, and not from the first clock tick.

```
task run_test;
  // ...
endtask : run_test

typedef enum bit [1:0] {NONE, NORMAL, SLOW, FAST} Mode;

program test(input logic clk, ready, pswitch, Mode pmode, ...);
  // ...
  sequence operational;
    @(posedge clk) pswitch && pmode == NORMAL ##1 !pswitch[*]
        ##0 ready;
  endsequence : operational

  initial begin
    @operational;
    run_test;
    $display("Test started");
    // ...
  end
endprogram : test
```

Fig. 11.14 Using sequence event control in programs

- The sequence may have many matches, but only the first one will have an effect in this specific case: when the sequence matches for the first time, the task `run_test` is executed, and control flow never returns to this point.

Sequence Event Controls in Modules and Interfaces. Sequence event control usage is not limited to programs only, it can also be used in modules and interfaces. Using sequence controls in modules may sometimes greatly simplify the design—it is similar to the situation with assertions, between their SVA specification and their manual implementation in procedural form. Unfortunately, industrial synthesis tools do not support this construct; therefore, the value of using it in modules and interfaces is only in nonsynthesizable (more abstract) models. However, using sequence event controls may be convenient in checkers.

11.3.2 Level-Sensitive Sequence Control

The execution of procedural code may be delayed until some event happens, using **wait** statement. The **wait** statement can also be used with the sequence method `triggered` (but not `matched`!).

For example, the code in Fig. 11.14 can be rewritten using the **wait** statement as shown in Fig. 11.15.

```
task run_test;
  // ...
endtask : run_test

typedef enum bit [1:0] {NONE, NORMAL, SLOW, FAST} Mode;

program test(input logic clk, ready, pswitch, Mode pmode, ...);
  // ...
  sequence operational;
    @(posedge clk) pswitch && pmode == NORMAL ##1 !pswitch[*]
        ##0 ready;
  endsequence : operational

  initial begin
    wait(operational.triggered);
    run_test;
    $display("Test started");
    // ...
  end
endprogram : test
```

Fig. 11.15 Using level-sensitive sequence event control in programs

When control flow reaches the **`wait`** statement, process execution is suspended until `operational.triggered` becomes true.

`wait` statement with sequence `triggered` is more verbose than sequence event control, but it may be more convenient when awaiting a Boolean expression containing a sequence `triggered` method to become true.

Example 11.33. Suspend the execution of the code until the command is complete (sequence `command_compete` has a match) or until an interrupt `intr` is asserted.

Solution:

```
wait(command_complete.triggered || intr);
```
□

11.3.3 Event Semantics of Sequence Match

Like assertions, sequences are also evaluated in the Observed region in much the same way. Sequence match points play a significant role in assertions as well as in other descriptions. Two situations are of particular importance. One is when a sequence match is used as a subexpression subsidiary to the assertion evaluation, whose result is used as a Boolean value true or false in the enclosing expression. The other situation is when it appears as an event control, likely used as a trigger for a process or a delayed statement.

Let us consider the first situation:

Example 11.34. A sequence match used as a subexpression

```
default clocking @(posedge clk);
endclocking
sequence req;
  start_req ##1 end_req;
endsequence: req
sequence end_ack;
  empty_slot[->1] ##1 no_conflict;
endsequence
sequence ack;
  enable ##[1:10] end_ack.triggered;
endsequence
a2: assert property (req |-> busy until_with ack.triggered);
```
□

In this example, sequences `req`, `end_ack` and `ack`, and assertion `a2` are all triggered by the same clock. The match point of sequence `ack` is at work in the consequent property of the assertion. Hence, there is an obvious contingency of sequence `ack` to the property, creating an order for evaluating the sequence prior to the assertion evaluation in the same time slot in which the clock occurs. Furthermore, sequence `ack` is dependent upon the match point of `end_ack`. The final order of evaluation is: `end_ack`, `ack`, `a2`.

Note that sequence `req` is simply substituted in `a2` and becomes part of the assertion property antecedent. The match point of `req` is not explicitly needed.

We should clarify here that the order of sequence evaluation is statically determined at compile time. Also, any cyclic dependency between sequences is semantically illegal, ensuring that a proper order of evaluation can always be found. This order remains constant throughout the evaluation.

A sequence match is determined in the Observed region.

The value of method `triggered` on a sequence is set in the Observed region as soon as it is evaluated. Thereupon, the method can be safely used in the Observed region and in the Reactive region. At the end of the time slot, the value of `triggered` is reset to false, and remains false until a match point of the sequence is detected in some future Observed region. Clearly, its use in the Active region is not advised as discussed earlier.

But, sequence match points do contribute to an important application in the Active region. This brings us to the second situation: using a sequence match point as event control in procedures.

Example 11.35. A sequence match used as a process trigger or a delay statement

```
module normal;
sequence begin_mode;
```

```
  @(posedge clk) (pswitch && pmode == normal ##1 !pswitch[*]) ##0
       ready;
endsequence : begin_mode
initial begin: I1
  @begin_mode setup(mode);
  $display("Mode started");
end
//...
endmodule
```
□

In this example, sequence `begin_mode` is used as event control in initial procedure `I1`. The event control subjects its execution to obtaining a match point of sequence `begin_mode`. Accordingly, the initial process gets suspended as it must wait until the Observed region where the sequence is evaluated. If the match point is attained there, the simulation control makes its way back to the Active region again following the normal course via the Reactive region. Now in the Active region, the initial procedure resumes to execute the `$display` statement.

Exercises

11.1. Two consecutive requests should be separated with four **`ack_wait`**, one acknowledgement (`ack`), and two `idle` cycles (in this order).

11.2. What is the difference between the assertion from Example 11.6 and the assertions below?

```
a1: assert property (start_ev |=> strong(next[->2:$] ##1 end_ev
    ));
a2: assert property (start_ev ##1 next[->2:$] |=> end_ev);
```

11.3. In this exercise, `e` is a Boolean expression, `m` and `n` are integer constants, `p` is a property. Implement the following PSL operators as SVA property definitions:

(a) *next_event e [n] p*
This property holds in the current clock tick iff `e` does not hold at least `n` times, starting at the current clock tick, or `p` holds at the `n`th occurrence of `e`.

(b) *next_event! e [n] p*
This property holds in the current clock tick iff `e` holds at least `n` times, starting at the current clock tick, and `p` holds at the `n`th occurrence of `e`.

(c) *next_event_a e [m:n] p*
This property holds in the current clock tick iff `p` holds at the `m`th through `n`th occurrences, inclusive, of `e`, starting at the current clock tick. If there are less than `n` occurrences of `e` then `p` holds on all of them, starting from the `m`th occurrence.

(d) *next_event_a! e [m:n] p*
This property holds in the current clock tick iff `e` holds at least `n` times, starting at the current clock tick, and `p` holds at the `m`th through `n`th occurrences, inclusive, of `e`.

(e) *next_event_e e [m:n] p*
This property holds in the current clock tick iff `p` holds at some occurrence of `e` among its `m`th through `n`th occurrences, inclusive, starting at the current clock tick, or there are less than `n` occurrences of `e`.

(f) *next_event_e! e [m:n] p*
This property holds in the current clock tick iff `p` holds at some occurrence of `e` among its `m`th through `n`th occurrences, inclusive, starting at the current clock tick, or there are less than `n` occurrences of `e`.

11.4. Write the following assertions:

(a) For each request, an acknowledgment should be sent from 2 to 5 times
(b) For each request, an acknowledgment should be sent 2 or 5 times

11.5. Implement the following assertion: In the transaction delimited by the `start_t` and `end_t`, there should be an even number of actions (`act`).

11.6. What is the meaning of the following assertions?

```
a1: assert property (a[=1]);
a2: assert property (strong(a[=1]));
a3: assert property (a[=1] |-> b);
```

Discuss their efficiency in simulation and in FV.

11.7. Implement the assertion from Example 11.14 without using operators from the **`intersect`** family.

11.8. Implement the assertion from Example 11.15 to ignore read and write requests happening simultaneously with the end of the transaction `end_t`.

11.9. During a memory transaction (delimited by `start_t` and `end_t`), the snoop request (`snoop_req`), and the credit update message (`credit_update`) must be sent in any order. The transaction must terminate in the clock tick when the later of these two events happen.

11.10. Which of the following assertions is equivalent to the assertion from Example 11.19?

```
a1: assert property(
      write |=> read[=1] intersect write[->1]);
a2: assert property(
      write |=> read[=1] ##1 1 intersect write[->1]);
a3: assert property(
      write |=> read[=1:$] intersect write[->1]);
a4: assert property(
      write |=> read[=1:$] ##1 1 intersect write[->1]);
a5: assert property(
      write |=> ##[*] read ##1 1[*] intersect write[->1]);
```

```
a6: assert property(
      write |=> ##[*] read ##1 1[+] intersect write[->1]);
a7: assert property(
      write |=> ##[+] read ##1 1[*] intersect write[->1]);
a8: assert property(
      write |=> ##[+] read ##1 1[+] intersect write[->1]);
```

11.11. Implement the following assertion: if a transaction contains at least two read requests, there should be at least three clock tick delay between it and the following transaction. Assume that the transactions cannot overlap.

11.12. Write the following assertions:

(a) After `start` is asserted, at least one of the following events should happen: two consecutive `read` or two consecutive `write`. When the first such event happens (e.g., if two consecutive `write` happen first then in the clock tick of the second `write`), `done` must be asserted.
(b) When after `start` is asserted, one of the following events happens for the first time: two consecutive `read` or two consecutive `write`, `done` must be asserted (e.g., if two consecutive `write` happen first then in the clock tick of the second `write`).
(c) After `start` is asserted either `read` or `write` request should arrive, and in the clock tick when the first of them arrives, `done` must be asserted.
(d) In the clock tick when one of the `read` or `write` requests arrives for the first time, `done` must be asserted.

11.13. Show that the following properties are not equivalent (Example 11.25):

(a) `a |-> b ##1 first_match(c[*] ##1 d)##1 e` and
`a |-> b ##1 c[*] ##1 d ##1 e`
(b) `a ##1 first_match(b[*] ##1 c)|-> d` and `a ##1 b[*] ##1 c |-> d`

11.14. Read transaction (delimited by `start_read` and `end_read`) may only be issued if a write transaction (delimited by `srart_write` and `end_write`) finished beforehand.

11.15. Implement the past temporal operator *before* for a Boolean argument (see Sect 11.2.1.1).

11.16. Implement the past temporal operator *backto* for Boolean arguments (see Sect 11.2.1.1).

11.17. Modify the sequence event control example in Sect. 11.3.1 to take the reset `rst` into account in the sequence.

Chapter 12
Clocks

The only reason for time is so that everything doesn't happen at once.

— Albert Einstein

Concurrent assertions are fundamentally temporal in nature. The evaluation of a concurrent assertion, and of its constituent subsequences and subproperties, evolves over time in a discrete way. *Clocks*, or, more precisely, *clocking events*, are the constructs that define the discretization of time. Clocking events form a rich subset of general SystemVerilog events. These include familiar edge events, such as **`posedge`** `clk`, declared events, as well as more general and complex event expressions.

In SVA, clocking events are declarations with scopes, not operators. As such, they do not have strengths. Rather, they determine the measurement of time and the times of evaluation of operators and expressions within their scopes. Within the scope of a clocking event, one unit, or *cycle*, of discrete time is measured from one occurrence of the clocking event to the next. Occurrences of a clocking event are also called *ticks of the clock*, or simply *clock ticks*. The intervals between successive clock ticks can be regular or irregular in length, but in all cases they constitute one unit of discrete time. This reckoning gives meaning to operators such as `##1`, `|=>`, and **`nexttime`**, whose semantics involves the notion of the "next point in time". The *leading clocking event* of a concurrent assertion, together with the context in which the assertion is written, determine when evaluation attempts of the assertion begin.

This chapter discusses the mechanics of declaring clocks and the rules that determine their scoping, including default clocking. Many concurrent assertions of practical interest are *singly clocked*, meaning that all parts of the assertion are

E. Cerny et al., *SVA: The Power of Assertions in SystemVerilog*,
DOI 10.1007/978-3-319-07139-8_12

governed by a single clocking event. Other concurrent assertions have portions that fall under the scopes of two or more clocking events and are called *multiply clocked.*[1]

12.1 Overview of Clocks

This section gives an intuitive overview of clocks based on examples.

Clocks, or, more precisely, *clocking events*, define the discretization of time within concurrent assertions. They are declarations with scopes, not operators. A clocking event for a concurrent assertion must not occur more than once per time step. If a clocking event occurs more than once in a time step, then the LRM defines no behavior for the assertion and a tool may issue an error. It is the assertion writer's responsibility to ensure that the clocks for assertions are "glitch-free" in this sense. Within the scope of a clocking event, one unit, or *cycle*, of discrete time is measured from one occurrence of the clocking event (i.e., clock tick) to the next. Since the clocking event must be glitch-free, one unit of discrete time is at least one time step. Clocks also influence the sampling of values within a concurrent assertion. In most cases the *sampled value* of a variable or net is the value from the Preponed region of a time step (see Sect. 4.4.3 and Appendix A). If a reference to a variable or net appears within the scope of a clocking event, then the time steps relevant for sampling are those in which the clocking event occurs.

12.1.1 Specifying Clocks

This section describes various ways to specify clocks for concurrent assertions, illustrated by singly clocked examples.

The concurrent assertion in Fig. 12.1 specifies an explicit clocking event control,[2] `@(posedge clk)`, in Line 3. The scope of the clocking event is the entire property expression `a |=> b`, so `a1` is an example of a *singly clocked* assertion. This implies that the references to `a` and `b` in Line 4 are evaluated using sampled values in time steps in which `posedge clk` occurs. The one cycle delay specified by `|=>` from its antecedent to its consequent is from one occurrence of `posedge clk` to the next. Finally, the *leading clocking event* of `a1` is also `posedge clk`. Since `a1` is a static concurrent assertion (i.e., one that is not in a procedural context), a new evaluation attempt of `a1` begins at each occurrence of `posedge clk`.

---

[1]The SystemVerilog LRM avoids the phrase "multiply clocked", using instead the grammatically suspicious adjectives "multiclock" and "multiclocked".

[2]We use the phrase "clocking event control" to emphasize the inclusion of the `@` symbol in the syntax.

```
module simple_clock(input logic clk, a, b);
   a1:  assert property(
      @(posedge clk)
      a |=> b
   );
endmodule
```

Fig. 12.1 Module with explicitly clocked concurrent assertion

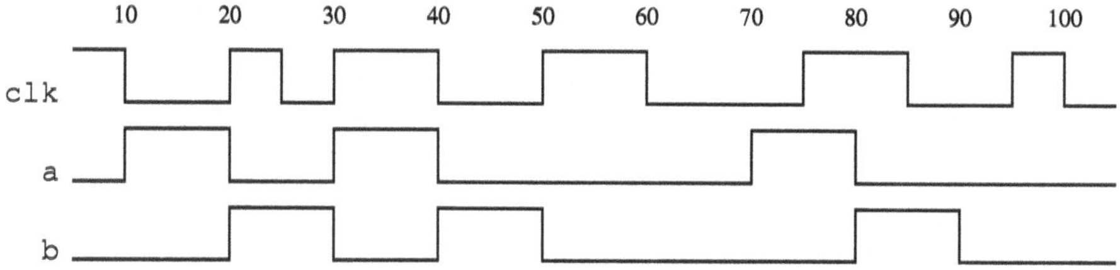

Fig. 12.2 Waveform for module `simple_clock`

Figure 12.2 shows a possible waveform for the signals in this example. The event **posedge** `clk` occurs at times 20, 30, 50, 75, and 95. The intervals from time 20 to 30, from time 30 to 50, etc. each constitute one cycle of discretized time, although their lengths vary in units of simulation time. This illustrates the fact that the clock ticks do not have to be regular as measured against simulation time. From among these clock ticks, the sampled value of `a` is `1'b1` only at times 20 and 75. Therefore, the evaluation attempts of `a1` that begin at times 30, 50, and 95 succeed vacuously. The attempt that begins at time 20 checks the sampled value of `b` at time 30 and finds it to be `1'b1`, so this attempt succeeds. The attempt that begins at time 75 checks the sampled value of `b` at time 95 and finds it to be `1'b0`, so this attempt fails. The fact that the sampled value of `b` is `1'b1` at the clock tick at time 50 is irrelevant for the evaluation of `a1`.

Clocking events form a rich subset of general SystemVerilog events and are specified using a limited event control syntax:

clocking_event ::=
  `@` identifier
| `@` `(` event_expression `)`

The event expression in the form `@(` event_expression `)` can be a familiar edge event, such as **posedge** `clk`, the name of a declared event, or a more general, and possibly complex, event expression. The identifier in the form "`@` identifier" can be the name of a declared event or the name of a clocking block, the latter specifying the clocking event of the referenced clocking block.

Figure 12.3 illustrates some of the different forms of the clocking event control syntax in a module with various explicit concurrent assertion clocking declarations. Assertion `a2` is clocked by an explicit edge event expression and is similar to `a1` in the previous example. Assertion `a3` is clocked by named event `e`. The **always**

```
1 module various_clocks(input logic clk1, clk2, a, b, c);
2    event e;
3    always @(negedge clk1) ->e;
4    clocking PCLK2 @(posedge clk2); endclocking
5    a2:  assert property(@(negedge clk1) a |=> b);
6    a3:  assert property(@e a[*2] |=> c);
7    a4:  assert property(@PCLK2 a |=> b);
8 endmodule
```

Fig. 12.3 Module with various explicit concurrent assertion clocking declarations

```
1 module various_clocks(input logic clk1, clk2, a, b, c);
2    default clocking NCLK1 @(negedge clk1); endclocking
3    clocking PCLK2 @(posedge clk2);
4    endclocking a2:  assert property(a |=> b);
5    a3:  assert property(a[*2] |=> c);
6    a4:  assert property(@PCLK2 a |=> b);
7 endmodule
```

Fig. 12.4 Module with default clocking

procedure in Line 3 triggers `e` at every occurrence of **negedge** `clk1`, so `a3` behaves equivalently to the following variant:

```
a3_v2:  assert property(@(negedge clk1) a[*2] |=> c);
```

The clocking event for `a4` is `PCLK2`, the name of the clocking block in Line 4, and so `a4` is clocked by the event **posedge** `clk2` of that clocking block.[3]

Often, many assertions within a module, interface, program, or checker share the same clock. In this situation, it is convenient to specify a *default* clocking block (see also Sect. 2.3.2). The module `various_clocks` from Fig. 12.3 is recoded in an equivalent way in Fig. 12.4. Line 2 declares `NCLK1` to be the default clocking for the module, with event **negedge** `clk1`. As a result, explicit clocking events can be omitted on assertions `a2` and `a3`: the default is understood to apply to them. The default can be overridden by an explicit clocking event, as in `a4`.

Default clocking applies to concurrent assertions, not to sequence and property declarations. This convention allows a sequence or property to be declared without clocks and to inherit the clock from the context in which it is instantiated. Figure 12.5 shows another equivalent encoding of module `various_clocks` illustrating this style. This encoding also dispenses with the declaration of clocking block `PCLK2`, putting the event expression **posedge** `clk2` directly in `a4`.

Clocks may also be declared within named sequence or property declarations. A clock in the declaration of a named sequence or property declaration applies to all

[3] According to the SystemVerilog LRM, Clause 14.13, the event associated with the clocking block name, here `PCLK2`, occurs in the Observed region, while **posedge** `clk2` occurs in the Active region. In singly clocked assertions that use sampled values from the Preponed region, there is no observable difference in behavior between the two forms.

```
module various_clocks(input logic clk1, clk2, a, b, c);
   default clocking NCLK1 @(negedge clk1); endclocking
   property p1; a |=> b; endproperty
   a2:  assert property(p1);
   a3:  assert property(a[*2] |=> c);
   a4:  assert property(@(posedge clk2) p1);
endmodule
```

Fig. 12.5 Module with default clocking and unclocked property declaration

```
module various_clocks(input logic clk1, clk2, a, b, c);
   default clocking NCLK1 @(negedge clk1); endclocking
   property p1; @(posedge clk2) a |=> b; endproperty
   a2:  assert property(a |=> b);
   a3:  assert property(a[*2] |=> c);
   a4:  assert property(p1);
endmodule
```

Fig. 12.6 Module with default clocking and clocked property declaration

```
module various_clocks(input logic clk1, clk2, a, b, c);
   default clocking NCLK1 @(negedge clk1); endclocking
   property p1(event ev = $inferred_clock);
      @ev a |=> b;
   endproperty
   a2:  assert property(p1);
   a3:  assert property(a[*2] |=> c);
   a4:  assert property(p1(.ev(posedge clk2)));
endmodule
```

Fig. 12.7 Module with default clocking and clocked property declaration with **`event`** argument

instances of the named sequence or property, overriding any clock from the context in which it is instantiated. Figure 12.6 illustrates this style with another equivalent encoding of module `various_clocks`. The named property `p1` can no longer be instantiated in `a2` because the clocking event **`posedge`** `clk2` in the declaration of `p1` would override the default clocking in the instance.

Another approach for specifying clocks in declarations of named sequences and properties is to pass the clocking events as arguments. This can be done with untyped arguments or with arguments of type **`event`**. The system function `$inferred_clock` can be used as a default actual argument. If no actual argument is passed to the formal in an instance, then `$inferred_clock` as default actual specifies that the clock from the instantiation context applies. Figure 12.7 illustrates this usage with a final equivalent coding of module `various_clocks`. `a2` instantiates `p1` without an actual, so `$inferred_clock` specifies that the default clocking applies. `a4` instantiates `p1` and passes the event expression **`posedge`** `clk2` to the event argument `ev`.

```
module multiply_clocked(input logic clk1, clk2, a, b, c);
   a5:  assert property(
      @(posedge clk1) a |=> @(posedge clk2) b
   );
endmodule
```

Fig. 12.8 Module with multiply clocked assertion

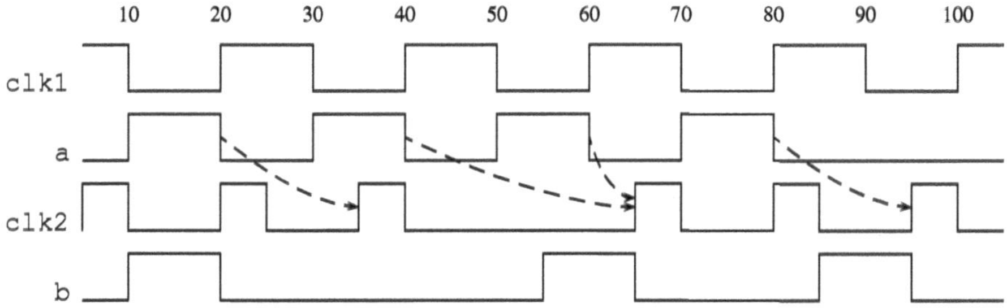

Fig. 12.9 Waveform for assertion `a5` in module `multiply_clocked`

12.1.2 Multiple Clocks

All examples of concurrent assertions in the preceding section were singly clocked. Figure 12.8 gives an example of a multiply clocked assertion. The leading clocking event for `a5` is **posedge** `clk1`, and the reference to `a` is within the scope of this clock. The reference to `b` is within the scope of **posedge** `clk2`.

The multiply clocked behavior of `a5` merits further explanation. Since the leading clock is **posedge** `clk1` and since `a5` is a static concurrent assertion, a new evaluation attempt of `a5` begins at each tick of **posedge** `clk1`. Let t_0 be such a time. If the sampled value of `a` at t_0 is `1'b0` (or `1'bx` or `1'bz`), then the attempt succeeds vacuously. Otherwise, the antecedent of `|=>` is matched at t_0, and evaluation of the consequent is obligated. Since the consequent is governed by a different clock, `|=>` does *not* specify advancement to the next tick of **posedge** `clk1` after t_0. Rather, `|=>` serves as a *synchronizer* between the two clocks. It specifies that evaluation of the consequent begin at the nearest tick of **posedge** `clk2` that is strictly after t_0. In that time step, the sampled value of `b` is checked, and if it is `1'b1`, then the overall attempt succeeds. Otherwise, the overall attempt fails.

Figure 12.9 shows a possible waveform for `a5`. An attempt of `a5` begins at every tick of **posedge** `clk1`. The sampled value of `a` is `1'b1` at times 20, 40, 60, and 80, so in each of these time steps the antecedent of `|=>` matches. The attempt beginning at time 20 looks for the nearest tick of **posedge** `clk2` that is strictly later than time 20. This clock tick is at time 35, where the sampled value of `b` is found to be `1'b0`, and so the attempt fails. Because the operator `|=>` has been used, it does not matter that **posedge** `clk2` occurs at time 20 since this occurrence is not strictly later. The attempts beginning at times 40 and 60 both find the nearest strictly future tick of **posedge** clk2 at time 65, where the sampled value of b is 1'b1, and so these attempts succeed. The attempt beginning at time 80 finds the nearest strictly future tick of **posedge** clk2 at time 95, where the sampled value of b is again 1'b1, and so it succeeds. As before, the fact that **posedge** clk2 occurs at time 80 is irrelevant because the operator |=> has been used.

The overlapping suffix implication |-> can also be used as a synchronizer between different clocks. Suppose that the following assertion is added to the module in Fig. 12.8:

```
a6:  assert property(
   @(posedge clk1) a |-> @(posedge clk2) b
);
```

If a match of the antecedent of |-> ends at time t_0, then the consequent will be checked at the nearest time greater than *or equal to* t_0 in which **posedge** clk2 occurs. Comparing with the waveform in Fig. 12.9, the attempt of a6 beginning at time 20 succeeds because **posedge** clk2 occurs at time 20 and the sampled value of b at that time is 1'b1. The attempts of a6 beginning at times 40 and 60 behave the same as the corresponding attempts of a5. Finally, the attempt of a6 beginning at time 80 fails because there is a tick of **posedge** clk2 at this time and the sampled value of b is 1'b0.

The operators ##1 and ##0 can be used as synchronizers between different clocks in sequences. The timing associated with ##1 as a synchronizer is the same as that of |=>. Here is an example:

```
sequence s1;
   @(posedge clk1) a[*2] ##1 @(posedge clk2) b;
endsequence
```

Referring again to Fig. 12.9, s1 matches over the intervals from times 20 to 65, 40 to 65, and 60 to 95. The following variant replaces ##1 with ##0:

```
sequence s2;
   @(posedge clk1) a[*2] ##0 @(posedge clk2) b;
endsequence
```

The timing associated with ##0 as a synchronizer is the same as that of |->. Therefore, the intervals in Fig. 12.9 over which s2 matches are from times 20 to 65 and 40 to 65. There is no match of s2 beginning at time 60 because the subsequence a[*2] matches ending at time 80 and there is an tick of **posedge** clk2 at this time with the sampled value of b equal to 1'b0.

##1 and ##0 are the only sequence operators that can be used as synchronizers between different clocks. For all other sequence operators, the operands must be singly clocked sequences clocked by the same clocking event. Here is an example of an illegal sequence declaration:

```
sequence s3_illegal;
   @(posedge clk) a[*2] within @(negedge clk) b[->1];
endsequence
```

This sequence is illegal because the operands of **within** are clocked by different clocking events.

```
1 property p2(event ev1, ev2, bit a, b);
2     @(ev1) a[*2] |=>
3     (
4         !a
5         and
6         @(ev2) b
7     );
8 endproperty
```

Fig. 12.10 Logical operator joining differently clocked properties

In addition to `|=>` and `|->`, the property operators `#=#` and `#-#` can be used as synchronizers between different clocks. The timing of `#=#` (resp., `#-#`) as a synchronizer is the same as that of `|=>` (resp., `|->`). **`if-else`** and **`case`** can also serve as synchronizers, with timing the same as that of `|->`. Here is an example:

```
a7:  assert property(
       @(ev1)
       if (a)
          @(ev2) b[*2]
       else
          @(ev3) c
    );
```

In `a7`, the scope of `ev1` is the condition `a` of the **`if-else`**. Assuming that `a` is of type **`bit`**, the following variant behaves equivalently to `a7` and explains how the timing of **`if-else`** as a synchronizer is the same as that of `|->`:

```
1 a7_v2:  assert property(
2     @(ev1)
3     ( a |-> @(ev2) b[*2])
4     and
5     (!a |-> @(ev3) c)
6 );
```

This encoding also illustrates some of the *clock flow* rules of clock scoping. The scope of `ev1` distributes to the two operands of **`and`** and flows into the parenthesized subproperties in Lines 3 and 5. As a result, `a` in Line 3 and `!a` in Line 5 are both under the scope of `ev1`.[4]

The LTL operators **`nexttime`**, **`always`**, **`s_eventually`**, **`until`**, and their variants can also be used as synchronizers. When this is done, the time advance specified by the LTL operator is determined by the *incoming* clock, not by the leading clock or clocks of the operands. See Sect. 12.2.5.1 for more details.

The logical property operators **`and`**, **`or`**, **`iff`**, and **`implies`** can be used to join differently clocked properties. Figure 12.10 gives an example. The scope of `ev1` includes the antecedent `a[*2]` of `|=>` and the operand `!a` of **`and`** in Line 4 **`and`** joins

[4]If `a` were of type **`logic`**, to preserve the semantics of **`if-else`** in the presence of X/Z values the negation of `a` in `a7_v2` would have to be written as `(!bit'(a) |-> @(ev3)c)`.

one operand clocked by `ev1` and one clocked by `ev2`. The antecedent of `|=>` matches if the sampled value of `a` is `1'b1` at two successive ticks of `ev1`. Suppose that such a match ends at t_0. Line 4 of the consequent says that at the nearest tick of `ev1` strictly after t_0, the sampled value of `a` must be `1'b0`. Line 6 of the consequent says that at the nearest tick of `ev2` strictly after t_0, the sampled value of `b` must be `1'b1`.

A multiply clocked concurrent assertion is required to have a unique leading clock. If the concurrent assertion is static (i.e., not within a procedural context), then it has implicit "always" semantics and the leading clock determines when new evaluation attempts of the assertion begin. If the concurrent assertion is procedural, then the leading clock determines when evaluation begins of an attempt that has matured from the procedural assertion queue (see Sect. 14.5). The following example is illegal:

```
a8_illegal:  assert property(
   @(ev1) a or @(ev2) b
);
```

The assertion is illegal because it has two leading clocks, `ev1` and `ev2`. There is a simple way to get around this restriction as shown in the modified example:

```
a8_illegal:  assert property(
   @(ev1 or ev2) 1'b1 |-> @(ev1) a or @(ev2) b
);
```

The restriction and the workaround forces the author of the assertion to be aware of the events that trigger the evaluation attempts. Note that the detection of simultaneous clocking events is guaranteed in simulators even if the events occur in different scheduling regions.

The remainder of this section discusses a few abstract, but practically motivated, examples of multiply clocked properties.

Example 12.1. Write an assertion to check that the time from any occurrence of `EV1` to the nearest strictly subsequent occurrence of `EV2` is at least `MINTIME` simulation time steps.

Solution: This encoding uses local variables (see Chap. 15) to capture timestamps for comparisons. Because of the use of timestamps, it is not so well suited for formal verification.

```
1 property p_mintime(event ev1, ev2, time mintime);
2    time basetime;
3    @(ev1) (1'b1, basetime = $time)
4    |=> @(ev2) $time >= basetime + mintime;
5 endproperty
6 a_EV1_EV2_MINTIME:  assert property(
7    p_mintime(.ev1(EV1), .ev2(EV2), .mintime(MINTIME))
8 );
```

The expectation is that `MINTIME` is a constant, perhaps a parameter, that has been coordinated with the simulation timescale. Line 2 declares the local variable `basetime`. Line 3 specifies that when `ev1` occurs, the value of `$time` is stored in `basetime`. According to Line 4, at the nearest strictly subsequent occurrence of `ev2`, the value of `$time` must be at least the sum of `basetime` and `mintime`. □

Example 12.2. Write an assertion to check that after an occurrence of event `ev_start`, event `ev_wait` cannot occur any earlier than the time step of the first occurrence of event `ev_enable`.

Solution: This solution assumes that in any time step these events will occur before the Observed region.

```
1  sequence s_ev(event ev);
2     @(ev) 1'b1
3     ##0 @(ev_enable or ev_wait) 1'b1;
4  endsequence
5  a_order:  assert property(
6     @(ev_start) 1'b1
7     |=> @(ev_enable or ev_wait) (
8        s_ev(ev_wait).triggered
9        ->
10       s_ev(ev_enable).triggered
11    )
12 );
```

The basic idea of this solution is as follows. If `ev_start` occurs, then advance to the nearest strictly subsequent occurrence of either `ev_enable` or `ev_wait`. In that time step, if `ev_wait` has occurred, then `ev_enable` must also have occurred.

In Line 6, the antecedent of `|=>` matches at an occurrence of `ev_start`. The consequent is clocked by the compound event expression "`ev_enable or ev_wait`", so it advances to the nearest strictly subsequent occurrence of either `ev_enable` or `ev_wait`. Lines 8 through 10 use the Boolean implication `->` to encode the check that if `ev_wait` has occurred in the current time step, then `ev_enable` must also have occurred in the current time step. The job of `s_ev` is to detect whether its event formal argument `ev` occurs. The detection is accomplished by applying sequence method `triggered` to instances of `s_ev` in Lines 8 and 10.

In Line 2, `s_ev` begins a match at an occurrence of its event formal argument `ev`. Line 3 is counterintuitive. It addresses the following restriction on the use of `triggered`: the ending clock of a sequence instance to which `triggered` is applied must be identical to the clock governing the context in which the application of `triggered` appears. In Lines 8 and 10, `triggered` is applied in a context clocked by `ev_enable or ev_wait`, so Line 3 ensures that `s_ev` ends on this clock. Line 3 does not actually cause any time advance for matches of the instances of `s_ev` in Lines 8 and 10. The reason is that the actual event arguments in these instances are `ev_wait` and `ev_enable`. If one of these events occurs in a time step, then a fortiori the compound event "`ev_enable or ev_wait`" occurs in that time step. □

12.2 Further Details of Clocks

This section delves into further details of specifying clocks, their scoping, and the use of multiple clocks.

12.2.1 Preponed Value Sampling

In general, references to variables and nets that appear in a concurrent assertion use *sampled values*, i.e., in most cases the values from the Preponed region of the time step (see Appendix A). The following are exceptions for this rule:

- Disable condition of **`disable iff`**.
- Clocking event expressions.
- Actual arguments passed to **`ref`** or **`const ref`** arguments[5] of subroutines attached to sequences.
- Assertion action blocks.

References in the other contexts above always use current values unless they appear within the system function `$sampled`. Since subroutines attached to sequences and action blocks execute in the Reactive region, this means that references in the last two contexts use Reactive region values.

If Preponed value sampling applies to a reference to a variable or net and the reference appears within the scope of a clocking event, then the time steps in which the reference is evaluated are those in which the clocking event occurs. References in the abort condition of an asynchronous abort (**`accept_on`** or **`reject_on`**) use sampled values, but the abort condition is not governed by a clock.[6]

As a simple example, consider the following:

```
1 a_strange_clk:  assert property(
2     @(posedge clk)
3     clk
4 ) else $error("FAIL:  clk=%b", clk);
```

For simplicity, assume that `clk` is of type **`bit`** and that it changes value at most once in any time step. The reference to `clk` within the clocking event in Line 2 uses the current value, while the reference in Line 3 uses the sampled value. In a time step in which **`posedge`** `clk` occurs, the sampled value will always be `1'b0`. Therefore, at each such clock tick, `a_strange_clk` will fail and the action block in Line 4 will execute. The reference to `clk` after the control string in the display statement

[5] **`ref`** and **`const ref`** both specify that the actual argument is passed by reference. External changes to the actual argument are visible to the subroutine. A **`ref`** argument can also be modified by the subroutine, while a **`const ref`** argument cannot.

[6] Technically, one could say that disable conditions and asynchronous abort conditions are not within the scope of any clock.

uses the current, Reactive region value. Since there was a tick of **posedge** `clk` in the current time step, prior to the Observed region, the value of `clk` in the Reactive region is `1'b1`. Therefore, there is a mismatch between the value in Line 3 that causes the assertion failure and the value written by the display statement. Changing the action block error call to

```
$error("FAIL:  clk=%b", $sampled(clk));
```

fixes the mismatch. The assertion remains counterintuitive, though, because of the relationship between Lines 2 and 3. Care must always be taken when interpreting assertions that reference the same variable in contexts where sampling is and is not used.

There are the following kinds of references within concurrent assertions that use current values as the sampled values:

- Local variables.
- **const** cast expressions or automatic variables in a concurrent assertion within procedural code.
- Free checker variables.

References within a **const** cast expression or to an automatic variable in a concurrent assertion within procedural code resolve to the values that existed when the assertion was placed in the procedural assertion queue. See Sect. 14.2 for more details. References to free checker variables use current values, which may reflect the result of randomization in the current time step. See 23.1 for more details.

Preponed value sampling in a concurrent assertion is not allowed to conflict with other sampling defined within a clocking block. In particular, if a clocking block input variable is referenced in a concurrent assertion, then the variable must be sampled with `#1step`[7] in the clocking block and the clock governing the reference in the assertion (if there is one) must be the same as that of the clocking block.

12.2.2 Default Clocking

A clocking block may be declared as the default within a given module, interface, program, or checker. There are two syntactic forms for specifying default clocking. One prepends the keyword **default** to the clocking block declaration, as in the following example:

```
default clocking PCLK @(posedge clk); endclocking
```

The other uses a separate top-level declaration to specify the default clocking, as in the following example:

```
clocking PCLK @(posedge clk); endclocking
...
default clocking PCLK;
```

[7] `#1step` specifies Preponed value sampling.

```
module various_clocks(input logic clk1, clk2, a, b, c);
   default clocking NCLK1 @(negedge clk1); endclocking
   clocking PCLK2 @(posedge clk2); endclocking
   a2:  assert property(a |=> b);
   a3:  assert property(a[*2] |=> c);
   module nested_1;
      default clocking PCLK2;
      a4:  assert property(a |=> b);
   endmodule
   module nested_2;
      a9:  assert property(a |=> @PCLK2 c);
   endmodule
endmodule
```

Fig. 12.11 Module and nested modules with default clocking declarations

The scope of a default clocking declaration is the entire module, interface, program, or checker in which it appears, including nested declarations of modules, interfaces, or checkers. A nested module, interface, or checker may, however, have its own default clocking declaration, which overrides a default from outside. The scope of a default clocking declaration does not descend into instances of modules, interfaces, or checkers.

The clocking event of a default clocking block will be called the *default clocking event*, or simply the *default clock*. Throughout the scope of a default clocking declaration, the default clock applies to all cycle delay operations whose clocking is not otherwise specified. In particular, the default clock serves as the leading clock of all concurrent assertions whose leading clock is not explicitly specified or otherwise inferred (see Sect. 14.2 for rules of inference of clocks for procedural concurrent assertions).

If the default clock is the leading clock for a concurrent assertion, then the rules of clock flow (see Sect. 12.2.4.1) determine what subsequent parts of the concurrent assertion are also clocked by the default clock. If the concurrent assertion has no explicit or otherwise inferred clocking event, then it is singly clocked by the default clock.

Figure 12.11 shows a variant of module `various_clocks` that illustrates these ideas. `a2` and `a3` are singly clocked by the default clock `NCLK1`. Module `nested_1` has its own default clock, so `a4` is singly clocked by `PCLK2`. Module `nested_2` inherits the default clock `NCLK1` from its parent. `a9` is multiply clocked. Its leading clock is the default clock `NCLK1`, but its consequent is clocked by `PCLK2`.

A default clock does not apply to declarations of sequences or properties. Clock scoping rules apply to instances of such declarations in the context of instantiation.

12.2.3 Restrictions in Multiply Clocked Sequences

The only synchronizers allowed in sequences are `##0` and `##1`. Therefore, the general form of a multiply clocked sequence s is

$$s = r_0 \;\#\#n_1\; r_1 \;\#\#n_2\; \cdots \;\#\#n_k\; r_k$$

where $k \geq 1$, each r_i, $0 \leq i \leq k$, is a singly clocked sequence, and each n_i, $1 \leq i \leq k$, is either `0` or `1`. We may assume that in this form r_i and r_{i+1} are differently clocked for each $0 \leq i < k$, since otherwise they could be combined into a larger singly clocked subsequence. Then the sequences r_i are the maximal singly clocked subsequences of s.

SVA requires that the maximal singly clocked subsequences of a multiply clocked sequence not admit empty match. This guarantees that each r_i has unambiguous starting and ending clock ticks for any match, thereby ensuring that there is a well-defined leading clock and that the clock changes for each synchronizer are well defined.

For example, the following multiply clocked sequence is illegal:

```
@(ev1) a[*] ##1 @(ev2) b
```

The maximal singly clocked subsequences are `@(ev1)a[*]` and `@(ev2)b`, and the former admits empty match. In this situation, we cannot be sure whether the leading clock is `ev1` or `ev2`, and this ambiguity is disallowed. Changing the sequence to

```
@(ev1) a[+] ##1 @(ev2) b
```

makes it legal. The first maximal singly clocked subsequence is now `@(ev1)a[+]`, which does not admit empty match. Now we can be sure that the leading clock of the sequence is `ev1` and that `##1` synchronizes between a tick of `ev1` and a tick of `ev2`.

12.2.4 Scoping of Clocks

In SVA, clocking event controls are declarations with scopes, not operators. As such, clocks have no strengths. The scoping rules for clocks have been designed to allow the scopes of clocks to extend intuitively through the structure of the assertions, sequences, and properties and to reduce the need for parenthesizing and repetition of clocking event controls.

There are actually two sets of rules that work together to determine how each part of a concurrent assertion is clocked. The first set of rules, called *clock flow* rules, defines how scopes of clocks descend from the outside in, beginning with the default clock or inferred clock, if it exists. A basic idea in clock flow is that the scope of a clocking event cannot flow across another clocking event control. In other words, the inner clock blocks and takes precedence over a clock flowing in from above or outside. The second set of rules defines the *set of semantic leading clocks* for a sequence or property expression. These rules work from the inside out and capture the notions that inner clocks take precedence over outer clocks and that some expressions require an incoming clock.

In the presence of instances of named sequences and properties, both sets of rules are understood to apply to the assertions, sequence expressions, and property expressions that result from expanding the instances.[8]

12.2.4.1 Clock Flow

The clock flow rules define how scopes of clocks descend from the outside in. They are intended to be intuitive and to reduce the need for parenthesizing and repetition of clocking event controls. Reliance on the rules can always be reduced by adding explicit clocking event controls, although doing so in sequence or property declarations may reduce their reusability. Here are the clock flow rules:

CF1 A default clock flows to every concurrent assertion in its scope.

CF2 An inferred clock for a procedural context (see Sect. 14.2) overrides a default clock and flows to every concurrent assertion in its scope.

CF3 Clock *c* flows out of `@(`*c*`)`.

CF4 A clock *c* that flows to a clocking event control `@(`*d*`)` does not flow across the clocking event control. Instead, the scope of *c* is halted by `@(`*d*`)`, and the scope of *d* begins after `@(`*d*`)`.[9]

CF5 A clock that flows to an instance of a named property flows into the body of the corresponding declaration. A clock that flows to an instance of a named sequence flows into the body of the corresponding declaration, whether or not a sequence method is applied to the instance. Furthermore, if the instance is of a sequence, then the clock also flows across the instance, regardless of whether a sequence method is applied. A clock in the body of a declaration does not flow out of an instance.

CF6 A clock that flows to a parenthesized subexpression (either a subsequence or a subproperty) flows into the subexpression. If the subexpression is a sequence, then the clock also flows across the parenthesized subexpression. A clock inside the subexpression does not flow out of the enclosing parentheses. This rule applies to parentheses enclosing a sequence to which one or more sequence match items are attached. Analogous rules apply to operators with explicit parentheses: **`strong`**`()`, **`weak`**`()`, **`first_match`**`()`.

CF7 A clock that flows to a maximal Boolean expression *b* governs *b* and flows across *b*. Analogous rules apply to Boolean repetitions *b*`[->n]`, *b*`[=n]`, etc.

[8]See the Rewriting Algorithms in Annex F.4 of [8].

[9]Rule CF2 can be thought of as a special case of CF4 if the inferred clock is understood to specify a clocking event control at the beginning of each of the concurrent assertions in its scope.

CF8 A clock that flows to one of the operators `##n`, `[*n]`, `|->`, `|=>`, `#-#`, and `#=#` flows across the operator. If the operator is not a synchronizer, then the clock also governs time advances associated with the operator. Analogous rules apply to ranged variants of `[*n]`.

CF9 A clock that flows to the left operand of one of the infix operators **or**, **and**, **intersect**, **within**, **throughout**, **iff**, **implies**, and **until** flows to the operator and distributes to (i.e., flows into) both operands. The clock also governs time advance for **until**. Analogous rules apply to the various variants of these operators.

CF10 A clock that flows to one of the prefix operators **not**, **nexttime**, **always**, and **eventually** flows to the operand of the operator. The clock also governs time advance in **nexttime**, **always**, and **eventually**. Analogous rules apply to all the variants of these operators.

CF11 A clock that flows to an **if-else** governs the test condition of the **if-else**. The clock also flows into each of the underlying properties of the **if-else**. Analogous rules apply to **case**.

CF12 A clock that flows to a **disable iff**, **accept_on**, or **reject_on** flows into the underlying property. The clock does *not* govern the reset condition.

CF13 A clock that flows to a **sync_accept_on** or **sync_reject_on** governs the abort condition and flows into the underlying property.

The following examples illustrate the clock flow rules.

Example 12.3. Analyze the clock flow in the following property:

```
@(ev1) a |=> b ##1 @(ev2) c
```

Solution: By *CF3*, `ev1` flows to `a`. By *CF7*, `ev1` governs `a` and flows to `|=>`. By *CF8*, `ev1` flows across `|=>` to `b`. By *CF7*, `ev1` governs `b` and flows to `##1`. By *CF8*, `ev1` flows across `##1` to `@(ev2)`. By *CF4*, the scope of `ev1` does not flow across `@(ev2)`. Therefore, `##1` is a synchronizer between `ev1` and `ev2`. By *CF3*, `ev2` flows to `c`. By *CF7*, `ev2` governs `c`. In summary, the property is equivalent to the following, in which each of the Booleans is explicitly clocked:

```
@(ev1) a |=> @(ev1) b ##1 @(ev2) c
```

□

Example 12.4. Analyze the clock flow in the following property:

```
@(ev1) a ##1 (b ##1 @(ev2) c) |=> d
```

Solution: By *CF3*, *CF7*, and *CF8*, `ev1` flows to and governs `a` and flows across `##1` to the parenthesized subsequence `(b ##1 @(ev2)c)`. By *CF6*, `ev1` flows into and across the parenthesized subsequence. Therefore, `ev1` flows to and across `|=>` (*CF8*), and so it flows to and governs `d` (*CF7*). Within the parenthesized subsequence, `ev1` flows to, governs, and flows across `b` (*CF7*), flows across `##1` (*CF8*), and ends at `@(ev2)` (*CF4*). Therefore, the `##1` within the parenthesized subsequence synchronizes between `ev1` and `ev2`. By *CF3*, `ev2` flows to and governs `c`, but `ev2` does not flow out of the enclosing parentheses by *CF6*. As a result, `|=>` is a synchronizer between `ev2` and `ev1`. In summary, the property is equivalent to the following, in which each of the Booleans is explicitly clocked:

```
@(ev1) a ##1 @(ev1) b ##1 @(ev2) c |=> @(ev1) d
```

□

Example 12.5. Analyze the clock flow in the following module:

```
module m1 (logic a, b, c, d, event ev1, ev2);
   default clocking EV1 @(ev1); endclocking
   sequence s4;  b ##1 @(ev2) c; endsequence
   a10:  assert property(a ##1 s4 |=> d);
endmodule
```

Solution: By *CF1*, `ev1` flows to `a10`, hence flows to and governs `a` (*CF7*) and flows across `##1` to the instance of `s4` (*CF8*). By *CF5*, `ev1` flows into the body of `s4` for this instance and also across the instance (*CF5*). Within the body of `s4`, `ev1` flows to and governs `b` (*CF7*), flows across `##1` (*CF8*), and stops at `@(ev2)` (*CF4*). `ev2` governs `c` (*CF3*, *CF7*), but `ev2` does not flow out of the instance of `s4` (*CF5*). `ev1` flows across `|=>` (*CF8*), and so it flows to and governs `d` (*CF7*). In summary, `a10` behaves the same as the property in the preceding example. □

After application of the clock flow rules, each Boolean expression that stands as a subsequence within a concurrent assertion must be governed by a clock. Otherwise, the assertion is not legal. The following example illustrates an illegal assertion:

```
module m2 (logic a, b, event ev1);
   a11_illegal:  assert property(
      (@(ev1) a) implies b
   );
endmodule
```

By *CF3* and *CF7*, `ev1` governs `a`, but by *CF6* `ev1` does not flow out of the enclosing parentheses. There is no default clock, so no clock governs `b`.

12.2.4.2 Semantic Leading Clocks

The rules of semantic leading clocks define how the leading clock or clocks of a sequence or property are determined from the inside out. One of the basic ideas of clock flow is that an outer clock is replaced by, rather than flowing through, an inner clock. This means that `@(`c`)@(`d`)` p behaves semantically the same as `@(`d`)` p. Syntactically, the leading clock of `@(`c`)@(`d`)` p appears to be c, but semantically it is d.

Another principle of concurrent assertions is that every subsequence, in particular every Boolean that stands as a subsequence, must be clocked. When examining semantic leading clocks from the inside out, though, there may be no clock at hand.

For example, in the presence of default clocking, the following concurrent assertion is legal:

```
c1:  cover property (a ##1 b);
```

When examining the underlying sequence `a ##1 b`, there is no clock at hand in the expression. Therefore, the definition of semantic leading clocks uses a device to indicate that a clock needs to be provided from outside, namely the *inherited* semantic leading clock.

The rules of semantic leading clocks propagate these ideas through the various sequence and property forms. They appear below and define the set LC of semantic leading clocks for a sequence or property. In the rules, b denotes a Boolean; n denotes a natural number; r, r_1, r_2 denote sequences; *item* denotes a sequence match item; p, p_1, p_2 denote properties; x denotes either a sequence or a property; and c denotes a clocking event.

LC1 If $inherited \in LC(x)$, then $LC(\texttt{@(}c\texttt{)}\ x) = \{c\} \cup (LC(x) - \{inherited\})$. Otherwise, $LC(\texttt{@(}c\texttt{)}\ x) = LC(x)$.

LC2 $LC(\texttt{(}x\texttt{)}) = LC(x)$.

LC3 : $LC(b) = LC(b\texttt{[->}n\texttt{]}) = LC(b\texttt{[=}n\texttt{]}) = \{inherited\}$. Analogous rules apply to variants of these operators.

LC4 $LC(b$ **throughout** $r) = \{inherited\} \cup LC(r)$.

LC5 $LC(r_1$ **and** $r_2) = LC(r_1) \cup LC(r_2)$. The same rule applies if **and** is replaced by any of **or**, **intersect**, and **within**.

LC6 $LC(r_1\ \texttt{\#\#}n\ r_2) = LC(r_1)$. Analogous rules apply to variants of ##n.

LC7 $LC(r\texttt{[*}n\texttt{]}) = LC(r)$. Analogous rules apply to variants of [*n].

LC8 $LC(\texttt{(}r\texttt{,}\ item\texttt{)}) = LC(r)$.

LC9 $LC($**first_match**$(r)) = LC(r)$. The same rule applies if **first_match**() is replaced by **strong**() or **weak**().

LC10 $LC($**not** $p) = LC(p)$.

LC11 : $LC(p_1$ **and** $p_2) = LC(p_1) \cup LC(p_2)$. The same rule applies if **and** is replaced by any of **or**, **iff**, and **implies**.

LC12 $LC(r \texttt{ |-> } p) = LC(r)$. The same rule applies if |-> is replaced by any of |=>, #-#, and #=#.

LC13 $LC($**nexttime** $p) = \{inherited\}$. The same rule applies if **nexttime** is replaced by **always** or **s_eventually** or by variants of any of these operators.

LC14 $LC(p_1$ **until** $p_2) = \{inherited\}$. Analogous rules apply to the variants of **until**.

LC15 If p is an **if-else** or **case** property, then $LC(p) = \{inherited\}$.

LC16 $LC($**accept_on**$(b)\ p) = LC(p)$. The same rule applies if **accept_on** is replaced by **reject_on** or **disable iff**.

LC17 $LC($**sync_accept_on**$(b)\ p) = \{inherited\}$. The same rule applies when **sync_accept_on** is replaced by **sync_reject_on**.

Rule *LC1* captures the fact that an outer clock applied to x is semantically significant if, and only if, something within x requires an incoming clock, as evidenced by the presence of *inherited* in $LC(x)$. Rule *LC3* says that Booleans require incoming clocks.

Rules *LC1* through *LC9* account for semantic leading clocks in sequences. They do not enforce the various restrictions on the clocking of sequences, such as those from Sect. 12.2.3. Rather, they allow for partial clocking of sequences that still requires an incoming clock, as in

```
(a[*2] and @(ev1) b[->1]) ##1 c
```

LC of this sequence is $\{inherited,$ `ev1`$\}$, where *inherited* records the fact that `a[*2]` requires an incoming clock. Since **`and`** is not a synchronizer for sequences, the incoming clock must be identical to `ev1` for the concurrent assertion in which this sequence appears to be legal.

Rules *LC1*, *LC2*, and *LC9* through *LC17* account for semantic leading clocks in properties. *LC13* and *LC14* capture the fact that the incoming clock governs time advancement in the temporal operators of the **`nexttime`**, **`always`**, **`s_eventually`**, and **`until`** families. *LC15* indicates that the condition of an **`if-else`** or **`case`** is governed by the incoming clock, while *LC16* indicates that the reset condition of an asynchronous abort is independent of the incoming clock.

Rule *LC17* reflects the definition in the LRM for synchronous aborts, namely, that the incoming clock governs the abort condition and serves as leading semantic clock. However, the LRM leaves ambiguous whether and how synchronous abort operators may be used as synchronizers.[10] Therefore, it is advised to use synchronous aborts only when there is at most one explicit (i.e., non-*inherited*) semantic leading clock of the underlying property and this clock is identical to the incoming clock.

The top-level property of a concurrent assertion is always required to have a single semantic leading clock after the resolution of clock scoping. If p is such a top-level property, then this means that $LC(p)$ must have one of the following forms:

- $\{c\}$. In this case, c is the unique, explicit semantic leading clock of p, and no incoming clock is required or has any effect.
- $\{inherited\}$. In this case, p has no explicit semantic leading clock and requires an incoming clock, either from default clocking or from a procedural context.
- $\{inherited, c\}$. In this case, c is an explicit semantic leading clock of p, but p also requires an incoming clock, which must be identical to c.[11]

[10] In fact, various rewrite rules in Annex F.5 of the LRM lead to the conclusion that, while the abort condition of a synchronous abort is governed by the incoming clock, the set of semantic leading clocks is determined from the underlying property, in contradiction to *LC17*.

[11] The LRM does not define precisely the criterion "identical", but through examples it indicates that syntactically identical events are "identical", while syntactically distinct, but semantically equivalent, events are not "identical".

Example 12.6. Compute the set of semantic leading clocks in the following assertion and determine any requirements on the context in which the assertion appears:

```
a12:  assert property(
   a or @(ev1) b and nexttime @(ev2) c
);
```

Solution:

$$
\begin{aligned}
& LC(\texttt{b and nexttime @(ev2) c}) \\
= \;& LC(\texttt{b}) \cup LC(\texttt{nexttime @(ev2) c}) && (LC11) \\
= \;& \{inherited\} && (LC3, LC13)
\end{aligned}
$$

Therefore by *LC1*,

$$LC(\texttt{@(ev1) b and nexttime @(ev2) c}) = \{\texttt{ev1}\}$$

By *LC3*, $LC(\texttt{a}) = \{inherited\}$, and so by *LC11*, the set of semantic leading clocks for the entire assertion is $\{\texttt{ev1}, inherited\}$. This means that the assertion must be in a context that guarantees an incoming clock, either by default clocking or by inference from a procedural context, and the incoming clock must be identical to `ev1`. □

12.2.5 *Finer Points of Multiple Clocks*

This section covers a few finer points regarding the use of multiple clocks in sequences and properties.

12.2.5.1 Clocking LTL Operators

When LTL operators **`nexttime`**, **`always`**, **`s_eventually`**, **`until`**, and their variants are used as synchronizers, it is important to remember that the time advance specified by the LTL operator is determined by the *incoming* clock, not by the leading clock or clocks of the operands. Consider the following:

```
module m3 (logic a, b, clk1, clk2);
   a13:  assert property(
      @(posedge clk1) a |-> nexttime @(posedge clk2) b
   );
   a14:  assert property(
      @(posedge clk1) a |-> ##1 @(posedge clk2) b
   );
endmodule
```

Assertions `a13` and `a14` look similar, but they behave differently. In both, the antecedent of `|->` matches whenever the sampled value of `a` is `1'b1` at an occurrence of **`posedge`** `clk1`. Suppose that this occurs at time t_0. In `a13`, **`posedge`** `clk1`

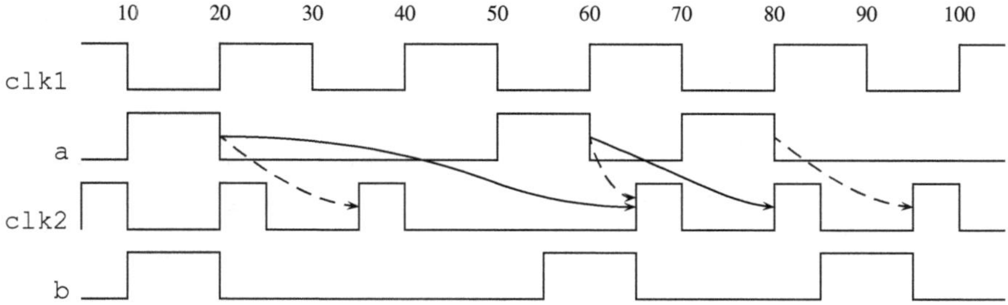

Fig. 12.12 Waveform for assertions `a13` and `a14` of module `m3`

flows to **`nexttime`**, and so the **`nexttime`** causes advance to the next occurrence of **`posedge`** `clk1` strictly after t_0 before looking for a concurrent or subsequent occurrence of **`posedge`** `clk2` at which to evaluate `b`. In `a14`, `##1` is a synchronizer between the occurrence of **`posedge`** `clk1` at t_0 and the earliest strictly subsequent occurrence of **`posedge`** `clk2`, where it evaluates `b`.

This difference is illustrated in the waveform of Fig. 12.12. The solid arrows represent evaluations of `a13`, while the dashed arrows represent evaluations of `a14`. The evaluation attempt of `a13` beginning at time 20 matches the antecedent of `|->` at time 20 and then advances to time 40 because of the **`nexttime`** clocked by **`posedge`** `clk1`. At time 40, the attempt begins looking for the next concurrent or future occurrence of **`posedge`** `clk2`, which is at time 65. At time 65, the sampled value of `b` is checked and found to be `1'b1`, so the overall evaluation passes. The evaluation of `a14` beginning at time 20 behaves differently. After matching the antecedent of `|->` at time 20, the synchronizer `##1` causes this evaluation to begin looking for the next strictly future occurrence of **`posedge`** `clk2`, which is at time 35. At time 35, the evaluation checks `b` and fails. The evaluation of `a13` beginning at time 60 advances to time 80 due to the **`nexttime`**. Since **`posedge`** `clk2` also occurs at time 80, `b` is checked at this time, and the evaluation fails. Again, the evaluation of `a14` beginning at time 60 behaves differently. This evaluation finds the next strictly future occurrence of **`posedge`** `clk2` at time 65, checks `b` there, and passes.

The structure of `a13` ensures that evaluation of **`nexttime`** always begins in a time step in which **`posedge`** `clk1` occurs and therefore is already *aligned* to the governing (i.e., incoming) clock for **`nexttime`**. It is possible to use **`nexttime`** as a synchronizer in a way that need not start in a time step aligned to its governing clock. In such a case, **`nexttime`** specifies both of the following temporal actions:[12]

[12]It may seem strange that **`nexttime`** specifies both alignment and advancement to the next tick. This behavior is aligned with PSL and is needed in order for the behavior of **`nexttime`** as a synchronizer to converge to the singly clocked behavior of **`nexttime`** under certain clock convergence scenarios. See Sect. 12.2.5.3 and Exercise 12.5.

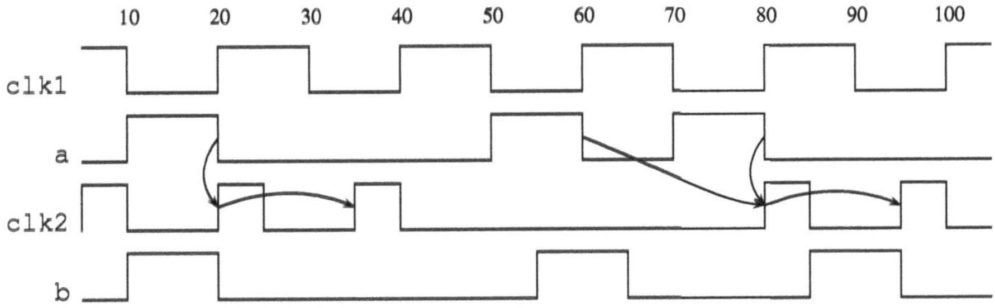

Fig. 12.13 Waveform for assertion `a13_v2` of module `m3_v2`

- First proceed to the nearest current or future tick of the governing clock (i.e., first *align* with the governing clock).
- From that point, advance to the next tick of the governing clock.

Here is an example illustrating such use of **`nexttime`**:

```
module m3_v2 (logic a, b, clk1, clk2);
   a13_v2:  assert property(
      @(posedge clk1) a |-> @(posedge clk2) nexttime b
   );
endmodule
```

Figure 12.13 shows a waveform with solid arrows representing the temporal actions of **`nexttime`** in the evaluation of `a13_v2`. Note that all signals are the same as in Fig. 12.12 except for `clk2`, which no longer has a **`posedge`** at time 65. For the evaluation attempt beginning at time 20, alignment with **`posedge`** `clk2` does not advance time. Advancing to the next tick carries the attempt to time 35, where it fails. For the evaluation attempt beginning at time 60, alignment with **`posedge`** `clk2` carries the attempt to time 80. For the attempt beginning at time 80, alignment with **`posedge`** `clk2` does not advance time. For both of these attempts, advancement to the next tick carries the attempt to time 95, where the result is pass. Because **`posedge`** `clk1` and **`posedge`** `clk2` both occur at times 20 and 80, these evaluation attempts behave the same as `a14`.

The operator **`nexttime`**`[0]` specifies only alignment with its governing (i.e., incoming) clock. For any property p, **`nexttime`**`[0]` p is equivalent to `1'b1 |->` p. This operator can be used, e.g., if p may have a leading clock different from the incoming clock and it is desired to ensure that alignment with the incoming clock occurs first. Examples of such usage are given in Sect. 17.1.

Similar considerations apply when other LTL operators are used as synchronizers: time advance specified by the LTL operator is always with respect to the clock governing (i.e., incoming to) the LTL operator, and evaluation of the LTL operator first specifies alignment to that clock in case its evaluation is not guaranteed to begin at such a point. See Exercise 12.4.

12.2.5.2 Unclocked Synchronizers and Logical Operators

The synchronizers **if-else**, **case**, and the LTL operators must always be within the scope of a clock because the incoming clock determines when the condition of the **if-else** or **case** is evaluated and when the time advance of an LTL operator occurs.

The other synchronizers are `##1` and `##0` for sequences and `|=>`, `|->`, `#=#`, `#-#` for properties. These operators can synchronize between clocks specified by their operands and are not themselves actually required to be within the scope of a clock. In a similar way, the logical operators **and**, **or**, **iff**, and **implies** can join differently clocked operands and do not themselves require a clock.

The situation of an unclocked synchronizer or logical operator occurs when the synchronizer or operator is the top-level operator of a concurrent assertion in a context where there is no incoming clock. Of course, the concurrent assertion must still have a single semantic leading clock.

The following example illustrates an unclocked synchronizer and an unclocked logical operator in static concurrent assertions within a module with no default clocking:

```
module m4 (logic a, b, c, event ev1, ev2);
   a15:  assert property((@(ev1) a) |=> (@(ev2) b));
   a16:  assert property((@(ev1) b) or (@(ev1) c));
endmodule
```

Both `a15` and `a16` have `ev1` as semantic leading clock. In `a15`, the synchronizer `|=>` is not within the scope of any clock, and in `a16` the logical operator **or** is not within the scope of any clock.

The following example is illegal because an operator that is not a synchronizer remains unclocked:

```
module m5 (logic a, b, event ev1, ev2);
   a17_illegal:  assert property(
      (@(ev1) a) |-> ##2 (@(ev2) b)
   );
endmodule
```

The operator `##2` is not a synchronizer and requires a clock to determine how time is advanced, but in this module it is not within the scope of a clock.

12.2.5.3 Continuity Under Clock Convergence

The semantics of multiple clocks has been designed so that it has a quality of *continuity* with respect to convergence of clocks. In other words, as the different clocks coordinated by synchronizers or logical operators become semantically equivalent, the behavior of the multiply clocked sequence or property converges to the behavior of the singly clocked sequence or property obtained by aligning all the clocks to a single clock. As an example of this idea, consider the following:

```
module m6 (logic a, b, event ev1, ev2);
   a15:  assert property((@(ev1) a) |=> (@(ev2) b));
   a18:  assert property(@(ev1) a |=> b);
endmodule
module m7;
   logic A,B;
   event EV;
   ...
   m6 m6_inst(.a(A), .b(B), .ev1(EV), .ev2(EV));
endmodule
```

The instance of `m6` within `m7` connects the same event `EV` to both of the ports `ev1` and `ev2`. Therefore, `ev1` and `ev2` are equivalent in this instance, and so `a15` and `a18` behave identically.

12.2.5.4 Sequence Methods

SystemVerilog provides two methods for detecting the endpoint of match of an instance of a named sequence: `triggered` and `matched`. `triggered` is discussed in detail in Sect. 11.2. Both of these methods may be used in multiply clocked sequences and properties, but their behaviors are very different.

Whenever `triggered` is used in a Boolean expression of a sequence, the clock governing that Boolean expression must be the same as the ending clock of the sequence instance to which `triggered` is applied. This restriction was illustrated in Example 12.2 from Sect. 12.1.2.

`matched` can, and should, be used when the ending clock of the instance to which it is applied is *different* than the clock governing the Boolean expression in which `matched` appears. `matched` serves as a synchronizer between these two clocks. Upon completion of a match of the underlying instance, this fact is stored until the earliest strictly subsequent time step in which there is a tick of the clock governing the context in which `matched` appears. In that time step, the value of the application of `matched` to the instance is true.

As an example, suppose that we need to check that if `dvalid` is true at an occurrence of **posedge** `dclk`, then the sequence `req ##1 ack` must have already completed a match clocked at **posedge** `rclk`, but the match must not have completed before the nearest strictly prior occurrence of **posedge** `dclk`, if such an occurrence exists. It does not matter when the match starts. This can be encoded using `matched` as shown in Fig. 12.14. Line 7 illustrates the syntax for applying method `matched` to an instance of a sequence.

Figure 12.15 shows a waveform for `a_matched`. There are two matches of the instance `s_req_ack(`**posedge** `rclk)`. The first is from times 20 to 40 and causes Line 7 to be true at time 55. The second is from times 60 to 80 and causes Line 7 to be true at time 105. As a result, the evaluation of `a_matched` that starts at time 55 succeeds. However, the evaluation of `a_matched` that starts at time 80 fails. The match of `s_req_ack(`**posedge** `rclk)` completing at time 80 is too late because it is not strictly before time 80. The other evaluations of `a_matched` succeed vacuously.

```
sequence s_req_ack(event ev);
   @(ev) req ##1 ack;
endsequence
a_matched:  assert property(
   @(posedge dclk)
   dvalid |->
   s_req_ack(posedge rclk).matched
);
```

Fig. 12.14 Assertion using `matched`

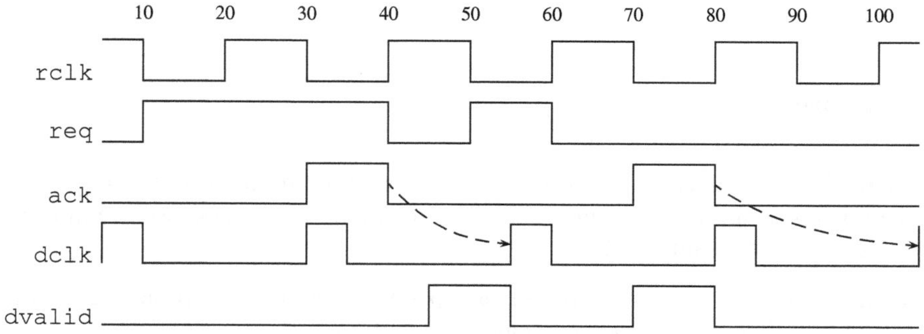

Fig. 12.15 Waveform for assertion `a_matched`

Note that if `s_req_ack(posedge rclk).matched` were used in two or more different clocking event scopes, then there are effectively as many different `.matched` end points because their validity ends at different clocking events.

12.2.6 Declarations Within a Clocking Block

Sequences and properties may be declared within a clocking block. No explicit clocking event control can be written in such a declaration. Instead, all instances of the named sequence or property are understood to be singly clocked by the clocking event of the clocking block. If a declaration of a sequence or property within a clocking block itself instantiates a sequence or property, then that instance must be singly clocked by a clock that is identical to the clock of the clocking block.

Concurrent assertions cannot be written within a clocking block. Therefore, to instantiate a named sequence or property that is declared within a clocking block, the clocking block must be named and the named sequence or property must be referenced hierarchically.

Here is an example:

```
module decl_in_clocking_block (logic a, b, c, clk);
   clocking PCLK; @(posedge clk);
      property p3; a |=> p4; endproperty
   endclocking
   property p4; b until c; endproperty
   a19:  assert property(PCLK.p3);
endmodule
```

Since `p4` is declared without clocks, the instance of `p4` within the declaration of `p3` is legal. The clock of `p3` is **`posedge`** `clk`, which flows into and clocks this instance of `p4`.

Exercises

12.1. Write an assertion to check that after an occurrence of event `ev_start`, there is no occurrence of event `ev_wait` until strictly after an occurrence of event `ev_enable` (cf. Example 12.2).

12.2. Without using the **`nexttime`** operator, rewrite assertion `a14` from Sect. 12.2.5.1 so that its behavior is identical to that of `a13`.

12.3. Using the waveform in Fig. 12.12, analyze the behavior of the following assertion:

```
a1:  assert property(
   @(posedge clk1) a |=> @(posedge clk2) nexttime b
);
```

12.4. Using the waveform in Fig. 12.12, analyze the behavior of the assertions in the following module for the evaluation attempts beginning at times 20 and 60:

```
module m (logic a, b, clk1, clk2);
   a1:  assert property(
      @(posedge clk1) a |->
      (@(posedge clk2) a) until (@(posedge clk2) b)
   );
   a2:  assert property(
      @(posedge clk1) a |-> @(posedge clk2) a[*] ##1 b
   );
endmodule
```

12.5. Consider the assertion

```
a1:  assert property(
   @(posedge clk1) a |-> @(posedge clk2) nexttime b
);
```

Create waveforms in which `clk2` converges to `clk1` from the right to explain why the synchronizing behavior of **`nexttime`** must both align and advance to the next tick in order for the multiply clocked semantics to converge to the singly clocked semantics. What happens if `clk2` converges to `clk1` from the left?

12.6. Analyze the clock flow in each of the following expressions.

1. `@(ev1) a |=> @(ev2) b until c.`
2. `@(ev1) a |=> (@(ev2) b) until c.`
3. `not(@(ev1) a) #-# (@(ev2) b) implies (@(ev3) c).`
4. `@(ev1) (a ##1 @(ev2) b) |=> if (c) @(ev3) d else e.`
5. `@(ev1) sync_accept_on(a) b until @(ev2) c and nexttime @(ev3) d.`

12.7. Compute the set of semantic leading clocks for each of the following properties.

1. `(@(ev1) a ##1 b[*]) |=> c.`
2. `if (a) @(ev1) p else q.`
3. `@(ev1) if (a) p else q.`
4. `@(ev1) a implies (@(ev2) b) or c.`
5. `(@(ev1) a) implies (@(ev2) b) or c.`
6. `accept_on(a) (@(ev1) b) and strong(c throughout @(ev2) d).`

12.8. Using rules of clocking, determine for each of the following whether it is legal, illegal, or its legality depends on the existence or nature of an incoming clock. For those in the last category, identify the conditions on an incoming clock to make the expression legal. Assume that a, b, c, etc. are Boolean expressions with no embedded clocking event controls.

1. `@(ev1) a ##1 b[*] |=> c.`
2. `(@(ev1) a ##1 b[*]) |=> c.`
3. `@(ev1) a within @(ev2) b[->1].`
4. `(@(ev1) a) within b[->1].`
5. `assert property(if (a) @(ev1) b else c);.`
6. `assert property(@(ev1) if (a) b else c);.`
7. `assert property(@(ev1) a implies (@(ev2) b) or c);.`

12.9. Suppose that Fig. 12.14 is modified as follows:

```
sequence s_req_ack;
   req ##1 ack;
endsequence
a_triggered:  assert property(
   @(posedge dclk)
   dvalid |->
   s_req_ack.triggered
);
```

Is the assertion legal? If no, why? If yes, interpret the behavior of the assertion on the waveforms in Fig. 12.15.

Chapter 13
Resets

The Metropolis should have been aborted long before it became New York, London or Tokyo.

— John Kenneth Galbraith

As the evaluation of a concurrent assertion evolves over time, certain conditions may occur upon which it is desired to stop the present evaluation attempt in a preemptive or abortive way. The prototypical example is the occurrence of design reset: most concurrent assertions should not continue evaluation across reset of the design. As a result, such preemptive or abortive conditions have come broadly to be termed *reset conditions*. It is cumbersome, at best, to instrument every step of a concurrent assertion with sensitivity to a reset condition. Therefore, SVA provides *reset* constructs with which reset conditions can be declared and their scopes specified.

This chapter covers declaration, scoping, and semantics of reset constructs. There are *abort* property operators which come in both synchronous and asynchronous forms, and in both passing and failing flavors. These are in addition to the existing asynchronous **`disable iff`** construct at the top-level of a concurrent assertion.

13.1 Overview of Resets

This section gives an intuitive overview of resets based on examples.

A *reset condition* is a condition upon which it is desired to stop evaluation of a concurrent assertion or subproperty in a preemptive or abortive way. The prototypical example is the occurrence of design reset: most concurrent assertions should not continue evaluation across reset of the design under test. Encoding a concurrent assertion to be sensitive throughout its evaluation to occurrence of a reset

E. Cerny et al., *SVA: The Power of Assertions in SystemVerilog*,
DOI 10.1007/978-3-319-07139-8_13

```
1 a_disable: assert property(
2    disable iff (reset)
3    @(posedge clk) a |=> b
4 ) else $error("FAIL");
```

Fig. 13.1 Simple concurrent assertion with a disable clause

condition is cumbersome, so SVA provides various *reset* constructs with which to declare reset conditions and specify their scopes.

A reset is *asynchronous* if the associated reset condition is checked at every time step during the evaluation of the underlying property. SVA provides three asynchronous resets: **disable iff**, **accept_on**, and **reject_on**. A reset is *synchronous* if the associated reset condition is governed by a clock and checked only in time steps in which the clocking event occurs. SVA provides two synchronous resets: **sync_accept_on** and **sync_reject_on**. Apart from **disable iff**, all of the resets are referred to collectively as *abort* operators.

13.1.1 Disable Clause

A *disable clause* is specified with the compound keyword **disable iff**. It defines a top-level asynchronous reset condition that applies throughout the evaluation of a concurrent assertion. The reset condition is called the *disable condition* of the disable clause. With the exception of overriding an incoming default disable condition (see below), all nesting of disable clauses is illegal.

The meaning of a disable clause is that the current (*not* the sampled) value of the disable condition is checked continuously throughout the evaluation of the underlying property of the concurrent assertion. If the disable condition is true in any time between the start of an evaluation attempt in the Observed region and the end of the evaluation attempt then the evaluation attempt is disabled. The evaluation of the property thus stops and neither passes nor fails. Instead, the overall result of the evaluation is *disabled*. If the disable condition neither is nor becomes true during the evaluation, then the evaluation either passes or fails according to the result of the evaluation of the underlying property.

Figure 13.1 gives a simple example. A disable clause may be specified either before or after an explicit clocking event control in a concurrent assertion, but it must precede all other terms of the underlying property. In this example, the disable condition is `reset`. The underlying property appears in Line 3 and is singly clocked by **posedge** `clk`.

Figure 13.2 shows a possible waveform for `a_disable`. The attempt that begins at time 20 is disabled by the transition to `1'b1` of `reset` at time 35. In the absence of the disable clause, this attempt would have failed at time 40, but, because it is preempted, no failure occurs and the failing action block in Line 4 does not execute.

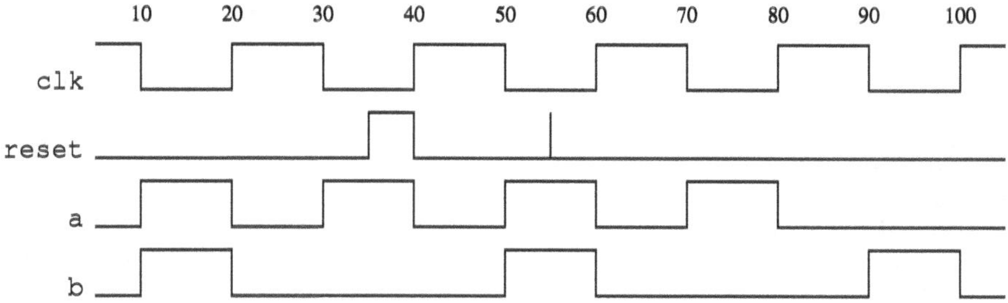

Fig. 13.2 Waveform for `a_disable`

The attempt that begins at time 40 is also disabled. Exactly when it is disabled depends on the behavior of `reset` at time 40. According to the waveform, the sampled value of `reset` at time 40 is `1'b1`, but a disable condition is evaluated using current values, not sampled values. If the value of `reset` remains `1'b1` until the beginning of the assertion evaluation attempt at time 40, then that attempt is preempted at time 40. Otherwise, the attempt is disabled by the brief, glitchy transition to `1'b1` of `reset` at time 55.[1] Again, because of the preemption, this attempt neither passes nor fails, even though, in the absence of the disable clause, the evaluation of the underlying property would have passed at time 60. The attempts that begin at times 60, 80, and 100 are not disabled. The first fails at time 80, the second passes at time 100, and the third passes vacuously at time 100.

To avoid preemption on glitches in the disable condition, such as the pulse at time 55 in Fig. 13.2, `$sampled` can be used to force evaluation using only sampled values. Suppose that Line 2 of Fig. 13.1 is changed to

```
disable iff ($sampled(reset))
```

The disable condition is still monitored continuously, but the use of `$sampled` means that, in any time step, only the sampled value of `reset` is relevant, and this value is persistent throughout the time step. As a result, the glitch in `reset` at time 55 is no longer visible to the disable condition. And the behavior at time 40 is more predictable: the sampled value of `reset` at time 40 is `1'b1`, which persists as the sampled value throughout the time step and preempts the attempt beginning at time 40.

A disable clause may be specified within the declaration of a named property. The disable clause must follow any local variable declarations (see Sect. 16.1). It may be specified either before or after an explicit clocking event control in the body of the property declaration, but it must precede all other terms of the underlying property. If a named property specifies a disable clause, then instances of the property must ensure that, after elaboration,[2] the disable clause is at the top-level

[1] It is presumed that this impulse is a glitch, although, technically, the graphical representation does not imply this without further information, e.g., concerning the timescale. A waveform tool can identify glitches unambiguously.

[2] See the Rewriting Algorithms specified in Annex F.4 of the LRM [8].

of each concurrent assertion in which it appears, preceded only by local variable declarations and leading clocking event controls. The following variant of the code in Fig. 13.1 behaves equivalently:

```
property p_disable;
   disable iff (reset) a |=> b;
endproperty
a_disable:  assert property(
   @(posedge clk) p_disable
) else $error("FAIL");
```

A disable clause may *not* be specified within the declaration of a named sequence.

13.1.1.1 Default Disable Condition

Like a clocking event, a disable condition may be shared by many concurrent assertions. In this situation, it is convenient to be able to specify a *default disable condition*. A default disable condition applies throughout the generate block, module, interface, or program in which it appears, including nested scopes except those with their own default disable condition. The default applies to all concurrent assertions within the scope that do not have disable conditions otherwise specified. It does not apply to declarations of named sequences or properties. The following example illustrates the syntax:

```
1  module m_default_disable(logic reset, a, b, clk);
2     default disable iff reset;
3     a_disable:  assert property(
4        @(posedge clk) a |=> b
5     )  else $error("FAIL");
6     a_override:  assert property(
7        disable iff (1'b0)
8        @(posedge clk) reset |=> !reset
9     );
10 endmodule
```

The default disable condition, `reset`, applies to `a_disable`. Since `a_override` has an explicit disable condition, the disable condition `1'b0` applies to it and overrides the default.

Nesting of disable conditions is only allowed when the inner disable condition overrides an incoming default disable condition. The overriding by the disable clause in Line 7 above is a legal example. Figure 13.3 shows an illegal example. Line 8 results in a nesting of disable conditions, the outer one from the concurrent assertion `a_disable` in Line 4 and the inner one from the instantiated property `p_disable` in Line 7. The outer disable condition is not a default. Therefore, the nesting is illegal, despite the fact that the two disable conditions are identical.

The disable condition can also be inferred from the instantiation context using the system function `$inferred_disable` which can be used in place of a default

```
module m_illegal_disable_nesting(logic reset, a, b, clk);
   default clocking PCLK @(posedge clk); endclocking
   property p_disable;
      disable iff (reset) a |=> b;
   endproperty
   a_disable:  assert property(
      disable iff (reset)
      p_disable
   )  else $error("FAIL");
endmodule
```

Fig. 13.3 Illegal nesting of disable conditions

argument value of a property declaration. This is especially useful when creating reusable properties for a library. For example,

```
property p_inferred_disable(x, y,
   event ck = $inferred_clock,
   logic rst = $inferred_disable);
   disable iff (rst) @ck x |=> y;
endproperty
module m_inferred_disable(logic reset, a, b, clk);
   default disable iff reset;
   default clocking @(posedge clk); endclocking
   a_inferred_disable:
      assert property(p_inferred_disable(a, b))
         else $error("FAIL");
endmodule
```

The property declaration `p_inferred_disable` uses the *inferred value functions* to specify that in the absence of the actual argument for `ck` or for `rst`, the value of the argument should be inferred from the instantiation context. In the case of `rst`, it can only be inferred from the default disable declaration. If there is none then the inferred value is `1'b0`. The assertion `a_inferred_disable` does not specify these arguments; therefore, the clocking event is inferred from the default clocking declaration as **`posedge`** `clk`, and the disabling condition is inferred from default disable declaration as `reset`.

13.1.2 Aborts

The term *abort* refers to resets specified by the following property operators: **`accept_on`**, **`reject_on`**, **`sync_accept_on`**, and **`sync_reject_on`**. The first two of these are *asynchronous* aborts, while the last two are *synchronous* aborts. Each of these property operators has two operands. The first is the *abort condition*, which is enclosed in parentheses, and the second is the underlying property governed by the abort operator. Here is an example of the syntax:

```
a_accept: assert property(
   @(posedge clk)
   accept_on (retry)
   a |=> b
) else $error("FAIL");
```

Fig. 13.4 Simple concurrent assertion with an asynchronous abort

```
1 a_simple_abort: assert property (
2    @(posedge clk)
3    start
4    |=>
5    accept_on(retry) check_trans_complete
6 );
```

Line 5 specifies an asynchronous **accept_on** with `retry` as abort condition. The underlying property is the instance `check_trans_complete`.

Aborts behave similarly to a disable clause in the way that they preempt evaluation of the underlying property, but there are a number of important differences:

- The scope of an abort condition is limited to its underlying property operand, not the entire concurrent assertion. A thread or subthread of evaluation does not become sensitive to the abort condition until it reaches the associated abort operator.
- Abort conditions are always checked using sampled values. Therefore, unlike a disable condition, an abort condition is not sensitive to glitches.
- If the sampled value of an abort condition is `1'b1` in any time step in which an evaluation is sensitive to it, then the evaluation of the underlying operand property is aborted. This rule applies even in the same time step that the underlying property evaluation would complete.
- An abort is a property, so the result of an evaluation is either pass or fail. An aborted evaluation results in pass for the "accept" operators and fail for the "reject" operators. This result applies only to the abort property itself. If the abort property is a subproperty, then this result must be combined with the results of other subevaluations in the usual ways to determine the overall result of the concurrent assertion evaluation.
- Abort operators may be nested arbitrarily.
- There are no default abort conditions.

13.1.2.1 Asynchronous Aborts

Figure 13.4 shows a simple concurrent assertion with an asynchronous abort. It is similar to the assertion `a_disable` of Fig. 13.1, but its abort condition is `retry`. The underlying property is `a |=> b`, and the entire concurrent assertion is clocked by **posedge** `clk`. At each occurrence of this clocking event, evaluation of the abort

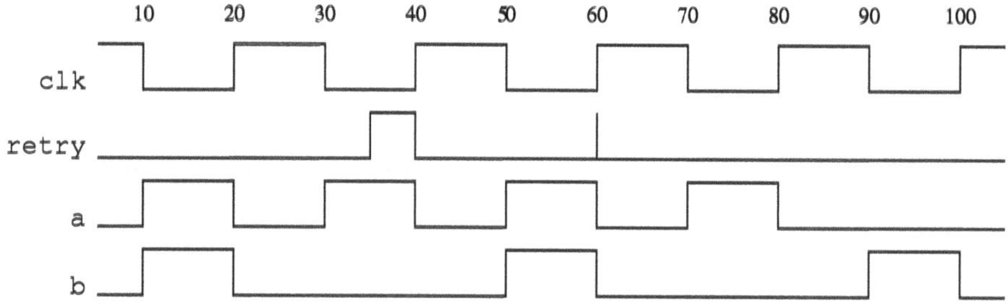

Fig. 13.5 Waveform for a_accept

property begins. This starts monitoring of the abort condition, using sampled values of retry, and also starts evaluation of the underlying property. Because the abort is asynchronous, the sampled value of retry is checked in every time step, including the first, that the evaluation of the underlying property is ongoing and has neither been aborted nor already completed on its own. If the sampled value of retry is 1'b1 in any of these checks, then the evaluation is aborted and passes in that time step. If the evaluation is not aborted, then it completes when the underlying property evaluation completes (i.e., at the next occurrence of **posedge** clk) and the result of the evaluation is the same as that of the underlying property.

The difference between Figs. 13.1 and 13.4 is that in the former the condition is not sampled and the outcome of evaluation when the condition is true is *disabled*, and in the latter the sampled value is used and the outcome is *success*.

Figure 13.5 shows an example waveform for a_accept. It is similar to Fig. 13.2 and will illustrate differences between an asynchronous abort and a disable clause. The evaluation attempt of a_accept that begins at time 20 is aborted in the first time step after time 35 and passes at that time. The evaluation does not abort at time 35 because sampled value of retry is 1'b0 in that time step. The sampled value of retry at time 40 is 1'b1, so the attempt beginning at time 40 immediately aborts and passes in that time step. The 0-width glitch on retry at time 60 does not affect any sampled value, so the evaluation beginning at time 60 is not aborted. This evaluation fails at time 80. The evaluation beginning at time 80 is also not aborted and passes at time 100.

Abort operators may be nested. The scope of the outer abort condition includes any nested abort property. The scope of the nested abort condition is limited to the underlying property of that abort operator. While evaluating the inner abort property, the outer abort condition takes precedence over the inner abort condition in case both conditions occur in the same time step.

Figure 13.6 shows a concurrent assertion with nested asynchronous aborts. The outer abort is an **accept_on** with abort condition retry whose scope is the entire property of the concurrent assertion. The inner abort is a **reject_on** with abort condition bad whose scope is the consequent of |=>. The entire assertion is singly clocked by **posedge** clk. The inner abort does not begin evaluation until after

```
a_accept_reject:  assert property(
   @(posedge clk)
   accept_on (retry)
   a |=> reject_on(bad) b[*2]
) else $error("FAIL");
```

Fig. 13.6 Concurrent assertion with nested asynchronous aborts

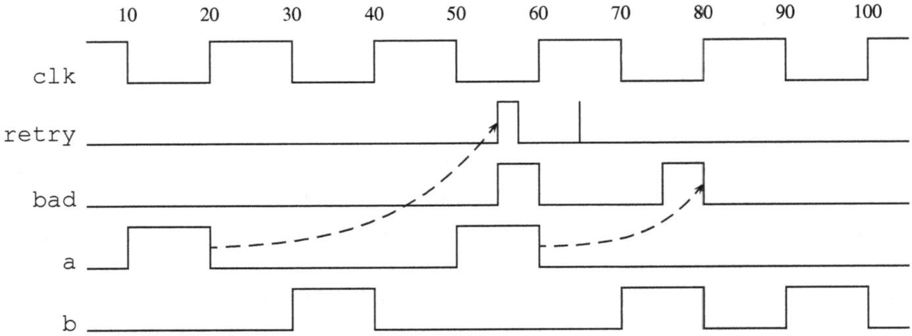

Fig. 13.7 Waveform for `a_accept_reject`

matching the antecedent `a` and advancing to the next occurrence of **`posedge`** `clk`, as specified by `|=>`. Only at that point does the evaluation become sensitive to the inner abort condition.

Figure 13.7 shows a waveform for `a_accept_reject`. The evaluation attempt that begins at time 20 begins executing the outer abort, becomes sensitive to the abort condition `retry`, and matches the antecedent of `|=>` at time 20. The evaluation then advances to time 40 and begins executing the inner abort. At that time, it becomes sensitive also to the abort condition `bad` and tests that the sampled value of `b` is `1'b1`. The evaluation then continues toward time 60 and encounters both `retry` and `bad` in the time step after time 55. In this situation, the outer abort condition takes precedence. Therefore, the evaluation of the outer abort property aborts and passes in that time step, and hence the overall evaluation of the concurrent assertion also passes. The evaluation that begins at time 60 starts similarly. The fact that the sampled value of `bad` is `1'b1` at time 60 is irrelevant because this evaluation is not yet sensitive to the inner abort condition. After matching the antecedent of `|=>`, the evaluation advances to time 80. The glitch on `retry` at time 65 is not observable by the abort operator. The evaluation does not abort in the time step after time 75 because, again, it is not yet sensitive to `bad`. At time 80, though, the evaluation becomes sensitive to `bad`. Since the sampled value of `bad` is `1'b1` at time 80, the evaluation of the inner abort property aborts and fails in that time step. This causes the consequent of `|=>` to fail. Therefore, the overall evaluation of the concurrent assertions fails at time 80 and the failing action block executes.

```
a_sync_accept:  assert property(
   @(posedge clk)
   sync_accept_on (retry)
   a |=> b
) else $error("FAIL");
```

Fig. 13.8 Concurrent assertion with synchronous aborts

```
a_sync_accept_reject:  assert property(
   @(posedge clk)
   sync_accept_on (retry)
   a |=> sync_reject_on(bad) b[*2]
) else $error("FAIL");
```

Fig. 13.9 Concurrent assertion with nested synchronous aborts

13.1.2.2 Synchronous Aborts

The synchronous abort operators **`sync_accept_on`** and **`sync_reject_on`** behave the same as their asynchronous counterparts with the exception that their abort conditions are only checked in time steps in which there is an occurrence of the clocking event.

Figure 13.8 shows a concurrent assertion with a synchronous abort. The assertion behaves like `a_accept` in Fig. 13.4 except that the abort condition `retry` is only checked in time steps in which **`posedge`** `clk` occurs. In the waveform in Fig. 13.5, the evaluation of the synchronous abort property for the attempt of `a_sync_accept` that begins in time 20 is aborted, but not in the time step after time 35 as in the case of the asynchronous abort. Instead, the synchronous abort occurs at time 40, where there is an occurrence of **`posedge`** `clk` and the sampled value of `retry` is `1'b1`. This illustrates the fact that if the sampled value of an abort condition is `1'b1` in the same time step that the underlying property evaluation would complete, the abort condition takes precedence and the evaluation is aborted. Since the abort is of the "accept" form, the evaluation of the synchronous abort property passes at time 40, hence there is an overall pass for the concurrent assertion.

Figure 13.9 shows the result of rewriting `a_accept_reject` from Fig. 13.6 using synchronous aborts. Figure 13.10 shows the same waveform as Fig. 13.7, but with arrows adjusted for the evaluations of `a_sync_accept_reject`. The pulse on `retry` beginning at time 55 is not relevant to the synchronous abort because it does not affect the sampled value at a tick of the clock. Therefore, the evaluation beginning at time 20 is aborted at time 60. The clock ticks at this time, and the sampled value of `bad` is `1'b1`. Therefore, the inner abort condition causes failure of the consequent of `|=>`, hence failure of the overall concurrent assertion, and the failing action block executes. The evaluation that begins at time 60 behaves the same as that of `a_accept_reject` because both evaluations become sensitive to `bad` at time 80, which is a tick of the clock at which the sampled value of `bad` is `1'b1`.

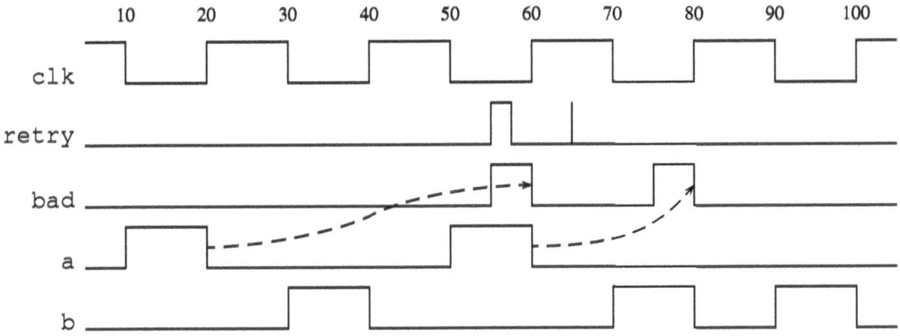

Fig. 13.10 Waveform for a_sync_accept_reject

13.2 Further Details of Resets

This section covers a few further details of specifying resets.

13.2.1 Generalities of Reset Conditions

The examples of reset conditions so far in this chapter have been simple references to variables or nets. Reset conditions can be general expressions, with the following provisos:

1. Reset conditions may not reference assertion local variables.
2. A disable condition may reference a sequence instance to which the sequence method `triggered` is applied. An abort condition may not make such a reference.
3. If a reset condition references a sampled value function other than `$sampled`, then the clock of the sampled value function must be explicitly specified.
4. Reset conditions may not contain instances of sequences to which the sequence method `matched` is applied.

The rationale for the first rule is that the meaning of such a reference may be unclear at the beginning of evaluation or as subevaluation threads create copies of local variables and assign independent values to them. The second rule reflects the fact that the sampled value of `triggered` is not useful—it is always `1'b0`. The third and fourth rules echo the fact that the asynchronous reset conditions are not governed by a clock.[3]

As an example of a more general reset condition, suppose that design reset is synchronous and occurs only if the sampled value of `reset` is `1'b1` in a time step

[3]These rules exist in the LRM [8], although they could be relaxed for synchronous aborts.

```
a19:  assert property(
   @(posedge clk)
   sync_accept_on(reset || $past(reset, , , @(posedge clk)))
   dOut == $past(dIn)
);
```

Fig. 13.11 Abort with a compound abort condition

```
1  a_abort_subproperties:  assert property(
2     @(posedge clk)
3     a |=>
4     (accept_on (retry) b[*2])
5     and
6     (reject_on (bad) c |=> !c)
7  );
```

Fig. 13.12 Assertion with multiple abort subproperties

in which `posedge clk` occurs. If we need to check that `dOut` is equal to last cycle's value of `dIn`, but only if no reset occurred in either cycle, this can be accomplished by the code in Fig. 13.11 (see also Exercise 13.3).

13.2.2 Aborts as Subproperties

Since an abort is a property, it participates in the determination of the result of evaluation of an enveloping property in the same way as other subproperties, regardless of whether the disposition of the abort is due to occurrence of the abort condition.

In the assertion of Fig. 13.12, the consequent of `|=>` in Line 3 is the conjunction of two abort subproperties.

Consider the waveform of Fig. 13.13. The attempt of `a_abort_subproperties` beginning at time 20 starts subevaluations of Lines 4 and 6 at time 40. The occurrence of `retry` in the time step after time 45 causes the evaluation of Line 4 to pass. The evaluation of Line 6 continues, and the occurrence of `bad` in the time step after time 55 causes the evaluation of Line 6 to fail. The overall assertion therefore fails. The attempt of `a_abort_subproperties` beginning at time 60 passes because Line 4 results in success at time 100 and Line 6 results in vacuous success at time 80.

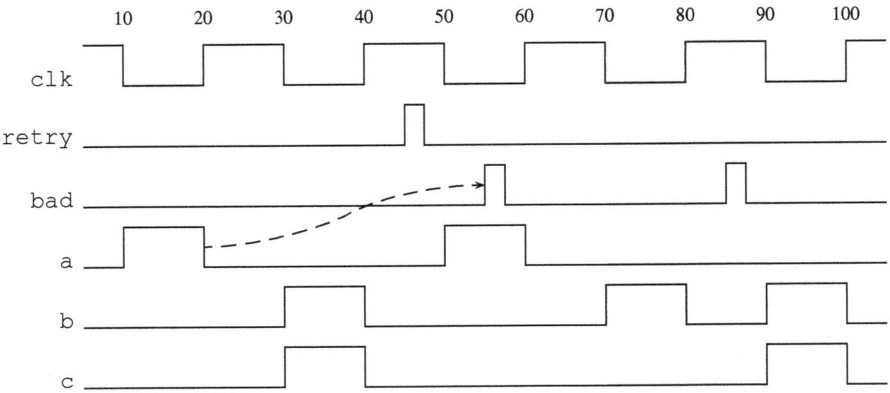

Fig. 13.13 Waveform for `a_abort_subproperties`

Exercises

13.1. Explain why assertions `a1` and `a2` below have the same passing and failing evaluation behavior:

```
a1:  assert property(@(posedge clk) sync_accept_on(a)
   b[*2] |=> c
);
a2:  assert property(@(posedge clk)
   !a throughout b[*2] |=> a || c
);
```

Rewrite the following coverage assertion without using **`sync_reject_on`**:

```
c1:  cover property(@(posedge clk) sync_reject_on(a)
   b[*2] ##1 c
);
```

13.2. In the module below, determine for each occurrence of a nested disable condition whether or not the nesting is legal. Also, for each concurrent assertion that is not involved with an illegal nested disable clause, identify the disable condition that governs it, if any.

```
module #(parameter BAD) m (
  logic a, b, c, clk, reset, retry, bad
);
   default clocking PCLK @(posedge clk); endclocking
   default disable iff reset;
   property p; disable iff (retry) b |=> c; endproperty
   a0:  assert property(a |=> b);
   a1:  assert property(disable iff (retry) a |=> b);
   a2:  assert property(p);
   a3:  assert property(a |=> p);
   generate if (BAD)
```

```
            begin: GEN_BAD
               default disable iff bad;
               a0:  assert property(a |=> b);
               a1:  assert property(disable iff (retry) a |=> b);
               a2:  assert property(p);
               a3:  assert property(a |=> p);
            end
         endgenerate
         module m_nested;
            default disable iff reset && bad;
            a0:  assert property(a |=> b);
            a1:  assert property(disable iff (retry) a |=> b);
            default disable iff retry;
            a2:  assert property(p);
            a3:  assert property(a |=> p);
         endmodule
      endmodule
```

13.3. Give an alternative encoding of the assertion in Fig. 13.11 that uses a simpler abort condition.

13.4. Explain the meaning of the following properties:

1. **accept_on**(a) **reject_on**(b) p.
2. **reject_on**(a) **accept_on**(b) p.

Chapter 14
Procedural Concurrent Assertions

My use of language is part and parcel of my message.

— Theo Van Gogh

A traditional way of writing assertions is to place them and consider them as procedural statements. Various programming languages already provide some syntactic forms to express assertions, either as first class language features or as language extensions [4, 50, 53, 60] expressed using pragmas or comments. Depending on the objectives of a language, assertions can vary from being simple Boolean checks that ensure the sanctity of variable values to being event or time based for expressing checks over temporality of values. We have already seen the immediate and deferred assertions in Sects. 4.2 and 4.3 of SystemVerilog that are written as procedural statements.

A concurrent assertion written in a procedure is called a *procedural concurrent assertion.*

Clearly, concurrent assertions are more complex than immediate or deferred assertions. The influence of clocks and synchronous delays in the assertion evaluation is appreciable. Such evaluations may require more than a single simulation time step, sometimes open ended with no predetermined time span. Yet, a concurrent assertion attempt behaves in a similar way to a task, that, once started, carries on its execution of statements with no predetermined time span. Another procedural statement that is similar in its behavior is `fork..join`, which starts executing parallel processes with individual threads of evaluation that possibly terminate without a coordinated end point between the threads.

E. Cerny et al., *SVA: The Power of Assertions in SystemVerilog*,
DOI 10.1007/978-3-319-07139-8_14

In this chapter, we describe how concurrent assertions can be placed in procedural code and how one can make use of the code context in which they are placed. There are many nuances that need to be understood for proper usage of procedural concurrent assertions. We discuss a commonly used form for replicating assertions by placing them in a for-loop. We describe in detail the simulation semantics, that is, how procedural assertions are evaluated in simulation. We describe the use of the **disable** statement to abort evaluations of procedural concurrent assertions that have been invoked but not yet committed. Finally, we explain how the instantiation of a checker in procedural code creates procedural concurrent assertions, from both static and procedural concurrent assertions in the checker declaration.

14.1 Using Procedural Context

Due to the temporal behavior of concurrent assertions, they are restricted to be placed either in an **always** procedure or an **initial** procedure. Contrary to functions and other constructs which must not incur any time delays, these procedures allow evaluations to continue past a single time unit. The **always** procedure may be of any kind, including **always_comb**, which has an implicit event expression for controlling the execution of the procedure. Concurrent assertions cannot be placed in a function, task, or a class.

> Concurrent assertions may only be placed in an **always** or **initial** procedure.

Let us start with a simple example of placing a concurrent assertion in an **always** procedure.

Example 14.1. A concurrent assertion in an **always** procedure:

```
module e1Unit(input logic clk, ...);
  logic i1, i2, d1, dout;
  always @(posedge clk) begin
    d1 <= i1|i2;
    dout <= d1;
    a1: assert property (nexttime[2] dout == $past(i1|i2, 2));
  end
  ...
endmodule
```

A new evaluation attempt of assertion `a1` is started each time a clock tick **posedge** `clk` occurs, and this clocking event is used to reckon time within `a1`. `a1` evaluates to make sure that the current value of `i1|i2` becomes the value of dout two clock ticks later. This reflects the intent of the cascade of non-blocking assignments to d1 and dout. The values of d1 and dout may be changing at successive clock ticks. Each evaluation attempt of a1 provides its own results, without interfering in the evaluation of other attempts. □

Placing an assertion in procedural code can greatly improve the understanding of the purpose of the procedural code. The surrounding code forms the context for and motivates the placement of the assertion. This naturally leads to greater readability and maintainability of the code and the assertion. When the assertion fails, debugging is improved because the context of the assertion is readily available. In the above example, it is clear that assertion a1 is placed to ensure the correctness of propagation of i1|i2 through d1 to dout. If the temporal propagation of values is violated, an error message is generated to indicate the precise point of failure. This is immensely useful to the user as the debugging of the failure is made by inspecting the values of the variables dout, i1, i2, and the intermediate quantity d1.

The context of a procedural concurrent assertion affects the semantics of its execution in two ways. First, the assertion is treated as a statement by the event simulation semantics to start an evaluation at the point it is reached in procedural execution. We shall say that a procedural concurrent assertion is *invoked* at the point when it is reached in the flow of procedural execution. Second, the leading clock of the assertion is inferred from the enclosing procedure and used to mark the passage of time, just as in any concurrent assertion. These two effects are illustrated in the following modification of Example 14.1.

> The clock of a procedural concurrent assertion is inferred from its context.

Example 14.2. A concurrent assertion under an enabling condition:

```
module e1Unit(input logic clk, ...);
  logic i1, i2, d1, dout, out_en;
  always @(posedge clk) begin
    d1 <= i1|i2;
    if (out_en) begin
      dout <= d1;
      a2_1: assert property (nexttime dout == $past(i1|i2, 2));
    end
    else
      a2_2: assert property (nexttime $stable(dout));
  end
  ...
endmodule
```

Assertion a2_1 gets invoked if the value of out_en is true. Otherwise a2_2 gets invoked. Both assertions infer the clocking event **posedge** clk from context. □

In an **initial** procedure, the flow of execution is initiated at the beginning of simulation, and is carried on until the end of the procedure, without ever returning to the initial point again. During this flow of execution, an assertion is invoked whenever it is reached. Embedding a concurrent assertion in an **initial** procedure is shown in Example 14.3.

Example 14.3. A concurrent assertion in an **initial** procedure:

```
module b1Unit(input bit clk, rst, ...);
  bit running;
  initial begin: B1
    if (!rst) begin
      a3: assert property (@(posedge clk) running until rst);
      ...
    end
  end
  ...
endmodule
```

Procedural block B1 is executed at time 0. Suppose that rst is initially low. At that point, assertion a3 is initiated. It waits until the clock tick **posedge** clk occurs, and then starts an evaluation attempt. No more new assertion attempts are started, but the initial attempt will continue its evaluation until completed in accordance with the normal assertion semantics. □

14.2 Clock Inferencing

When an assertion leading clock is not explicitly specified, this clock is inferred from the assertion context. For concurrent assertions outside procedural code, the clock inference rules have been discussed in Chap. 12. For procedural concurrent assertions, the inferred clock is derived from the preceding event control statement in the procedure. In the examples seen so far, the inferred clock has been the entire event expression of the event control. The rules for clock inference follow synthesis guidelines to provide for more general extraction of an inferred clock. There is some subtlety to the rules, and in some cases the form of the procedure will not admit any inferred clock.

> Clock inferencing closely follows common design synthesis guidelines.

We say that an event expression is *inferable* if the entire expression has one of the following forms:

I1 An event variable or a clocking block event.

I2 edge_op expr1 [**iff** expr2], where edge_op is one of **posedge**, **negedge**, or **edge**.[1]

We say that an inferable event expression *E* is *valid* for an **always** or **initial** procedure provided is satisfies the following:

V1 The procedure has one and only one event control, and *E* is an expression appearing in that event control.
V2 There is no other blocking timing control in the procedure.
V3 If *E* is an event variable or a clocking block event, then *E* does not appear in the body of the procedure, except in a clocking event or within an assertion statement.
V4 If *E* has the form edge_op expr1 [**iff** expr2], then no term in expr1 appears in the body of the procedure, except in a clocking event or within an assertion statement.
V5 If *E* has form edge_op expr1, then a larger event expression *E* **iff** expr2 does not appear in the event control of the procedure.

Now we can state the *Rule of Clock Inferencing*:

Event expression *E* is the clock inferred for an **always** or **initial** procedure provided *E* is the one and only valid inferable event expression for the procedure. If the number of valid inferable event expressions for the procedure is zero or greater than one, then no clock is inferred for the procedure.

If no clock is inferred, then clocks may be explicitly specified for the assertions. Another alternative is to declare a default clock, which provides the clock for assertions that would otherwise remain unclocked.

The following examples illustrate some details of clock inferencing.

Example 14.4. Clock inferencing with **iff** in the event control:

```
always @(posedge clk iff en) begin
  d1 <= i1|i2;
  a4: assert property (d1 |=> i3|i4);
  dout <= f_ecap(d1);
end
```

The entire expression **posedge** `clk` **iff** `en` is inferred as the clock, not just the subexpression **posedge** `clk`. According to *V5*, the smaller event expression **posedge** `clk` is not valid. □

[1]Note that while **edge** v is semantically the same as **posedge** v **or** **negedge** v, these forms are not interchangeable for clock inferencing. **edge** v is inferable, while **posedge** v **or** **negedge** v is not.

Example 14.5. No clock inferred – **logic** variable without an edge operator:

```
logic clk;
...
always @(clk) begin
  d1 <= i1|i2 ;
  a5: assert property (d1 |=> i3|i4);
  dout <= f_ecap(d1);
end
```

`clk` is just a **logic** variable, without any edge operator, so it is not an inferable event expression. Therefore, `clk` is not inferred as the leading clock for assertion `a5`. □

Example 14.6. Event variable inferred as clock:

```
event ev;
...
always @(ev) begin
  d1 <= i1|i2;
  a6: assert property (d1 |=> i3|i4);
  dout <= f_ecap(d1);
end
```

`ev` is an event variable and is inferred as the clock for `a6`. `ev` would also be inferred if it were a reference to a clocking block event. □

Example 14.7. No clock inferred—a variable in the event expression is used in the procedure:

```
always @(posedge(e1|e2) iff !reset) begin
  d1 <= i1|i2;
  e1N = ~e1;
  a7: assert property (d1 |=> i3|i4);
  dout <= f_ecap(d1);
end
```

Because term `e1` is used in the blocking assignment, Rule *V4* specifies that **posedge**`(e1|e2)` **iff** `!reset` is not valid. This restriction, however, does not apply to terms in the right-hand operand expression of operator **iff**. Term `reset` could be used in the **always** procedure without affecting validity. □

> Terms in the right-hand operand expression of **iff** may be used freely in the procedural block without affecting validity.

Example 14.8. Terms referenced in an assertion statement do not affect validity:

```
always @(posedge(e1|e2)) begin
  d1 <= i1|i2;
  a8_1: assert property (d1 |=> i3|i4);
  a8_2: assert property (@(negedge e1) !e2);
```

```
  dout <= f_ecap(d1);
end
```

By *V4*, the references to terms `e1` and `e2` in the assertion `a8_2` do not affect validity, so **posedge**`(e1|e2)` is inferred as the leading clock for assertion `a8_1`. Such a reference can be in the body property or in an action block of the assertion statement. □

> Terms from an event expression may be used within assertion statements in the procedural block without affecting validity.

When the event control of a procedure consists of a list of event expressions, separated either by comma or by operator **or**, then the Rule of Clock Inferencing requires that there be only one event expression from the list that is inferable and valid for the procedure. Otherwise, no clock is inferred from the procedure.

Example 14.9. No clock inferred—two valid inferable event expressions:

```
always @(posedge e1 or posedge e2) begin
  d1 <= i1|i2 ;
  a9: assert property (d1 |=> i3|i4);
  dout <= f_ecap(d1);
end
```

The Rule of Clock Inferencing is violated. □

> A clock is inferred only if a single valid inferable event expression is specified in the event control.

In the following example, asynchronous reset is specified as an event expression. Even though the event control has two inferable event expressions, only one is valid.

Example 14.10. A common use of asynchronous reset event:

```
always @(posedge e1 or posedge reset) begin
  if (reset)
    d1 <= 0;
  else begin
    d1 <= i1|i2;
    a10: assert property (d1 |=> i3|i4);
    dout <= f_ecap(d1);
  end
end
```

The clock **posedge** e1 is inferred for assertion a11 since it is the only valid inferable event expression. By *V4*, event expression **posedge** reset is not valid due to the use of term reset inside the procedure. □

In the example below, the additional event control @(**posedge** e2) violates *V1* and prevents any clock inferencing for assertion a11.

Example 14.11. No clock inferred—additional event control:

```
always @(posedge e1) begin
  d1 <= i1|i2 ;
  @(posedge e2) egL <= sL & sL1;
  a11: assert property (d1 |=> i3|i4);
  dout <= f_ecap(d1);
end
```
□

Only one event control may appear in the procedure for clock inferencing.

In the example below, the inclusion of a delay statement #6 violates *V2* and bars clock inferencing for assertion a12.

Example 14.12. No clock inferred—presence of a delay statement:

```
always @(posedge e1) begin
  d1 <= i1|i2 ;
  #6;
  a12: assert property (d1 |=> i3|i4);
  dout <= f_ecap(d1);
end
```
□

Whether a concurrent assertion is placed in an **always** procedure or an **initial** procedure, the assertion is invoked, like any other statement, only if the simulation execution reaches the assertion statement. From that moment onward, the temporal assertion evaluation is driven by its leading clock as if the assertion were placed outside its enclosing procedure. If the leading clock is the same as the contextually inferred clock, then the evaluation for that clock tick happens in the same time step as its invocation. Otherwise, the evaluation waits further until the leading clock occurs. We discuss the precise semantics of scheduling procedural concurrent assertions later in this chapter.

14.3 Using Automatic Variables

An important consideration in constructing a procedural concurrent assertion is the presence of variables declared in the procedure that are part of the simulation execution reaching the assertion statement. Static variables are treated differently than the automatic variables. Even though the values of static variables are in progress during the simulation execution, references to those variables in the assertion expression use the Preponed region values, following the same paradigm as all other concurrent assertions. Using Preponed region values is essential to obtaining a deterministic result. This is due to the fact that continuing temporal evaluation from one clock tick to the next takes place in its own thread of execution, which is apart from the execution of the enclosing procedural block.

The value of an automatic variable is captured at the time a procedural concurrent assertion is invoked.

The semantics for an automatic variable in a procedural concurrent assertion is, however, to capture its value at the time the assertion is invoked, i.e., when procedural execution reaches the assertion. The captured value is used for the variable throughout that evaluation attempt of the assertion.

In the next examples, we assume that `store` and `ptr` are static variables and that `ev` is an event variable.

Example 14.13. Automatic variable in a procedural assertion and a timing problem:

```
always @(ev) begin
  automatic dataT d = pipeline.pop_front();
  store[ptr] = d;
  a13: assert property (@(posedge clk) store[ptr] == d);
  ptr = next_ptr(ptr);
end
```

Each time `a13` is invoked, the value of automatic variable `d` is recorded at that point, and this value is used to compare to `store[ptr]` at **posedge** `clk`. Since `store` and `ptr` are static variables, their references in `a13` use Preponed region values. This causes a timing problem because `ptr` will typically have been updated before the time step of the next **posedge** `clk`, so the value of `d` is compared to the wrong element of `store`. □

The following variant solves the timing problem by using another automatic variable to capture the value of `ptr` for use in the assertion.

Example 14.14. Automatic variables in procedural assertion:

```
always @(ev) begin
  automatic dataT d = pipeline.pop_front();
  automatic ptrT ptr_copy = ptr;
  store[ptr] = d;
  a14: assert property (@(posedge clk) store[ptr_copy] == d);
  ptr = next_ptr(ptr);
end
```

Since both d and ptr_copy are automatic, their values are captured at the invocation of a14. The reference to store still uses Preponed region values in the time step of **posedge** clk, but ptr_copy holds the value that ptr had when the assignment to store was made in the preceding line of the procedure. □

Another good option exists to control whether a procedural assertion uses Preponed region values of static variables or captures their values at invocation of the assertion. By using **const** cast, as in **const**'(v), the value of variable v is captured when procedural execution reaches the assertion, and this value is used throughout that evaluation attempt, regardless of the way v may be updated. In this way, a static variable can be treated as automatic in assertion evaluation without having to create an auxiliary automatic variable.

Example 14.15. Use of **const** cast for a static variable:

```
always @(ev) begin
  automatic dataT d = pipeline.pop_front();
  store[ptr] = d;
  a15: assert property (@(posedge clk) store[const'(ptr)] == d);
  ptr = next_ptr(ptr);
end
```

By using **const** cast for static variable ptr, the behavior of assertion a15 is the same as a14. □

The same rules apply to references to automatic variables and **const** cast expressions in the action blocks of procedural concurrent assertions. The values of automatic variables and **const** cast expressions are captured when the assertion is invoked, and these values are used in the assertion evaluation and when any action blocks execute for that attempt. Recall that at the time of execution of action blocks, references to static variables use Reactive region values, not Preponed region values.

14.4 Assertions in a For-Loop

So far we have seen how clocks and values are applied to procedural assertions. In this section, we show a form in which an assertion is replicated and activated under nested conditions. This form is well suited and natural to express a variety of scenarios and is based on dynamic values of the surrounding procedural conditions.

Example 14.16. Replication of assertions using for-loop:

```
logic treg;
logic [3:0] dreg;
logic [7:0] tr;
//...
always @(posedge clk) begin
  if (treg) begin
    for  (int i = 0; i < 4; i++) begin
```

```
      dreg[i] <= tr[i + 1];
      c16: cover property (dreg[i] ##[1:8] tr[i + 1]);
    end
  end
end
```

Four evaluation attempts of `c16` are initiated when the for-loop gets executed, each with a different value of index `i`, from `0` to `3`. Following their invocations, the inferred clock, **posedge** `clk`, drives each evaluation attempt to progress and finish independently of the others. Since variable `i` is an automatic variable, its value is captured separately at each invocation of `c16` and used throughout the respective evaluation attempt. Thus, `c16` behaves like four covers:

```
always @(posedge clk) begin
  if (treg) begin
    for  (int i = 0; i < 4; i++) begin
      dreg[i] <= tr[i + 1];
    end
    c16_0: cover property (dreg[0] ##[1:8] tr[1]);
    c16_1: cover property (dreg[1] ##[1:8] tr[2]);
    c16_2: cover property (dreg[2] ##[1:8] tr[3]);
    c16_3: cover property (dreg[3] ##[1:8] tr[4]);
  end
end
```

The four covers `c16_0`, `c16_1`, `c16_2`, and `c16_3` are started whenever the condition `treg` is true for a clock tick of **posedge** `clk`. An FV tool may actually split the single assertion into four equivalent assertions as shown here. This technique is well suited for formal verification, where analysis of assertion behaviors generally accounts for evaluation attempts differently than simulation. □

In the following example, the number of for-loop iterations is controlled by signal `count`, resulting in a varying number of assertion invocations from clock tick to clock tick. Again, the value of the automatic variable `i` is captured whenever `c17` is invoked.[2]

Example 14.17. A variable for-loop index replicating of assertions:

```
always @(posedge clk)  begin
  if (treg) begin
    for  (int i = 0; i < count; i++) begin
      dreg[i] <= tr[i + 1];
      c17: cover property (dreg[i] ##[1:8] tr[i + 1]);
    end
  end
end
```

□

[2]FV tools may not be able to handle assertions in procedural loops that cannot be statically unrolled.

14.5 Event Semantics of Procedural Concurrent Assertions

Before delving into the details of semantics, let us first review event simulation semantics that are important for the execution of procedural concurrent assertions. The event semantics of assertion simulation is mostly carried out in three regions: Active region, Observed region, and Reactive region. Although scheduling evaluation, performing evaluation, and detecting events take place in all three regions, each region has a unique role in these activities for assertions. The majority of actual evaluation of assertion expressions takes place in the Observed region, the scheduling is largely performed in the Active region, and the Reactive region is used for processing the action blocks. The role of the regions and their order is depicted in Fig. 3.2.

To support procedural concurrent assertion evaluation within the event simulation semantics, two new semantic objects are introduced: *procedural assertion queue* and *matured assertion queue*.

A procedural assertion queue is used as a temporary holding place for instances of assertions that have been invoked, together with any captured values of variables. These assertion instances *pend* in the procedural assertion queue until it is determined in the Observed region that they should mature for evaluation. Thereupon, the assertions are transferred to the matured assertion queue to await the arrival of their respective leading clocks. Each procedure, such as **`always`** or **`initial`**, that contains procedural concurrent assertions has its own procedural assertion queue. There is only one matured assertion queue for all procedures. While in a procedural assertion queue, prior to maturing, an instance of an assertion can get purged, as we will illustrate later in the examples.

Note that these semantic objects have no explicit representation in the language. Their sole purpose is to explain the semantics of procedural concurrent assertions.

Here are the roles of the two queues for processing an assertion:

1. In the Active region, when procedural execution reaches a concurrent assertion, an instance of the assertion is entered in the procedural assertion queue of the process, together with any captured values of variables for that instance.
2. During the execution in the Active region, an assertion instance waiting in the procedural assertion queue may get purged.
3. In the Observed region, all surviving assertion instances from the procedural assertion queues *mature* and are transferred to the matured assertion queue.
4. If the leading clock of a matured assertion instance did trigger in the Active region, then the evaluation attempt of that assertion begins and the assertion instance is removed from the matured assertion queue to follow its normal course of evaluation. Otherwise, the assertion instance waits in the matured assertion queue until its leading clock triggers in some future time step.

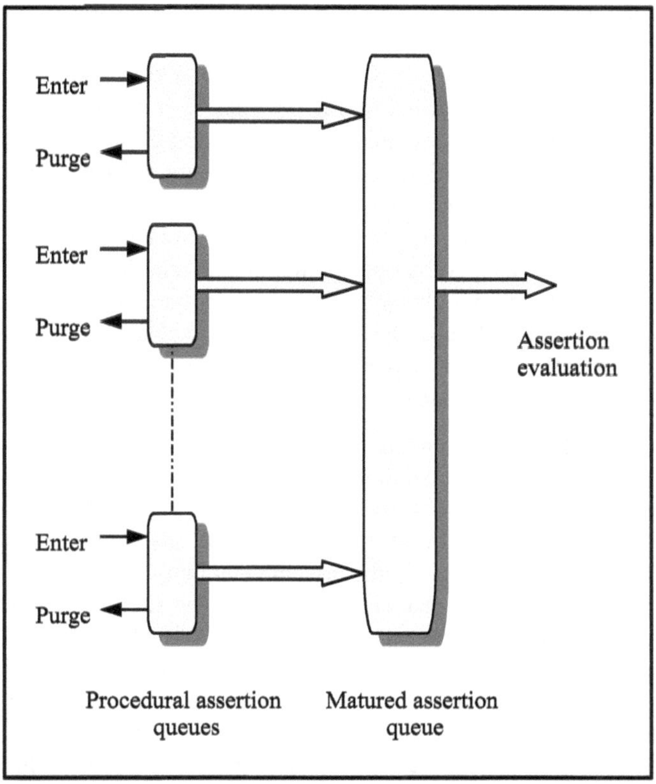

Fig. 14.1 The role of queues for procedural assertions

> Each procedure containing a concurrent assertion uses its own procedural assertion queue to enter and purge the assertion attempts.

Figure 14.1 illustrates the role of the two queues.

We will see later how an assertion gets removed from the procedural assertion queue. First, let us follow an example to see how events cause assertion evaluation.

Example 14.18. Implicit events triggering assertion evaluation:

```
always @(*) begin: B1
  r11 <= v11;
  if (c1_long) begin
    r12 <= v12 & v13;
    a18: assert property (@(posedge clk) r11 |=> r12);
  end
end
```

When **always** procedure block B1 gets triggered because of its sensitivity to implicit events, the nonblocking assignment to r11 gets scheduled. Next, if c1_long is false, no further action takes place. Otherwise, nonblocking assignment to r12 is scheduled. An instance of assertion a18 is now placed in the procedural assertion queue associated with B1. Assume that no other implicit events occur in the Active region. Eventually the simulation control moves to the Observed region. In the Observed region, the instance of a18 is still in the procedural assertion queue, so it is moved to the matured assertion queue. If the leading clock **posedge** clk occurred in the Active region, then an evaluation attempt of a18 is initiated in the current time step in the Observed region. Otherwise, a18 stays in the matured queue pending a tick of **posedge** clk in a future time step. Once an evaluation attempt matures, it continues its evaluation at every clock tick until it completes, independently of any other attempt that may be initiated. □

Now, let us see how an extraneous assertion attempt is blocked from evaluation. Suppose in Example 14.18 that multiple right-hand terms of the nonblocking assignments are updated in the Active region of a time step. This causes multiple occurrences of the implicit event for B1. Assume also that c1_long is true throughout the time step. To be specific, assume that first v11 gets a new value, and then later in the Active region processing v12 get a new value. When v11 changes, B1 gets triggered, and, as above, an instance of a18 is placed in the procedural assertion queue for B1. Then, when v12 changes, B1 is again triggered. This second trigger of B1 causes all pending instances in the procedural assertion queue for B1 to be purged. Since we are assuming c1_long remains true, the second procedural execution of B1 causes an instance of a18 to be placed in the procedural assertion queue for B1. In general, whenever a procedural block is triggered, all assertions pending in the procedural assertion queue for that block are purged. No matter how many times B1 is triggered in the Active region, only the instance of a18 from the last execution of B1 can remain in the procedural assertion queue at the end of the Active region. The one surviving instance subsequently matures in the Observed region and awaits arrival of a tick of **posedge** clk to continue evaluation. By purging pending assertions each time the procedural block is triggered, the effect of zero-width glitches within a single time step is eliminated.

Triggering a procedural block causes immediate purging of its procedural assertion queue. This prevents multiple invocations of the same procedural concurrent assertion due to zero-width glitches.

Furthermore, the semantics also takes care of the situation where multiple invocations are made to an assertion within a for-loop. The procedural assertion queue for a block is purged only on triggering of that block. Multiple executions of a statement within the block, as in the body of a loop, do not themselves cause purging of the queue.

Example 14.19. Event semantics for assertions in a for-loop:

```
always @(*) begin
  if (c1_long) begin
    for (int i=0; i<4; i++) begin
      dreg[i] <= tr[i+1];
      a19: assert property (@(posedge clk) dreg[i] |=> r12);
    end
  end
end
```

In this example, if the for-loop executes then four instances of assertion `a19` are entered in the procedural assertion queue. Each assertion instance is considered distinct for the purpose of evaluation and has its own value of the automatic loop index variable `i`. □

14.6 Things to Watch Out For

There are many subtle situations that can arise due to unscrupulous placement of procedural concurrent assertions. Interspersing assertions with timing control statements (i.e., delay control, event control, or **wait** statements) can be especially tricky and counterintuitive because of flushing of the queue. In general, the procedural assertion queue is flushed any time execution of the procedure resumes after being suspended, not just when the procedure is re-invoked from the beginning.

> The procedural assertion queue is flushed any time execution of the associated procedure resumes after being suspended, not just when the procedure is re-invoked from the beginning.

The next example illustrates interspersing procedural assertions with a delay control statement.

Example 14.20. Assertions interleaved with a positive delay control statement:

```
always @(*) begin: B1
   r11 <= v11 | v12;
   a20_1: assert property (@(posedge clk) v11 |=> v12);
   #5;
   r12 <= v12 & v13;
   a20_2: assert property (@(posedge clk) r11 |=> r12);
end
```

Consider a time step in which block `B1` starts evaluating. An instance of assertion `a20_1` enters the procedural assertion queue for `B1`. The evaluation of the delay control #5 blocks further evaluation in the Active region. In the Observed region the instance of a20_1 is transferred to the matured assertion queue, pending the arrival of its clock. Due to the blocking effect of the delay control statement, multiple triggerings of B1 in the same time step cannot occur since the process is suspended.

The behavior of a20_2 is quite different. After the delay of five units of time, the procedural block restarts its evaluation in the Active region, and an instance of a20_2 enters the procedural assertion queue for B1. If B1 is not triggered again prior to the Observed region in the same time step, then a20_2 is transferred to the matured assertion queue. Otherwise, triggering B1 causes assertion a20_2 to be purged from the procedural assertion queue. Execution of B1 would continue until the delay control statement, without rescheduling a20_2 in that time step. □

The following variant of Example 14.20 switches the delay control from #5 to #0, significantly changing the assertion evaluation behavior.

Example 14.21. Assertions interleaved with a #0 delay control statement:

```
always @(*) begin: B1
  r11 <= v11 | v12;
  a21_1: assert property (@(posedge clk) v11 |=> v12);
  #0;
  r12 <= v12 & v13;
  a21_2: assert property (@(posedge clk) r11 |=> r12);
end
```

When B1 is triggered, an instance of a21_1 is placed in the procedural assertion queue for B1. Execution of #0 suspends the procedure and schedules its resumption in the Inactive region of the current time step. When the Active region processing finishes, Inactive region events are moved to Active and the procedure resumes. At that point, the instance of a21_1 is flushed from the procedural assertion queue. Because of this flushing, a21_1 can *never* survive to the Observed region to mature. After the #0 delay, an instance of a21_2 is placed in the queue. That instance will mature provided B1 is not triggered again before the Observed region. □

Another peculiar situation transpires when the event control of the **always** procedure is different than the clock of the assertion, as in the example below.

Example 14.22. An assertion with a different clock than its **always** procedure:

```
always @(posedge clk) begin
  i3 <= lb1 && lb8;
  i4 <= lb2 && lb11;
end
always @(posedge e1)  begin
  d1 <= i1 + i2 ;
  a22: assert property (@(posedge clk) d1 |=> (i3 || i4));
  dout <= f_ecap(d1);
end
```

The number of evaluations of `a22` is determined by the rate of occurrences of event **posedge** `e1` and of the assertion clock, **posedge** `clk`. There is one instance of `a22` scheduled per occurrence of event **posedge** `e1`, even if the assertion clock occurs more frequently. This seems reasonable. The same rule applies if **posedge** `e1` occurs much more frequently than the assertion clock. In that case, `a22` is placed into the matured assertion queue more than once, waiting for the next occurrence of **posedge** `clk`, and all those instances get evaluated in parallel for the same occurrences of the assertion clock. Indeed, the evaluation behavior of `a22` is the same as the following multiply clocked, non-procedural concurrent assertion:

```
a22_non_proc:  assert property (
  @(posedge e1) 1 |=>
  @(posedge clk) d1 |=> (i3 || i4)
);
```

□

> When the leading clock of a procedural concurrent assertion is not inferred, multiple redundant evaluation attempts may be possible.

We have seen how the procedural assertion evaluation framework can filter out superfluous assertion evaluations due to the transient value changes in signals in the Active region. This glitch protection can be lost if the signals change in the Reactive region rather than the Active region. One such case is shown below.

Example 14.23. An assertion trigger due to a change in Reactive region:

```
1  program prg();
2    integer i;
3    initial
4      for (i=0; i<8 ; i++) begin
5        mod.v11= 1; #5;
6        mod.v11 = 0;
7      end
8  endprogram
9
10 module mod(input sig2,clk);
11   bit  v11,r11,r12;
12   wire v12;
13   assign v12 = sig2;
14   always @(posedge v11 or posedge v12) begin: B1
15     r11 = v11 | v12;
16     a23: assert property (@(posedge clk) v11 |=> v12);
17     r12 = v11 & v12;
18   end
19 endmodule
```

When the event **posedge** `v12` occurs due to the **assign** statement in Line 13, block `B1` is evaluated. As expected, an instance of assertion `a23` is entered in the procedural assertion queue, with a transfer to the matured assertion queue in the Observed region, assuming no more triggers of `B1` in this Active region. Now, in the Reactive region, when variable `v11` is assigned `1`, it causes the initiation of the Active region once again, re-triggering block `B1`. The procedural assertion queue for `B1` is purged, but the previous instance of `a23` has already matured, so the purging has no effect. Then another instance of assertion `a23` is entered in the procedural assertion queue. Finally, that assertion evaluation is also transferred to the matured queue when the Observed region is entered for the second time. At that point, there are now two instances of `a23` in the matured assertion queue that await their clock tick. □

14.7 Dealing with Unwanted Procedural Assertion Evaluations

As we learnt from Example 14.23, there are some situations where redundant evaluations of assertions occur. The user is provided with an explicit means to stop such cases. The procedural assertion queue can be purged entirely or for a specific assertion with the **`disable`** statement. But, the evaluation attempts already advanced to the matured queue or the ongoing attempts that were started in previous time steps are not affected.

The example below modifies Example 14.23 by adding a **`disable`** statement for the assertion to prevent the extra evaluation of `a24` in case both **`posedge`** `v11` and **`posedge`** `v12` occur in the same time step. Crafting the conditions for such a **`disable`** statement can be delicate.

Example 14.24. A **`disable`** statement purging an assertion evaluation:

```
program prg();
  integer i;
  initial
    for (i=0; i<8 ; i++) begin
      mod.v11= 1;
      #5;
      mod.v11 = 0;
    end
endprogram
module mod(input bit sig2,clk);
  bit v11,r11,r12;
  wire v12;
  assign v12 = sig2;
  always @(posedge v11 or posedge v12) begin: B1
    r11 = v11 | v12;
    a24: assert property (@(posedge clk) v11 |=> v12);
    r12 = v11 & v12;
    if (v11 && !$sampled(v11) && v12 && !$sampled(v2))
      disable a24;
```

```
  end
endmodule
```

The entry of assertion a24 is removed by the **disable** statement under the condition that both v11 and v12 are high and **posedge** events have been seen on both in the time step so far. □

> Evaluation attempts already in the matured queue or ongoing from previous time steps are not affected by the **disable** statement.

In some cases, purging of all assertions for a block may be appropriate. This can be accomplished by executing a **disable** statement with the block name as its target. Bear in mind that such a **disable** applies to all code in the block, not just the procedural concurrent assertions.

Example 14.25. A **disable** statement purging all assertion evaluations in a block:

```
always @(*) begin: B2
  a25_1: assert property (@(negedge clk) !(dreg & tr));
  if (c1_long) begin
    for (int i=0; i<4; i++) begin
      dreg[i] = tr[i+1];
      treg = flogic (dreg[i]);
      a25_2: assert property (@(posedge clk) dreg[i] |=> r12);
    end
  end
  if (retry_cond) disable B2;
end
```

Assume that retry_cond signals a condition under which all instances of assertions of block B2 should be purged from the procedural assertion queue. This can be accomplished with the **disable** targeting the whole block B2. If retry_cond is true, then the instance of a25_1 and all instances of a25_2 are purged from the procedural assertion queue. Otherwise, the instances mature and are evaluated normally. By placing the **disable** at the end of the block, it doesn't disturb any of the assignments made in the block. □

To conclude this section, we mention that a formal verification tool may extract the procedural assertions from the procedures as shown in Example 14.16, including the enabling conditions, and evaluate them as regular concurrent assertions. Such extraction may create differences between the simulation and formal verification semantics. For more details on formal verification, see Chap. 21.

14.8 Procedural Checker Instances

Checkers are generalizations of concurrent assertions that also serve as containers for verification code that supports the assertions. As such, it is desirable to be able to instantiate checkers wherever concurrent assertions can be written, including within procedures. Supporting this capability comes with some challenges to define how various checker constructs interact with the surrounding procedural context. The instantiation rules formulated for static checker instances hold also for procedural checker instances, but there are additional rules applicable to procedural instances, as well as considerations for their effective use. These are the subjects of this section.

Static Assertions

Static assertions in a checker declaration (i.e., those checker assertions that are not in the scope of any checker procedure) become procedural assertions as the result of a procedural instantiation of the checker.

Example 14.26. Assertion `a1` is static in checker `check1`:

```
checker check1(a, event clk = $inferred_clock);
  a1: assert property (@clk a);
endchecker : check1
```

The instantiation `c1` of checker `check1` in module `m`

```
module m(input logic clock, b, en, ...);
  // ...
  always @(posedge clock) begin
    if (en) begin
      // ...
      check1 c1(b);
    end
  end
endmodule : m
```

is conceptually[3] equivalent to

```
module m(input logic clock, b, ...);
  // ...
  always @(posedge clock) begin
   if (en) begin
      // ...
    a1: assert property (@(posedge clock) b);
    end
  end
endmodule : m
```

That is, the static assertion `a1` in checker `check1` behaves as a procedural assertion in checker instantiation `c1` in module `m`. □

---

[3]We use the term "conceptually" because of the hierarchical assertion naming in the checker, as explained in Sect. 9.3.2.

Checker Procedures

Straightforward inlining cannot work for checker procedures: it is illegal to have nested procedures in SystemVerilog. One should regard checker procedures as if they were instantiated outside the procedural code. Therefore, always procedures in checkers are sort of screens "protecting" their contents from the procedural code at the place of the checker instantiation.

Example 14.27. Consider the following checker `check` instantiated in module `m`.

```
checker check(a, event clk);
  always_ff @clk
    a1: assert property (@clk a);
endchecker : check
module m(input logic clock, b, en);
  // ...
  always @(posedge clock) begin
    if (en) begin
      // ...
      check mycheck(b, negedge clock);
    end
  end
endmodule : m
```

This checker instantiation is conceptually equivalent to the following code:

```
module m(input logic clock, b, en);
  // ...
  always @(posedge clock) begin
    if (en) begin
      // ...
    end
  end
  always_ff @(negedge clock)
    a1: assert property (@(negedge clock) b);
endmodule : m
```

The **always_ff** procedure in the checker screens assertion `a1` from the direct effect of the **always** procedure in the module: assertion `a1` is controlled by **negedge** `clock` and does not depend on the value of `en`. This behavior is different from the behavior of static assertion `a1` in checker `check1` from Example 14.26. □

Example 14.28. Assertion `a1` in the following checker is instantiated in the scope of an **initial** procedure.

```
checker check(a, event clk = $inferred_clock);
  initial
    a1: assert property (@clk s_eventually a);
endchecker : check
module m(input logic clock, b, en);
  // ...
  always @(posedge clock) begin
    if (en) begin
      // ...
```

```
      check mycheck(b, negedge clock);
    end
  end
endmodule : m
```

This checker instantiation is conceptually equivalent to the following code:

```
module m(input logic clock, b, en);
  // ...
  always @(posedge clock) begin
    if (en) begin
      // ...
    end
  end
  initial
    a1: assert property (@(negedge clock) s_eventually b);
endmodule : m
```

Even though the checker is instantiated in an **always** procedure of module m, assertion a1 is monitored only once. □

> Monitoring of procedural checker assertions is done according to the placement of the assertions in the checker, not according to the placement of the checker instantiation in a module or in an interface.

Checker Instantiation in Procedural Loops

Procedural loops are just a particular case of procedural code, and thus all the rules of checker instantiation in procedural code are applicable here too. Note, however, that if a checker actual argument depends on a loop variable, the loop variable should have automatic lifetime or **const**' cast should be applied to the actual argument.

Example 14.29. Consider the following checker instantiation:

```
checker check(a, b, event clk = $inferred_clock);
  a1: assert property(@clk a |=> b);
endchecker : check
module m(input logic clock, logic [7:0] req, ack);
  // ...
  always @(posedge clock) begin
   for (int i = 0; i < 8; i ++)
      if (i != 3) begin
        // ...
        check mycheck(req[i], ack[i]);
      end
  end
endmodule : m
```

According to the instantiation semantics in procedures, the checker instantiation is conceptually equivalent to

```
module m(input logic clock, logic [7:0] req, ack);
  // ...
  always @(posedge clock) begin
   for (int i = 0; i < 8; i ++)
      if (i != 3) begin
        // ...
        a1: assert property(req[i] |=> ack[i]);
      end
  end
endmodule : m
```

It is important that the loop variable `i` be automatic. If the loop in the module `m` were written as:

```
always @(posedge clock) begin
  int i; //Static lifetime
  for (i = 0; i < 8; i ++)
    // ...
```

the resulting assertion would be multiply and redundantly evaluated as

```
  assert property(req[$sampled(i)] |=> ack[$sampled(i)]);
```

This is unlikely to have been intended (see also the discussion about assertion instantiation in loops in Sect. 14.4). If for some reason it is necessary to have the loop variable with static lifetime, the **const'** cast should be explicitly applied:

```
check mycheck(req[const'(i)], ack[const'(i)]);
```

□

Example 14.30. Consider now a checker that contains procedural assertions:

```
checker check(a, b, c, event clk = $inferred_clock);
  default clocking @clk; endclocking
  always_ff @clk begin
    a1: assert property(a);
    a2: assert property(b |=> c);
  end
endchecker : check
module m(input logic clock, ok, logic [7:0] req, ack);
  // ...
  always @(posedge clock) begin
   for (int i = 0; i < 8; i ++)
      if (i != 3) begin
        // ...
        check mycheck(ok, req[i], ack[i]);
      end
  end
endmodule : m
```

The checker instantiation is conceptually equivalent to the following code:

```
module m(input logic clock, ok, logic [7:0] req, ack);
  // ...
  always @(posedge clock) begin
   for (int i = 0; i < 8; i ++)
      if (i != 3) begin
```

```
        // ...
      end
  end
  always_ff @(posedge clock) begin
    a1: assert property(@(posedge clock) ok); // Legal
    a2: assert property(@(posedge clock) req[i] |=>
                                           ack[i]);  // Illegal
  end
endmodule : m
```

There is only one instance of each procedural checker assertion `a1`, `a2` regardless of the checker instantiation in the loop. Even though the instantiation of assertion `a1` is meaningful, the instantiation of `a2` is not because variable `i` is not visible in the context where the procedural checker assertion is rewritten. □

> The code in checker **always** procedures should not depend on procedural loop control variables.

Procedural Checkers with Checker Variables

The behavior of procedural checkers containing checker variables complies with the rules described thus far: the **always** procedure in a checker remains a separate process after instantiation.

Example 14.31. Consider the following instantiation of checker `stable_for_two_ticks` defined in Example 9.19:

```
module m(input logic clock, reset, logic [7:0] en, ...);
  default disable iff reset;

  logic cond;
  logic [7:0] driver;
  // ...
  always @(posedge clock) begin
    for (int i = 0; i < 7; i++) begin
      if (en[i] && cond) begin
        driver[i] <= ...;
        stable_for_two_ticks check_driver(driver[i]);
      end
    end
  end
endmodule : m
```

This is roughly equivalent to the following code:

```
module m(input logic clock, reset, logic [7:0] en, ...);
  default disable iff reset;
  logic cond;
  logic [7:0] driver;
  bit toggle = 1'b0;
```

```
  // ...
  always @(posedge clock) begin
    for (int i = 0; i < 7; i++) begin
      if (en[i] && cond) begin
        driver[i] <= ...;
        a1: assert property (!toggle |-> $stable(driver[i]));
      end
    end
  end
  always @(posedge clock)
    toggle <= reset ? 1'b0 : !toggle;
endmodule : m
```

As usual, the real names `toggle` and `a1` have a different hierarchy than in the inlined version of the checker. Also, the non-blocking assignment to checker variable `toggle` is performed in the Reactive region, and not in the Active region as in modules.

The resulting assertion `a1` remains in the scope of the **for**-loop. In fact, this assertion is checked for all indices `i` for which `en[i] && cond` is true. The assignment of the checker variable `toggle` is performed in a separate process, and does not depend on the values of `i` and the condition `en[i] && cond`.

Checker `stable_for_two_ticks` may be safely instantiated in a procedural loop because the checker variable does not depend on the loop control variables. Otherwise, the checker instantiation in the loop would be illegal. □

> If a checker is instantiated in procedural loop, its variables should not depend on the loop control variables.

Clock Inference in Checker Procedures

Consider the following checker:

```
checker mycheck(a, event clk);
  always_ff @clk begin
    a1: assert property (a);
  end
endchecker : mycheck
```

Will a clock be inferred for `a1` from the checker procedural context of the **`always_ff`**? It depends on the actual argument passed to the formal argument `clk`. The Rule of Clock Inferencing cannot be applied to `@clk`. It must instead be applied after the substitution of the actual argument for `clk` from the instance of `mycheck`. If, for example, the actual argument of `clk` is **`posedge`** `clock`, then **`always`** `@clk` becomes **`always`** `@(`**`posedge`** `clock)`. From the Rule of Clock Inferencing it follows that in this case the assertion infers its clocking event from the **`always`** procedure. Suppose now that the actual argument of `clk` is just `clock`, where `clock` is not an event variable or clocking block event, hence not an inferable event expression. Then **always** @clk becomes **always** @clock after substitution, and the assertion clock cannot be inferred from the procedural context of the **always**. The assertion might still get its clock otherwise, as from a default clock that applies in the context of the checker declaration.

Example 14.32. In the following checker, the inferred clock for assertion a1 is **posedge** clk.

```
checker check1(a, clk);
  always_ff @(posedge clk)
    a1: assert property(a);
endchecker : check1
```

This inference is due to the structure of the checker itself. The event control of the checker procedure is an edge expression **posedge** clk. □

Example 14.33. Consider checker check2 and its instantiations c1 through c5:

```
checker check2(a, event clk = $inferred_clock);
  always_ff @clk
    a2: assert property(a);
endchecker : check2
module m(input logic clock, b, ...);
  // ...
  check2 c1(b, posedge clock);
  check2 c2(b, clock); // Illegal
  always @(edge clock) begin: B1
    check2 c3(b);
  end
  always @(edge clock or negedge b) begin: B2
    check2 c4(b, edge clock);
    check2 c5(b); // Illegal
  end
endmodule : m
```

Assume that no default clock applies to the checker or within the module.

The clock of assertion a2 inferred in the instance c1 is **posedge** clock since this is the actual argument passed to clk and it is inferable.

Instance c2 will not compile since the actual argument corresponding to clk is clock, which is not inferable. In the absence of a default clock, c2.a2 remains unclocked.

The clock **edge** clock is inferred for procedural block B1, so this clock is passed through $inferred_clock as default actual argument to clk in instance c3. **edge** clock is then inferred for the **always_ff** procedure in check2 and is the clock for c3.a2.

No clock is inferred for procedural block B2 because **edge** clock and **negedge** b are both valid inferable event expressions, so uniqueness is violated in the Rule of Clock Inferencing. The instance c5 is therefore illegal because there is no actual argument for clk. □

Exercises

14.1. Add some procedural assertions to represent your understanding of the intent of following code examples.

1.
```
module m #(W=8, D=32) (
   input logic clk,
   input logic [W-1:0] d_in,
   output logic [W-1:0] d_out
);
   logic [D*W-1:0] sh;
   assign d_out = sh[D*W-1:(D-1)*W];
   always @(posedge clk) begin
      sh <= {sh[(D-1)*W-1:0], d_in};
   end
endmodule : m
```

2.
```
typedef logic[31:0] natT;
module m (
   input logic clk, rst_n, load, incr,
   input natT a_in, b_in,
   output natT d_out
);
   natT a, b;
   assign d_out = a;
   always @(posedge clk or negedge rst_n) begin
      if (!rst_n) begin
         a <= 0;
         b <= 0;
      end else if (load) begin
         a <= a_in;
         b <= b_in;
      end else if (incr) begin
         a <= a + b;
         b <= a;
      end
   end
endmodule : m
```

14.2. For each of the following procedures, determine whether or not a clock is inferred. If so, give the inferred clock. If not, explain why not. Assume the following declarations throughout:

```
logic clk, en, rst_n;
logic [7:0] a, b, d_in;
event ev;
clocking CLK @(posedge clk); endclocking
```

1.
```
always @(edge clk) begin
   a <= d_in;
   A_STABLE: assert (a != d_in)
```

```
        else $display("d_in=%h, clk=%b", d_in, clk);
end
```

2.
```
always @(edge clk) begin
   if (clk)
      a <= a_in;
   else
      b <= a_in;
end
```

3.
```
always @((posedge clk iff en) or negedge rst_n) begin
   if (!rst_n)
      a <= 0;
   else if (!en) $error("en expected");
   else
      a <= d_in;
end
```

4.
```
always @((posedge clk iff en) or negedge rst_n) begin
   if (!en) $error("en expected");
   else
      a <= d_in;
end
```

5.
```
always @(posedge clk) begin
   a <= d_in;
   @(negedge clk) b <= d_in;
end
```

6.
```
always @(ev iff en) begin
   a <= d_in;
end
```

7.
```
always @(CLK or negedge rst_n) begin
   if (!rst_n)
      a <= 0;
   else
      a <= a_in;
end
```

8.
```
always @(CLK or (ev iff en)) begin
   a <= a_in;
end
```

14.3. A shift register `sh` has data width `W` and depth `D`. Whenever `load` is true, the entries of `sh` shift up one index, with new data shifted into entry `sh[0]` from `d_in` and the last data in entry `sh[D-1]` shifted out to `d_out`. Code for the shift register is shown below:

```
logic clk, load;
logic [W] d_in, d_out;
logic [D][W] sh;
```

```
always @(posedge clk) begin
   if (load) begin
      sh[0] <= d_in;
      for(int i=1; i<D; i++)
         sh[i] <= sh[i-1];
      d_out <= sh[D-1];
   end
end
```

1. Write a procedural concurrent assertion to check that if `load` is true, then the current value in position `j` of the shift register is the next value in position `j+1`, where `j` ranges from `0` to `D-2`.
2. Write a procedural concurrent assertion to check that after `D+1` occurrences of `load`, `d_out` holds the value that was originally in `d_in`. [Hint: Use **const** cast to capture the value of `d_in` when your assertion is invoked.]

14.4. In Example 14.18, what assertion evaluation results in a time step in which `c1_long` changes value from true to false? Be sure to consider that the change in `c1_long` may be neither the first nor the last event in the time step that triggers procedural block `B2`.

14.5. Consider the following variant of Example 14.21:

```
always @(*) begin: B1
  a <= a_in;
  a_1: assert property (@(posedge clk) a == const'(a_in));
  @(edge clk) b <= b_in;
  a_2: assert property (@(posedge clk) b == const'(b_in));
end
```

Describe scenarios in which triggering `B1` causes:

1. An instance of `a_1` that gets purged and an instance of `a_2` that matures.
2. An instance of `a_1` that matures and an instance of `a_2` that gets purged.
3. Instances of `a_1` and `a_2` that both mature.
4. Instances of `a_1` and `a_2` that both get purged.

14.6. Consider the following checker declaration:

```
checker chk(a, b, event ev = $inferred_clock);
   a1: assert property(@ev a);
   always_ff @ev begin
      a2: assert property(b);
   end
endchecker : chk
```

For each of the following instances of checker `chk`, either give a rewrite that is conceptually equivalent and eliminates the checker instance or explain why the checker instance is illegal.

1.
```
always @(posedge clk) begin
   A <= A_in;
```

```
      B <= B_in;
      chk chk_1(A, B);
   end
```

2.
```
   always @(posedge clk) begin
      for (int i=0; i<8; i++) begin
         A[i] <= A_in[i];
         chk chk_2(A[i], B);
      end
      B <= B_in;
   end
```

3.
```
   always @(posedge clk) begin
      automatic logic T <= B_in;
      A <= A_in;
      chk chk_3(A, T);
   end
```

4.
```
   always @(posedge clk or negedge B_in) begin
      A <= A_in;
      B <= B_in;
      chk chk_4(A, B);
   end
```

5.
```
   always @(posedge clk or negedge B_in) begin
      A <= A_in;
      B <= B_in;
      chk chk_5(A, B, posedge clk or negedge B_in);
   end
```

6.
```
   always @(posedge clk or negedge rst_n) begin
      A <= A_in;
      B <= B_in;
      chk chk_6(A, B);
   end
```

Chapter 15
An Apology for Local Variables

Local color has a fatal tendency to remain local; but it is also true that the universal often borders on the void.

— DuBose Heyward and Hervey Allen

Local variables are a powerful feature of SystemVerilog Assertions that enable an assertion to capture the value of an expression at a specified point in its evaluation and store that value for later reference, perhaps after further modification. This feature makes the encoding of many assertions much easier and helps to eliminate the need for auxiliary state machines to support assertions.

A local variable must be declared within the declaration of a named sequence or property, and the scope of a local variable does not extend outside the sequence or property in which it is declared. Local variables are, therefore, not a first-class construct of SVA. Each evaluation attempt of a named sequence or property has its own copies of the local variables declared within it. In this sense, local variables are "local" to these individual evaluation attempts.

This chapter gives an intuitive introduction to local variables based on examples. For each example, an alternative encoding is shown that does not use local variables. By comparing the encodings, the reader should gain an appreciation for the semantics and, in most cases, the benefits of local variables. Throughout this chapter, we assume that all assertions are clocked at `posedge clk` and that there is a default clocking specification.

The examples described in this chapter can be implemented naturally as checkers. However, in order to make the study of local variables independent of checkers, we have rendered all the examples as modules or module fragments. Section 9.1 contains a detailed discussion of both module- and checker-based implementations of the sequential protocol of Sect. 15.2. Exercise 15.11 explores checker-based implementation of the protocols of this chapter.

E. Cerny et al., *SVA: The Power of Assertions in SystemVerilog*,
DOI 10.1007/978-3-319-07139-8_15

15.1 Fixed Latency Data Pipeline

To get started, suppose that there is a fixed latency data pipeline whose data checking requirement is specified by the following English:

1. `start` is a signal of type **`logic`**. `dataIn` and `dataOut` are signals of type `dataType`.
2. `LATENCY` is a positive integer parameter.
3. Whenever `start` is high, `dataIn` is valid.
4. The value of `dataIn` when `start` is high must equal the value of `dataOut` `LATENCY` cycles later.

The specification can be encoded without using local variables as shown in Fig. 15.1.

By using `$past(dataIn,LATENCY)`, one should expect performance in simulation and formal verification to be similar to that of encoding a cascade of `LATENCY` delay variables of type `dataType`. If `LATENCY` equals three, then the cascade of delay variables is

```
dataType dataIn_D1, dataIn_D2, dataIn_D3;
always @(posedge clk) begin
   dataIn_D1 <= $sampled(dataIn);
   dataIn_D2 <= dataIn_D1;
   dataIn_D3 <= dataIn_D2;
end
```

and the reference to `$past(dataIn,LATENCY)` is like a reference to `dataIn_D3` (see also the discussion of `$past` in Sect. 7.2.1.2).

Using a local variable, the pipeline data check can be encoded as shown in Fig. 15.2. The local variable `data` of type `dataType` is declared on Line 2 within

```
a_pipeline_data_check: assert property (
   start
   |->
   ##LATENCY dataOut == $past(dataIn, LATENCY)
);
```

Fig. 15.1 Encoding of pipeline data check without local variables

```
1 property p_pipeline_data_check;
2    dataType data;
3    (start, data = dataIn)
4    |->
5    ##LATENCY dataOut == data;
6 endproperty
7 a_pipeline_data_check: assert property (p_pipeline_data_check);
```

Fig. 15.2 Encoding of pipeline data check with a local variable

the declaration of property `p_pipeline_data_check`. Local variable declarations follow the same format as other variable declarations in SystemVerilog. Line 3 is an example of attaching a local variable assignment to a Boolean expression. The Boolean `start` is separated from the local variable assignment `data = dataIn` by a comma, and the two are enclosed in parentheses. `(start, data = dataIn)` is a sequence (not a Boolean) with the following meaning:

- The value of `start` is tested when evaluation of the sequence begins.
- If the value of `start` is high, then the value of `dataIn` is assigned to the local variable `data` and the sequence matches at that point.
- Otherwise, the sequence fails to match and no assignment to the local variable `data` occurs.

If `start` is high when evaluation of `p_pipeline_data_check` begins, then the value of `dataIn` is assigned to `data` and the antecedent of the implication `|->` in Line 4 matches. Therefore, the consequent specified in Line 5 must match. Line 5 says that `LATENCY` cycles should be advanced and then the value of `dataOut` must equal the value stored in the local variable `data`. In summary, when `start` is high, the value of `dataIn` is captured in the local variable `data`, and this value is compared `LATENCY` cycles later to the value of `dataOut`.

From a data storage perspective, the simulation performance of the encoding with the local variable should never be worse than that of the encoding using `$past`. This is because at most `LATENCY` threads of evaluation of `p_data_pipeline_LATENCY` can be active simultaneously, each with its own copy of the local variable `data`. The storage needed for the local variable encoding varies in direct proportion to the frequency of occurrences of `start`, while the storage needed for the encoding using `$past` is fixed by the parameter `LATENCY`.

15.2 Sequential Protocol

Now let us switch from a fixed latency pipeline to a protocol that is *sequential* in the sense that its transactions do not overlap. This is the same protocol that was discussed in Sect. 9.1. We repeat here, with minor modifications, the sequential protocol description and the module-based implementation without local variables.

In the sequential protocol, there is not a constant latency from `dataIn` to `dataOut`. Instead, a Boolean signal `complete` determines when `dataOut` is valid. Here is the English description:

1. `start` and `complete` are signals of type **`logic`**. `dataIn` and `dataOut` are signals of type `dataType`.
2. Whenever `start` is high, `dataIn` is valid. Whenever `complete` is high, `dataOut` is valid.

```
a_no_start: assert property (
   start |=> !start throughout complete[->1]
);
a_no_complete: assert property (
   complete |=> !complete throughout start[->1]
);
initial
   a_initial_no_complete: assert property (
      !complete throughout start[->1]
   );
```

Fig. 15.3 Encoding of control part of sequential protocol

3. If `start` is high, then the value of `dataIn` at that time must equal the value of `dataOut` at the next strictly subsequent cycle in which `complete` is high.
4. If `start` is high, then `start` must be low in the next cycle and remain low until after the next strictly subsequent cycle in which `complete` is high.
5. `complete` may not be high unless `start` was high in a preceding cycle and `complete` was not high in any of the intervening cycles.

The last two English rules specify that the protocol is sequential. Let us say that a Boolean *occurs* if it is high. Then these rules say that a second `start` cannot occur until after the `complete` for the first `start` occurs, and an occurrence of `complete` corresponds to the nearest preceding occurrence of `start`. A transaction spans the set of cycles from an occurrence of `start` to its corresponding occurrence of `complete`, and two transactions do not overlap.

Note that this specification decomposes into a control part, which ensures the sequential pairing of occurrences of `start` and `complete`, and a data part, which checks the data correspondence between `dataIn` and `dataOut` for each such pair. The control part of the specification does not involve data or local variables and can be encoded as shown in Fig. 15.3.

Assertion `a_no_start` checks Rule 4. Rule 5 is checked by `a_no_complete` and `a_initial_no_complete`. The first two assertions are symmetric in `start` and `complete`, while the last is not. Since the last assertion is within an **`initial`** procedure, only one evaluation attempt of `a_initial_no_complete` is begun at the first occurrence of the clocking event. That evaluation checks that there is no occurrence of `complete` until after the first occurrence of `start`.

Now let us move to the data part of the specification and begin without using local variables. Since the latency from an occurrence of `start` to the next subsequent occurrence of `complete` is not fixed, `$past` will not work. The value of `dataIn` at an occurrence of `start` needs to be stored somewhere, though, since otherwise it is lost and the comparison with `dataOut` cannot be made. Since the protocol is sequential,

```
1 dataType last_dataIn;
2 always @(posedge clk)
3    if ($sampled(start))
4       last_dataIn <= $sampled(dataIn);
5 a_seq_data_check: assert property (
6    start ##1 complete[->1]
7    |-> dataOut == last_dataIn
8 );
```

Fig. 15.4 Encoding of sequential protocol data check without local variables

```
1 property p_seq_data_check;
2    dataType data;
3    (start, data = dataIn) ##1 complete[->1]
4    |-> dataOut == data;
5 endproperty
6 a_seq_data_check: assert property (p_seq_data_check);
```

Fig. 15.5 Encoding of sequential protocol data check with a local variable

one auxiliary storage variable can be used to hold the value of `dataIn` from the nearest preceding occurrence of `start`. Figure 15.4 shows such an encoding.[1]

Using local variables, we can follow the same data capture idiom from the pipeline data check. The encoding is shown in Fig. 15.5. Note the similarity between Lines 6 and 7 from Fig. 15.4 and Lines 3 and 4 from Fig. 15.5.[2] The encoding with local variables avoids the auxiliary modeling code to define how `last_dataIn` is updated, and it makes clear the timing of the data capture because the local variable assignment `data = dataIn` is attached to the Boolean `start`.

15.3 FIFO Protocol

Next, let us generalize the sequential protocol to a FIFO (i.e., in-order) protocol by allowing multiple occurrences of `start` prior to the next occurrence of `complete` and multiple occurrences of `complete` before the next occurrence of `start`. The occurrences of `start` and `complete` pair up by order, so that for each $n \geq 1$, the

[1]The use of `$sampled` specifies that sampled values of `start` and `dataIn` are used in the **`always`** procedure, maintaining consistency with the implicit use of sampled values in `a_seq_data_check`. See the detailed discussion in Sect. 9.1.1.

[2]The antecedent written in Line 6 of Fig. 15.4 is more verbose than the one in Line 22 of Fig. 9.4. This has been done to help highlight the similarities of the code in Figs. 15.4 and 15.5.

```
1 bit [0:$clog2(MAX_OUTSTANDING)] outstanding;
2 initial
3    outstanding <= '0;
4 always @(posedge clk)
5    outstanding <= outstanding + $sampled(start - complete);
```

Fig. 15.6 Encoding of the number of outstanding transactions

nth occurrence of `start` pairs with the nth occurrence of `complete`. Here are the English rules:

1. `start` and `complete` are signals of type **`logic`**. `dataIn` and `dataOut` are signals of type `dataType`.
2. Whenever `start` is high, `dataIn` is valid. Whenever `complete` is high, `dataOut` is valid.
3. `MAX_OUTSTANDING` is a positive integer parameter.
4. `start` may be high if and only if `complete` is not high and the number of preceding occurrences of `start` minus the number of preceding occurrences of `complete` is less than `MAX_OUTSTANDING`.
5. `complete` may be high if and only if `start` is not high and the number of preceding occurrences of `start` minus the number of preceding occurrences of `complete` is positive.
6. For all $n \geq 1$, at the nth occurrence of `complete`, the value of `dataOut` must equal the value of `dataIn` at the nth occurrence of `start`.

As with the sequential protocol, this specification decomposes into a control part, governing the signals `start` and `complete`, and a data part. The control part of the specification must keep track of the difference between the number of preceding occurrences of `start` and the number of preceding occurrences of `complete`. Let us call this difference the *number of outstanding transactions*. One way to keep track of this number is to encode an auxiliary variable to store it. Figure 15.6 shows how this can be done.

Line 1 references the `$clog2` system function, which returns the ceiling of the base-2 logarithm of its argument. This declaration ensures that `outstanding` has enough bits to store the number `MAX_OUTSTANDING` (see Exercise 15.3). (Question: Why is it important that `outstanding` be able to store `MAX_OUTSTANDING`?) The **`always`** procedure updates `outstanding`, incrementing it whenever `start` occurs and decrementing it whenever `complete` occurs. Of course, if `start` or `complete` does not obey the control part of the specification, then `outstanding` may overflow or underflow.

Using `outstanding`, the control part of the FIFO protocol specification can be encoded as shown in Fig. 15.7. Without using local variables, the various values of `dataIn` for the outstanding transactions need to be stored in some data structure. A bounded queue of maximum size `MAX_OUTSTANDING` is a good choice because of the in-order pairing of corresponding occurrences of `start` and `complete`. Figure 15.8

```
a_start_valid: assert property (
   start |-> !complete && outstanding < MAX_OUTSTANDING
);
a_complete_valid: assert property (
   complete |-> !start && outstanding > 0
);
```

Fig. 15.7 Encoding of control part of FIFO protocol

```
1 dataType dataQ[$:MAX_OUTSTANDING-1] = {};
2 always @(posedge clk)
3    if ($sampled(start))
4       dataQ.push_back($sampled(dataIn));
5    else if ($sampled(complete))
6       dataQ.pop_front;
7 a_fifo_data_check: assert property (
8    complete |-> dataOut == dataQ[0]
9 );
```

Fig. 15.8 Encoding of FIFO protocol data check without using local variables

shows the declaration and management of such a queue, as well as the simple data check assertion that references it.

Note the use of the built-in queue methods `push_back` in Line 4 and `pop_front` in Line 6. Because of this update policy, the data needed for comparison with `dataOut` at the next occurrence of `complete` is always in `dataQ[0]`, which is referenced in the assertion in Line 8. The execution of `pop_front` in Line 6 does not lose the data needed for the comparison in Line 8 because the reference to `dataQ[0]` within the assertion is to the sampled value, which is not affected by the execution of the queue method in the same time step.

Using local variables, the same data capture idiom we have seen for the pipeline and sequential protocol examples continues to work. The challenge is to determine when the corresponding `complete` occurs. To accomplish this, we capture not only `dataIn` when `start` occurs, but also the value of `outstanding`. The value of `outstanding` determines how many occurrences of `complete` must be skipped before arriving at the occurrence of `complete` at which the data comparison should be performed. As occurrences of `complete` are observed, they are accounted for so that the assertion detects when the corresponding `complete` occurs. An encoding following this approach is shown in Fig. 15.9.

Property `p_fifo_data_check` has two local variables, `data` declared in Line 2 and `numAhead` declared in Line 3. The type of `numAhead` is the same as that of `outstanding`. In general, any number of local variables may be declared within the declaration of a named sequence or property. In Line 4, two local variable assignments are attached to the Boolean `start`. The first stores the value of `dataIn` in the local variable `data`, just as we have seen before. The second stores the value of `outstanding` in the local variable `numAhead`. If multiple local variable assignments need to be performed on successful test of a Boolean, then the assignments are simply separated by commas and are performed in the order in which they are written. When Line 4 completes, `numAhead` holds the number of occurrences of `complete` that need to be skipped before arriving at the occurrence of `complete`

```
1  property p_fifo_data_check;
2     dataType data;
3     bit [0:$clog2(MAX_OUTSTANDING)] numAhead;
4     (start, data = dataIn, numAhead = outstanding)
5     ##1 (numAhead > 0 ##0 complete[->1], numAhead--)[*]
6     ##1 (numAhead == 0 ##0 complete[->1])
7     |->
8     dataOut == data;
9  endproperty
10 a_fifo_data_check: assert property (p_fifo_data_check);
```

Fig. 15.9 Encoding of FIFO protocol data check using local variables

at which the data check will be performed. Lines 5 and 6 cause the evaluation to advance to the cycle of the data check and are discussed in detail below. Finally, Line 8 performs the simple data comparison.

Line 5 does the job of advancing the evaluation through the occurrences of `complete` that need to be skipped. This line deserves careful study. It begins with `##1`, which simply advances to the cycle after the occurrence of `start`. The rest of the line is a repetition of zero or more occurrences of the sequence

```
(numAhead > 0 ##0 complete[->1], numAhead--)
```

Let us call this sequence the *skipping sequence*. The top-level structure of the skipping sequence attaches the local variable assignment `numAhead--` to the subsequence

```
numAhead > 0 ##0 complete[->1]
```

In general, local variable assignments may be attached to any sequence that does not admit an empty match. The assignment `numAhead--` uses the decrement operator `--` and behaves the same as `numAhead = numAhead - 1`. The subsequence to which it is attached begins with the Boolean condition `numAhead > 0`, which is true if and only if there remain occurrences of `complete` that need to be skipped. This Boolean is fused via `##0` to the sequence `complete[->1]`, which advances to the next occurrence of `complete`. Each match of the skipping sequence in the repetition therefore behaves as follows:

- Confirm that `numAhead > 0`, hence that there remains at least one occurrence of `complete` that needs to be skipped.
- Advance to the next occurrence of `complete` by matching `complete[->1]`.
- Decrement `numAhead`.

Because `numAhead` is decremented for each match in the repetition, the skipping sequence is matched at most a number of times equal to the value that was stored in `numAhead` in Line 4. In fact, as explained below, Line 6 forces the skipping sequence to be matched exactly this number of times. It is possible that this number is zero, meaning that no occurrences of `complete` need to be skipped. In this case, no matches of the skipping sequence are possible, and the zero repetition case is used to match Line 5.

Line 6 does two jobs. The first is to prevent the repetition of Line 5 from stopping early. This is done by enforcing the Boolean condition `numAhead == 0`. If the evaluation tries to proceed from Line 5 to Line 6 when `numAhead > 0`, then the Boolean `numAhead == 0` will be false and the evaluation must revert to Line 5 to attempt another repetition of the skipping sequence. The second job of Line 6 is to advance to the next occurrence of `complete` by matching `complete[->1]`. This occurrence is the one that corresponds to the occurrence of `start` in Line 4 and at which the data check should be performed.

As an intuitive summary, the matching of Lines 5 and 6 accomplishes the following:

- While `numAhead` is positive, advance to the next occurrence of `complete` and decrement `numAhead`.
- When `numAhead` becomes zero, advance to the next occurrence of `complete` and stop.

Managing a local variable counter such as `numAhead` within a repetition as shown in this example may seem daunting at first. After fully understanding a few such patterns, the reader will acquire the skill and confidence to put them into practice.

Exercise 15.6 explores the storage requirements of the encodings of the FIFO protocol data check with and without local variables.

15.4 Tag Protocol

In this section, we switch from an in-order protocol to an out-of-order protocol in which two occurrences of `complete` do not have to be in the same order as the corresponding occurrences `start`. Additional data are needed to determine which occurrence of `complete` corresponds to a given occurrence of `start`. This protocol uses a tag to do the matching. The signal `tagIn` is valid with an occurrence of `start` and determines the tag of the transaction. The tag is active for that transaction while the transaction is outstanding. The signal `tagOut` is valid with an occurrence of `complete`, and matching of the values of `tagIn` and `tagOut` is used to define the correspondence. While a tag is active for a transaction, it must not be reused by another transaction. Here are the English rules:

1. `start` and `complete` are signals of type **`logic`**. `dataIn` and `dataOut` are signals of type `dataType`. `tagIn` and `tagOut` are signals of type `tagType`.
2. Whenever `start` is high, `dataIn` and `tagIn` are valid. Whenever `complete` is high, `dataOut` and `tagOut` are valid.
3. In each cycle, each tag value is either active or inactive, according to the following rules:
   - Every tag value begins inactive.
   - If there is no occurrence of `start` or `complete` in a cycle, then no tag value changes state in the next cycle.

```
bit active[tagType];
always @(posedge clk)
   if ($sampled(start)) begin
      a_no_tag_reuse: assert final (
         !active.exists($sampled(tagIn))
      );
      active[$sampled(tagIn)] <= 1'b1;
   end
   else if ($sampled(complete)) begin
      a_comp_tag_ok: assert final (
         active.exists($sampled(tagOut))
      );
      active.delete($sampled(tagOut));
   end
```

Fig. 15.10 Encoding of tag protocol control check

- A tag value becomes active the cycle after an occurrence of `start` at which `tagIn` held that value.
- A tag value becomes inactive the cycle after an occurrence of `complete` at which `tagOut` held that value.

4. `start` may be high if and only if `complete` is not high and the value of `tagIn` is inactive in that cycle.
5. `complete` may be high if and only if `start` is not high and the value of `tagOut` is active in that cycle.
6. An occurrence of `start` corresponds to an occurrence of `complete` if and only if the following conditions are all satisfied:
   - The occurrence of `start` is strictly before the occurrence of `complete`.
   - The value of `tagIn` at the occurrence of `start` equals the value of `tagOut` at the occurrence of `complete`.
   - There is no earlier occurrence of `complete` satisfying both of the two preceding conditions.
7. If an occurrence of `start` corresponds to an occurrence of `complete`, then the value of `dataIn` at that occurrence of `start` equals the value of `dataOut` at that occurrence of `complete`.

The control part of the tag protocol specification is more complicated than in our previous examples. As for the FIFO protocol, auxiliary variables can help in encoding the control part of the specification. In the tag protocol, we need to keep track of which tags are active. An associative array of bits can be used to do this as shown in Fig. 15.10. This encoding uses final assertions to perform the control checks specified in Rules 4 and 5. Note that the value associated to a tag in the associative array `active` is not important, only whether or not the tag exists in the array.

```
dataType data[tagType];
always @(posedge clk)
   if ($sampled(start)) begin
      a_no_tag_reuse: assert final (
         !data.exists($sampled(tagIn))
      );
      data[$sampled(tagIn)] <= $sampled(dataIn);
   end
   else if ($sampled(complete)) begin
      a_comp_tag_ok: assert final (data.exists($sampled(tagOut))
         );
      a_data_check: assert final (
         $sampled(dataOut) == data[$sampled(tagOut)]
      );
      data.delete($sampled(tagOut));
   end
```

Fig. 15.11 Encoding of tag protocol control and data check

```
property p_tag_data_check;
   tagType tag;
   dataType data;
   (start, tag = tagIn, data = dataIn)
   ##1 (complete && tagOut == tag)[->1]
   |-> dataOut == data;
endproperty
a_tag_data_check: assert property (p_tag_data_check);
```

Fig. 15.12 Encoding of tag protocol data check using local variables

Without using local variables, a similar associative array can be used to store the data values for the active tags. This array keeps track of which tags are active and their associated data, so the `active` array is no longer needed in this approach. Another final assertion can perform the data check specified in Rule 7. Such an encoding is shown in Fig. 15.11.

The data check can be encoded using local variables in a style similar to that of the FIFO data check with local variables, as shown in Fig. 15.12. At `start`, the property captures the values of both `tagIn` and `dataIn` in the local variables `tag` and `data`. The property then advances to the nearest occurrence of `complete` at which `tagOut` equals the value stored in `tag`, and at that point compares `dataOut` to the value stored in `data`. This encoding still relies on the final assertions in Fig. 15.10 to perform the control checks.

Associative arrays are convenient for encoding the checks of the tag protocol, but because of their dynamic and unbounded nature, they will typically not be supported in formal verification tools. To provide an encoding for formal verification, the auxiliary variables should be declared in a way that is explicitly bounded. If the number of tag values is not too large, then the `active` associative array can be

```
bit active[0:MAX_TAG];
initial
   active <= '0;
always @(posedge clk)
   if ($sampled(start))
      active[$sampled(tagIn)] <= 1'b1;
   else if ($sampled(complete))
      active[$sampled(tagOut)] <= 1'b0;
a_no_tag_reuse: assert property (start |-> !active[tagIn]);
a_comp_tag_ok: assert property (complete |-> active[tagIn]);
```

Fig. 15.13 Encoding of tag protocol control check without associative arrays

recoded as a bounded array. For simplicity, suppose that the tag values range from `0` to `MAX_TAG`. Then the array of active bits can be encoded as shown in Fig. 15.13. There we use concurrent assertions to illustrate a coding style suitable for formal verification.[3] The code of this figure, together with that of the data check using local variables shown in Fig. 15.12, gives a complete encoding of the tag protocol checks suitable for formal tools.

If the number of tag values times the number of bits needed to store a data value is also not too large, then a similar bounded array of elements of type `dataType` can be used to store the data of the active transactions, replacing the associative array of Fig. 15.11.

In practice, it may be that the number of tag values or the product of the number of tag values times the number of bits needed to store a data value will be too large for static allocation and formal verification. If the number of active transactions can be bounded by a number significantly smaller than the number of tags, then smaller arrays that store the tag and data values for each active transaction can be encoded. Such arrays are essentially RTL implementations of bounded associative arrays.

Suppose that the positive integer parameter `MAX_ACTIVE` is an upper bound for the number of simultaneously active transactions. Figure 15.14 shows an encoding of the tag protocol control checks assuming this bound. Line 1 defines `indexType`, which can store values from `0` through `MAX_ACTIVE` (cf. Exercise 15.3). An array `tags` indexed from `0` to `MAX_ACTIVE` is used to store the tags of the active transactions. Each element of the array has a validity bit (`valid`) and a tag data field (`activeTag`). If the validity bit is `1'b1`, then the tag data field stores the value of an active tag. If the validity bit is `1'b0`, then the tag data field is not meaningful and that position in the array is considered free. The data for an element are organized in a structure (Line 2) with type name `extendedTagType` (Line 5). In Line 6 the array of structures is declared with a declaration assignment initializing its elements to zero. The expression `'{default: '{1'b0, '0}}` is an assignment pattern that says that each element in the array is initialized to value `'{1'b0, '0}`, which in turn specifies that the `valid` member of the structure gets value `1'b0` and the `activeTag` member of the structure gets value `'0`. The assignment pattern accomplishes initialization

[3]A formal tool may also accept procedural concurrent assertions or final assertions.

```
typedef bit[0:$clog2(MAX_ACTIVE)] indexType;
typedef struct {
   bit valid;
   tagType activeTag;
} extendedTagType;
extendedTagType tags[0:MAX_ACTIVE] = '{default: '{1'b0, '0}};

function automatic indexType freeIndex;
   for (indexType i=0; i < MAX_ACTIVE; i++)
      if (!tags.valid[i]) return i;
   return MAX_ACTIVE; // no free index
endfunction : freeIndex

function automatic indexType tagIndex (tagType tag);
   for (indexType i=0; i < MAX_ACTIVE; i++)
      if (tags.valid[i] && tags.activeTag[i] == tag)
         return i;
   return MAX_ACTIVE; // tag not found
endfunction : tagIndex

always @(posedge clk)
   if ($sampled(start))
      tags[freeIndex] <= '{1'b1, $sampled(tagIn)};
   else if ($sampled(complete))
      tags.valid[tagIndex($sampled(tagOut))] <= 1'b0;

a_no_tag_reuse: assert property (
   start
   |-> tagIndex(tagIn) == MAX_ACTIVE // tagIn not found
);
a_comp_tag_ok: assert property (
   complete
   |-> tagIndex(tagOut) < MAX_ACTIVE // tagOut found
);
```

Fig. 15.14 Encoding of tag protocol control check with bound on number of active transactions

of the array without explicitly enumerating values or iterating over indices.[4] We allocate `MAX_ACTIVE+1`, and not just `MAX_ACTIVE`, array elements and reserve the top element as a sink in case our assumed bound on the number of active transactions is violated.

The auxiliary function `freeIndex` returns the smallest index that is not currently storing a tag, if there is one. Otherwise, it returns `MAX_ACTIVE`. The auxiliary function `tagIndex` returns the smallest index of the entry that is currently storing

[4]The initialization could be skipped because the `valid` members are of data of type **`bit`**, which is initialized by default to `1'b0`. The explicit initialization clarifies the intention. The initial values of the `activeTag` members are not important. Note also that only the first `MAX_ACTIVE` entries of the array need be initialized.

a tag and whose `activeTag` field is equal to the function's argument `tag`. If the tag protocol control checks are satisfied, then there will be at most one such index. If there is no such index, then the function returns `MAX_ACTIVE`. These functions have been declared with **`automatic`** lifetime to guard against undesired contention on their internal loop counter variables in the event of multiple calls in the same time step.

A similar array can be encoded to store the data values for the active transactions and used to encode the tag protocol data check in a way that is suitable for formal verification. See Exercise 15.9. The complexity of the data management in such arrays is striking in comparison with the simplicity of the data check using local variables.

15.5 FIFO Protocol Revisited

This section shows an alternative encoding of the FIFO protocol control and data checks that uses only local variables.

The encoding with local variables from Sect. 15.3 makes use of only one nonlocal variable, namely, `outstanding` as defined in Fig. 15.6. Thus, the present encoding

```
typedef bit [0:$clog2(MAX_OUTSTANDING)] counterType;
property p_fifo_data_check(local input counterType numAhead);
   dataType data = dataIn;
   ##1 (numAhead > 0 ##0 complete[->1], numAhead--)[*]
   ##1 (numAhead == 0 ##0 complete[->1])
   |->
   dataOut == data;
endproperty
property p_fifo_all_checks;
   counterType outstanding, nextOutstanding = '0;
   (
      (start || complete)[->1],
      outstanding = nextOutstanding,
      nextOutstanding += start - complete
   )[+]
   |->
   if (start) (
      (!complete && outstanding < MAX_OUTSTANDING)
      and p_fifo_data_check(.numAhead(outstanding))
   ) else ( // complete
      !start && outstanding > 0
   );
endproperty
initial
   a_fifo_all_checks: assert property (p_fifo_all_checks);
```

Fig. 15.15 Encoding of all FIFO protocol checks using only local variables

needs to keep track of the number of outstanding transactions internally by using local variables. This encoding is shown in Fig. 15.15.

The style of this encoding is somewhat more complicated than those shown in Sect. 15.3, although each of its parts corresponds to a part of the encoding using local variables that appears in that section. This encoding uses a **typedef** on Line 1 to provide the type `counterType` that simplifies the declarations on Lines 2 and 10. The top-level property `p_fifo_all_checks` is responsible for accounting for the number of outstanding transactions, performing the control checks, and calling property `p_fifo_data_check` to perform the data check. The accounting of the number of transactions is done in the repetition in Lines 11–15, and this approach, which will be explained in more detail below, assumes that there is only one evaluation attempt of `p_fifo_all_checks` running. For this reason the assertion `a_fifo_all_checks` on Line 24 appears in an **initial** procedure. See Exercise 15.7.

The data checking property `p_fifo_data_check` is very similar to the property of the same name from Sect. 15.3, so we focus on the differences. In Line 2, `numAhead` is declared as a formal argument. The keywords **local input** in this declaration specify that the formal argument `numAhead` is in fact a local variable that will receive an initial value from its actual argument. `p_fifo_all_checks` computes the value of `outstanding` in Line 13 and passes it in to `numAhead` through the instantiation in Line 19. In all other respects, `numAhead` behaves the same here as it did in Sect. 15.3. In Line 3, the local variable `data` is declared, and the declaration includes a declaration assignment to the value of `dataIn`. The meaning of this assignment is that whenever evaluation of `p_fifo_data_check` begins, the copy of `data` for that evaluation begins with the value of `dataIn` at that time. Line 17 guarantees that `p_fifo_data_check` is called only when `start` occurs, so `dataIn` is valid whenever the declaration assignment of Line 3 is performed. The rest of `p_fifo_data_check` is identical to the encoding from Sect. 15.3 and behaves the same.

The structure of `p_fifo_all_checks` is an implication whose antecedent is an unbounded repetition (Lines 11–15) and whose consequent enforces the control checks (Lines 18 and 21) and the data check (Line 19). There are two local variables, `outstanding` and `nextOutstanding`, both declared in Line 10. The reason for having two local variables will be explained below. Note that `nextOutstanding` is declared with a declaration assignment to the value `'0`. This assignment corresponds to the nonblocking assignment on Line 3 of Fig. 15.6.

The repetition in Lines 11–15 uses a pattern similar to the decrementing counter from Line 5 of Fig. 15.9. In this case, successive iterations of the repetition match each time a `start` or `complete` occurs. The assignments in Lines 13 and 14 are executed for each iteration and are performed in the order that they appear. The assignment to `nextOutstanding` is analogous to Line 5 of Fig. 15.6 and accomplishes the basic accounting of the number of outstanding transactions.

The reason for using the two local variables is that when `outstanding` is declared as a static variable as in Sect. 15.3, references to it from within sequences and properties resolve to the sampled value for that cycle. On the contrary, when outstanding is declared as a local variable, then any assignment to it is immediately visible to subsequent parts of the sequence or property. Thus, the present encoding uses nextOutstanding to compute the new number of outstanding transactions, and outstanding is assigned the old value of nextOutstanding just before nextOutstanding is updated. The result is that the local variable outstanding has the same timing as the static variable from Sect. 15.3. Of course, the local variable outstanding can be eliminated from p_fifo_all_checks (see Exercise 15.8). An alternative encoding using recursive properties also avoids the use of two local variables (see Fig. 17.11).

15.6 Tag Protocol Revisited

This section shows two alternative encodings of the tag protocol control and data checks using local variables. One uses a single-bit auxiliary static variable, and the other requires no auxiliary static variable. These encodings also do not rely on the existence of the small bound, MAX_ACTIVE, on the number of active transactions.

15.6.1 Tag Protocol Using a Single Static Bit

An encoding of the tag protocol control and data checks using local variables and a single-bit auxiliary static variable is shown in Fig. 15.16. It is unlikely that this encoding is supported in current formal verification tools because assignment to the static variable is done within a task that is called from within a property. Nevertheless, there is nothing in the encoding that is essentially beyond the capabilities of formal verification tools. Its simplicity and elegance stand in contrast to the more cumbersome data management of the encodings that rely on MAX_ACTIVE (e.g., as shown in Fig. 15.14 and explored in Exercise 15.9).

Property p_start_and_data_checks captures the values of tagIn and dataIn in the local variables tag and data whenever start occurs (Line 11). Lines 14–16 check that, beginning in the next cycle, there is not another occurrence of start with the same tag until after the nearest occurrence of complete with the same tag. These lines enforce the rule that an active tag cannot be reused, as specified in Rule 4 of the tag protocol. When the corresponding complete occurs, Line 19 compares dataOut with the local variable data. This comparison performs the data check specified in Rule 7 of the tag protocol.

It remains to explain how this encoding performs the check specified in Rule 5: complete may occur only if tagOut is an active tag. To gain some insight into why checking Rule 5 is more difficult than checking the Rule 4, it is helpful to note that there is an incomplete symmetry between start and complete in the tag protocol. The symmetry can be described by the following statements:

```
1  bit complete_justified;
2  always @(posedge clk)
3     complete_justified = 1'b0;
4  task t_justify_complete;
5     @(negedge clk)
6     complete_justified = 1'b1;
7  endtask : complete_justified
8  property p_start_and_data_checks;
9     tagType tag;
10    dataType data;
11    (start, tag = tagIn, data = dataIn)
12    |=>
13    (
14       !(start && tagIn == tag)
15       throughout
16       (complete && tagOut == tag)[->1]
17    )
18    ##0 (1'b1, t_justify_complete)
19    ##0 dataOut == data;
20 endproperty
21 a_start_and_data_checks: assert property (
22    p_start_and_data_checks
23 );
24 a_complete_check: assert property (
25    complete |=> complete_justified
26 );
```

Fig. 15.16 Encoding of all tag protocol checks using local variables and a single-bit static variable

- Once `start` occurs for a given tag, another `start` may not occur for that tag until after `complete` occurs for that tag.
- Once `complete` occurs for a given tag, another `complete` may not occur for that tag until after `start` occurs for that tag.

These statements can be encoded in a straightforward way using local variables: the first is represented in Lines 11 through 17 of Fig. 15.16; for the second, see Exercise 15.10. The asymmetry is that initially there can be a `start` on any tag, but there cannot be a `complete` on a tag until after an occurrence of `start` on that tag. Accounting for the tags for which no `start` has occurred is the challenging part, and doing so explicitly is essentially encoding a data structure to represent all the tags in a particular state, similar to the data structures discussed in Sect. 15.4.

The novelty of the present encoding is that it entirely avoids explicit representation of such a data structure. Instead, it makes use of the fact that there is one evaluation thread of property `p_start_and_data_checks` tracking each active tag. When a `complete` occurs, we need to determine whether or not there is such a thread tracking the tag value in `tagOut`. SystemVerilog provides no way to query such information from a set of threads of evaluation. Instead, `p_start_and_data_checks` is encoded so that the relevant thread, if it exists, announces itself. The medium of communication is the static bit `complete_justified`.

The communication mechanism works as follows. In every time step in which **posedge** `clk` occurs, the communication bit is cleared by writing the value `1'b0` into `complete_justified` in Line 3. This assignment does not change the sampled value of `complete_justified` in that time step. If `complete` occurs and there is a thread of evaluation with `tag == tagOut`, then that thread will finish match of Lines 14–16 and execution will proceed to Line 18, where `t_justify_complete` is called. Line 18 illustrates attachment of a subroutine call to a sequence. The sequence in this case is just the Boolean `1'b1`. Like local variable assignments, subroutine calls may appear in such a comma-separated list, and they are scheduled to execute in the Reactive region in the order that they appear. When it executes, the task `t_justify_complete` advances to **negedge** `clk` (Line 5) and then writes the value `1'b1` into `complete_justified` (Line 6). The value `1'b1` will remain in `complete_justified` until after the Preponed region of the next time step in which **posedge** `clk` occurs. Thus, a thread announces itself as justifying a `complete` by causing the sampled value of `complete_justified` to be `1'b1` in the time step of the next occurrence of **posedge** `clk`. Later in that time step, the value `1'b0` will again be written into `complete_justified` by Line 3, clearing the communication bit.

The assertion `a_complete_check` simply looks for an announcement by requiring that `complete_justified` be high the cycle after each occurrence of `complete`. Because the reference to `complete_justified` in Line 25 uses the sampled value, it is able to see the announcement even though the communication bit is also cleared in the same cycle. This accomplishes the validation of `complete` as specified in Rule 5 of the tag protocol.

This encoding shows management of `complete_justified` by clearing it at **posedge** `clk` (Line 3) and setting it at **negedge** `clk` (Line 5). Other schemes will work provided the value `1'b1` set in task `t_justify_complete` persists to the Preponed region of the time step of the next occurrence of **posedge** `clk`.

15.6.2 Tag Protocol Using Only Local Variables

In the previous solution, the mechanism for announcing the existence of a justifying thread relies somewhat delicately on the SystemVerilog scheduling semantics in the way the static bit `complete_justified` is updated. Another approach is to code an auxiliary sequence that will match exactly when a justifying thread exists and finishes matching Lines 14–16 of Fig. 15.16. The existence of a justifying thread is then detected by reference to the endpoint of match of this sequence using the sequence method `triggered` (see Sect. 11.2.1). No auxiliary static bit is needed. Such an encoding is shown in Fig. 15.17.

The auxiliary sequence is `s_start_and_complete`. It simply mimics the temporal patterns of Lines 11–18 of Fig. 15.16. In order to maximize the sharing

```
1  sequence s_start_and_complete;
2     tagType tag;
3     (start, tag = tagIn)
4     ##1 (
5        !(start && tagIn == tag)
6        throughout
7        (complete && tagOut == tag)[->1]
8     );
9  endsequence
10 property p_start_and_data_checks;
11    dataType data;
12    (start, data = dataIn)
13    |->
14    s_start_and_complete
15    ##0 dataOut == data;
16 endproperty
17 a_start_and_data_checks: assert property (
18    p_start_and_data_checks
19 );
20 a_complete_check: assert property (
21    complete |-> s_start_and_complete.triggered
22 );
```

Fig. 15.17 Encoding of all tag protocol checks using only local variables

of sequential code between the auxiliary sequence and the correctness property `p_start_and_data_checks`, the latter has been reorganized to instantiate the former in Line 14 as part of the consequent of the operator `|->`. The result is equivalent to the version of the property in the previous encoding. Sequence method `triggered` is applied to the instance of the auxiliary sequence in Line 21. This method converts the sequence instance into a Boolean that is true in any cycle in which the sequence finishes a match and is false otherwise, accomplishing the communication of the announcement.

Because there are two instances of the auxiliary sequence, this solution incurs a nominal doubling of the local variable storage for tags as compared to the previous encoding. A tool could mitigate or eliminate this storage overhead by recognizing that the antecedent `start` of `|->` in Line 12 is not restrictive on the possible matches of `s_start_and_complete` in the consequent.

Exercises

15.1. Show that if the three assertions of Fig. 15.3 hold, then the following assertion also holds:

```
a_mutex: assert property (!(start && complete));
```

15.2. Suppose that the sequential protocol is relaxed to allow an occurrence of `complete` to be concurrent with the corresponding occurrence of `start`. If this happens, then `start` may occur again in the subsequent cycle.

1. Modify the precise English description of the protocol to account for this relaxation.
2. Modify the encodings given in Figs. 15.3, 15.4, and 15.5 to align them with this relaxation.

15.3. Show that for any positive value of `MAX_OUTSTANDING` that does not overflow `$clog2`, the declaration of `outstanding` in Fig. 15.6 has enough bits to store the number `MAX_OUTSTANDING`. Show that if `MAX_OUTSTANDING` is not a power of two, then this declaration has more than enough bits to store the number `MAX_OUTSTANDING`. Find a declaration that always allocates the minimum number of bits required to store the number `MAX_OUTSTANDING`.

15.4. Will the behavior of `p_fifo_data_check` in Fig. 15.9 change if the `##1` in Line 5 is changed to `##0`? If so, does the new behavior correctly advance to the cycle at which the data comparison of Line 8 needs to be performed? What happens if the `##1` in Line 6 is changed to `##0`?

15.5. The encoding of `p_fifo_data_check` in Fig. 15.9 uses the decrementing local variable counter `numAhead`. Give an alternative encoding that uses an incrementing local variable counter. [Hint: Use three local variables, one to capture `dataIn`, another to capture `outstanding`, and the other to serve as the incrementing counter.]

15.6. Assume that storage for the queue declared in Line 1 of Fig. 15.8 is allocated at compile time. (This is likely an accurate assumption for a formal verification tool, although it might not be accurate for a simulation tool.) Compare the storage requirements for the encodings of `a_fifo_data_check` with local variables (Fig. 15.9) and without local variables (Fig. 15.8). Consider the storage requirements for the local variables if `outstanding` remains small compared to `MAX_OUTSTANDING` and if `outstanding` becomes close to `MAX_OUTSTANDING`.

15.7. What will happen if the assertion `a_fifo_all_checks` of Fig. 15.15 is written as a module item instead of within an **`initial`** procedure?

15.8. Show how to recode `p_fifo_all_checks` from Fig. 15.15 to eliminate the local variable `outstanding`.

15.9. Suppose that the product of the number of tag values times the number of bits needed to store a data value in the tag protocol is too large to allocate storage for the data associated with every tag. Enhance the encoding shown in Fig. 15.14 to store also the data for each active transaction. Show how to encode the tag protocol data check without using local variables.

15.10. Write a property using only local variables that checks that once `complete` occurs for a given tag, another `complete` may not occur for that tag until after `start` occurs for that tag.

15.11. Implement protocol specifications provided in this chapter as checkers. [Hint: A checker-based sequential protocol implementation was discussed in detail in Sect. 9.1.]

Chapter 16
Mechanics of Local Variables

I think I can safely say that nobody understands quantum mechanics.

— Richard Feynman

The previous chapter introduced local variables by illustrating their use in intuitive and realistic examples of increasing richness. This chapter covers the mechanics of declaring, assigning, and referencing local variables in a more complete way. Input and output with local variables, and behavior of local variables with LTL operators, multiple clocks, and resets are discussed.

Local variables are divided into two kinds. A *body local variable* is one that is declared in the body of a named sequence or property, whereas an *argument local variable* is one that is declared as a formal argument of a named sequence or property. The second kind is especially helpful for writing recursive properties (see Chap. 17) in which local variable values need to be passed into recursive property instances. In most respects, local variables of both kinds behave the same, and the term *local variable* is used to refer to one of either kind. Another usability feature enables initialization of a local variable to be specified in the local variable's declaration.

Certain sequence and property operators (e.g., **`or`**, **`and`**, **`always`**) cause the evaluation of an assertion to fork into subevaluation threads. Depending on the operator and its context, the forked subevaluation threads may or may not later join. Local variables have been designed to work well in this multithreaded setting. When evaluation forks, each of the subevaluation threads receives its own copy of all the local variables together with all the values currently stored in them. As the subevaluations continue, each thread independently manages its copies of the local variables.

Forked subevaluation threads of property operators never join back, so the local variables in the forked threads remain independent. For a sequence operator, the nature of the operator (**`or`** vs. **`and`**, e.g.) determines how the subevaluations join

E. Cerny et al., *SVA: The Power of Assertions in SystemVerilog*,
DOI 10.1007/978-3-319-07139-8_16

when the sequence matches. A system of rules defines which local variables may be referenced after match of that sequence. These rules ensure that a local variable that might have inconsistent values in a thread of evaluation ensuing from the match may not be referenced after the match until an unambiguous value is stored into it. The rules can be checked at compile time and do not depend on the particular values stored in the local variables in the evaluation threads.

There are mechanisms for passing values of local variables into instances of named sequences or properties and out of instances of named sequences. Argument local variables enable improved type checking and self-documentation of code intent for these mechanisms.

We continue to assume that, unless otherwise specified, all assertions are clocked at **posedge** `clk` and there is a default clocking specification.

16.1 Declaring Body Local Variables

Body local variables are declared immediately after the header of the named sequence or property in which they appear. Here is an example:

```
sequence s1;
   logic l_a, l_b[4];
   dataType l_data;
   @(posedge clk)
   ...
endsequence
```

This example declares `l_a` to be a local variable of type **logic**, `l_b` to be a local variable that is an unpacked array of four elements of type **logic**, and `l_data` to be a local variable of the user-defined type `dataType`. Body local variable declarations precede the main sequence or property of the declaration. In particular, they precede any clocking event or **disable iff** specified in the declaration.

The form of a body local variable declaration is a special case of the form of a SystemVerilog variable declaration. The data type of a local variable declaration must be explicit and must be one of the types allowed in assertion Boolean expressions (see Clause 16.6 of the LRM [8]). The following example shows some illegal local variable declarations.[1]

```
property p_illegal_loc_var_decl;
   logic l_a, [3:0] l_b; // packed dimension not in data type
   l_c;                  // no explicit data type
   bit l_d [];           // dynamic array type not allowed
   ...
endproperty
```

[1]Note that dynamic array elements can appear in expressions as long as the type of the expression is cast compatible with an integral type. However a local variable itself cannot be of dynamic array type.

Unlike other SystemVerilog variables, local variables have no default initial values. A body local variable may optionally be declared with a *declaration assignment*. Declaration assignments are also called *initialization assignments* because in each evaluation they provide the initial values to the associated local variables. Consider the following example:

```
sequence s2(logic start, b[4]);
   logic l_a = 1'b0, l_b[4] = b;
   dataType l_data;
   @(posedge clk)
   start ##1 ...
endsequence
```

The local variables `l_a` and `l_b` have declaration assignments. The expression on the right-hand side of a declaration assignment can be any expression that may be assigned to the local variable. It need not be constant. For a given evaluation of `s2`, the declaration assignments are performed in the first time step in which there is an occurrence of the leading clocking event, **posedge** `clk`. At that point, `l_a` is assigned the value `1'b0` and `l_b` is assigned the value of the unpacked array formal argument `b`. The local variable `l_data`, however, has no value at that point and is said to be *unassigned*. If evaluation of `s2` begins in a time step in which **posedge** `clk` has occurred, then the declaration assignments are performed at that point.[2]

Evaluation of the right-hand side of a declaration assignment follows the rules for evaluation of the right-hand side of an ordinary local variable assignment, as discussed in Sect. 16.3. The delay, as necessary, of the performance of a declaration assignment until occurrence of the leading clocking event of the sequence or property implies that these assignments are semantically equivalent to ordinary local variable assignments performed at the appropriate points. For example, `s2` above is semantically equivalent to the following variant:

```
sequence s2_v2(logic start, b[4]);
   logic l_a, l_b[4];
   dataType l_data;
   @(posedge clk)
   (start, l_a = 1'b0, l_b[4] = b) ##1 ...
endsequence
```

The local variable declaration assignments have been moved and attached to the first Boolean expression, `start`, of the sequence. Alternate interpretation is prefixing the original sequence using `##0`:

```
sequence s2_v2(logic start, b[4]);
   logic l_a, l_b[4];
   dataType l_data;
```

[2]In a singly clocked setting, where all timing is aligned to the same clocking event, evaluation of a sequence or property always begins in a time step in which the clocking event has occurred.

```
  @(posedge clk)
   (1, l_a = 1'b0, l_b[4] = b) ##0 start ##1 ...
endsequence
```

Or, in the case of a property by using `(1, l_a = 1'b0, l_b[4] = b)|-> ...` or `(1, l_a = 1'b0, l_b[4] = b)#-# ...`.

Declaration assignments are performed in the order that they appear, so there can be dependency of the right-hand side of a later assignment on a local variable whose declaration assignment is performed previously. It is illegal, though, for the right-hand side of the later assignment to depend on a local variable for which there is no declaration assignment. Here is an example of legal declaration assignments illustrating such a dependency:

```
property p1(byte data);
  byte l_byte = data, l_byteMasked = l_byte & mask;
  ...
endproperty
```

When the declaration assignments are performed, `l_byte` is first assigned the value of `data`, and then `l_byteMasked` is assigned the bitwise-and of the value just assigned to `l_byte` and the value of `mask`. The following example shows illegal declaration assignments:

```
property p1_illegal;
  byte l_byte, l_byteMasked = l_byte ^ mask;
  ...
endproperty
```

The declaration assignments are illegal because the right-hand side of the assignment to `l_byteMasked` references `l_byte`, for which there is no declaration assignment.

The timing of declaration assignments involves some subtlety in the presence of multiple clocks. The key principle is that a declaration assignment is always performed after aligning with a leading clocking event. This ensures that the value stored by the declaration assignment comes from a known sampling point. If a property has multiple leading clocking events,[3] then separate local variables are created for each leading clocking event and the corresponding declaration assignments are performed upon reaching alignment with each of those various clocking events.

Figure 16.1 shows declaration of a property with a local variable declaration assignment and multiple leading clocks. Suppose evaluation of `p_mult_leading_clks` begins at time t_0. At that time, two copies of `l_v` are created. One copy is used in the evaluation of the subproperty on Line 3 and the other copy is used in the evaluation of the subproperty on Line 5. The assignment `l_v = e` for the first copy is performed in the first time step concurrent or subsequent to t_0 in which **posedge** `clk1` occurs. Similarly, the assignment

[3] According to SystemVerilog 2012, only properties can have multiple leading clocking events. Sequences always have a unique leading clocking event and so do assertions.

```
property p_mult_leading_clks;
   byte l_v = e;
   (@(posedge clk1) a1 until b1 == l_v)
   and
   (@(posedge clk2) a2 until b2 == l_v);
endproperty
```

Fig. 16.1 Property with local variable declaration assignment and multiple leading clocks

`l_v = e` for the second is performed in the first time step concurrent or subsequent to t_0 in which **posedge** `clk2` occurs. The behavior of `p_mult_leading_clks` is therefore equivalent to the following variant, in which the declaration assignment has been eliminated:

```
property p_mult_leading_clks;
   byte l_v;
   (@(posedge clk1) (1'b1, l_v = e) #-# a1 until b1 == l_v)
   and
   (@(posedge clk2) (1'b1, l_v = e) #-# a2 until b2 == l_v);
endproperty
```

16.2 Declaring Argument Local Variables

The preceding section presented the following example:

```
sequence s2(logic start, b[4]);
   logic l_a = 1'b0, l_b[4] = b;
   dataType l_data;
   @(posedge clk)
   start ##1 ... // assume no further reference to b
endsequence
```

The body local variable `l_b` has a declaration assignment whose entire right-hand side is a reference to the like-typed formal argument `b`. The declaration assignments are performed when the evaluation reaches alignment with **posedge** `clk`, and at that point the value in the formal argument `b` is assigned to `l_b`. Assume that the body of `s2` makes no further reference to `b`. Then the sole use of this formal argument is to provide the local variable `l_b` its initial value, which is determined from the actual argument expression associated with `b` in the relevant instance of `s2`.

In such a case, it is convenient to be able to declare the formal argument itself as a local variable. SystemVerilog provides this capability in *argument local variables*.[4] Under the assumption that the only reference to `b` in `s2` is in the declaration assignment to `l_b`, the following variant is semantically equivalent to `s2`:

---

[4] In the LRM[8], argument local variables are called local variable formal arguments.

```
sequence s2_v3(logic start, local input logic l_b[4]);
   logic l_a = 1'b0;
   dataType l_data;
   @(posedge clk)
   start ##1 ...
endsequence
```

The keyword **local** specifies that l_b is an argument local variable, while the direction **input** specifies that l_b will receive its initial value from the associated actual argument expression, after casting to the type of l_b. The keyword **local** prevents preceding data type information from applying to the associated formal argument, so the data type **logic** must be repeated for l_b.

The semantic equivalence of s2 and s2_v3 is exact. In general, an argument local variable of direction **input** behaves exactly like a body local variable of the same type, together with a dummy formal argument of the same type, where the body local variable has a declaration assignment whose entire right-hand side is a reference to the dummy formal argument. As a result, an **input** argument local variable gets its initial value at the same time step in which body local variable declaration assignments are performed.

Argument local variables are considered to precede body local variables. If a sequence or property has both **input** argument local variables and body local variables with declaration assignments, then the initialization assignments of the **input** argument local variables are performed first. It is legal for the right-hand side of a body local variable declaration assignment to reference an **input** argument local variable. Figure 16.2 shows a variant of property p1 from the preceding section illustrating this capability.

Argument local variables can also be declared of direction **output** or **inout**, but only in a sequence declaration. An argument local variable of a property must be of direction **input**. An **output** argument local variable outputs its value to the actual argument whenever the sequence matches. The actual argument must itself be a local variable. An **inout** argument local variable behaves as a combination of an **input** and an **output** argument local variable—it receives its initial value from the actual argument and also outputs its value back to the actual argument whenever the sequence matches. The actual argument must again be a local variable.

Figure 16.3 gives an example of a sequence s_arg_dirs with argument local variables of all three directions and a property p_arg_dirs that instantiates it. The instance of s_arg_dirs on Line 12 passes the expression e_i to l_s_i and passes the local variables l_p_io and l_p_o to l_s_io and l_s_o, respectively. In the evaluation of s_arg_dirs, l_s_i gets its initial value from e_i, while l_s_io gets

```
property p1_v2(local input byte l_byte);
   byte l_byteMasked = l_byte & mask;
   ...
endproperty
```

Fig. 16.2 Body local variable declaration assignment referencing argument local variable

```
1  sequence s_arg_dirs(
2     local input  byte l_s_i,
3     local inout  byte l_s_io,
4     local output byte l_s_o
5  );
6     ... // l_s_o must be assigned in the body
7  endsequence
8
9  property p_arg_dirs;
10    byte l_p_io, l_p_o;
11    (start, l_p_io = e_io)
12    |-> s_arg_dirs(e_i, l_p_io, l_p_o)
13    |-> results_ok(l_p_io, l_p_o);
14 endproperty
```

Fig. 16.3 Sequence with argument local variables of all directions

its initial value from `l_p_io`, which is assigned the value of `e_io` in Line 11. `l_s_o` gets no initial value. Whenever `s_arg_dirs` matches, the values of `l_s_io` and `l_s_o` are output to `l_p_io` and `l_p_o`, respectively, and these values are used in the check of `results_ok` in Line 13. Further details on **`output`** and **`inout`** argument local variables will be discussed in Sect. 16.5.

The following rules apply in declaration of argument local variables: If a direction is specified for an argument, then the keyword **`local`** must also be specified, and if the keyword **`local`** is specified, then the data type must be explicitly specified, including any packed dimensions. Unpacked dimensions may also be specified. Since the arguments of `s_arg_dirs` have distinct directions, they must each specify the keyword **`local`** and therefore must also repeat the data type **`byte`**.

It is allowed to specify the keyword **`local`** without a direction, and in this case the direction **`input`** is understood. Thus, the following variant is semantically equivalent to `s_arg_dirs`:

```
sequence s_arg_dirs_v2(
   local        byte l_s_i,
   local inout  byte l_s_io,
   local output byte l_s_o
);
   ...
endsequence
```

Consecutive declarations of argument local variables can share the same specifications of the keyword **`local`**, the direction, and the data type (including any packed dimensions), but only if the following conditions are satisfied:

1. The first of the declarations specifies **`local`**, specifies the direction either explicitly or implicitly, explicitly specifies the data type (including any packed dimensions), and does not specify any unpacked dimensions.
2. Subsequent declarations specify only the formal argument identifier. No packed or unpacked dimensions may be specified together with the subsequent identifiers.

Here is an example of legal declarations illustrating this capability and the restrictions:

```
sequence s3(
   local input byte l_a, l_b,
   local input byte l_c[8],
   local input byte l_d,
   local inout byte l_e,
   local input bit  l_f,
   local input bit  [0:3] l_g, l_h
);
   ...
endsequence
```

`l_a` and `l_b` are both of direction **`input`** and data type **`byte`**. `l_c` is of direction **`input`** and is an unpacked array of eight bytes. Because of the unpacked dimension, the declaration of `l_c` cannot share the keyword **`local`** and direction with either the preceding or the subsequent declaration. Because the direction of `l_e` does not match the preceding direction, the declaration of `l_e` must specify the keyword **`local`** and the direction and data type. `l_g` and `l_h` are both packed vectors of 4 bits. Even though `l_g` has the same direction and base data type as `l_f`, it has different packed dimension and so must have its own specification of the keyword **`local`**, direction, and data type.[5]

An **`input`** argument local variable may be declared with an optional default actual argument, which can be any expression that may be assigned to the argument local variable. An **`output`** or **`inout`** argument local variable may not be given a default actual argument because the actual argument must specify the local variable that will receive the output value. The syntax for a default actual argument is the same as that for ordinary formal arguments of sequences or properties (see Sect. 8.2). The default actual argument serves as the actual argument in any instance of the sequence or property that does not otherwise specify the actual argument. As usual, names in the default actual argument expression resolve in the context of the sequence or property declaration, not in the context of its instantiation. Here is a modification of property `p1` that specifies a default actual argument:

```
property p1_v3(local input byte l_byte = data);
   byte l_byteMasked = l_byte & mask;
   ...
endproperty
```

The default actual argument for `l_byte` is `data`.

[5]The rules of sharing components of declarations are more stringent than necessary. Future versions of SystemVerilog may relax them.

16.3 Assigning to Local Variables

The fundamental capability provided by local variables is to enable an assertion to capture the value of an expression at a specified point in its evaluation and store that value for later reference, perhaps after further modification. This section discusses rules for assigning to local variables. The declaration and initialization assignments introduced in Sects. 16.1 and 16.2 are particular cases of assignments to local variables. They must follow the rules presented here.

A local variable assignment may be attached to a subsequence of a named sequence or property. The local variable assignment is written after the subsequence, separated by a comma, and the pair is enclosed in parentheses. The subsequence must not admit an empty match. Whenever the subsequence matches in the course of evaluation of the named sequence or property, the local variable assignment is performed. The result of attaching a local variable assignment to a subsequence is always a sequence, even if the subsequence were itself a Boolean. Here is a simple example:

```
property p_ttype_vs_data;
   transType l_ttype;
   (start, l_ttype = ttype)
   ##1 (dataValid within complete[->1])
   |-> ttypeAllowsData(l_ttype);
endproperty
```

The local variable assignment `l_ttype = ttype` has been attached to the Boolean subsequence `start`. Whenever `start` is tested and evaluates to true, the Boolean subsequence matches and the local variable assignment is performed, capturing the value of `ttype` for later reference in the call to the function `ttypeAllowsData`. This property checks that if `dataValid` occurs after `start` and not later than `complete`, then the transaction type specified in `ttype` at the time of `start` is one that allows data as encoded in `ttypeAllowsData`.

Multiple local variable assignments may be attached to a single subsequence. The local variable assignments are performed in order whenever the subsequence matches.

```
sequence s_compare_two_data_and_parity;
   dataType l_data;
   parityType l_parity;
   start
   ##1 (dataValid[->1], l_data = data, l_parity = parity)
   ##1 dataValid[->1]
   ##0 data == l_data && parity == l_parity;
endsequence
```

This sequence compares the values of `data` and `parity` from the first two occurrences of `dataValid` after `start`. Two assignments have been attached to the first goto subsequence `dataValid[->1]`: `l_data = data` and `l_parity = parity`. The local variables capture the values of `data` and `parity` from the first occurrence

```
property p_data_and_parity;
   dataType l_data;
   parityType l_parity;
   start ##1
   (
      (dataValid, l_data = data, l_parity = parity)
      within complete[->1]
   )
   |-> parityOK(l_data, l_parity);
endproperty
```

Fig. 16.4 Data and parity check

of `dataValid` and hold them for comparison with the corresponding values at the second occurrence of `dataValid`. If the corresponding values are equal, then the overall sequence matches.

In general, the evaluation of the right-hand side of a local variable assignment follows the rules for evaluation of expressions within a concurrent assertion. After resolving the terms of the expression through elaboration (including argument passing, module instantiation, bind instantiation, etc.), sampled values are used for those terms that are not local variables, while current values are used for terms that are local variables. These rules apply to declaration assignments for body local variables and initialization assignments for argument local variables.

Each evaluation attempt of a named sequence or property gets its own, private copy of all the declared local variables for use within that evaluation. Local variables are thus "local" to an individual thread of evaluation, and variables for one thread of evaluation cannot be referenced in another thread of evaluation. The examples presented so far in this chapter have required only one copy of each local variable per evaluation attempt. In general, evaluation of a sequence or property may involve branching into subevaluations. Part of the power of local variables is that their semantics includes automatic allocation of additional copies when needed to store multiple values arising from such subevaluations. Consider the example in Fig. 16.4. This property checks that for every occurrence of `dataValid` that happens after `start` and not later than `complete`, the values of `data` and `parity` that are present with `dataValid` satisfy the condition encoded in the function `parityOK`. The local variable assignments are attached to the Boolean subsequence `dataValid`, and the resulting sequence is the first operand of the **`within`** operator. The **`within`** sequence is itself part of the antecedent of `|->`. The semantics of `|->` requires that every match of its antecedent result in a check of its consequent. Because `dataValid` may occur at multiple points within the interval of matching of `complete[->1]`, there may be multiple matches of the antecedent, and each such match will have its own copy of the local variables to store the values of `data` and `parity` from the particular point that `dataValid` occurred for that match.

Figure 16.5 shows a possible waveform for this property. Only the values of `data` and `l_data` are shown, since the timing of the local variable capture for `l_parity` is the same. `start` occurs at time 20 and `complete` occurs at time 90. Between them there are three occurrences of `dataValid`, at times 30, 50, and 70. Therefore,

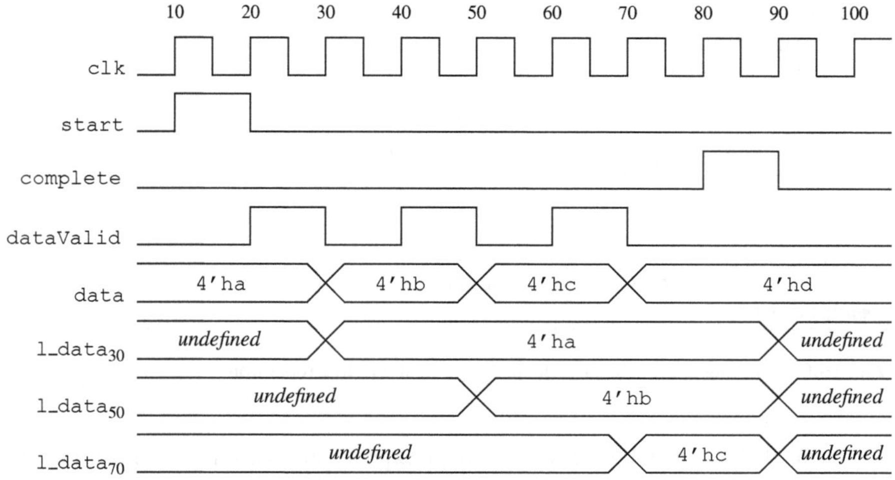

Fig. 16.5 Waveform for data and parity check

the evaluation of `p_data_and_parity` that starts at time 20 will obtain three copies of `l_data`, one for each of the three occurrences of `dataValid`. These copies are shown as $\texttt{l\_data}_{30}$, $\texttt{l\_data}_{50}$, and $\texttt{l\_data}_{70}$ in the waveform, along with the values of `data` that they capture.

16.3.1 Assignment Within Repetition

If the operand of a sequence repetition is a subsequence with a local variable assignment attached, then the local variable assignment executes on each iterative match of the subsequence. Such assignments can be used to count the number of iterations or to compute aggregate values based on increments that are observable at the successive matches of the subsequence.

Figure 16.6 shows an example of counting the number of iterations. [6] When `start` occurs, the property stores the transaction type in the local variable `l_ttype` (Line 4). The property checks that the number of occurrences of `dataValid` that happen after `start` and not later than `complete` is allowable for the transaction type according to the function `numBeatsOK`. The number of occurrences of `dataValid` is counted in the local variable `numBeats`. `numBeats` is initialized to zero in its declaration on Line 3. The various occurrences of `dataValid` are matched by the goto `dataValid[->1]` within the repetition on Line 6, and `numBeats` is incremented

[6]For formal verification, the **`int`** type of `numBeats` should be replaced with the type of smallest bitwidth needed for the counter.

```
1  property p_ttype_vs_beats;
2     transType l_ttype;
3     int numBeats = 0;
4     (start, l_ttype = ttype) ##1
5     (
6        (dataValid[->1], numBeats++)[*]
7           ##1 !dataValid[*]
8        intersect
9        complete[->1]
10    )
11    |-> numBeatsOK(l_ttype, numBeats);
12 endproperty
```

Fig. 16.6 Counting using a local variable assignment within a repetition

for each of these matches. The increment expression `numBeats++` illustrates the fact that local variable assignments may be specified by increment and decrement expressions. They may also be specified by general operator assignments (`+=`, `&=`, etc.). The subsequence of Lines 6 and 7 is intersected with the goto `complete[->1]` of Line 9 to ensure that only the relevant occurrences of `dataValid` are counted. The zero repetition option for `[*]` in Line 7 allows the last `dataValid` to be concurrent with `complete`. The zero repetition option for `[*]` in Line 6 allows for the possibility that there is no occurrence of `dataValid`.

16.3.2 Sequence Match Items

A local variable assignment attached to a sequence is an example of a *sequence match item*, i.e., an item to be performed upon match of the sequence. The other kind of sequence match item is a subroutine call. A subroutine called as a sequence match item can be a task, a task method, a void function, a void function method, or a system task.

Sequence match items may be attached to any sequence that does not admit empty match. The first match item is separated from the sequence by a comma, and further match items may be written as a comma-separated list. The sequence and list of match items are enclosed in parentheses. Whenever the sequence matches, the match items are processed in the order of the list. Local variable assignments are performed immediately in the Observed region. Subroutine calls are scheduled for execution in the Reactive region in the order that they appear. The assertion evaluation does not wait on or get information back from a subroutine.

Arguments passed to a subroutine call must be passed either by value as inputs or by reference (**`ref`** or **`const ref`**). Local variables may be passed only by value and must flow to the point of the subroutine call (see Sect. 16.4.1). Actual arguments passed by value are evaluated in the Observed region like other expressions in

```
1  property p_data_and_parity_v2;
2     dataType l_data;
3     parityType l_parity;
4     start ##1
5     (
6        (
7           dataValid,
8           l_data = data, l_parity = parity,
9           $display("time=%0d data=%h parity=%h",
10             $time,l_data,l_parity)
11       )
12       within complete[->1]
13    )
14    |-> parityOK(l_data, l_parity);
15 endproperty
```

Fig. 16.7 Property with subroutine call attached to a sequence

assertions: current values are used for local variables, whereas sampled values are used otherwise. Actual arguments passed by reference are evaluated using Reactive region values when the subroutine executes.

A common use of subroutine calls as sequence match items is to export information from the assertion evaluation thread, especially values of local variables. The export can be for debugging, to communicate with other parts of a testbench, or to populate a coverage model. The capability to place the subroutine call as a sequence match item is essential to get visibility to the local variables, which cannot be referenced from an action block. Figure 16.7 shows a variant of `p_data_and_parity` illustrating this usage. In each time step that `dataValid` is matched in Line 7, the local variable assignments are performed in Line 8 and then the `$display` is called. The arguments to $display are inputs. Their values are computed in the Observed region, using the values assigned to the local variables in Line 8 and the current value of `$time`.

As another example, suppose that `startCode` is a signal of type `startType` that is valid with `start`, while `endCode` is a signal of type `endType` that is valid with `complete`. Suppose that we want to collect coverage on the pairs of `startCode` and `endCode` values that occur for transactions of the sequential protocol (see Sect. 15.2). This can be done using a covergroup and calling its sample method as a sequence match item. Figure 16.8 shows an encoding. The signature of the covergroup `sample` method is declared in Line 2, and the covergroup is instantiated in Line 7. The sequence `s_SEC` has a local variable to capture the value of `startCode`. Once `complete` is reached in Line 13, the covergroup sample method is called in Line 14. The signal `endCode` is valid at this time, while the `startCode` has been stored in `l_startCode`. See Sect. 18.2.3 for further discussion of covergroups and Sect. 18.2.4 for more examples of this kind.

```
covergroup cg_SEC_type
   with function sample(startType startCode, endType endCode);
   SC: coverpoint startCode;
   EC: coverpoint endCode;
   SEC: cross SC, EC;
endgroup
cg_SEC_type cg_SEC = new();
sequence s_SEC;
   startType l_startCode;
   (start, l_startCode = startCode)
   ##1
   (
      complete[->1],
      cg_SEC.sample(l_startCode, endCode)
   );
endsequence
c_SEC:  cover property(s_SEC);
```

Fig. 16.8 Collecting coverage using a local variable and a subroutine call

16.4 Referencing Local Variables

A local variable that is assigned a value may be referenced within the same named sequence or property. Local variables can be referenced in expressions such as:

- Boolean expressions.
- Bit-select and part-select expressions.
- Array indices.
- Arguments of task and function calls.
- Arguments of sequence and property instances.
- Expressions assigned to local variables.

Local variables cannot be referenced in the following kinds of expressions:

- Expressions that are required to be compile-time constants, such as *n* in each of the following operators: `##n`, `[*n]`, `[->n]`, `[=n]`. Local variables also may not be referenced in the constant expressions of ranged forms of these operators and bounded property operators.
- Clocking event expressions.
- The reset expression of a **`disable iff`**.
- The abort condition of a reset operator (**`accept_on`**, **`sync_accept_on`**, and the reject forms of these operators).
- An argument expression to a sampled value function (`$rose`, `$fell`, `$past`, etc.).

A local variable that is unassigned may not be referenced.

Each evaluation attempt of a sequence or property has its own, private copies of the local variables declared in that sequence or property. Further copies of the local variables may be created as the evaluation evolves, branching into subevaluation

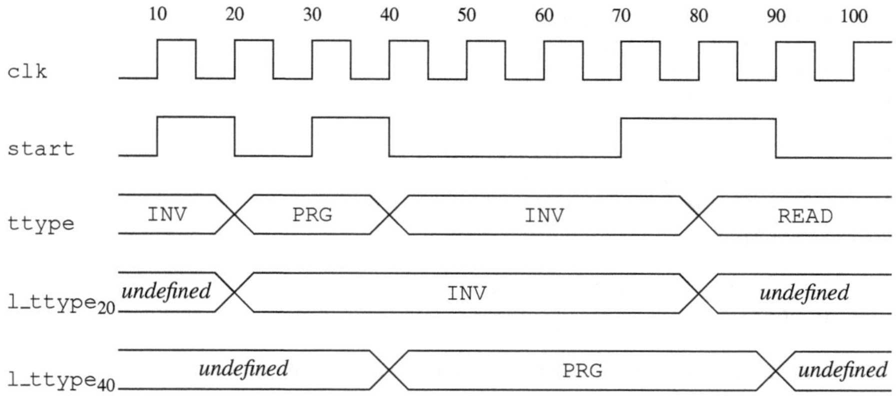

Fig. 16.9 Waveform for transaction type check

threads. In general, a local variable assigned in a thread of evaluation may be referenced later in that thread or in a descendent of that thread. However, SystemVerilog provides no mechanism for externally referencing or cross-referencing a local variable. One evaluation attempt cannot reference the copies of a local variable from another evaluation attempt. Similarly, a subevaluation thread cannot reference updates to a sibling's copy of a local variable. If the siblings are subsequence evaluations that later join (e.g., subevaluations of operands of sequence **and** or **intersect**), then structural rules described below determine whether or not the thread of evaluation that proceeds from the join can reference the updates from one of the siblings. Hierarchical references to local variables are illegal.

As a simple example to illustrate these ideas, suppose ttype is valid with start and that if there is an occurrence of start with ttype == INV or ttype == PRG, then at the next two occurrences of start, ttype must not have the same value that it has at the current occurrence of start. This check can be coded as

```
property p_ttype_check;
  transType l_ttype;
  (start && (ttype == INV || ttype == PRG), l_ttype = ttype)
  |=> start[->1:2]
  |-> ttype != l_ttype;
endproperty
```

Consider the waveform shown in Fig. 16.9. The evaluation attempt of property p_ttype_check that begins at time 20 observes an occurrence of start together with ttype == INV, and so it stores INV in its copy of the local variable l_ttype. This copy of the local variable is represented by the row labeled l_ttype$_{20}$ in the waveform. Another evaluation attempt of p_ttype_check begins at time 40, observes an occurrence of start together with ttype == PRG, and stores PRG in its copy of the local variable, which is labeled l_ttype$_{40}$ in the waveform. The evaluation that begins at time 20 sees only l_ttype$_{20}$, not l_ttype$_{40}$. At time 40, this thread observes the first subsequent occurrence of start, and the check

```
1  property p_ttype_vs_beats;
2     transType l_ttype;
3     int numBeats = 0;
4     (start, l_ttype = ttype) ##1
5     (
6        (dataValid[->1], numBeats++)[*]
7           ##1 !dataValid[*]
8        intersect
9        complete[->1]
10    )
11    |-> numBeatsOK(l_ttype, numBeats);
12 endproperty
```

Fig. 16.10 Illustrating local variable flow, reference, and reassignment

`ttype != l_ttype` in Line 5 compares `PRG != INV` and succeeds. At time 80, though, this thread fails since Line 5 compares `INV != INV`, and so the overall evaluation beginning at time 20 fails. The evaluation that begins at time 40 succeeds since its two checks of Line 5 compare `INV != PRG` at time 80 and `READ != PRG` at time 90.

16.4.1 *Local Variable Flow*

To understand more thoroughly where a local variable may be referenced and what value will be yielded, we need to explore in more detail how the scope of the local variable extends into various subsequences and subproperties and how the value stored in the local variable is carried along and changed in the corresponding subevaluations. A local variable is said to *flow into* a subsequence or subproperty if it is assigned (i.e., has a value) when evaluation of the subsequence or subproperty begins. A local variable is said to *flow out of* a subsequence if it is guaranteed to be assigned (i.e., to have a value) upon reaching the end of a match of the subsequence.

In general, if a local variable flows into a subsequence or subproperty, then the local variable may be referenced within that subsequence or subproperty provided the local variable remains assigned at the point of reference. (See Sect. 16.4.2 for a discussion of how a local variable may become unassigned.) A reference to a local variable yields the latest value assigned to it in the evaluation or subevaluation that reaches the point of the reference. The latest assignment may have occurred in the current time step. A local variable reference is always to the current value stored in the variable.

As an example of local variable flow, reference, and reassignment, let us revisit the property `p_ttype_vs_beats`, which is copied here in Fig. 16.10. The property has two local variables: `l_ttype` has no declaration assignment, while `numBeats` has a declaration assignment initializing its value to `0`. Therefore, `l_ttype` is unassigned and does not flow into the body of the property in Line 4, while numBeats does flow into Line 4. If start evaluates to true in Line 4, though, then l_ttype is assigned the value of ttype. Therefore, both l_ttype and numBeats flow into Line 5. The subsequence of Lines 5–10 makes no reference to and no reassignment of l_ttype. Therefore, l_ttype flows with value unchanged into Line 11, where it is referenced as an argument to the call to function numBeatsOK. In contrast, the subsequence on Line 6 makes reference to and reassigns numBeats for each consecutive match of dataValid[->1] in the repetition. The reference to numBeats is implicit in the increment operator ++. As a result, numBeats flows out of the subsequence of Lines 6 and 7. Because numBeats is reassigned in only one of the operands of the **intersect** operator appearing on Line 8, numBeats also flows out of the entire subsequence of Lines 5–10. Therefore, numBeats flows with its last reassigned value into Line 11, where it is also referenced as an argument to the call to function numBeatsOK.

The rules of local variable flow have been designed to be intuitively reasonable and also computable at compile time, rather than varying dynamically with the course of evaluation. The rules do not depend on the specific value stored in a local variable at a particular point, only on whether the local variable is guaranteed to be assigned some value at that point. Because the rules depend only on the structure of the sequences and properties and where the assignments to the local variables occur, they can be checked at compile time.

Below are the rules of local variable flow for declaration, instance, sequence, and property forms. In these rules, v, w stand for local variables; r, s stand for sequences; p, q stand for properties; e stands for an expression; and b stands for a Boolean expression.

DF A local variable declared in a named sequence or property flows into the body sequence or property expression of that declaration iff it is assigned in an initialization assignment.

IF1 A local variable that flows into an instance of a named sequence or property does not flow into the body sequence or property expression in the declaration of that instance. The value of the local variable may be passed into the instance through an argument (see Sect. 16.5.1).

IF2 A local variable that flows out of the body sequence expression of the declaration of a named sequence[7] does not flow out of an instance of the named sequence. If the local variable is an untyped formal argument or an argument local variable of direction **output** or **inout**, then its value may be passed out of the instance (see Sect. 16.5.2).

IF3 v flows out of b iff v flows into b. Analogous rules apply when b is replaced by Boolean repetitions b[->n], b[=n], etc.

---

[7]More precisely, that flows out of the sequence expression that results from the body sequence expression of the declaration by substituting actual arguments from an instance for formal arguments, as described in the rewriting algorithms of Annex F.4 of the LRM [8].

IF4 v flows out of `(r, v = e)`. If w flows into `(r, v = e)`, then w flows into r. w flows out of `(r, v = e)` iff w flows out of r. If e references w, then w must flow out of r.

IF5 If v flows into r `##n` s, then v flows into r. If v flows out of r, then v flows across `##n` into s. v flows out of r `##n` s iff v flows out of s. Analogous rules apply to the variants of the binary concatenation operator. The flow rules for unary concatenation are obtained from those for binary concatenation by replacing r with `1'b1`.

IF6 If v flows into r **`or`** s, then v flows into both r and s. v flows out of r **`or`** s iff v flows out of both r and s.

IF7 If v flows into r **`and`** s, then v flows into both r and s. v flows out of r **`and`** s iff either v flows out of r and there is no assignment to v in s or v flows out of s and there is no assignment to v in r. Analogous rules apply to **`intersect`** and **`within`**.

IF8 If v flows into b **`throughout`** r, then v flows into both b and r. v flows out of b **`throughout`** r iff v flows out of r.

IF9 If v flows into **`first_match`**`(r)`, then v flows into r. v flows out of **`first_match`**`(r)` iff v flows out of r.

IF10 v flows out of r `[*0]` iff v flows into r `[*0]`. The value of v does not change as a result of empty match of r `[*0]`.

IF11 If v flows into r `[*n]`, where n is positive, then v flows into the first iteration of r. v flows out of r `[*n]` iff v flows out of r, in which case v flows into and out of each iteration of r. If v does not flow out of r, then v does not flow into any iteration of r after the first. Analogous rules apply to ranged forms of the repetition operator. If the lower range is zero, then the flow rule is obtained by decomposing r `[*0:n]` as r `[*0]` **`or`** r `[*1:n]` (n positive or `$`).

PF1 If v flows into **`strong`**`(r)`, then v flows into r. An analogous rule applies to **`weak`**.

PF2 If v flows into **`not`** p, then v flows into p.

PF3 If v flows into p **`or`** q, then v flows into both p and q. Analogous rules apply to **`and`**, **`implies`**, and **`iff`**.

PF4 If v flows into r `|->` p, then v flows into r. If v flows out of r, then v flows across `|->` and into p. Analogous rules apply to `|=>`, `#-#`, and `#=#`.

PF5 If v flows into **`if`** `(b)` p **`else`** q, then v flows into b, p, and q. An analogous rule applies to **`case`**.

PF6 If v flows into **`nexttime`** p, then v flows into p. Analogous rules apply to **`s_nexttime`** and to the indexed forms of these operators.

PF7 If v flows into **`always`** p, then v flows into p. Analogous rules apply to **`s_always`** and to the ranged forms of these operators.

PF8 If v flows into p **`until`** q, then v flows into both p and q. Analogous rules apply to **`s_until`**, **`until_with`**, and **`s_until_with`**.

PF9 If v flows into **`s_eventually`** p, then v flows into p. Analogous rules apply to the ranged form of this and the weak **`eventually`** operator.

```
1  property p_flow_analysis;
2     byte l_v;
3     (
4        (a[->1], l_v = e)
5        within
6        b[->1]
7     )
8     #=#
9     (
10       (c == l_v)
11       until
12       (
13          (d == l_v)
14          and
15          nexttime(c != l_v)
16       )
17    );
18 endproperty
```

Fig. 16.11 Property illustrating local variable flow

PF10 If v flows into **`disable iff`** `(`b`)` p, then v flows into p. Analogous rules apply to **`accept_on`**, **`reject_on`**, **`sync_accept_on`**, and **`sync_reject_on`**.

As an example of analyzing local variable flow using these rules, consider the property in Fig. 16.11. By Rule *DF*, `l_v` does not flow into Line 3. By Rule *IF4*, `l_v` flows out of Line 4, and so by Rule *IF7*, `l_v` flows out of Line 7. Rule *PF4* then says that `l_v` flows into Line 9, and so by Rule *PF8*, `l_v` flows into both Line 10 and Line 12. This implies that the reference to `l_v` in Line 10 is legal. By Rule *PF3*, `l_v` flows into both Line 13 and Line 15. This implies that the reference to `l_v` in Line 13 is legal. Finally, by Rule *PF6*, `l_v` flows into the expression `c != l_v`, so the reference to `l_v` in Line 15 is also legal.

16.4.2 Becoming Unassigned

Once assigned, local variables only become unassigned upon match of subsequences formed from certain sequence operators. Code in which local variables become unassigned is of poor style and should be avoided.

Figure 16.12 shows an example in which the local variable `l_v` becomes unassigned upon match of an **`and`** subsequence. The semantics of the **`and`** sequence operator requires the subevaluations of the operands to join together once both have matched. This requires all local variables that are assigned to have consistent, unambiguous values in the continuing thread of evaluation. There are assignments to `l_v` in both operands of the **`and`** (Lines 4 and 6), so `l_v` may not have a single

```
1  sequence s_and_ambiguous;
2    bit l_v;
3    (
4      (a[->1], l_v = e)
5      and
6      (b[->1], l_v = f)
7    ) // SF5: l_v does not flow out, becomes unassigned
8    ##1 c == l_v;  // illegal
9    reference to l_v endsequence
```

Fig. 16.12 Local variable becomes unassigned due to ambiguity of value after **and**

```
1  sequence s_and_unambiguous;
2    bit l_va, l_vb;
3    (
4      (a[->1], l_va = e)
5      and
6      (b[->1], l_vb = f)
7    )
8    ##1 c == l_va || c == l_vb;
9  endsequence
```

Fig. 16.13 Assigning to distinct local variables in the operands of **and**

```
1  sequence s_or_ambiguous;
2    bit l_v;
3    ( // DF: l_v does not flow in
4      (a[->1], l_v = e)
5      or
6      b[->1]  // SF1: l_v does not flow out
7    )  // SF4: l_v does not flow out, becomes unassigned
8    ##1 c == l_v;  // illegal reference to l_v
9  endsequence
```

Fig. 16.14 Local variable becomes unassigned due to ambiguity of value after **or**

value upon match of the **and**. According to Rule *IF7*, `l_v` does not flow out of the **and** subsequence, and as a result `l_v` becomes unassigned in Line 7. The reference to `l_v` in Line 8 is therefore illegal.

SF5 applies similarly to sequence operators **intersect** and **within**. Therefore, when coding local variable assignments in operands of sequence **and**, **intersect**, or **within**, do not make assignments to a single local variable in more than one operand. The ambiguous example from Fig. 16.12 can be recoded using two local variables as shown in Fig. 16.13. Note, in particular, that the comparison to `c` in Line 8 has been disambiguated in Fig. 16.13 as the disjunction `c == l_va || c == l_vb`.

Figure 16.14 shows a different problem occurring with assignments in operands of a sequence **or**. In this example, `l_v` is not assigned before the **or**, and it is assigned in only one operand of the **or** (Line 4). The semantics of sequence **or** specifies that if either operand subsequence matches, then there is a match of the

```
1  sequence s_or_unambiguous;
2     bit l_v = 1'b0;
3     (
4        (a[->1], l_v = e)
5        or
6        b[->1]
7     )
8     ##1 c == l_v;
9  endsequence
```

Fig. 16.15 Local variable assigned before **or**

```
1  sequence s_nested_ambiguity;
2     bit l_v = 1'b0;
3     (
4        (a[->1], l_v = e)
5        or
6        (
7           (b1[->1], l_v = f1)
8           and
9           (b2[->1], l_v = f2)
10       )  // SF5: l_v does not flow out
11    )     // SF4: l_v does not flow out, becomes unassigned
12    ##1 c == l_v;  // illegal reference to l_v
13    endsequence
```

Fig. 16.16 Nested structure results in local variable becoming unassigned

or, and evaluation continues for each such match. After Line 7, it is not known which operand of the **or** matched, and so it is not known whether `l_v` has been assigned. According to Rule *IF6*, `l_v` does not flow out of the **or**, and `l_v` becomes unassigned in Line 7. The reference to `l_v` in Line 8 is therefore illegal.

The situation in Fig. 16.14 can be remedied by ensuring that `l_v` will be assigned a value no matter which operand of the **or** matches. This can be done by assigning a value to `l_v` prior to the **or** or by assigning a value to `l_v` in the second operand (Line 6). Figure 16.15 uses a declaration assignment to provide a value to `l_v` prior to the **or**.

The flow rules must be applied recursively to sequences and properties with nested structure. Figure 16.16 shows an example in which the second operand of an **or** subsequence is itself an **and** subsequence. According to *IF7*, `l_v` does not flow out of the **and** (Line 10). Even though `l_v` is assigned before the **or** (Line 2) and is assigned within each operand of the **or**, *IF6* then implies `l_v` does not flow out of the **or** and becomes unassigned in Line 11. The reference to `l_v` on Line 12 is therefore illegal.

16.4.3 Multiplicity of Matching with Local Variables

Without local variables, multiple matches of a sequence over the same interval of a trace can usually be treated as the same.[8] For example, consider the following assertion:

```
a_mult_match:  assert property(
   (a[*1:2] or b[*1:2]) |-> c
);
```

For each evaluation attempt, the antecedent sequence can either not match or match over one or two cycles. If a and b are not mutually exclusive, then each match could be of multiplicity either one or two. Checking of the consequent is obligated for each match of the antecedent, but the consequent evaluation is the same for multiple matches over the same interval. Therefore, a tool can simply keep track of which intervals are matched by the antecedent and not distinguish multiple matches over the same interval.

With local variables, the values stored in them can differentiate matches over the same interval of a trace. If there are subsequent references to the local variables, then tools must keep track of the different values in the local variables for multiple matches over the same interval. The following code illustrates this situation:

```
property p_mult_match_loc_var;
   byte l_v;
   ((a[*1:2], l_v = e) or (b[*1:2], l_v = f)) |-> g != l_v;
endproperty
a_mult_match_loc_var:  assert property(p_mult_match);
```

Property `p_mult_match_loc_var` has the same possibilities for matching of its antecedent as assertion `a_mult_match`. However, because different expressions are assigned to the local variable `l_v` in the two operands of **or**, this local variable can distinguish multiple matches over the same interval. Since `l_v` is referenced in the consequent, tools must keep track of all of the various multiple matches of the antecedent and the associated values of `l_v`. The checking performed by `a_mult_match_loc_var` is equivalent to the following assertion:

```
a_mult_match_2:  assert property(
       (a |-> g != e)
   and (b |-> g != f)
   and (a[*2] |-> g != $past(e))
   and (b[*2] |-> g != $past(f))
);
```

[8]When collecting coverage for a **cover sequence** assertion statement, all matches must be counted with the appropriate multiplicities.

16.5 Input and Output with Local Variables

This section describes the mechanisms provided by SystemVerilog for passing values of local variables into instances of named sequences and properties and out of instances of named sequences. Special rules apply when passing values into an instance of a named sequence to which a sequence method (`triggered` or `matched`) is applied. There is no notion of passing values of local variables out of instances of named properties.

16.5.1 Input with Local Variables

In general, any local variable that is assigned a value at the point of instantiation of a named sequence or property may have its value passed into that instance simply by referencing the local variable as an entire actual argument expression of the instance or as a proper subexpression of an actual argument expression of the instance. The local variable itself cannot be referenced from within the instance, so the instance does not track changes to the local variable or to copies of it that might occur in the instantiating context. Rather, the value of the local variable passed into the instance remains constant throughout the evaluation of the instance.

The following simple example illustrates input of a local variable to an instance of a named property:

```
1  property p_no_repeat_ttype(
2     transType varying_ttype, captured_ttype
3  );
4     start[->1:2]
5     |-> varying_ttype != captured_ttype;
6  endproperty
7  property p_ttype_check;
8     transType l_ttype;
9     (start && (ttype == INV || ttype == PRG), l_ttype = ttype)
10    |=> p_no_repeat_ttype(
11       .varying_ttype(ttype), .captured_ttype(l_ttype)
12    );
13 endproperty
```

This example is an equivalent recoding of the property `p_ttype_check` from Sect. 16.4. It uses an auxiliary property, `p_no_repeat_ttype`. In the instance of `p_no_repeat_ttype` in Lines 10–12, the signal `ttype` is passed to the formal argument `varying_ttype`, while the local variable `l_ttype` is passed to the formal argument `captured_ttype`. During an evaluation of this instance of `p_no_repeat_ttype`, the formal argument `captured_ttype` behaves as a constant equal to the value of the local variable `l_ttype` at the time the evaluation begins. Referring back to the waveform in Fig. 16.9, the evaluation of `p_ttype_check` beginning at time 20 causes an evaluation of `p_no_repeat_ttype` to begin at time 30. This evaluation continues until time 80, and throughout it the formal argument `captured_ttype` has the value `INV`. By contrast, argument `varying_ttype` tracks changing values of the actual argument `ttype` because `ttype` is a signal rather than a local variable. Therefore, the evaluation of `p_no_repeat_ttype` that begins at time 30 fails at time 80, where it observes the second occurrence of `start` together with `varying_ttype` and `captured_ttype` both equal to `INV`.

Note that the preceding example is also equivalent to the following:

```
1  property p_no_repeat_ttype(
2     transType varying_ttype, captured_ttype
3  );
4     start[->1:2]
5     |-> varying_ttype != captured_ttype;
6  endproperty
7  property p_ttype_check;
8     transType l_ttype;
9     (start && (ttype == INV || ttype == PRG), l_ttype = ttype)
10    |=>
11    (
12       (1'b1, l_ttype = ttype)    // no effect on instance below
13       and
14       p_no_repeat_ttype(
15          .varying_ttype(ttype), .captured_ttype(l_ttype)
16       )
17    );
18 endproperty
```

In this variant, `l_ttype` is modified in Line 12. Flow rules *PF4* and *IF7* say that the value of `l_ttype` that flows into Line 14 is determined by the assignment to `l_ttype` in Line 9. The assignment in Line 12 does not affect the instance of `p_no_repeat_ttype` in Lines 14–16. This emphasizes the fact that changes to a local variable in the instantiating context that do not flow into an instance do not affect the evaluation of the instance (cf. Exercise 16.7).

It can be useful for the value of a local variable passed into an instance to be further modified within the evaluation of the instance. The example of Fig. 16.17 passes a local variable to an argument local variable of direction **`input`**. It modularizes the encoding of the FIFO protocol data check from Fig. 15.9 so that an instance of sequence `s_skip` accomplishes the skipping of occurrences of `complete` to get to the occurrence that is relevant for the current evaluation attempt of `p_fifo_data_check`. The local variable `numAhead` is assigned in Line 10, and its value is then passed to the argument local variable `numToSkip` in Line 11.

SystemVerilog also allows untyped formal arguments to be l values in assignments and thereby function as local variables. The sequence `s_skip` from Fig. 16.17 can be recoded as follows:

```
sequence s_skip(numToSkip);
   (numToSkip > 0 ##0 complete[->1], numToSkip--)[*]
   ##1 (numToSkip == 0 ##0 complete[->1]);
endsequence
```

```
sequence s_skip(
    local input bit [0:$clog2(MAX_OUTSTANDING)] numToSkip
);
    (numToSkip > 0 ##0 complete[->1], numToSkip--)[*]
    ##1 (numToSkip == 0 ##0 complete[->1]);
endsequence
property p_fifo_data_check;
    dataType data;
    bit [0:$clog2(MAX_OUTSTANDING)] numAhead;
    (start, data = dataIn, numAhead = outstanding)
    ##1 s_skip(.numToSkip(numAhead))
    |->
    dataOut == data;
endproperty
a_fifo_data_check: assert property (p_fifo_data_check);
```

Fig. 16.17 FIFO protocol data check using a subsequence

The usual rules of passing actual arguments to untyped formal arguments apply in this case: the bit width of `numToSkip` for a given instance is determined by the corresponding actual argument, and there is no type checking to ensure compatibility between the formal and actual argument.

Exception for Sequence Methods

There is an exception to all of the preceding discussion of this section: it does not apply to a local variable passed to an instance of a named sequence to which a sequence method (`triggered` or) is applied. SystemVerilog does not allow the value of a local variable passed into such an instance to be observed by the instance.[9] This means that the declaration of the sequence must not reference the corresponding formal argument unless the formal argument has been assigned as a local variable within the sequence body. This restriction is important for the preservation of forward progress of time and causality because, in general, it is not known how the time of assignment to a local variable in the instantiating context relates to the start time of a match of the sequence to which a sequence method is applied. For example, the following contrived code is not legal:

[9]In addition, a local variable passed to a sequence instance to which a sequence method is applied must be the entire actual argument passed to the corresponding formal argument and must not be passed to an argument local variable of direction **`input`** or **`inout`**.

```
1 sequence s_3_in_a_row(bit signal, goal);
2    (signal == goal)[*3];
3 endsequence
4 property p_illegal_causality;
5    bit l_v;
6    (start, l_v = a)
7    |-> s_3_in_a_row(.signal(b), .goal(l_v)).triggered;
8 endproperty
```

In Line 7, this code attempts to pass the value of local variable `l_v` into an instance of sequence `s_3_in_a_row` to which sequence method `triggered` is applied. The intended meaning of the code is that if `start` occurs, then `b` must equal `a` now and `b` must now have held this value for at least three cycles, including the current one. This encoding is forbidden by SystemVerilog because it requires evaluation of `s_3_in_a_row` to be aware of the future value of `a` when looking for a match to satisfy the instance to which `triggered` is applied (cf. Exercise 16.8).

16.5.2 Output with Local Variables

Values of local variables can be output only from instances of named sequences. There is no notion of passing values of local variables out of instances of named properties because:

- The evaluation of a property or subproperty is terminal in the sense that there is no ensuing thread of assertion evaluation.
- In general, property evaluation has no well-defined finite endpoint.

There are two mechanisms for passing the value of a local variable out of an instance of a named sequence. The first is to pass the value out through an argument local variable of direction **`output`** or **`inout`**. The following rules apply to named sequences with argument local variables of direction **`output`** or **`inout`** and their instances:

1. The entire actual argument bound to an argument local variable of direction **`output`** or **`inout`** must itself be a local variable to whose type the type of the formal argument can be cast. The actual argument is called the *receiver local variable* for the corresponding formal argument.
2. In a given sequence instance, a local variable may not be referenced more than once as a receiver local variable.
3. The structure of the declaration of the named sequence must guarantee that each argument local variable of direction **`output`** or **`inout`** be assigned at the completion of a match of the instance of the named sequence.[10]
4. The instance of the named sequence must not admit empty match.

[10] In particular, each argument local variable of direction **`output`** or **`inout`** must flow out of every match of the sequence expression that results from the body sequence expression of the declaration

```
1  sequence s_cnt_occurrences(bit a, local output int num_a);
2     (1'b1, num_a = a)
3     ##1 (a[->1], num_a++)[*]
4     ##1 !a[*];
5  endsequence
6  property p_num_dataValid_check;
7     transType l_ttype;
8     int num_dataValid;
9     (start, l_ttype = ttype)
10    ##1
11    (
12       s_cnt_occurrences(.a(dataValid), .num_a(num_dataValid))
13       intersect
14       complete[->1]
15    )
16    |->
17    num_dataValid_OK(l_ttype, num_dataValid);
18 endproperty
19 a_num_dataValid_check: assert property (p_num_dataValid_check);
```

Fig. 16.18 Checking the number of occurrences of `dataValid` using local variable output

If these conditions are satisfied, then at the endpoint of any match of the sequence instance, the values of the argument local variables of direction **`output`** or **`inout`** are cast-converted and assigned to the corresponding receiver local variables. If there are multiple matches of the sequence instance, then each match results in an ensuing thread of evaluation in the instantiation context with its own copies of the receiver local variables, and for each such match the receiver local variables are assigned the values of the corresponding argument local variables for that match. Rule 2 ensures that the result of these assignments is independent of the order in which they are performed. Rules 3 and 4 ensure that each receiver local variable is assigned upon match of the instance of the sequence.

As an example, suppose that we need to count the number of occurrences of `dataValid` that are strictly after `start` and not later than `complete`, perhaps so that a validity check can be performed based on this number and the value of `ttype` that appeared with `start`. The code in Fig. 16.18 shows how this can be done using a sequence, `s_cnt_occurrences`, with an argument local variable of direction **`output`**.[11] `s_cnt_occurrences` monitors the argument `a` of type **`bit`** and accumulates the number of occurrences of `a` in the **`output`** argument local variable `num_a`. The sequence begins in Line 2 by assigning the value of `a` to `num_a`, which accounts for an occurrence of `a` in the first cycle of match of the sequence. The sequence continues in Line 3 with repetition of going to the next occurrence of `a` and incrementing `num_a`. This accounts for subsequent occurrences

---

by substituting actual arguments for formal arguments, as described in the rewriting algorithms of Annex F.4 of the LRM [8].

[11] For formal verification, the **`int`** type of `num_a` and `num_dataValid` should be replaced with the type of smallest bitwidth needed for the counters.

of a. The sequence ends with !a[*], which allows any number of cycles without occurrences of a at the end. Within property p_num_dataValid_check, an instance of s_cnt_occurrences, which binds dataValid to a and num_dataValid to num_a, is intersected with the goto complete[->1]. The intersection ensures that the instance of s_cnt_occurrences only counts those occurrences of dataValid that occur after start and not later than complete. In Line 17, a function (or property) num_dataValid_OK is called to check that the number of occurrences of dataValid is allowable for the given transaction type captured in l_ttype in Line 9.

The second mechanism for passing out values of local variables uses untyped formal arguments. This mechanism is discouraged unless the flexibility of untyped arguments is required. If an untyped formal argument of a named sequence is used an l value in a local variable assignment, then that formal argument is understood to represent a local variable. The entire actual argument passed to such a formal argument must be a local variable (the receiver local variable). The formal argument behaves like an untyped version of an argument local variable of direction either **output** or **inout**. The following variant encoding of the sequence s_cnt_occurrences exhibits this style and can be substituted in Fig. 16.18 to achieve an equivalent effect:

```
sequence s_cnt_occurrences(bit a, untyped num_a);
   (1'b1, num_a = a)
   ##1 (a[->1], num_a++)[*0:$]
   ##1 !a[*0:$];
endsequence
```

The value of a local variable can be output from an instance of a named sequence to which a sequence method (triggered or matched) is applied. The rules above apply, with the additional conditions that argument local variables of direction **input** or **inout** cannot appear in the declaration of such a named sequence, and the sequence instance must be a maximal expression (i.e., there is no negation or other expression operator applied to it.)

Exercises

16.1. A default actual argument for an **input** argument local variable may reference preceding arguments in the port list of the sequence or property provided that none of the arguments referenced is an **output** argument local variable. Explain the rationale for this restriction.

16.2. For each of the following sequences, explain why the syntax is not legal and give an example of a legal sequence that, arguably, captures the same intent as the illegal one.

1.
```
sequence s1(bit a, b, c);
    bit l_b;
    (a, l_b = b) throughout c[->1];
endsequence
```

2.
```
sequence s2(bit a, b, c);
    bit l_b;
    a throughout (c, l_b = b)[->1];
endsequence
```

3.
```
sequence s3;
    byte l_numBeats;
    (start, l_numBeats = numBeats) ##1
    (
       (dataValid[->1])[*l_numBeats] ##1 !dataValid[*]
       intersect
       complete[->1]
    );
endsequence
```

16.3. Explain why the following is not legal:

```
sequence s_illegal(local input bit l_a, l_b = l_a);
  bit l_c, l_d = l_c;
  (t, l_c = l_a) ##1 {l_a,l_b} != {l_c,l_d};
endsequence
```

16.4. For each of the following declarations, analyze the local variable flow and determine whether any references to local variables are illegal. Use rules of local variable flow to justify your analysis.

1.
```
sequence s1;
    byte l_v;
    a ##1 (b, l_v &= e);
endsequence
```

2.
```
property p2;
    byte l_v;
    (a, l_v = e)
    |->
    s_nexttime s_eventually (b == l_v);
endproperty
```

3.
```
property p3;
    byte l_v;
    (a, l_v = e)
    implies
    s_nexttime s_eventually (b == l_v);
endproperty
```

4.
```
property p4;
    bit [1:0] l_v = e;
```

```
    a |=>
    case(l_v[0])
       1'b1:  s_eventually(b == l_v);
       1'b0:  c != l_v;
    endcase;
endproperty
```

5.
```
sequence s5;
    byte l_v;
    (
       ( (a[->1], l_v = e) and b[->1] )
       or
       (c, l_v = f)[*1:2] ##1 !c
    )
    ##1 (d == l_v);
endproperty
```

16.5. The nonoverlapped followed-by operator `#=#` is defined such that for a sequence `R` and a property `P`, the passing or failing of the following two properties is equivalent:

1. `R #=# P.`
2. `not (R |=> not P).`

Use flow rules to show that local variable flows through these two properties is also equivalent.

16.6. Use rules of local variable flow to show that `v` flows out of `R[*0:1]` iff both of the following two conditions are satisfied:

1. `v` flows into `R[*0:1]`.
2. If `v` flows into `R`, then `v` flows out of `R`.

What can you say about the conditions for `v` to flow out of `R[*0:n]`, where `n` is positive? What about `R[*]`?

16.7. Consider the following variant of `p_ttype_check`:

```
1  property p_ttype_check;
2     transType l_ttype;
3     (start && (ttype == INV || ttype == PRG), l_ttype = ttype)
4     |=>
5     (
6        (1'b1, l_ttype = ttype)
7        #-#
8        p_no_repeat_ttype(
9           .varying_ttype(ttype), .captured_ttype(l_ttype)
10       )
11    );
12 endproperty
```

Explain the local variable flow. Does the assignment in Line 6 affect the instance of `p_no_repeat_ttype`? Compare or contrast the behavior of this property with the encoding that results by replacing `#-#` with **and**.

16.8. Give a legal encoding of the intent of `p_illegal_causality` from Sect. 16.5.1.

16.9. Recode the example of Fig. 16.18 so that the antecedent of `|->` in property `p_num_dataValid_check` is replaced by an instance of a named sequence to which method `triggered` is applied. In this instance, the local variables `l_ttype` and `num_dataValid` should both be bound to **`output`** argument local variables.

Chapter 17
Recursive Properties

There is repetition everywhere, and nothing is found only once in the world.

— Johann Wolfgang von Goethe

SystemVerilog allows named properties to be recursive. A named property is *recursive* if its declaration instantiates itself. More generally, a set of named properties may be *mutually recursive*, which means that there is a cyclic dependency in the way that they instantiate themselves and one another. Recursion provides a very flexible framework for coding properties. In general, from a flow diagram for a desired check an encoding can be created in which certain nodes of the flow diagram correspond to named properties. If the flow diagram contains cycles, then some of the named properties will be recursive or mutually recursive. This situation occurs, for example, if the check involves retry scenarios. Unlike recursion in programming languages, property recursion does not return. It simply specifies that a thread of evaluation should begin executing the named property again. As a result, there is no stack associated with property recursion, and infinite property recursion is possible on an infinite trace. As a theoretical matter, an instance of a recursive property can be rewritten to avoid recursion.[1] However, for complex properties, it is often simpler to write and maintain a recursive encoding, either because it is more succinct or because the assertion writer can think about the properties in a more procedural way.

[1] An instance of a recursive property can be rendered as an alternating automaton. See [40] for a sketch; restrictions on the actual arguments to recursive instances play a role. LTX augmented with regular expressions is known to be as expressive as alternating automata (see [12], e.g.).

E. Cerny et al., *SVA: The Power of Assertions in SystemVerilog*,
DOI 10.1007/978-3-319-07139-8_17

17.1 Overview of Recursion

This section gives an intuitive overview of recursion based on examples.

A named property is *recursive* if its declaration instantiates itself. Figure 17.1 shows a simple example. The instance `my_always(p)` behaves equivalently to **`always`** `p`. Whenever evaluation of `my_always(p)` begins, Line 2 causes alignment with the incoming clock, followed by evaluation of `p`. Line 4 has a recursive instance and causes evaluation of `my_always(p)` to begin again after advancing to the next occurrence of the incoming clock. The use of **`nexttime`**`[0]` in Line 2 ensures that the first evaluation of `p` begins after alignment with a tick of the incoming clock.[2] From the beginning of an evaluation of `my_always(p)`, property `p` is checked starting at each tick of the incoming clock. Indeed, the following is an unrolling of the instances of the recursive form:

```
    nexttime[0] p
and nexttime[1] p
and nexttime[2] p
...
```

Checking this is exactly the same as checking **`always`** `p`. In particular, the recursion need not end: on a trace with infinitely many ticks of the incoming clock, there are infinitely many recursive evaluations.

Suppose now that we want to get the effect of **`always`**`[low:high] p` in a situation where `low` and `high` are not constants. In such a case, **`always`** cannot be used. Instead, we can use a recursive property and capture the values of the expressions defining the range bounds into local variables when the property begins. Figure 17.2 shows an encoding. The recursive property `ranged_always_recur` is called from the wrapper property `ranged_always`, which passes its arguments through and passes `0` as actual argument to `cnt`. The wrapper is not necessary, but it hides `cnt`, simplifying instantiation and usage. In the recursive property, `low`, `high`, and `cnt` are argument local variables. As a result, the values of the actual arguments passed to `low` and `high` are captured when evaluation of the wrapper begins. If `low` and `high` were not local variables, then changes to their actual arguments would

```
1 property my_always(property p);
2    (nexttime[0] p)
3    and
4    nexttime my_always(p);
5 endproperty
```

Fig. 17.1 Recursive encoding of **`always`**

[2]Alignment with a tick of the incoming clock is necessary for equivalence of `my_always(p)` and **`always`** `p` in case `p` has one or more leading clocks that differ from the incoming clock. See Sect. 12.2.5.1 for more on **`nexttime`**`[0]`.

```
1  property ranged_always(int unsigned low, high, property p);
2     ranged_always_recur(low, high, 0, p);
3  endproperty
4  property ranged_always_recur(
5     local input int unsigned low, high, cnt,
6     property p
7  );
8     if (cnt <= high)
9     (
10       (if (cnt >= low) p)
11       and
12       nexttime ranged_always_recur(low, high, cnt+1, p)
13    );
14 endproperty
```

Fig. 17.2 Recursive encoding of ranged **`always`** allowing non-constant range bounds

```
1  property my_until(property p, q);
2     (nexttime[0] q)
3     or
4     ((nexttime[0] p) and nexttime my_until(p, q));
5  endproperty
```

Fig. 17.3 Recursive encoding of **`until`**

affect the meaning of the range while evaluation of the recursive property was underway (see Exercise 17.1). `cnt` is an incrementing counter that keeps track of the number of cycles that have elapsed. Line 8 checks that `cnt` has not yet exceeded `high`. When it does, the recursion ends. When `cnt` reaches `low`, checking of `p` is caused by Line 10. The recursive instance in Line 12 increments `cnt` in the actual argument expression.

Figure 17.3 shows a recursive encoding of **`until`**. Each time evaluation of `my_until` begins, either property `q` must evaluate to true (Line 2) or both property `p` must evaluate to true and also evaluation of `my_until` must start again in the next cycle (Line 4). The use of **`nexttime`**`[0]` in Lines 2 and 4 ensures that the first evaluations of `q` and `p` begin after alignment with a tick of the incoming clock. Satisfaction of `q` results in satisfaction of `my_until` without evaluation of the recursive instance in Line 4, and so it allows the recursion to stop. Until `q` is satisfied, Line 4 must be satisfied, which implies that `p` must be satisfied. It is possible that `q` is never satisfied, in which case `p` must be satisfied always. Thus, `my_until(p,q)` has the same semantics as `p` **`until`** `q`.

Suppose that we need to check a ranged form of `p` **`until`** `q`, in which satisfaction of `q` must occur within a range `[low:high]` of cycles from the start of the evaluation. Let us denote this property `p` **`until`**`[low:high]` `q`. Note that this is not legal SVA code, just a notation. The property `p` **`until`**`[low:high]` `q` is satisfied iff there exists k, `low` $\leq k \leq$ `high`, such that `q` is satisfied in the kth cycle from the start of evaluation and `p` is satisfied in every prior cycle from the start of evaluation. Figure 17.4 shows a recursive encoding of this property.

```
1  property ranged_until(int unsigned low, high, property p, q);
2     if (low > high) 1'b0 // required k does not exist
3     else
4        ranged_until_recur(low, high, 0, p, q);
5  endproperty
6  property ranged_until_recur(
7     local input int unsigned low, high, cnt,
8     property p, q
9  );
10    if (cnt == high) q // last chance to satisfy q
11    else
12    (
13       (q and (cnt >= low))
14       or
15       (p and nexttime ranged_until_recur(low,high,cnt+1,p,q))
16    );
17 endproperty
```

Fig. 17.4 Recursive encoding of ranged **until**

```
property even;
   (nexttime[0] p) and nexttime odd;
endproperty
property odd;
   q and nexttime even;
endproperty
```

Fig. 17.5 Mutually recursive encoding of even–odd checks

Line 2 requires `low` to be at most `high`. If `low` exceeds `high`, then the required k does not exist and `ranged_until` fails. Otherwise, `ranged_until_recur` is called in Line 4. This recursive property is an adaptation of the ideas of the encoding of ranged always from Fig. 17.2 to the recursive form of `my_until` in Fig. 17.3. `cnt` is an incrementing counter of the cycle number, and it is initialized to `0` by the actual argument in Line 4. Lines 13–15 mimic Lines 2–4 of `my_until`. Line 13 adds the condition `cnt >= low`. This ensures that satisfaction of `q` discharges the evaluation only if `cnt` is in the range `[low:high]`. Line 10 requires `q` to be satisfied no later than the last cycle of this range.

For the next example, suppose that we want to check that `p` holds in every even tick of the incoming clock and `q` holds in every odd tick of the incoming clock, both reckoned from the start of evaluation.

Figure 17.5 shows an encoding with two *mutually recursive* properties. Property `even` checks `p`, after alignment with the incoming clock, and in the next cycle calls property `odd`. Property `odd` checks `q` and in the next cycle calls property `even`. Figure 17.6 shows a single recursive property that performs the same check as property `even`.

```
property even_odd (property p, q);
   (nexttime[0] p) and nexttime even_odd(q, p);
endproperty
```

Fig. 17.6 Recursive encoding of even–odd checks

```
property even_odd_stall(property p, q);
   if (stall) nexttime even_odd_stall(p, q)
   else
      p and nexttime even_odd_stall(q, p);
endproperty
```

Fig. 17.7 Recursive encoding of even–odd checks with stall

```
1 property p_fifo_data_check;
2    dataType data;
3    bit [0:$clog2(MAX_OUTSTANDING)] numAhead;
4    (start, data = dataIn, numAhead = outstanding)
5    ##1 (numAhead > 0 ##0 complete[->1], numAhead--)[*]
6    ##1 (numAhead == 0 ##0 complete[->1])
7    |->
8    dataOut == data;
9 endproperty
```

Fig. 17.8 Encoding of FIFO protocol data check using local variables

Suppose now that we want to modify the check so that cycles in which `stall` is true are skipped—i.e., no new check of `p` or `q` is started in such a cycle and it does not count for the reckoning of even and odd cycles. Figure 17.7 shows an encoding.

The remaining examples of this section give recursive encodings of several checks for the FIFO protocol from Sect. 15.3. Recall that the FIFO protocol requires `dataIn` at an occurrence of `start` to equal `dataOut` at the corresponding occurrence of `complete`. According to FIFO ordering, the nth occurrence of `start` corresponds to the nth occurrence of `complete`. An auxiliary variable `outstanding` (see Fig. 15.6) keeps track of the number of outstanding transactions, i.e., the difference between the number of preceding occurrences of `start` and the number of preceding occurrences of `complete`. The encoding of the FIFO protocol data check property from Fig. 15.9 is repeated here for reference in Fig. 17.8.

A drawback of this encoding is that the management of the local variable `numAhead` in Lines 5 and 6 involves a somewhat tricky pattern. Figure 17.9 shows a recursive encoding of the same property.

`p_fifo_data_check_recur` has argument local variables `data` and `numAhead`. The instance of this property in Line 2 initializes `data` to the value of `dataIn` and `numAhead` to the value of `outstanding`. Line 8 advances to the nearest strictly future occurrence of `complete`. At that point, if `numAhead` is zero, then `dataOut`

```
1  property p_fifo_data_check;
2     start |-> p_fifo_data_check_recur(dataIn, outstanding);
3  endproperty
4  property p_fifo_data_check_recur(
5     local input dataType data,
6     local input bit [0:$clog2(MAX_OUTSTANDING)] numAhead
7  );
8     ##1 complete[->1]
9     |->
10    if (numAhead == 0)
11       dataOut == data;
12    else
13       p_fifo_data_check_recur(data, numAhead--);
14 endproperty
```

Fig. 17.9 Recursive encoding of FIFO protocol data check

is compared to `data` (Line 11). Otherwise, the evaluation recurs, decrementing `numAhead` in the actual argument in Line 13. The overall encoding of Fig. 17.9 occupies more lines than the nonrecursive encoding, but the pattern in the body of the recursive property (Lines 8–13) may be more accessible than the trickier pattern in Lines 5 and 6 of the nonrecursive encoding.

Figure 17.10 is a copy of Fig. 15.15 for reference. It shows the encoding from Sect. 15.5 of all of the FIFO protocol checks using only local variables. This encoding is subtle in the use of two local variables of type `counterType` in `p_fifo_all_checks`. Because `outstanding` is managed as a local variable and updated in the repetition of Lines 11–15, it is necessary to have it shadow `nextOutstanding` so that the appropriate value will still be available for use in the consequent (Lines 17–22).

With recursion, there is more flexibility in where local variables are updated. As a result, the shadow variable can be eliminated. A recursive encoding is given in Fig. 17.11. Property `p_fifo_data_check_recur` is the same as in Fig. 17.9, except that the user-defined type `counterType` is used for `numAhead`. The main recursive property is `p_fifo_all_checks_recur`. It has a single argument local variable, `outstanding`, which is initialized to `'0` in the instance in Line 32. Line 16 advances to the next occurrence of `start` or `complete`. Lines 18–21 of the consequent update `outstanding` and, in the next cycle, recur with the new value of `outstanding`. This part of the code keeps the ongoing check running. Rules of local variable flow (see Sect. 16.4) ensure that the new value of `outstanding` from Line 19 is not visible in Lines 23–28. This is how the shadow variable is eliminated. Lines 23–28 perform the same checks as Lines 17–22 of the nonrecursive encoding. Again, the overall number of lines is a bit greater in the recursive case, but the code patterns are easier to understand.

```
typedef bit [0:$clog2(MAX_OUTSTANDING)] counterType;
property p_fifo_data_check(local input counterType numAhead);
   dataType data = dataIn;
   ##1 (numAhead > 0 ##0 complete[->1], numAhead--)[*]
   ##1 (numAhead == 0 ##0 complete[->1])
   |->
   dataOut == data;
endproperty
property p_fifo_all_checks;
   counterType outstanding, nextOutstanding = '0;
   (
      (start || complete)[->1],
      outstanding = nextOutstanding,
      nextOutstanding += start - complete
   )[+]
   |->
   if (start) (
      (!complete && outstanding < MAX_OUTSTANDING)
      and p_fifo_data_check(.numAhead(outstanding))
   ) else ( // complete
      !start && outstanding > 0
   );
endproperty
initial
   a_fifo_all_checks: assert property (
      p_fifo_all_checks
   );
```

Fig. 17.10 Encoding of all FIFO protocol checks using only local variables

17.2 Retry Protocol

This section presents a write protocol with retry. The protocol involves complex checks that are handled well with recursive properties. The protocol is a variant of the retry protocol from Sect. 16.12.17 of the SystemVerilog 2012 LRM. The LRM version includes transaction tags, which allow multiple write transactions to be in flight concurrently, distinguished by their tags. Tags have already been discussed at a high level in the Tag Protocol of Sect. 15.4. The tags are orthogonal to the retry, so they have been abstracted away for simplicity. As a result, the present protocol is sequential, meaning that a new transaction cannot start while the previous transaction remains in flight. We assume that the protocol is clocked at **posedge** clk, as specified by **default clocking**. Here are the English rules:

1. start, dataValid, complete, and retry are signals of type **logic**. data is a signal of the integral type dataType.
2. Start of a write transaction is signaled by occurrence of start.
3. A write transaction has between one and MAX_BEATS data beats, where MAX_BEATS is a positive integer parameter. Data for a single beat is of type dataType.

```
typedef bit [0:$clog2(MAX_OUTSTANDING)] counterType;
property p_fifo_data_check_recur(
   local input dataType data,
   local input counterType numAhead
);
   ##1 complete[->1]
   |->
   if (numAhead == 0)
      dataOut == data;
   else
      p_fifo_data_check_recur(data, numAhead--);
endproperty
property p_fifo_all_checks_recur(
   local input counterType outstanding
);
   (start || complete)[->1]
   |->
   (
      (1'b1, outstanding += start - complete)
      |=> p_fifo_all_checks_recur(outstanding)
   )
   and
   if (start) (
      (!complete && outstanding < MAX_OUTSTANDING)
      and p_fifo_data_check_recur(dataIn, outstanding)
   ) else ( // complete
      !start && outstanding > 0
   );
endproperty
initial
   a_fifo_all_checks:  assert property (
      p_fifo_all_checks_recur('0)
   );
```

Fig. 17.11 Recursive encoding of all FIFO protocol checks using only local variables

4. At an occurrence of `start`, expected data for the associated write transaction is available in the array `dataModel`. `dataModel` is an unpacked array of `MAX_BEATS` entries, each of type `dataType`. It is declared as

   ```
   dataType dataModel [MAX_BEATS];
   ```

   The beats of the transaction must transfer the data in the sequence `dataModel[0]`, `dataModel[1]`, ...
5. Subsequent to `start`, an occurrence of `dataValid` signals a data beat. Data beats do not have to be consecutive. In a cycle in which `dataValid` occurs, the data for the beat is carried on the signal `data`.
6. The last data beat for the data transfer is signaled by occurrence of `complete` together with `dataValid`. `complete` is meaningful only together with `dataValid`. If `complete` occurs without `dataValid`, then it is ignored. `dataValid` may not occur in the cycle after the last data beat.

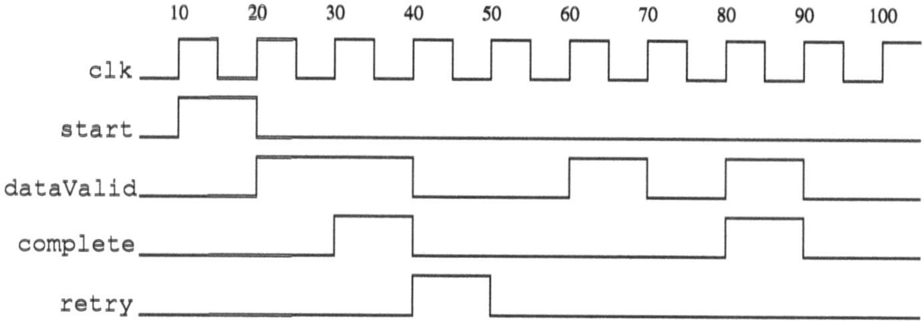

Fig. 17.12 Waveform for retry protocol

7. At any time subsequent to `start` and no later than the cycle after the last data beat, an occurrence of `retry` signals that the data transfer is forced to retry. This means that no further data beats in the current sequence may be transferred, and the data transfer must begin the sequence again, starting with `dataModel[0]`. The transaction itself does not restart after retry, only the data transfer. Specifically, the transaction does not reassert `start` and `dataModel` is not observable after `retry`. There is no limit to the number of times the data transfer may be forced to retry.
8. The overall write transaction completes in the cycle after the last data beat provided `retry` does not occur in that cycle. A write transaction is said to be *in flight* beginning in the cycle after `start` and continuing up to and including the cycle that the transaction completes.
9. Write transactions must be sequential. More precisely, `start` must not occur while a write transaction is in flight.
10. If `dataValid`, `complete`, or `retry` occurs while no write transaction is in flight, then it is ignored.

Figure 17.12 shows a waveform for the control signals of the retry protocol. A transaction starts at time 20. Two data beats occur at times 30 and 40, and `complete` is signaled at time 40. At time 50, `retry` occurs, so the data transfer must start again. Two data beats are repeated at times 70 and 90, and `complete` is signaled at time 90. Since `retry` does not occur again, the transaction completes at time 100.

Note that the protocol makes no requirement on the number of data beats in a retried data transfer. A retried data transfer might have fewer or more data beats than a previous data transfer, even if the previous data transfer signaled `complete`. However, for each beat that occurs in a retried data transfer, the data must match the value predicted by `dataModel`.

Figure 17.13 shows a flow diagram to check the retry protocol. The flow begins with `start`. In the cycle of `start`, `dataModel` is captured. The flow then advances to the next cycle and invokes synchronous reject on `start`. If `start` occurs while the check is ongoing, then the check fails (rule 9). The counter `i` is then initialized

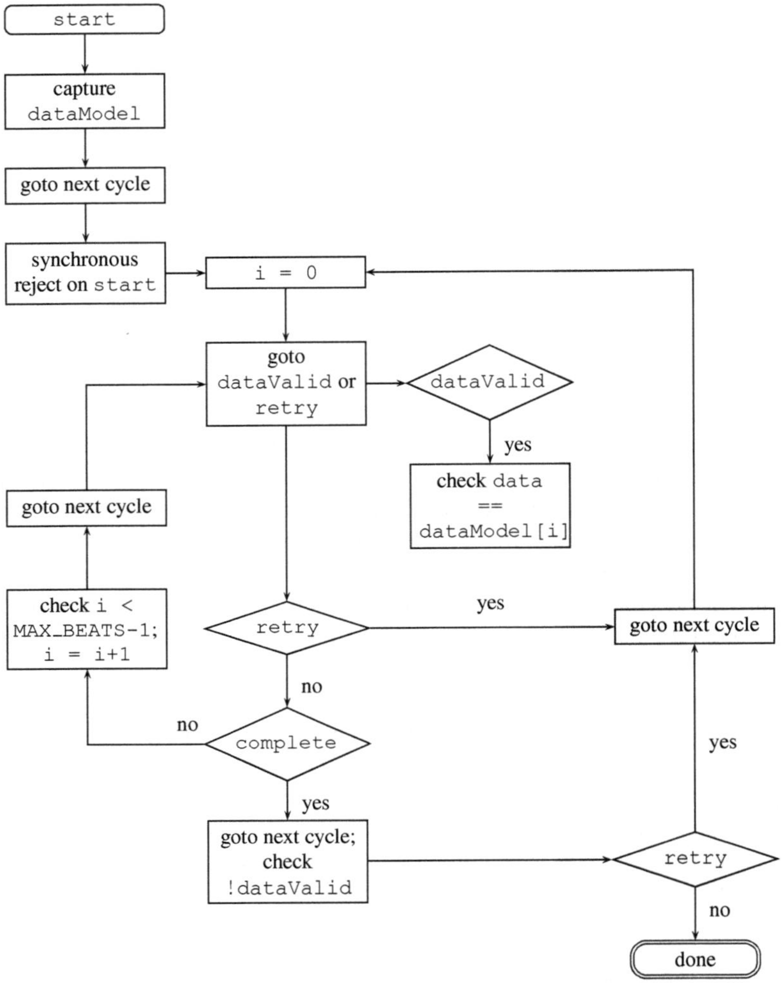

Fig. 17.13 Flow diagram for retry protocol checks

to zero, indicating that the check expects the first data beat. The flow then advances to the nearest occurrence of `dataValid` or `retry`. If `dataValid` occurs, the flow checks that `data` equals `dataModel[i]`. If `retry` occurs, the flow advances to the next cycle and resets `i` to zero to begin the data transfer again. If `complete` has not also occurred, then the flow increments `i`, checks that the result is not too big in comparison with `MAX_BEATS`, advances a cycle, and returns to look for the next `dataValid` or `retry`. If `complete` has also occurred, then the data transfer is complete and the flow advances one more cycle, where it checks that `dataValid` does not occur and checks for `retry` at the last opportunity. If `retry` does not occur,

```
1  property p_retry_check;
2     dataType l_dataModel [MAX_BEATS];
3     (start, l_dataModel = dataModel)
4     |=>
5     sync_reject_on(start) p_retry_check_recur(dataModel, 0);
6  endproperty
7  property p_retry_check_recur(
8     local input dataType l_dataModel [MAX_BEATS],
9     local input int unsigned i
10 );
11    (dataValid || retry)[->1]
12    |->
13    (dataValid |-> data == l_dataModel[i]) and
14    if (retry)
15       nexttime p_retry_check_recur(l_dataModel, 0)
16    else if (complete)
17       nexttime (
18          !dataValid and
19          if (retry)
20             nexttime p_retry_check_recur(l_dataModel, 0)
21       )
22    else (
23       (i < MAX_BEATS-1) and
24       nexttime p_retry_check_recur(l_dataModel, i+1)
25    );
26 endproperty
27 a_retry_check:  assert property (p_retry_check);
```

Fig. 17.14 Recursive encoding of retry protocol checks

then the flow is done. Otherwise, the flow advances a cycle and resets `i` to zero to begin the data transfer again.

Figure 17.14 shows a recursive encoding of the retry protocol checks. The encoding follows closely the flow diagram of Fig. 17.13. The outer property `p_retry_check` corresponds to the whole diagram, beginning from the `start` figure. The recursive property `p_retry_check_recur` corresponds to the subflow that is downstream from the node "goto `dataValid` or `retry`". There are three cycles in the flow diagram, and these correspond to the three recursive instances of `p_retry_check_recur`. The cycle on the left of the flow diagram corresponds to the recursive instance on Line 24. The smaller cycle on the right of the flow diagram corresponds to the recursive instance on Line 15. And the larger cycle on the right of the flow diagram corresponds to the recursive instance on Line 20. Note that management of the counter `i` is accomplished in the actual arguments to the instances of `p_retry_check_recur`. With careful study of these correspondences, the reader should gain an appreciation for the simplicity of encoding flow diagrams into sets of properties, where cycles correspond to recursion.

```
property p_check_data_xfer(
   local input dataType l_dataModel [MAX_BEATS]
);
   int unsigned j = 0;
   (
      (!retry && !(dataValid && complete))
      throughout dataValid[->1],
      j++
   )[*]
   ##1 (!retry && !dataValid)[*]
   ##1 (dataValid, j++)
   |->
   (
      j <= MAX_BEATS
      and data == l_dataModel[j-1]
      and if (!retry && complete)
         nexttime !dataValid
   );
endproperty
property p_check_retried_data_xfer(
   local input dataType l_dataModel [MAX_BEATS]
);
   (
      ( !(dataValid && complete) throughout retry[->1] )
      or
      (
         (!retry && !(dataValid && complete))[*]
         ##1 dataValid && complete
         ##0 first_match(##[0:1] retry)
      )
   )[+]
   |=> p_check_data_xfer(l_dataModel);
endproperty
property p_retry_check_non_rec;
   dataType l_dataModel [MAX_BEATS];
   (start, l_dataModel = dataModel)
   |=>
   sync_reject_on(start)
   (
      p_check_data_xfer(l_dataModel)
      and
      p_check_retried_data_xfer(l_dataModel)
   );
endproperty
a_retry_check_non_rec: assert property(p_retry_check_non_rec);
```

Fig. 17.15 Nonrecursive encoding of retry protocol checks

The remainder of this section provides an alternative, nonrecursive encoding of the retry protocol checks. The encoding is shown in Fig. 17.15. The main property, `p_retry_check_non_rec`, is similar to `p_retry_check`. It captures the data model into a local variable when `start` occurs. Evaluation of the consequent begins in the next cycle, where **`sync_reject_on`**`(start)` is invoked and two properties are instantiated and conjoined.

The first property is `p_check_data_xfer`, which checks the current data transfer up to its completion or to an occurrence of `retry`. The antecedent of `|->` in Lines 5 to 11 is crafted to match at each occurrence of `dataValid` in the current data transfer that is not strictly subsequent to an occurrence of `retry` or an occurrence of `dataValid && retry`. Lines 5–9 match zero or more occurrences of `dataValid`, none together with `complete`. Throughout match of these lines, `retry` must not occur. Lines 10 and 11 match one more occurrence of `dataValid`. In these lines, `retry` must not occur strictly before `dataValid`. The consequent of `|->` checks that the counter `j` has not grown too large and that `data` is correct. Also, it checks that if `dataValid` and `complete` occur without `retry`, then `dataValid` does not occur in the next cycle.

The second property is `p_check_retried_data_xfer`. This property is responsible for restarting `p_check_data_xfer` each time a data transfer is forced to retry. The overall antecedent of `|=>` is a repetition of one or more matches of the sequence in Lines 24 to 30. That sequence encodes the condition that a data transfer is forced to retry. The occurrence of `retry` may be strictly before `dataValid && complete`, which matches Line 24. Or the occurrence of `retry` may be concurrent with or in the cycle after `dataValid && complete`. These cases match Lines 27 to 29. The antecedent has been arranged to avoid multiple matches, which could result in redundant checking of the consequent. The consequent of `|=>` simply instantiates `p_check_data_xfer` to begin checking the data transfer again.

It should be clear that the sequential conditions in the antecedents of these two properties are complex and prone to error in comparison with the recursive encoding.

17.3 Restrictions on Recursive Properties

The coding flexibility afforded by recursive properties is tempered by four restrictions on their use. The restrictions are intended to avoid recursive declarations and instances that are problematic or whose semantics involves subtleties that are beyond the scope of the current semantic framework for recursion. This section quotes the restrictions from the LRM and elaborates briefly on the rationale for them.

17.3.1 Negation and Strong Operators

RESTRICTION 1: The negation operator **not** and the strong operators **s_nexttime**, **s_eventually**, **s_always**, **s_until**, and **s_until_with** cannot be applied to any property expression that instantiates a recursive property.

In view of the Rewriting Algorithms in Annex F.4 of the LRM, this restriction also forbids the application of these operators to an expression that, through instances of nonrecursive properties, ultimately depends on an instance of a recursive property.

Restriction 1 avoids the subtle interplay between negation and recursion. Consider the following example:

```
property confounding;
   nexttime not confounding;
endproperty
```

A naïve approach to the semantics of this property leads to contradiction. For simplicity, consider the unclocked semantics (see Chap. 22) of `confounding` over an infinite trace $w = \ell^{\omega}$, where ℓ is a normal letter (i.e., $\ell \notin \{\top, \bot\}$). Then also $w^{1..} = \ell^{\omega} = w$ and $\bar{w} = w$. Naïvely,

$w \models$ `confounding`
  iff $w \models$ **nexttime not** `confounding`
  iff $w^{1..} \models$ **not** `confounding`
  iff $[w^{1..} = w]$
    $w \models$ **not** `confounding`
  iff $[\bar{w} = w]$
    $w \not\models$ `confounding`

Clearly, something is wrong with this argument.[3] SVA currently avoids the problem by imposing Restriction 1.

The strong operators are related to negation because negation interchanges weak and strong. For example, **s_eventually** can be defined as a derived operator according to

s_eventually p $\equiv$ **not always not** p

[3][40] provides an approach to negation and recursion that allows their free interplay. Additional information is provided in the declarations of recursive forms that enables the interpretation of satisfaction using either a co-Büchi or Büchi acceptance criterion. With this approach, negation need not result in language complementation. For co-Büchi acceptance, one gets $w \not\models$ `confounding` and $w \not\models$ **not** `confounding`. For Büchi acceptance, one gets $w \models$ `confounding` and $w \models$ **not** `confounding`. [40] gives sufficient conditions on recursive forms to ensure that co-Büchi and Büchi acceptance are equivalent and that negation results in language complementation. Of course, `confounding` does not satisfy these conditions.

Creating a framework for recursion that interacts well with the strong operators involves dealing with negation and liveness in recursive forms. Again, SVA currently avoids these issues through Restriction 1.

17.3.2 Disable Clause

RESTRICTION 2: **disable iff** cannot be used in the declaration of a recursive property.

disable iff cannot be nested, so its use within a recursive property declaration is forbidden. If a disable clause is needed, it should be put in a wrapper property around the recursive property or in an assertion statement. For example, the following is illegal:

```
property illegal_disable(logic b, property p);
   disable iff (b)
   p and nexttime illegal_disable(b, p);
endproperty
```

The effect of the disable can be obtained using a wrapper, as in

```
property always_with_disable(logic b, property p);
   disable iff (b)
   my_always(p);
endproperty
```

where `my_always` is the recursive property from Fig. 17.1.

17.3.3 Time Advance

RESTRICTION 3: If `p` is a recursive property, then, in the declaration of `p`, every instance of `p` must occur after a positive advance in time. In the case of mutually recursive properties, all recursive instances must occur after positive advances in time.

Briefly, recursion must occur after a positive advance in time. This restriction is intended to avoid recursive forms that get "stuck" at a single point in time. For example, the following is illegal:

```
property illegal_stuck(property p);
   p and nexttime[0] illegal_stuck(p);
endproperty
```

`nexttime[0]` does not guarantee an advance in time, and infinitely many evaluations of `p` are specified beginning in the same time step.

17.3.4 Actual Arguments

RESTRICTION 4: For every recursive instance of property `q` in the declaration of property `p`, each actual argument expression `e` of the instance satisfies at least one of the following conditions:

- `e` is a formal argument of `p`.
- No formal argument of `p` appears in `e`.
- `e` is bound to a local variable formal argument of `q`.

This restriction is intended to avoid problematic recursive instances that result from passing compound actual argument expressions. Such instances can lead to an explosion of distinct actual argument expressions as the recursion unfolds. Such explosion has undesirable consequences for the complexity and tractability of checking the properties.

As an example of a problematic recursive instance, consider the following example that violates Restriction 4:

```
property illegal_arg(longint unsigned u, v, w);
   u == v*w and nexttime illegal_arg(u, v+w, v*w);
endproperty
```

Since the formal arguments of `illegal_arg` are not local variables, they are treated as reference arguments. This means that successive recursive instances require composition of the compound actual argument expressions. Tracking the evolution of the recursion, the comparison `u == v*w` expands to

cycle	comparison
0	`u == v*w`
1	`u == (v+w)*(v*w)`
2	`u == ((v+w)+(v*w))*((v+w)*(v*w))`
3	`u == (((v+w)+(v*w))+((v+w)*(v*w)))*` `(((v+w)+(v*w))*((v+w)*(v*w)))`

and so on, where the arithmetic is performed modulo 2^{64}, as specified by the unsigned `longint` type of the formal arguments and the rules for bitwidths in expressions. Even though the data type is bounded, the management of these expressions quickly gets out of hand.[4]

[4] See [22] for an example, due to D. Bustan, that shows how a recursive form violating the restrictions can represent a language that is not omega-regular.

Having seen how recursion can get out of hand when Restriction 4 is violated, let us consider the intuition for why it does not when at least one of the three conditions is satisfied by each actual argument expression in a recursive instance.

If the first condition is satisfied, then the actual argument `e` is itself a formal argument of the declaration of `p`. In this situation, `e` is simply being passed into the recursive instance of `q`, modified at most by being cast to the type of the associated formal argument of `q`. Iterated castings can involve some complexity, but it is benign compared to expression explosion described above.

If the second condition is satisfied, then the actual argument `e` makes no reference to any formal argument of `p`. As a result, `e` is composed of references to local variables of `p` and references to entities outside of the declaration of `p`. Those entities could be static variables, named sequences or properties, functions, etc. In any case, whatever the local variables or external entities are, the meanings of references to them are the same every time the recursive instance of `q` is encountered. Of course the value stored in a local variable or an external static variable, e.g., may be different, but the way in which these terms are combined to form `e` is not changing. There is, thus, no expression explosion associated with `e`.

Finally, if the third condition is satisfied, then the formal argument to which `e` is passed is an argument local variable. Argument local variables do not have the expression explosion problem because they behave like value arguments, not reference arguments. At each recursive instance, the value of the actual argument expression is computed and stored in the corresponding argument local variable, after which the form of the actual argument can be forgotten.

Exercises

17.1. The following is a variant of the encoding of ranged **always** from Fig. 17.2:

```
property ranged_always(int unsigned low, high, property p);
   ranged_always_recur(low, high, 0, p);
endproperty
property ranged_always_recur(
   int unsigned low, high,
   local input int unsigned cnt,
   property p
);
   if (cnt <= high)
   (
      (if (cnt >= low) p)
      and
      nexttime ranged_always_recur(low, high, cnt+1, p)
   );
endproperty
```

In this encoding, the arguments `low` and `high` of the recursive property are not local variables. How does this difference affect the meaning of the property?

17.2. Recode the property `ranged_always` of Fig. 17.2 as a single recursive property with argument local variables `low` and `high` and no other local variables. [Hint: Manage `low` and `high` as decrementing counters instead of using the incrementing local variable `cnt`.]

17.3. Recode the property `ranged_until` of Fig. 17.4 as a single recursive property with argument local variables `low` and `high` and no other local variables. [Hint: Manage `low` and `high` as decrementing counters instead of using the incrementing local variable `cnt`.]

17.4. Figure 17.4 gives a recursive encoding of `p until [low:high] q` for general properties `p` and `q`. Give a nonrecursive encoding under the restriction that `p` and `q` are Booleans. Try to find a nonrecursive encoding for general properties `p` and `q`.

17.5. Write a recursive property (or a set of mutually recursive properties) to check that properties `p0`, `p1`, and `p2` hold in cycles from the start of evaluation that are congruent to 0, 1, and 2 modulo 3, respectively.

17.6. Write a nonrecursive property to perform the check of property `even_odd` from Fig. 17.6. Write a nonrecursive property to perform the check of property `even_odd_stall` from Fig. 17.7.

17.7. Explain why `p_fifo_data_check_recur` in Fig. 17.9 becomes illegal if `##1` is deleted from Line 8. Assuming that `##1` is deleted from Line 8, explain how and why the property can be repaired by adding **`nexttime`** in Line 10. [Hint: Use the fact that `start` and `complete` are mutually exclusive.]

17.8. Modify the recursive encoding of the retry protocol in Fig. 17.14 so that there is a failure if `dataValid`, `complete`, or `retry` occurs while no write transaction is in flight. Also, check that there is never an occurrence of `complete` without `dataValid` while a write transaction is in flight.

17.9. Make the same modifications specified in Exercise 17.8 for the nonrecursive encoding in Fig. 17.15.

17.10. Let `a` and `b` be Booleans. Determine whether any of the following declarations or instances violates the restrictions on recursive properties.

```
property p1;
   a |-> p2;
endproperty
property p2;
   p4 and (b |-> p1);
endproperty
property p3;
   disable iff (reset)
   p4 and nexttime(a |-> p2);
endproperty
property p4;
   reject_on(bad)
   a |=> not p1;
endproperty
```

Which properties are recursive and which are nonrecursive?

17.11. Let a and b be Booleans. Determine whether any of the following declarations or instances violates the restrictions on recursive properties.

```
property p1(sequence s, property p);
   s |=> p;
endproperty
property p2(property q);
   q or p1(a, q or p1(b, p2(q)));
endproperty
property p3(sequence s);
   s |=> p2(weak(s)) and p3(s ##1 s);
endproperty
```

Which properties are recursive and which instances are recursive?

17.12. Explain why the following code is illegal:

```
property fib(int unsigned a, b, n, sig);
   if (n > 0)
      sig == a and nexttime fib(b, a+b, n-1, sig);
endproperty
```

How can the declaration be modified to make it legal, while preserving the intent of the original code?

Chapter 18
Coverage

Verification may not ever be complete, but we should know what was verified.

— Unknown.

An important mechanism for determining whether design validation has sufficiently verified the design on hand is to collect "coverage" information, both structural and functional. This chapter describes how assertions can be used to gather functional coverage information using **`cover property`** and **`cover sequence`** statements. It is mainly suitable to collect information about the occurrences (or not) of some sequences of events. SystemVerilog provides another mechanism for collecting coverage, called *covergroups*. They are particularly suitable for gathering information about the occurrence of data patterns and their cross correlation. Often, it is important to detect a particular sequence of events and then initiate collecting coverage on data patterns. This can be achieved by combining assertion coverage with that of covergroups.

As we have seen, assertions and assumptions provide a precise way to state functional specification of a design and its environment. While these assertion statements may nonvacuously pass in our tests, we may still find that the design contains errors. Why is that? The answer is quite simple: In simulation, we may have not exercised all functional modes of the design, and the bugs may be hiding there. In formal verification which is exhaustive with respect to each assertion, the problem may lie in the fact that the assumptions representing the behavior of the environment are more strict than the actual usage of the design, or because the search space was restricted due to memory limitations. In either situation, we should have means to determine the extent to which the design functionality has been exercised. There are essentially two main methods for measuring this extent: *Code Coverage* and *Functional Coverage*. Code coverage is concerned with measuring the percentage of lines of code executed, which conditional blocks were executed and caused by which conditions, etc. These measures indicate how much the implementation was

E. Cerny et al., *SVA: The Power of Assertions in SystemVerilog*,
DOI 10.1007/978-3-319-07139-8_18

structurally exercised. It does not tell us whether some functionality is missing. Functional coverage, however, derives for the most part its coverage targets from the functional specification of the design.

18.1 Immediate and Deferred Coverage

The simplest form of functional coverage measurement can be obtained using immediate and deferred **`cover`** statements. Like the immediate and deferred **`assert`** statements, they can be placed in procedural code. The immediate **`cover`** is useful in high-level functional models and test benches, while the deferred form is preferred in RTL models because of its ability to filter out 0-width glitches. Given that the argument to a **`cover`** statement is a Boolean expression and is associated with the design model, it is mostly suitable to measure whether some specific expression values, or their combinations have been encountered in the model. That is, like immediate and deferred assert statements, they are closer to the implementation level than to the purely functional level.

The syntactic form of the immediate and deferred cover statements is as follows:

```
cover ( expression ) statement_or_null
```
and
```
cover #0 ( expression ) statement_or_null
```
and
```
cover final ( expression ) statement_or_null
```
[1]

Notice that the action block is limited to an optional pass action block since there is no notion of failure. In addition, as in deferred assertions, the action block of a deferred cover statement is limited to a single subroutine call; that is, it cannot be a block of statements. In the case of deferred final cover, like in deferred final asserts, the action block may contain only passive statements. The coverage database will record the total number of evaluations of the statement and the number of times it succeeded (i.e., the `expression` evaluated true). The coverage database and the analysis and presentation tools built around it play an important role in assessing the quality of verification of the design. The tools usually provide means to merge coverage information from different tests, indicate which areas of the design have not yet been verified (or have no coverage collection statements), display the trend in coverage over different tests and time, and present the information in a concise and graphical manner.

In immediate covers, the statement in the pass action block executes immediately upon success of the cover statement. In deferred cover statements, however, the pass statement is limited to a single subroutine and is scheduled to execute in the Reactive

---

[1] In fact, in both cases of deferred covers this should specify "subroutine call or null" due to the restrictions placed on action blocks in deferred assertions.

region following the evaluation of the cover statement. Why is that so different from the simpler immediate cover? This is due to the glitch filtering mechanism inherent in deferred assertions. In any given simulation time step, if a particular cover instance evaluates several times without exiting a scheduling region (Active or Reactive), only the last result upon entry to the following Observed region is reported. The pass action subroutine is scheduled in the subsequent Reactive region in the case of deferred observed covers, and in the Postponed region in the case of deferred final covers. For more details on simulation semantics, see Sect. 3.3. The semantics of immediate and deferred cover statements is the same as for the corresponding assert statements. See also Sect. 4.7.

18.2 Sequence and Property Coverage

Concurrent assertion coverage statements come in two forms: **cover sequence** and **cover property**. Let us first consider the former.

18.2.1 Sequence Coverage

The syntactical form of this statement is defined in Sect. 4.7.

The body of **cover sequence** must be a sequence expression; that is, it may only contain an expression constructed using clock specifications, Boolean expressions, sequence instances, sequence operators, and possibly sequence match items. It may optionally contain a **disable iff** specification. A clocking event is optional if it can be inferred from the context. The optional pass action is executed when the sequence matches. In this case, the pass action may be any procedure and is not limited to a single subroutine call.

Example 18.1. Detect that `a` is followed by `b` within 100 clock ticks.

```
default clocking ck @(posedge clk); endclocking
int my_count = 0;
seq_cov: cover sequence ( disable iff (reset)
   a ##[1:100] b
) begin
   my_count++;
   $display("match of seq_cov at time %t", $time);
end
```

□

In simulation, an evaluation attempt is started at every tick of the clocking event, **posedge** `clk`. If `a` is true, then the evaluation will continue searching for all occurrences of `b` being true within 100 clock ticks. For each such occurrence of `b` a *match* is recorded in the coverage database. It is important to note that each match of `b` will be recorded as a coverage success for `seq_cov`. The simulator will also record how many attempts were started. What happens if either `a` is false at the beginning of an attempt or `b` is false throughout the range? In that case, there is no match and the coverage count is not incremented in the database. Thus, unlike those of **`assert`** statements, *failures* of covers are ignored.

The body of **`cover sequence`** may have a **`disable iff`** (*expression*) at the top. As with **`assert property`** statements, the evaluation of the expression is asynchronous to the evaluation of the sequence and uses the current values of the variables appearing in the *expression*. Whenever the expression is true, all the evaluation attempts currently in flight or starting are disabled and record no match. In other words, in cover statements, **`disable iff`** behaves similarly to **`reject_on`** (Chap. 13), but unlike **`reject_on`** it does not use sampled values and a success is reported as "disabled".

In the above Example 18.1, whenever the sequence `a ##[1:100] b` matches, and provided that `reset` is false, the increment of `my_count` and the `$display` statement are scheduled to execute in the Reactive region.

Despite its powerful expressive power, one has to be careful when writing such **`cover sequence`** statements. This is because the total coverage count of the matches includes multiple matches for a single attempt whenever they occur, and thus get mixed with matches from other attempts in the total coverage count. In the above example, consider a situation where `a` occurs at the first and then at every fifth clock tick, and `b` at every tenth clock tick. Suppose that there are 101 clock ticks. How many matches will be recorded? The attempts associated with the first two occurrences of `a` will match each ten times, on all the occurrences of `b`. The subsequent two occurrences of `a` will match nine times, etc. It can be easily seen that the total number of matches will considerably exceed the number of attempts. Furthermore, the same occurrence of `b` being true will account for several matches. When looking at the final count in the coverage database, the result may be difficult to interpret.

A more useful case of coverage determination is whether `a` was followed by a `b` within 100 clock ticks without any other intervening occurrence of `a`. The result of the coverage in Example 18.1 does not provide this information.

The coverage database for coverage on a sequence contains the number of evaluation attempts started and the total number of sequence matches. This total number of matches does not distinguish among the attempts, hence even if the total number of matches exceeds the total number of evaluation attempts, there could be attempts that had no match. Simulation tools may optionally provide a count of first matches, in which case it is possible to see whether there were evaluation attempts that were disabled or had no match.

The question is then where is the **`cover sequence`** statement useful? Our experience suggests that it is useful when an action needs to be taken for all the matches. For example to trigger evaluation of some tasks or increment counts used elsewhere in the test bench.

Efficiency Tip. Unless necessary, avoid using **cover sequence** on sequence expressions that may result in multiple matches, such as those containing delay and repetition ranges or sequence disjunction (**or**).

If the total number of matches is not of interest, the cover statement can be stated using **cover property** as discussed next (Sect. 18.2.2).

18.2.2 Property Coverage

The syntax of this statement is **cover property**(*property_spec*) *statement_or_null* We have seen a description of **assert property** in Sect. 4.4. Similarly, the body of **cover property** may consist of a **disable iff** specification, a clocking event, and a property expression. If the property expression is a sequence, it is *promoted* to a property, meaning that the first occurring match of the sequence is transformed into a success of the property for that particular evaluation attempt, and any further evaluation within that attempt is curtailed.

The coverage from Example 18.1 is rewritten using **cover property** in the following example.

Example 18.2.

```
default clocking ck @(posedge clk); endclocking
int my_count = 0;
prop_cov1: cover property ( disable iff (reset)
   a ##[1:100] b
) begin
   my_count++;
   $display("success of prop_cov1 at time %t", $time);
end
```
□

In this case, `my_count` will be incremented and the `$display` statement will be executed for only the first match of the sequence in any evaluation attempt. The total number of recorded successes will thus be less than or equal to the number of evaluation attempts of the **cover property**. Contrast this with the behavior of **cover sequence**, for which the total number of recorded matches can exceed the number of evaluation attempts.

Property successes are classified as *vacuous* and *nonvacuous* (see Sect. 10.6 for discussion of vacuous and nonvacuous successes). Vacuous and nonvacuous successes are recorded separately in the coverage database. Therefore, the information obtained from the coverage database consists of the number of evaluation attempts started, the number of attempts in which the property succeeded nonvacuously, and the number of attempts in which the property succeeded vacuously. Attempts that were *disabled* are not counted as either vacuous or nonvacuous successes. A simulation tool may optionally provide a separate count for disabled attempts.

While much can be expressed using sequences only, property operators can provide additional expressive power. The combination of (simple) sequences and property operators helps to state clearly the intent of the coverage statement.

Example 18.3. Cover with a property expression:

```
prop_cov2: cover property ( disable iff (reset)
   @(posedge clk) a |-> s_eventually[1:100] b
) begin
   my_count++;
   $display("success of prop_cov2 at time %t", $time);
end
```
□

The disadvantage of this form is that when a is false, there is a vacuous success of the property. Coverage of vacuous successes is not useful. The property can be reformulated using the *followed-by* operator, which is better suited for this purpose.

Example 18.4.

```
prop_cov3: cover property ( disable iff (reset)
   @(posedge clk) a #-# s_eventually[1:100] b
) begin
   my_count++;
   $display("success of prop_cov3 at time %t", $time);
end
```
□

This example illustrates two ideas: (1) the use of the operator *followed-by* `#-#` for concatenating sequences with properties (the Boolean a on the left-hand side of `#-#` is a simple sequence), and (2) the use of strong bounded eventuality requiring the simple sequence b (implicitly cast to property) to occur within 100 clock ticks after a and due to its strength not to record a hit if there are not enough clock ticks to complete an evaluation attempt at the end of simulation.

Efficiency Tip. Use **`cover property`** unless all matches are absolutely needed.

> Avoid using implications in **`cover property`** statements. Instead, either use sequences or replace the implication with a followed-by operator.

The followed-by operator is necessary when the consequent is expected to be a property or when you wish to indicate clearly the trigger sequence in the antecedent. Followed-by is particularly useful in checker libraries where the **`checker`** arguments may not be restricted to sequences only.

It is often of interest to obtain information on the conditions under which a particular coverage scenario occurred. For instance, in the above example it may be of interest to know which of the delays between a and b occurred. A simple solution is to place the coverage statement in a generate loop spanning the 100 possible delay values:

Example 18.5.

```
for (genvar i = 1; i <= 100; i++) begin : loop_delays
  prop_cov4: cover property ( disable iff (reset)
    @(posedge clk) a #-# s_nexttime[i] b
  ) $display(
    "prop_cov4 success, time %t, delay %0d", $time, i
  );
end : loop_delays
```
□

There will be 100 **cover property** statements to evaluate, each one triggering on a and then searching for the occurrence of b at the exact time specified by **s_nexttime**[i]. For small delay range values this is an acceptable solution. However, for larger values, like in this example, it rapidly degrades the performance in simulation. For FV, the performance impact may not be as heavy, as the individual covers become targets for reachability analysis.

For simulation there is an alternative, simpler solution. We can combine the power of temporal properties to detect patterns of signals over time with covergroups to record data and characteristics of the signal patterns. First we provide a brief introduction to the **covergroup** construct.

18.2.3 Covergroup

Coverage on properties and sequences is an excellent mechanism for detecting the occurrence of some specific series of Boolean values. It is not as useful for collecting data values, delays, etc. For that purpose the **covergroup** object is available in SystemVerilog. Not only does it allow collecting information from simple temporal sequences, but its main power is in collecting and correlating information from multiple data points. The information sampling is triggered either by some clocking event or by calling the sample method of the **covergroup**.

We do not present a detailed account of all the features of the **covergroup** object. The reader can find further details in the LRM and also in [19, 59]. We only introduce a small set of features by means of an example to illustrate the usage of covergroups jointly with cover properties to achieve an efficient implementation of coverage collection with data. The coverage collection is triggered by the successes of a **cover property**, and data collected from within the evaluation attempt of the property is stored using a **covergroup**. First let us see an example of a **covergroup** where sampling is triggered by a clock.

Example 18.6.

```
covergroup cover_delay @(posedge clk);
   dl_pt: coverpoint delay {
      bins delays [100]: {[10:110]};
   }
   dt_pt: coverpoint data;
   dlXdt: cross dt_pt, dl_pt;
endgroup
cover_delay cover_delay_inst = new();
```
□

The **covergroup** definition is named `cover_delay`. The **covergroup** tracks two variables, `delay` and `data`. They are identified by using the keyword **coverpoint** and by labeling them `dl_pt` and `dt_pt`, respectively. The variable values are read whenever the clocking event `@(`**posedge** `clk)` occurs and are recorded into individual bins according to their values.

For `dl_pt`, there is an array `delays` of 100 bins into which the occurrence counts of values of `delay` in the range `[10:110]` are maintained. In other words, when the recorded value of `delay` is `10` the bin `delays[0]` is incremented; if it is, say, 100 then the bin `delays[90]` is incremented; etc.

For `dt_pt`, the bins are allocated automatically. By default, there are at most 64 bins, but that value can be changed by an optional specification `auto_bin_max=number` of the **covergroup**. If the value range of the sampled variable is less than the specified maximum, the number of bins is that of the variable range. If the range is larger than the maximum, then the variable values are uniformly distributed over the bins.

The cross correlation `dlXdt` of the values of `delay` and `data` is specified using the keyword **cross**. This defines a set of pairs of values consisting of the Cartesian product of the sets of bins of **coverpoint**s of `dl_pt` and `dt_pt`. In this way, the user can observe the correlated pairs of values of `data` and `delay` that occurred during the simulation.

The clocking event can be replaced by an interface definition of the `sample` method as follows:

```
covergroup cover_delay
   with function sample(int unsigned delay, logic [7:0] data);
```

In this case, the coverage is triggered when the method `sample` is called. The actual arguments must be type-compatible with the formal arguments of the method. Note that the formal argument names match exactly those of the **coverpoint** variables `delay` and `data`. In this way, the same **covergroup** instance can read different variables passed as actuals to instances of the `sample` function and collect the coverage information in the same bins. Using the method is especially helpful when new instances of the variables are created over and over, as in the example in the following section where local variables are passed to a **covergroup** instance. When no clocking event or sample method is explicitly specified, the default `sample` method is available to trigger sampling of **coverpoint** variable values.

There are many variations on how to define parameterized covergroups, bins, crosses, conditional selection, etc. They are beyond the scope of this book, but the information provided here should be sufficient to illustrate their combined power with assertions.

18.2.4 Combining Covergroups and Assertions

We can use a **covergroup** to collect information on the delays in Example 18.5. The generate loop in that example can be disposed of as follows:

Example 18.7.

```
default clocking ck @(posedge clk); endclocking
covergroup delay_cg
   with function sample(int unsigned delay);
   dl_pt: coverpoint delay {
      bins delays [100]: {[1:100]};
   }
endgroup
delay_cg delay_cg_inst = new();
property p_delay_coverage;
   int unsigned ticks_l;
   disable iff (reset)
   (a, ticks_l = 0)
   #-# (ticks_l < 100, ticks_l++)[+]
   ##1 (
      b,
      delay_cg_inst.sample(ticks_l),              // match item
      $display("prop_cov4 success, time %t, delay %0d",
         $time, ticks_l)                          // match item
   );
endproperty
prop_cov4: cover property(p_delay_coverage);
```
□

The main components of this form of coverage collection are:

- **covergroup** `delay_cg` definition that specifies a **coverpoint** on the formal argument of the method `sample`.
- A **covergroup** instance `delay_cg_inst` that creates the actual coverage object.
- A cover on property `p_delay_coverage`, which uses a local variable `ticks_l` of the same type as the formal argument of the `sample` method to count the number of clock ticks until `b` matches.
- When `b` matches, `sample` is called to classify the current value of the local variable in the coverage database.

Whenever `prop_cov4` is triggered by the occurrence of `a`, the local variable `ticks_l` is initialized to 0. `ticks_l` is incremented at each clock tick thereafter until `b` occurs, provided it is within 100 clock ticks. When `b` matches, sampling of `tick_l` by the **covergroup** takes place. The coverage database thus has two entries, one for prop_cov4 that records the total number of successes of the property and a second one for the cover point dl_pt of the **covergroup** instance delay_cg_inst. dl_pt has 100 bins, one for each value in the delay range [1:100].

Note that in this formulation of the property, the property operator followed-by #-# could have been replaced by the overlapping sequence concatenation ##0 because the consequent of #-# is a sequence. The effect is the same.

When should one use one or the other? We believe that this depends on the property to be covered. If the consequent is a sequence, use ##0 and ##1. If it is a property, use #-#, and #=#, respectively. Followed-by is also useful even with sequences when it is desired to identify clearly the trigger condition (the antecedent of followed-by) in the **cover property** statement.

We cannot simply substitute **s_eventually** in place of the sequence operators. The reason is that **s_eventually** provides no way to increment the local variable when the clock advances. We can still extend the formulation with **s_eventually** by adding extra code to keep a global count of clock ticks outside the assertion. We initialize the local variable to the current count of the global counter when a matches, and when b matches we pass the difference between the global counter and the local variable value to the sample function call. This is shown in the next example.

Example 18.8.

```
default clocking ck @(posedge clk); endclocking
int unsigned tick_counter = 0;
covergroup delay_cg with function sample (int unsigned delay);
   dl_pt: coverpoint delay {
      bins delays [100]: {[1:100]};
   }
endgroup
delay_cg delay_cg_inst  new();
always @ck
   tick_counter <= tick_counter+1;
property p_delay_coverage;
   int unsigned start_tick;
   disable iff (reset) @(posedge clk)
   (a, start_tick = tick_counter)
   #-# s_eventually[1:100] (
      b,
      delay_cg_inst.sample(tick_counter-start_tick),
      $display("prop_cov5 success, time %t, delay %0d",
         $time, tick_counter-start_tick)
   );
endproperty
prop_cov5: cover property(p_delay_coverage);
```
□

The compression of many generated **cover property** statements into one together with a **covergroup** can be applied in other situations too. For example, it is common to write the same cover statement on all bits of a vector. Instead, we can use a generalized **cover property** on the vectors, save the bit indices that were triggered in a local variable, and finally update the database using a **covergroup**.

This is illustrated in the next example, where first we show the coverage collection using a generate loop and then rewrite it using a **covergroup**.

Example 18.9. On 32-bit vectors, cover that a rising transition on `x[i]` is eventually followed by `y[i]` asserted.

Solution: Using **generate**:

```
bit [31:0] x, y; bit clk;
default clocking ck @(posedge clk); endclocking
for (genvar i=0; i<32; i++) begin : loopi
   prop_cov6: cover property (
      !x[i] ##1 x[i] ##1 y[i][->1]
   );
end : loopi
```

Solution: Using a **covergroup**:

```
default clocking ck @(posedge clk); endclocking
covergroup cg_vect
   with function sample (bit [31:0] covered);
   vct: coverpoint covered {
      bins x0 = {covered[0]};
      ... // enumerate all bit positions
      bins x31 = {covered[31]};
   }
endgroup
cg_vect cg_inst = new();
property p_vector_cov;
   bit [31:0] rose_x, covered;
       (1, rose_x = x)
   ##1 (1, rose_x = x & ~rose_x)
   ##0 (|rose_x, covered = 0)
   ##1 (
      (y & rose_x & ~covered)[->1],
       covered |= (y & rose_x)
   )[+]
   ##0 (covered == rose_x, cg_inst.sample(covered));
endproperty
prop_cov7: cover property (p_vector_cov);
```
□

Property `p_vector_cov` is much more complex than the one used in `prop_cov6`. There are two local variables: `rose_x` to record the bits of `x` that had a rising transition at the start of the property; and `covered`, which maintains the set of bits of `y & rose_x` that have been covered so far. The final condition, `covered == rose_x`, indicates that all the rising transitions recorded in `rose_x` have been covered by high values on the corresponding bits of `y`. When that occurs, the `covered` bits are sent to the **covergroup** instance and recorded.

There are two issues with this encoding that need some remedy. First, if for one of the bits of `x` that had a rising transition the counterpart bit of `y` is never asserted, then no coverage is recorded. It is better always to sample when some match occurs inside the repetition loop. The second issue is in the declaration of the **covergroup**.

There is no way to "generate" a set of bins in a loop; they must be enumerated. To remedy this issue, a task can be created that decodes which bits are set in `covered`, and then calls method `sample` for each such bit, passing its index. The task is then called from the property in place of the `sample` method. The **covergroup** must be modified so that the index into the vector is the **coverpoint** variable and it has as many bins as the width of the vector. Both of these improvements are left as an exercise at the end of the chapter.

18.3 Covergroups in Checkers

Checkers can be used to encapsulate coverage collection statements. Both assertion **cover** statements and **covergroup** constructs are allowed in checkers. We show an example of using **cover** statement in checkers in Example 24.4. In this section, we focus on covergroups.

Example 18.10. Check that all states of the FSM are visited. Let the FSM states be described by the following enumeration:

```
typedef enum {
    INIT,
    IDLE,
    SEND,
    WAIT,
    RECEIVE,
    TO
} state_t;
```

The following checker collects coverage information about FSM states.

```
checker cover_fsm(
   state_t state,
   event clk = $inferred_clock,
   untyped rst = $inferred_disable
);
   covergroup cg_state @clk;
      coverpoint state iff (!rst);
      option.per_instance = 1;
   endgroup : cg_state
   cg_state cg = new();
endchecker : cover_fsm
```

This checker contains a covergroup `cg_state` triggered by the clocking event `@clk`. Coverage is collected in bins defined by the values of the enumeration type `state_t`, but only when `rst` is low. Recall that for actual coverage collection to take place, it is required to instantiate the covergroup using the **new** operator. □

It is possible to use checker variables in covergroups contained in the same checker.

Example 18.11. Cover the time interval distribution between request `req` and acknowledgment `ack`.

Solution:

```
checker req_ack_window2(
   req, ack,
   event clk = $inferred_clock,
   untyped rst = $inferred_disable
);
   default clocking @clk; endclocking
   default disable iff rst;
   int unsigned n = 0;

   always_ff @clk
      if (rst || ack)   n <= 0;
      else if (req)     n <= 1;
      else if (n == 0) n <= 0;
      else              n <= n + 1;

   covergroup cg_win @(clk);
      coverpoint n iff (ack && !rst);
   endgroup : cg_win
   cg_win cg = new();
endchecker : req_ack_window2
```

Capture of the value of `n` in the covergroup happens when the clocking event `@clk` takes place, i.e., before `n` is reset to 0 by nonblocking assignment.

We mentioned earlier that for efficiency reasons in FV, checker variables should have the smallest possible size. In this example, `n` is defined as an unsigned **`int`**, which is of size 32 bits. One reason for this declaration is that we do not know the maximal size of the time window, yet we need to reserve a large enough upper bound. More importantly, the goal of this checker is to collect coverage information in the covergroup in simulation, where **`int`** variables are efficient. The checker is not intended for use in FV. □

In the following section, we examine effects of strong and weak properties in **`cover property`** statements.

18.4 Coverage on Weak and Strong Properties

An important enhancement to SystemVerilog Assertions is the explicit notion of property *strength* (see Chaps. 10 and 21). In simulation, the impact of strength is seen on the result of property evaluation at the end of simulation, or more specifically, when there is no further clock tick. In the case of **`cover property`** statements, an evaluation attempt of a strong property for which there are not enough clock ticks to reach a definitive decision yields the result *not covered*. This is unlike in **`assert property`** statements, where such a situation leads to a failure of the evaluation attempt.

When a sequence `s` is implicitly promoted to a property by using it in the statement **cover property**(s), the interpretation is as **cover property** (**strong**(s)). It means that if the first match of the sequence is not reached when the last clock tick occurs, that attempt is not counted as covered. Recall that the statement **assert property**(s) interpretation is as **assert property** (**weak**(s)), meaning that if a match is not reached when clock ticks stop, the result is a success of the assertion.

An interesting situation occurs if a cover property statement is placed in a conditional branch of an always procedure, as shown in the following example.

Example 18.12. Cover in an always procedure

```
always @(posedge clk) begin
   if (!en) ... some code...;
   else begin
      c: cover property(a ##1 b);
      ... some other code ...
   end
end
```
□

The **cover property** is sampled using **posedge** clk, but an evaluation attempt will only start when control reaches the location of `c` in the procedure. That may or may not ever happen, hence the overall effect of the cover within the procedural code is similar to having written a static cover containing the property **s_eventually**((en && a)##1 b). The eventuality occurs when the procedure does execute the corresponding branch.

The next and final section provides two more complex examples.

18.5 Examples

Example 18.13. Recall the specification from Chap. 1, Fig. 1.6:

> The system consists of a transmitter and a receiver connected by a point-to-point duplex channel. The transmitter sends to the receiver packets and gets an acknowledgment from the receiver upon the packet receipt. The packet contains a header and a body. The header consists of 8 bits, and the two most significant bits contain information about the transaction type: *data* (10), *control* (01), or *void* (00). The remaining six bits of the header contain the transaction tag in case of a data transaction, and are 0 in case of a control transaction. For void packets the tag field may contain any value. The packet body consists of three bytes; these bytes contain raw data for data transactions and commands for control transactions . . ..
> Upon receipt of a data or a control packet the receiver sends back to the transmitter an acknowledgment signal. The acknowledgment consists of 7-bits: the most significant bit is set to 1, and the remaining 6 bits contain the tag of the received packet. If a void packet is received, its contents are ignored and no acknowledgment is sent . . .
> The transmitter is not allowed to send a new packet before an acknowledgment is received. If timeout is reached, the transmitter sends the same packet again. If after three retries it does not get an acknowledgment, it asserts the error signal and requires a manual reset.

We can now enhance the checker in Fig. 1.6 to include collection of coverage information to make sure that the behavior of the design is sufficiently exercised by simulation tests. The enhancements are shown in Fig. 18.1.

Property `tx_rx_ack` describes a situation in which a packet sent of the type `kind_t` having a particular tag eventually receives an acknowledgment. The correspondence between the sending packet tag and the acknowledged tag is assured by using the local variable `tag` in the property. It is assigned the tag of the transmitted packet at the time when `sent` is true and then checked when an acknowledgment arrives. Only when the tags match does the property succeed, and then the coverage count in the database is incremented. □

The next example illustrates another important point regarding the combined power of **`cover property`** and the **`covergroup`** construct. It shows that this scheme can save on coding effort, reduce simulation time, and consolidate reporting. The example uses a simplified $N \times N$ switch.

Example 18.14. A switching device has $N \leq 256$ 8-bit input ports and $N \leq 256$ 8-bit output ports arranged in packed arrays:

```
logic [N-1:0] [7:0] dataIn;
logic [N-1:0] [7:0] dataOut;
```

A packet consists of 256 8-bit bytes. Each packet enters the switch on one of the input ports and leaves the switch on one of the output ports. Packet data is transmitted one byte at a time across a port. The first byte of a packet contains the source ID where the packet originated. The second byte contains the destination ID where the packet is to be routed after it leaves the output port. A 1-bit signal is associated with each input and output port indicating the start of a packet. Vector **`bit`** `[N-1:0] startIn` is used for input ports, while **`bit`** `[N-1:0] startOut` is used for output ports. Bit `i` of either port is set to one when the first byte arrives. At all other times the bits are set to 0. The incoming data are all synchronized to **`posedge`** `clkIn`, the outgoing data are synchronized to **`posedge`** `clkOut`. There is a delay of at least 2 `clkIn` cycles before a packet can emerge on an output port. For simplicity, we assume more specifically that the first byte of a packet will appear on one of the output ports at the first posedge of clkOut at or after the posedge of clkIn when the second byte of the packet appeared on its input port.

It is required to construct a coverage collection system such that correlated information is collected about which source ID and destination ID appeared in a packet, together with the input port and output port through which the packet was routed in the switch.

Discussion: A simple approach is to define a property that characterizes one path of a packet from one input port to one output port, and use a generate loop to build as many instances of a cover property as there are possible combinations of paths through the switch and IDs. This is shown in Fig. 18.2.

While the definition of this coverage model is quite concise and simple to encode, it will impose a heavy burden on the simulator and will most likely produce a coverage report that is long and difficult to analyze. Even with a small $N = 4$, the

```
typedef enum {
   info = 2'b10, control = 2'b01, void = 2'b00, forbid = 2'b11
} kind_t;
typedef logic [5:0] tag_t;
typedef logic [23:0] data_t;
typedef struct {
   kind_t kind; tag_t tag; data_t data;
} packet_t;
typedef struct { logic ack_received; tag_t tag; } ack_t;

checker spec (
   packet_t tx_packet, // Packet to be transmitted
   rx_packet,          // Last received packet
   logic sent,         // Packet sent
   ack_t ack,          // Acknowledge
   logic timeout,      // Timeout active
   logic err,          // Error signal
   event clk,          // System clock
   logic rst           // Reset
);
default clocking @clk; endclocking
default disable iff rst;

...same as before...

// coverage of different packet types
let packet_sent_type(packet_t packet, kind_t kind) =
   sent && packet.kind == kind;
cov_kinds_tx_info: cover property (
   packet_sent_type(tx_packet, info));
cov_kinds_tx_void: cover property (
   packet_sent_type(tx_packet, void));
cov_kinds_tx_control: cover property (
   packet_sent_type(tx_packet, control));

// coverage of packet types transmission and acknowledgment.
property tx_rx_ack(kind_t kind);
   tag_t tag;
   (sent && tx_packet.kind == kind, tag = tx_packet.tag)
   #=# s_eventually(ack.ack_received && ack.tag == tag);
endproperty

cov_sent_ack_info: cover property(tx_rx_ack(info));
cov_sent_ack_void: cover property(tx_rx_ack(void));
cov_sent_ack_control: cover property(tx_rx_ack(control));

endchecker : spec
```

Fig. 18.1 System specification with coverage

```
parameter N = 4;
property path_cover(inIdx, outIdx, sourceId, destId);
    @(posedge clkIn)
        (startIn[inIdx] && sourceId == dataIn[inIdx]) ##1
            (destId == dataIn[inIdx])
        ##0
    @(posedge clkOut)
        (startOut[outIdx] && dataOut[outIdx] == sourceId) ##1
            (dataOut[outIdx] == destId);
endproperty

generate
  for (genvar pIn = 0; pIn < N; pIn++) begin : PORT_IN
    for (genvar pOut = 0; pOut < N; pOut++) begin : PORT_OUT
      for (genvar sId = 0; sId < 256; sId++) begin : SOURCE_ID
        for (genvar dId = 0; dId < 256; dId++) begin : DEST_ID
          cover_path:
          cover property(path_cover(pIn, pOut, sId, dId));
        end
      end
    end
  end
endgenerate
```

Fig. 18.2 Coverage using generate loops

packet path coverage for this 4×4 switch will generate $4 \times 4 \times 256 \times 256 = 2^{20}$ cover properties, all running at the same time.

Perhaps it is not necessary to track all source and destination IDs, hence the number of cover properties can be reduced. Still, the overhead may be large. A better and more elegant solution can be obtained by combining N cover property statements, one for each output port of the switch, with a **covergroup** that collects the coverage of paths. This is shown in Fig. 18.3.

The combination of N cover properties with a single **covergroup** provides a speed up of up to 2^{18} for a 4×4 switch over the simplistic solution using nested generate loops. Furthermore, the coverage report generated for the **covergroup** will nicely summarize which paths were covered and how many times. This is due to the **cross** statement in the **covergroup**. □

Exercises

18.1. For Example 18.13, write a **cover property** that records that a retry occurred.

18.2. For Example 18.13, write several **cover property** statements that record the number of retries that occurred (from 1 to 3).

```
parameter N = 4;
covergroup pathCg with function sample
 (bit [7:0] inIdx, outIdx, logic [7:0] sourceId, destId);
  cross inIdx, outIdx, sourceId, destId;
endgroup
pathCg pathCg_inst = new();
task samplePathInfo(
  bit [N-1:0] inIdx_lv,
  bit [7:0] outIdx,
  logic [N-1:0][7:0] sourceId_lv, destId_lv);  logic [7:0] outSourceId_lv, outDestId_lv;
  int i;
  for (i=0; i<N; i++) begin
    if (inIdx_lv[i] && (sourceId_lv[i] == outSourceId_lv)) begin
                    && (destId_lv[i] == outDestId_lv)
      pathCg_inst.sample(i, outIdx, sourceId_lv[i], destId_lv[i]);
      break;
    end
  end
endtask
property path_cover(outIdx);
  logic [N-1:0][7:0] sourceId_lv, destId_lv;
    bit [N-1:0] inIdx_lv;  logic [7:0] outSourceId_lv;
    @(posedge clkIn)
      ((|startIn), inIdx_lv = startIn,
                   sourceId_lv = dataIn) ##1
        (1'b1, destId_lv = dataIn)
       ##0
    @(posedge clkOut)
      (startOut[outIdx], outSourceId_lv = dataOut) ##1
         (1'b1, samplePathInfo(inIdx_lv, outIdx,
                               sourceId_lv, destId_lv,
                               outSourceId_lv, dataOut));
endproperty
generate
  for (genvar i = 0; i < N; i++) begin : PORT_OUT
    cover_path: cover property(path_cover(i));
  end
endgenerate
```

Fig. 18.3 Coverage of packet paths using a cover property and a cover group

18.3. Write a single **`cover property`** that records the same information as Exercise 18.2 with the help of a **`covergroup`**.

18.4. Write a **`cover property`** and a **`covergroup`** that record which tags were used with which tx packet kind.

18.5. Modify the **`cover property`** and the **`covergroup`** in Example 18.8 to remedy the two issues raised there.

18.6. Write two **`cover property`** statements that record whether between two conditions, `a` and `b`, where `b` follows `a` by some indefinite number of clock ticks, the condition `c` occurred and not occurred, respectively.

18.7. Modify the code in Fig. 18.1 to add a **`covergroup`** recording the kinds of the transactions that are witnessed. Fold the three **`cover property`** statements into a single **`cover property`** that collects this coverage.

18.8. The **`covergroup`** solution for Example 18.14 shown in Fig. 18.3 uses a strategy of running `N` parallel threads, one for each output channel. Each thread captures into local variables the packet start vector `startIn`, all bytes of `dataIn` for two consecutive cycles of `clkIn`, and the one byte of `dataOut` for the channel being tracked in the subsequent cycle of `clkOut`. At the next cycle of `clkOut`, the thread passes all this data, together with another byte of `dataOut` for its channel, to `samplePathInfo` to sort out which coverage data paths ending at the tracked channel have been witnessed.

1. Discuss the redundancy of data across the various threads running in parallel.
2. What redundancy is there, if any, in the processing of the calls to task `samplePathInfo` across the parallel threads?
3. Write a new encoding that uses the same **`covergroup`**, but which only runs a single thread for all the data channels. Make the single thread collect all the data from `dataIn` for two consecutive cycles as before, and make it collect all the bytes of `dataOut` in the subsequent cycle of `clkOut`. At the next cycle of `clkOut`, pass this data, together with all the bytes of `dataOut`, to a call to a new task that analyzes the data path coverage for all channels.
4. Discuss the savings in data capture with the new encoding and compare the processing complexity for its single task call with the `N` parallel calls to `samplePathInfo` in Fig. 18.3.

Chapter 19
Debugging Assertions and Efficiency Considerations

If everything seems to be going well, you have obviously overlooked something.

— Steven Wright

Properties and sequences allow us to describe complex behaviors in a very compact declarative form. That form is quite different from the procedural style used for writing RTL and other design models as well as test benches. Thus, assertions may also need a different style for debugging them. Issues related to the run time and memory overheads for complex temporal assertions also need to be addressed. The same behavior may be expressed using different assertions. Each may have different efficiency in formal verification and simulation. We discuss both debugging and efficiency in this chapter.

There are two kinds of situations to consider (see also [19]):

- A failure for an assertion from a checker library or a user-written one. Failure of an assertion from a library usually points to its incorrect usage or a problem in the design under verification, assuming that the library was validated. User-written assertion failure may be due to a design error or an incorrect formulation of the assertion.
- A failure for an assertion under development or during regression that needs to be analyzed. Failure of an assertion under development usually means a failure during specific simulation tests created to validate the assertion before use in verifying a design. Failure during regression is more likely to be due to an error in the design, assuming that the assertion was validated.

Developing assertion-based checkers uses similar techniques as for developing custom assertions. The main difference is that the testing and documentation must be quite extensive as demonstrated, for example, in the Accellera OVL checker library [10]. In either situation, effective means must be provided to pinpoint the source of the problem. Verification tool vendors often provide various mechanisms

E. Cerny et al., *SVA: The Power of Assertions in SystemVerilog*,
DOI 10.1007/978-3-319-07139-8_19

to view and debug assertions, however, in this book we are tool-agnostic and assume only that tools generally provide waveforms for viewing with some marking that identifies the start time and the end time of an evaluation attempt. To gather a more insightful view of the failure, we rely on means within the SV language to provide us with further information on the progress of the failing attempt. Debugging is usually done in simulation, even though the assertion may be developed for or may have failed in formal verification.

In the following sections, we address two scenarios: one, for debugging an assertion during its development, and the other, for debugging a failing assertion in a regression test for a design.

19.1 Debugging an Assertion Under Development

The starting point of any debugging effort while developing a custom assertion is a good requirement specification that states the trigger conditions and the sequence of signal combinations that must hold following the trigger. Based on this information, a simple test bench should be developed. If the assertion is complex, a random test bench is preferable, since a completely exhaustive test may be impractical. This guideline is similar to developing a test bench for verifying a design. While inspecting the results from simulating the assertion with the test bench, we must be careful to identify and verify any unwanted vacuous successes and incomplete evaluations.

The next step is to change the test bench to generate erroneous situations that induce assertion failures, while avoiding the acceptable ones as much as possible. This step is much harder, because the number of possibilities of failure may be quite large and the resulting test bench may unavoidably contain acceptable situations in which the assertion succeeds. It is these successes that must be scrupulously analyzed for validity.

Suppose that either an invalid success or failure is detected. Now, one needs to isolate the particular invalid attempt so as to not clutter the debugging information by data from other attempts, and to observe the progress of the evaluation of that attempt. Since most simulators provide information about the start time of a failing/succeeding attempt, that time can be used to control the starting and stopping of the assertion using the control system tasks `$assertoff` and `$asserton` (see Sect. 7.3). Assuming that the test bench is repeatable, the assertion should be stopped from time 0 using `$assertoff` till just before the start time of the attempt of interest at which point it should be started using `$asserton` for just one clock tick, and thereupon stopped again using `$assertoff`. This will start exactly the single attempt of interest.

Once the invalid attempt is isolated, we can instrument the assertion by adding local variables for collecting data, using match items in sequences to assign and display signal values, and use action blocks to display any additional information about the failure or success of the assertion. The number of possibilities is quite large to describe it in all generality; therefore, we shall illustrate the process using an example.

Suppose that we need to debug an assertion written according to the following requirements: When `req` becomes asserted for one clock cycle it is associated with a transaction id `req_tag`, the acknowledgment `ack` is also associated with a similar id `ack_tag` that establishes correspondence with the request having the same id. `ack` must arrive no later than 18 cycles after the `req`. Acknowledgments can arrive out of the requesting order.

Let us assume that the assertion has the following form:

Example 19.1.

```
property p;
  logic [3:0] tag;
   @(posedge clk) (req, tag = req_tag)
   |=>
   s_eventually [1:18] ack && (ack_tag == tag);
endproperty
a: assert property (p);
```
□

Clock ticks occur at times 1, 3, 5, ... An unexpected failure happened for the attempt starting at 105, ending at 141. That is, the failure occurred at the limit of 18 clock cycles after `req` was received. Is the failure genuine?

Making association between the failure and the sequence of values is laborious, so we opt for instrumenting the assertion and the test bench.

Example 19.2.

```
default clocking ck @(posedge clk);
endclocking
initial begin // add to the test
  $assertoff();
  #104;
  // start assertion just before
  // the start time of failure
  $asserton();
  #2;
  // stop assertion right after
  // the failing attempt at time 105
  $assertoff();
end

property p;
  logic [3:0] tag;
  (
    req, tag = req_tag, // make sure it triggered
    $display("[%t] req asserted, tag %0d", $time, tag)
  )
  |=>
  s_eventually [1:18] ack && (ack_tag == tag);
endproperty
a: assert property (p);
```

```
property p_cover;
  logic [3:0] tag;
  (
    req, tag = req_tag,
    $display("[%t] req asserted, tag %0d", $time, tag)
  )
  ##[*]
  (
    ack && tag == ack_tag,
     display("[%t] ack asserted, ack_tag %0d", $time, ack_tag)
  );
endproperty
c: cover property (p_cover);
```
□

The additional code stops the assertion at the beginning, starts it for one clock tick, and then again stops it. We added `$display` to the assertion to make sure that it does trigger (and fail) as observed originally. Finally, we added a cover property that searches for `ack` with the matching tag indefinitely from the time of the occurrence of `req`. The objective is to see whether `ack` ever arrives.

After running the simulation, we observe that the assertion and the cover triggered as expected, the assertion failed and the cover matched very quickly on the next clock tick at time 107. Assuming that the test is generating legal situations only, it suggests that the assertion missed the arrival of `ack` associated with the matching `req`. By examining the assertion, we can see that in fact the bounded eventuality starts checking only 2 cycles after the `req` arrival due to the one cycle delay introduced by `|=>`, and then another cycle by the lower bound of 1 in **`s_eventually`**. An easy correction is as follows:

Example 19.3.

```
property p;
  logic [3:0] tag;
  (req, tag = req_tag)
  |=>
  s_eventually [0:18] ack && (ack_tag == tag);
endproperty
a: assert property (p);
```
□

The solution brings out a question: Can the acknowledgment arrive with 0 delay? The requirement specification did not mention anything about the earliest arrival. This issue has to be clarified. If the answer is that the acknowledgment can be generated combinatorially with 0 clock tick delay, then `|=>` should be replaced by `|->` in the assertion. What should happen if `ack` arrives at the same time as `req` and with the same `ack_tag` value as the value of `req_tag`? If this situation is illegal, then the assertion should be modified to reject it as shown in the next example.

Example 19.4.

```
property p;
  logic [3:0] tag;
  (req, tag = req_tag)
```

```
  |->
  (
    s_eventually [1:18] ack && (ack_tag == tag)
    and
    !(ack && (ack_tag == tag))
  );
endproperty
a: assert property (p);
```

□

Or even more simply as

```
property p;
  logic [3:0] tag;
  (req, tag = req_tag)
  |->
  strong(
    !(ack && (ack_tag == tag)) ##1
    ##[1:18] ack && (ack_tag == tag)
  );
endproperty
a: assert property (p);
```

□

An alternative that may be more efficient with formal verification tools is to separate the two cases into two independent assertions.

Example 19.5.

```
property p1;
  logic [3:0] tag;
  (req, tag = req_tag)
  |->
  s_eventually [1:18] ack && (ack_tag == tag);
endproperty
property p2;
  logic [3:0] tag;
  @(posedge clk) (req, tag = req_tag)
  |->
  !(ack && (ack_tag == tag));
endproperty
a1: assert property (p1);
a2: assert property (p2);
```

□

Finally, what if the lower bound is greater than 1 clock cycle, e.g., 3? A possible solution to eliminate the unwanted early arrivals is to use a bounded always operator as follows:

Example 19.6.

```
property p;
  logic [3:0] tag;
  (req, tag = req_tag)
  |->
  (s_eventually [3:18] ack && (ack_tag == tag))
  and
  (s_always [1:2] !(ack && (ack_tag == tag)));
```

```
endproperty
a: assert property (p);
```

This could again be written using a strong sequence as follows:

```
property p;
  logic [3:0] tag;
  (req, tag = req_tag)
  |->
  strong(
    !(ack && (ack_tag == tag))[*2]
    ##[1:16]
    (ack && (ack_tag == tag))
  );
endproperty
a: assert property (p);
```

□

An option is to separate the two clauses into two independent assertions as shown in Example 19.5.

In the following section, we briefly discuss debugging assertion failures that occur during a test of a design.

19.2 Debugging Assertion Failures from a Test

Without specialized debugging tools that vendors may provide, a similar technique to the one in the preceding section can be used to debug assertion failures as a result of running a test. The failure can be due to either an incorrect assertion or an error in the design. One difference in debugging is that there may be many assertions (that fail or not) and that rerunning the test may become quite demanding on resources and time. In this scenario, we do not construct a new test bench, but continue with the same failing test.

Assuming that rerunning the test either with its original design or from a saved signal dump is possible, we can concentrate on the particular assertion failure. If there are more than one failure of the same assertion in the run, we start with the first one that is not due to some clearly apparent reason like neglecting to stop the assertion during the reset phase.

There are three possible ways to approach the debugging problem:

1. To concentrate on that failure, we should stop all assertions at time 0, and then start only the one of interest just before the clock tick associated with the start time of the failing attempt. We will thus need to specify the complete path to the assertion in the call to `$asserton` as well as to the subsequent `$assertoff` that shuts it off just after the attempt started. These calls can be placed in a new top-level module that is used only for this assertion control. We also instrument the assertion following the ideas shown in Example 19.2.
2. An alternative is to shut off all assertions and add a copy of the instrumented failing assertion into the new top-level module. The actual arguments then must

be stated as hierarchical references to the signals in the original assertion. This approach has the advantage that we do not modify the design in any way.

3. Alternately, we could make the new control module with ports that correspond to the signals used in the assertion, and instrument the same assertion but referring to these ports. We bind the new control module to the module instance that contains the failing assertion. Care must be taken to control just the instrumented assertion using the `$asserton` and `$assertoff` calls.

19.3 Efficiency Considerations

Depending on whether an implementation of assertions in simulation takes the attempt-based view, different forms of assertions expressing the same requirements may have different compile time and run time performance. Each simulator may have different forms of implementation; nevertheless, there are some general situations that can be exposed. We base this exposition on three abstract implementations of the assertions:

1. Assertion is an observer that issues only the fail result with no information about attempt start time of failure. No failure in simulation means success.
2. Assertion issues only the fail result with information about the earliest or latest attempt start time of the failure at a specific time (more than one attempt can reach a failure at the same time by the same condition). No failure in simulation means success.
3. Assertion is evaluated by maintaining information about all attempts, their start times and fail/pass times.

An abstract implementation of the first kind may be achieved by compiling the assertion into a single automaton of the negated property, as typically done for formal tools (Sect. 21.4.1). When the automaton is evaluated it may only provide information on the first failure detected and its failure time and perhaps the earliest/latest start time as indicated by the 2nd bullet above. Only on Boolean assertions of the form **`assert property`** `(expr);` both times would be exact.

For the third kind of implementation, it is possible to compile the assertion into processes that interpret the syntactic form of the property. An evaluation attempt would start at every tick of the leading clock, noting its time and then, when it succeeds or fails, it would report the result and the start and fail times.

Naturally, these are descriptions of abstract implementations, and a particular simulator may have a mixture of approaches. For example, there may be a different algorithm to evaluate Boolean concurrent assertions than for evaluating complex or recursive properties (Chaps. 10 and 17). Local variables may be another dimension to the implementation spectrum (Chap. 15).

When local variables are involved, the purely automaton-based evaluation may not be feasible for at least two reasons. First, determinizing the automaton may not always be practically feasible, and second, evaluating a nondeterministic automaton requires keeping track of which transitions are to be combined with the same set of local variable values.

Compile Time Performance

In general, the complexity of the automaton needed in implementations (1) and (2) can grow exponentially with the size of the property as measured by the number of operators and expressions in the property. However, implementation (3) based on the interpretation of the syntactic structure remains linear in size. This implies that form (1) for really complex properties involving nested property and sequence operators may take prohibitive time to generate. The final representation may also consume large amounts of memory, especially when large delay ranges are involved or there is a large set of choices due to an **or** operation or large ranges. This is one of the important efficiency considerations for formal verification to limit the size of ranges of delays (##) and repetitions of all kinds (`[* [..]`) (see Sect. 21.4.1).

Run Time Performance

In general, evaluation based on automata, especially if the automata are determinized, will yield higher performance, because there is minimum work involved in evaluating an expression and then advancing to the next state. When failure is detected, the simulator just reports the result and aborts further evaluation.

For the evaluation that keeps track of attempts, the simulator must maintain information about attempt start times, the local variable values if any, as well as information about multiple threads of evaluation within an attempt. This consumes both time and memory.

In the following, we examine several typical cases that may have different run time performance based on the algorithm scheme used for the implementation. In the preceding chapters, we often raised performance issues as "efficiency tips". We thus revisit some of them and provide explanation why it is so based on the abstract implementations.

- Fixed delay or repetition values `##N` as well as `[*N]` for some large `N`.

In an automaton-based implementation, large values used in these operators that also include `[->N]`, `[=N]` create at least `N` states with transitions arranged in a similar way as in counters. If used within other operators such as **intersect**, **within**, **and**, **or** the memory requirements both at compilation and at run time can rapidly grow.

In a process-based implementation, compilation may not be costly, but if such structures are part of an assertion that can create many overlapping attempts or threads then the result can be a growing number of concurrent processes and/or data structures that have to be allocated and evaluated at run time.

- Ranges `##[M:N] b` as well as `[*M:N]`, for some large `N > M`

The problem is similar to the preceding case, yet somewhat worse because of the implied nondeterminism representing a disjunction of fixed delay or repetition ranges. In this case, not only the value of N is of importance but also the span N-M+1 of the range itself. The consequences on compilation and run time performance are significant.

- Unbounded delay at the beginning of an antecedent `##[*] s |-> p`

The interpretation of this property reads as follows: Upon each occurrence of sequence `s`, property `p` must hold. If this property is used in an assertion outside an initial procedure, the assertion will evaluate the whole implication at every clock tick.

In an implementation that does not keep track of attempts and reports failures only, it may not cause problems since the multitude of evaluation threads that have equivalent next states is collapsed into one evaluation. However, the assertion evaluation may not complete in simulation if it does not fail in one of the evaluation attempts of `p`. The presence of a local variable further complicates the task of reducing the number evaluation threads, as the value of the local variable may not be the same in all threads.

The situation is quite different in process-based evaluation that does keep track of attempts. Here, if `s` and `p` are of temporal nature spanning several clock ticks, there can be a rapid accumulation of processes, each doing essentially the same thing, but needing to keep track of the attempt start times. Notice that unless `p` fails all attempts will keep evaluating till the end of simulation.

Possible solutions are:

- Place the assertion into an **`initial`** procedure thus creating only one evaluation attempt. Only the first failure will be reported.
- Remove `##[*]` from the antecedent because it is redundant for assertions that trigger at every clock tick.

Another problematic case is the occurrence of large delays in `$past`:

- Sampled value function `$past(exp, N)`, for some large `N`.

The sampled value function can be viewed as a shift register of length `N`. The first stage is loaded by the sampled value of `exp` at every clock tick. The last stage provides the value of the function `$past`. Therefore, the number of state bits is equal to `$bits(exp)* N`. This may create a large state space for formal verification (Chap. 21). For simulation, the issue is more related to the cost of updating the `N` registers.

- Using `$rose(exp, $global_clock)` or `$rising_gclk(exp)`

As the names of these functions imply, they differ by the clock tick at which they indicate that the least significant bit of `exp` changed to `1'b1`. `$rose` evaluates to true at the clock tick when the least significant bit of `exp` has risen, while the future value function `$rising_gclk` evaluates to true at the tick that precedes the tick when the signal rises, i.e., when the least significant bit of `exp` is to rise.

Future value functions are more efficient in FV because in the automata representation of the overall property, the next-state is already encoded in the automaton. If past-value functions, e.g., `$rose` are used, then the implied past-value register is created independently of the property automaton and thus adds one state bit to the overall state space. If many such functions are used, it may impact the performance of FV. For future-value functions, no such extra registers are needed. Furthermore, the future value functions often simplify the formulation of stability properties, make them more easily understandable, e.g., Example 7.31.

In simulation however, the effect on performance can be quite different. Simulation cannot know the value a signal will have at the next clock tick, it can only evaluate the present and store the past values. Therefore, if a future value function is used in a property, the compiler must shift the entire property evaluation by one global clock tick into the past. Furthermore, the reported failure times must be adjusted to values as if the evaluations were actually based on the future value functions, i.e., shifted by one clock period of the global clock. However, in general that global clock period is not known and must be computed by the simulator. This processing adds overhead in simulation.

- Ranges in **`always`** `[M:N] p` as well as **`eventually`** `[M:N] p`, for some large `N > M` (see Sect. 10.5).

As before, large upper bounds `N` and large span `N-M+1` increase the state space, and, in an attempt-based evaluation, if the assertion retriggers while previous evaluations are still in progress, the run time performance can be significantly affected.

- Trigger by level `req |-> p` or by value change `$rose(req)|-> p`

In many situations, `$rose(req)` is a more efficient form as it only triggers evaluation on a change.

As mentioned earlier, the use of sampled value functions implies additional registers. However, to avoid false firing at time 0, it may be necessary to shift the antecedent as `##1 $rose(req)` (see Sect. 7.2.1).

- Property **`and`** versus sequence **`and`** in `seq_1` **`and`** `seq_2` (see Sect. 10.2)

In an automata-based implementation, sequence **`and`** requires performing intersection of the argument sequences. Depending on the complexity of these sequences, the resulting automaton may be quite large, thus requiring more memory to represent, as well as slowing down the compilation. Therefore, if **`and`** is the top-level operator in the consequent property then it is more efficient in both simulation and formal verification to replace the sequence **`and`** with a property **`and`**. It is also likely that the verification tool, formal or simulator, does the replacement automatically.

- Properties ending with open-ended intervals such as `##[M:$] s` and **`s_eventually`** `p`.

An open-ended interval in a property implies that the tool will be searching for satisfying the arguments `s` and `p`, till the end of evaluation. If it cannot be satisfied in simulation it will run till the end. Depending on the strength of the operator, it will report a success (weak property) or failure (strong property). That is, the property cannot fail (if at all) until the end of simulation. If the evaluation of such an open-ended operator is repeatedly retriggered, this will cause accumulation of attempts and threads. In simulation, open-ended intervals should be replaced by a reasonably bounded range, while in FV, open ranges are much preferred because they add only few states to the automaton.

Exercises

19.1. Assertion `a: assert property(en ##1 !y[+] |-> x)` had attempts start at 1, 2, 3, 4 all of which succeeded at time 5, but the attempt that started at time 5 failed at time 6. What could be the problem? Is it a problem in the design or in the assertion formulation? Explain each of the cases.

19.2. Assertion `a: assert property(trig |-> s_eventually(sig));` failed at the end of simulation for several attempts. What is the reason for the failure and what could be the remedy(ies)?

19.3. The example in Chap. 17, Fig. 17.14 fails. How would you debug the failure in simulation?

19.4. If a checker like the `assert_handshake` in Chap. 24 fails, how could you approach debugging the failure (a) if you have access to the source code of the checker, and (b) if you do not have access to the source code.

19.5. Suppose that the assertion in Exercise 19.1 failed in formal verification by model checking. What means you may have to debug it?

19.6. Will property `req |-> s_eventually ack` be efficiently evaluated in simulation if it is required to provide full attempt information (start and fail/success times)? Does it depend on the protocol? That is, when `req` is a single clock tick pulse vs. when `req` should hold asserted until and including the assertion of `ack`? How would it perform in formal verification?

19.7. In the preceding problem, if `req` is to remain asserted until `ack` is asserted, is it important to you that an assertion using the preceding property reports all the start times of attempts that succeeded at the same time when `ack` is asserted? Should it report the earliest or the latest such start time only?

19.8. Suppose that `req` must remain asserted until and including `ack` is asserted, how could you modify the property in Exercise 19.6 to trigger only once for a given `req` - `ack` pair?

19.9. Can you identify other properties that may have impact on simulation performance depending on the form of evaluation and the amount of detail provided about the start and fail times of succeeding and failing attempts of the associated assertion(s)?

Part V
Formal Verification

Chapter 20
Introduction to Assertion-Based Formal Verification

The man of science has learned to believe in justification, not by faith, but by verification.

— Thomas Huxley

In this and the following chapter, we probe deeper into the principles of formal assertion-based verification: its methods of application, formal semantics of assertions, and underlying models and algorithms. In this chapter our objective is to familiarize the reader with the terminology as well as the methodologies that have proven to be indispensable for many design groups.[1]

There is a common opinion that only experts can do formal verification (FV), but nobody claims that to simulate an RTL design one has to be an expert in simulation. Indeed, it is not that difficult to run a simulator, but in some forms it is not more difficult to run an FV tool either. Therefore, even people without special expertise can carry out FV to some extent.

To run lightweight FV on an RTL block, it is only necessary to formulate adequate assumptions constraining the inputs of the block. This is the trickiest and the most effort consuming step in the verification. After that, running a formal verification tool is not very different from running a simulation.

In contrast, an exhaustive formal verification of a design is a full-time job. It requires model reduction and pruning, often writing abstract models for parts of the design, checking specification completeness, iterative refinements, and algorithm tuning. Perhaps in the future, if FV tool capacity drastically grows or FV-friendly design methodologies are developed, exhaustive FV will become automated, but currently it is not. Therefore, exhaustive verification is performed

[1]This chapter and the next discuss special questions of formal verification and may be skipped. However, this chapter explains further the nature of assertions, and it should be useful even for readers who are interested only in simulation.

E. Cerny et al., *SVA: The Power of Assertions in SystemVerilog*,
DOI 10.1007/978-3-319-07139-8_20

only for the most critical blocks, where correctness is crucial and simulation is too unreliable. Examples include multipliers, dividers, other arithmetic units, arbiters, coherency managers, branch predictors, critical controllers, bus protocols, etc.

Verifying that assertions hold on a design is the primary purpose of FV, yet checking coverage is also useful for several reasons:

- To make sure that the FV model is not overconstrained, i.e., that the assumptions are not too strong and allow meaningful model behavior.
- To assist dynamic validation. If a coverage point is proven to be unreachable in FV, there is no point in trying to construct test cases for it. The benefit of checking coverage points in FV may be very significant.
- To evaluate new FV algorithms and tools. It is difficult to evaluate a new algorithm by trying to discover assertion violations in a real mature design. Hitting tough coverage points is more meaningful because the intended design behavior is known.

20.1 Counterexample and Witness

If an assertion fails in FV, the FV tool reports a *counterexample*, often abbreviated as CEX, a sequence of input stimuli leading to the assertion failure. For convenience, the tool usually also shows the values of relevant internal signals. Suppose that a right shift operator `>>` were specified instead of the left shift operator `<<` on Line 8 of Fig. 1.7. The FV tool could produce the counterexample shown in Table 20.1. Indeed, we can see that at clock cycle 3 our assertion fails: `shift_reg` is `8'b00000000`, while `$past({shift_reg[6:0],shift_reg[7]})` is `8'b00000010`.

Note that some values in the counterexample are not important. For example, the value of `val` is important only at clock cycle 1, at other clock cycles it may assume any value without affecting the result. The FV tool may explicitly report a *don't care* (x) value in this case.

Cover statements may also be checked in FV: if a coverage point can be hit, the FV tools report a *witness*. A witness is a sequence of input stimuli leading to the coverage point hit, while satisfying all the specified assumptions. Otherwise, if the coverage point cannot be hit FV tools report that this coverage point is

Table 20.1 Counterexample for `check_shift`

Clock cycle	set	rst	val	shift_reg
0	1'b0	1'b1	8'b00000000	8'b00000000
1	1'b1	1'b0	8'b00000001	8'b00000000
2	1'b0	1'b0	8'b00000000	8'b00000001
3	1'b0	1'b0	8'b00000000	8'b00000000

Table 20.2 Witness for `cov_shift`

Clock cycle	set	rst	val	shift_reg
0	1'b0	1'b1	8'b00000000	8'b00000000
1	1'b1	1'b0	8'b10000000	8'b00000000
2	1'b0	1'b0	8'b00000000	8'b10000000
3	1'b0	1'b0	8'b00000000	8'b00000001

unreachable under the specified assumptions. It may also be impossible to hit the coverage point if the FV tool runs out of memory or allocated run time.

Consider the following cover statement for the shift register in Fig. 1.7:

```
cov_shift: cover property (@(posedge clk) disable iff (rst)
    shift_reg == 8'b10000000 && !set
    ##1 shift_reg == 8'b000000001);
```

The cover statement states that we wish to cover two consecutive clock cycles such that in the first cycle `shift_reg` has the value `8'b10000000` and `set` is inactive, and in the second cycle `shift_reg` has the value `8'b000000001`. A possible witness is shown in Table 20.2.

As with assertion counterexamples, coverage point witnesses may contain don't care values. In our example, all values of `val` are actually don't cares except for the one in clock cycle 1.

20.2 Complete and Incomplete Methods

So far we discussed *complete* FV methods: these methods report for each assertion whether it passes or fails. Of course, this is the best thing to have, but because of capacity limitations, *incomplete* FV methods are often used. These methods have three possible outcomes: *passed*, *failed*, or *unknown*. Strictly speaking, even the methods we called complete can also report an unknown status for assertions when they time out or exceed memory limitations. The difference is that the complete methods are intended to find the exact solution, while the incomplete methods may give up even when computing resources are still available. Usually incomplete FV methods are much faster than complete ones.

Typically, incomplete methods are *bounded* FV methods: given a *verification bound* n they check that there is no assertion violation with a counterexample shorter than n cycles starting from an initial state. The initial state need not be the reset state. These methods do not guarantee that an assertion is correct, but only that it cannot be violated "too soon". Bounded FV methods are widely used, and they can provide good confidence in design correctness. For example, if the design is pipelined, then the bound n equal to or a little bigger than the depth of the pipeline is usually sufficient.

20.3 Approximation

Sometimes *approximation* [18] is used in formal verification. While exact methods should always return accurate results, approximation-based methods can return an inaccurate result due to approximation error: either a *false negative*, in which case failure is reported for an assertion that should pass, or a *false positive*, in which case success is reported for an assertion that should fail. According to the error types introduced by the approximation, one distinguishes between *overapproximation*, which may introduce false negatives, and *underapproximation*, which may introduce false positives. Overapproximation is *sound*, meaning that a result of success is always accurate (i.e., free of approximation error). Underapproximation is *safe*, meaning that a result of failure is always accurate.

20.3.1 Overapproximation

Overapproximation occurs either automatically as part of the verification strategy of FV tools, in which case it is transparent to the users, or manually when a design model is abstracted. The abstract model is usually simpler than the original one and allows more behaviors. If the assertion is proven on this abstract model, it also holds on the original model. If the assertion fails on the abstract model, the counterexample may be *spurious*, i.e., impossible in the original model. Manual checking of whether counterexamples are spurious may be difficult, especially when there are many of them. Therefore, overapproximation may lead the user to ignore assertion failures, thus missing true bugs. The benefit of model reduction that enables a successful FV run has its cost in the analysis and elimination of spurious counterexamples. This is discussed in more detail below.

Overapproximation may also happen inadvertently, for example, when one or more assumptions is missing. In this case the model allows for more unintended behaviors, hence false negatives are likely to occur. This is why a validated thorough system specification is important.

Another common case of overapproximation happens when we try to formally verify only a part of a bigger model by removing some subcomponents or blocks of statements. Figure 20.1 shows a toy module generating signal `req`: `req` is asserted only when the system state is `idle`. Assertion `req_when_idle` checks that if `idle` is asserted, then in the next clock cycle `req` is asserted unless `rst` happens.

Let us assume that we wish to verify a smaller model. We manually delete the assignment statement on Line 3. What we obtained is an overapproximated model. Now the assertion will fail since `idle` may assume any value at any time.[2] This toy

[2]It is possible to argue that since `idle` is now unassigned it will keep the value `X` all the time. This is true in simulation, but FV tools usually consider undriven variables, like `idle`, as a free variable which may assume any value at any time.

```
module reqgen(input logic busy, clk, rst, output logic req);
  wire idle;
  assign idle = !busy;

  always @(posedge clk or posedge rst) begin
    if (rst) req <= 1'b0;
    else if (busy) req <= 1'b0;
    else  req <= 1'b1;
  end
  req_when_idle: assert property (
    @(posedge clk) disable iff (rst) idle |=> req);
endmodule : reqgen
```

Fig. 20.1 Request generator

example illustrates a very important problem encountered in FV: on the one hand it is desired to reduce the model to fit the capacity of FV tools, and on the other hand, it cannot be done by a naïve, mechanical deletion. The boundary around the missing model part must be carefully characterized by adding relevant assumptions. In our case, the following assumption needs to be added to define the behavior of `idle`:

```
idle_when_not_busy: assume final (idle == !busy);
```

Of course, in this toy example we just replaced an assignment by an assumption, and the abstract version can hardly be more efficient than the original one. In realistic cases, however, model abstraction may bring significant performance improvement because the benefit of the model reduction outweighs the cost of the boundary assumptions.

20.3.2 Underapproximation

Underapproximation is very commonly used, although most people do not realize when they are using it. The most common example is simulation: we check the model behavior only on a given simulation trace, while other possible traces remain unchecked. Simulation is safe: if the assertion fails in simulation, the model is definitely wrong (presuming, of course, that the assertion is written correctly and no assumption is violated). There may be false positives, however: if the assertion passes in simulation, it does not mean that the design is correct. Therefore, simulation is not sound.

Another common example of underapproximation is bounded verification: if the assertion fails within the verification bound, a bug is discovered and can be analyzed using the generated counterexample. If the bug cannot be stimulated within the bound, no assertion violation is reported and we cannot conclude anything about the validity of the assertion.

There is nothing wrong with underapproximation. It is convenient because all failures are correct and no spurious counterexamples are reported. One should, however, keep in mind that the underapproximation is not sound: if no bugs are reported the model is not necessarily correct. Of course, everybody understands this when running simulation, but it is possible to get confused with FV. Therefore, FV tools usually issue an appropriate message in case underapproximation is used, such as "the assertion has not been violated up to bound 50 clock cycles", rather than "the assertion passed".

Underapproximation happens inadvertently when we overconstrain the model by writing stronger assumptions than intended. If we add the following assumption to the original model in Fig. 20.1

```
always_busy: assume final (busy);
```

assertion `req_when_idle` will hold trivially, or *vacuously*, because `idle` is never asserted in this case. This situation is dangerous because we may believe that everything is checked, but essentially nothing is verified. Many FV tools report assertion vacuity to help spot such cases. Unfortunately, overconstraining is usually less obvious, and it cannot always be discovered by automatic tools. Therefore, it is important to validate the assumptions, as discussed in Sect. 20.5. As mentioned earlier, trying to hit coverage points also helps to discover overconstraining.

Empty Model

An extreme case of model overconstraining occurs when there are contradictory assumptions added to the original model, such as adding

```
always_busy: assume final (busy);
```

and

```
always_idle: assume final (idle);
```

to Fig. 20.1.

It is possible to think that in this case all the assertions will fail in FV, but the opposite is true—all of them will pass, vacuously. This is because the hypothesis of the FV proof is that all the assumptions are satisfied, but this hypothesis does not hold due to the contradiction in the assumptions. We have in fact created an *empty model*—the entire constrained model contains no state due to the contradiction.

An empty model may occur not only when two assumptions are mutually contradictory. There may be a larger set of contradictory assumptions in which no two of them are mutually contradictory. For example,

```
m1: assume property (@(posedge clk) a |=> c);
m2: assume property (@(posedge clk) b |=> !c);
m3: assume property (@(posedge clk) a && b);
```

Assumption `m1` states that if `a` is true then `c` will be true in the next clock cycle. Assumption `m2` states the same thing about `b` and `!c`, while assumption `m3` states

```
1 logic a;
2 always @(posedge clk) a <=!a;
3 a_stable: assume property (@(posedge clk) nexttime $stable(a));
```

Fig. 20.2 Empty model example

that a and b are always true. Therefore, overall the set of assumptions implies that c and !c should be true simultaneously! Clearly, this is an impossible situation.

An empty model may also occur when assumptions contradict the behavior of the design, as shown in Fig. 20.2. Here, the statement in Line 2 causes a to toggle every clock cycle, while assumption a_stable requires that a remain stable all the time. Therefore, it is recommended that every assumption has *some* design input or a free variable[3] in its support. Otherwise, the assumption is likely to collide with the behavior of the design.

Many FV tools have the capability to report an empty model, but usually their ability is limited to several basic cases, as exhaustive empty model discovery is very costly.

20.3.3 Pruning

Since FV tool capacity is limited, it is important to reduce the model size. We have mentioned that this operation should be done carefully. A blind removal of a part of the model will likely result in numerous false negatives. *Pruning* (see, e.g., [58]) provides a more controlled model size reduction for FV purposes.

The main pruning directives are *set* and *free*. *set* assigns a constant value (for example, 0 or 1) to a given signal, while *free* disconnects a signal from its fan-in. There is no standard SystemVerilog support for pruning directives,[4] hence FV tools usually provide some custom directives for model pruning. With these directives, the user can prune a model without actually changing the design. For instance, to verify data propagation through a queue, it may be sufficient to watch the propagation of a single bit, while all other bits may be set to zero. Furthermore, if there is some complex logic defining the behavior of a signal, and an assertion should hold for every value of this signal, there is no need to keep this complex logic. It is possible to free the signal. For example, to verify a pipeline stage that performs addition, the verification does not depend on the functionality of the preceding stages. Therefore, the logic driving the data incoming to this stage can be safely pruned.

All FV tools automatically prune most of the irrelevant parts of the model. Manual pruning is needed only for those signals that affect the checked assertion

[3]Recall that free variables can take any value of its type at any time

[4]The SystemVerilog **force** statement is the best candidate, but it is usually ignored by FV tools.

or the assumptions in some indirect way that the automatic pruning algorithms cannot determine. The verification engineer must thus have detailed knowledge of the design.

Note that the pruning directives relate to different types of approximation:

- *Setting an input signal* is underapproximation—we eliminate some model behaviors, which can lead to false positives but not to false negatives. Setting input signals is thus safe but not sound.
- *Freeing any signal* in the model is an overapproximation—it can only introduce more behaviors than allowed by the original model. Therefore, it can only introduce false negatives, but not false positives. Freeing is sound but not safe.
- *Setting an internal signal* may forbid some behaviors of the original model, but may also introduce new behaviors (for example, when in the original model a signal is toggling, and now it is set to a constant value). This kind of pruning may introduce both false positives and false negatives, and it is neither safe nor sound.

Often there are additional pruning directives provided by the tools, such as black-boxing parts of the model, but they are beyond the scope of this book.

20.4 Formal Verification Flows

We now have the necessary background to discuss formal verification flows in RTL design and verification. There are many possible scenarios; we focus on three of them:

- Exhaustive verification of model specification.
- Lightweight verification.
- Early RTL verification.

20.4.1 Exhaustive Verification of Model Specification

Figure 20.3 represents a block diagram of the typical FV flow for exhaustively verifying model compliance to its specification.

The flow starts with writing the specification for the design. The specification consists of a set of assumptions on the primary inputs and a set of assertions describing the expected behavior under these assumptions. The specification is checked first for sanity and debugging in simulation. If there is no simulation environment available, it is possible to check parts of the specification on manually created traces. Upon completion of sanity checks and debug, a formal verification tool is applied. There may be three outcomes of checking a specific assertion:

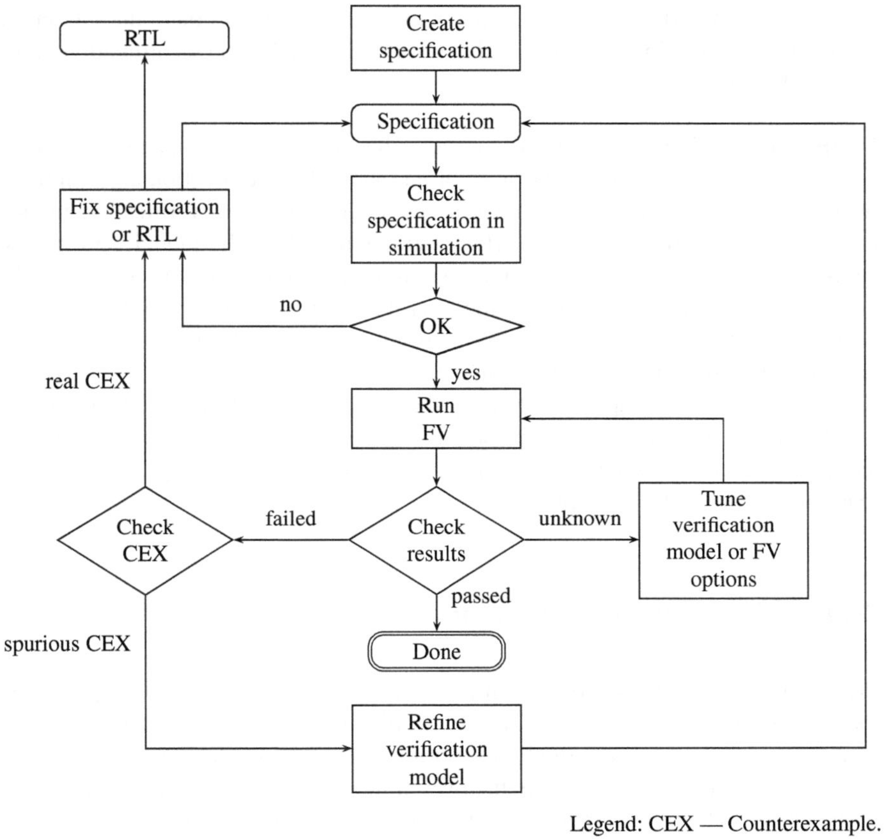

Fig. 20.3 Exhaustive formal verification flow

Success: The assertion is true for any input sequences for which all the assumptions hold. In this case, it is tempting to say that the model is correct, but the assertion may pass because of overconstraining assumptions or because the assertion itself is too weak (e.g., vacuous).

Failure: A counterexample (CEX) is generated consisting of an explicit or implicit input sequence for which all assumptions hold, but on which the assertion fails. There may be several reasons for assertion failure. For example, the assertion may be wrong. If the assertion is correct, the model may be underconstrained, e.g., the applied abstraction may be too coarse. If it is not clear whether the counterexample is real or spurious, the assertion failure should be reproduced in simulation. This may be challenging since it requires propagating the counterexample from the inputs of the FV model to the inputs of the simulation model. Usually, this task is done manually or with the support of debugging tools. A fully automatic solution for the problem of counterexample propagation is as hard as formal verification of the simulation model. If the assertion appears to be correct, the problem may lie in the design. Further probing using simulation should then lead to the identification of the source of the problem.

Unknown: The verification result is inconclusive. This can happen because of timeout, memory overflow, or because of an incomplete verification algorithm. To obtain conclusive results, it becomes necessary to refine the verification model: to use a more aggressive or a smarter abstraction, reduce the model size, or add auxiliary assertions. These auxiliary assertions are called *lemmas*: they may be easier to prove, and when proved, they can be used as assumptions to prove the original assertion. Another possibility is to tune the FV tool options—to choose a specific FV algorithm, adjust its parameters, etc.

Note that it is also important to verify the completeness of the specification, that is, whether the assertions fully represent the desired behavior and assumptions. We leave this rather sophisticated problem out of the scope of this book. The interested reader may consult for example [32].

20.4.2 Lightweight Verification

Figure 20.4 represents a block diagram of the lightweight FV flow for verifying local assertions inserted in the RTL code. Unlike in the exhaustive verification flow where the goal is to prove formally the correctness of the model, here the objective is to obtain a greater confidence in the overall model correctness and to detect bugs in the design. The lightweight verification flow is much less effort-consuming than the exhaustive verification flow.

The main purpose of this flow is bug hunting, hence only assertion failures are investigated. Failure investigation in this case is much simpler than in exhaustive verification; spurious counterexamples (CEX) are due only to missing assumptions. There is no need for model refinement or tool tuning.

Lightweight verification is usually faster than the exhaustive flow because only small parts of the RTL design affect the behavior of the local assertions.

20.4.3 Early RTL Verification

Another application of FV is early RTL verification [66]. The flow is essentially the same as in lightweight verification, but it runs in the early stages of the design using fast verification algorithms, such as bounded FV with small verification bounds. The goal of this flow is to clean up obvious bugs quickly, before the simulation environment is ready. It is well known that building a simulation environment is a complex task, and using FV in the early stages of the design allows starting verification earlier, thus reducing the time to market.

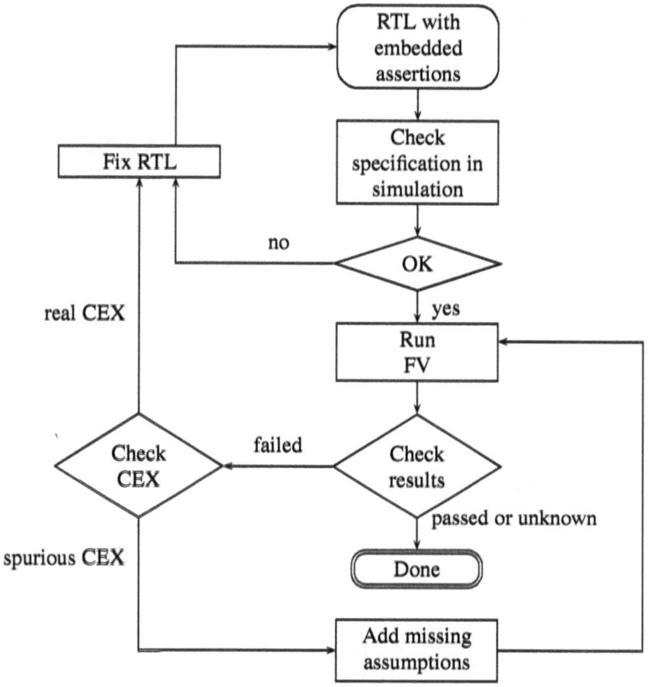

Fig. 20.4 Lightweight formal verification flow

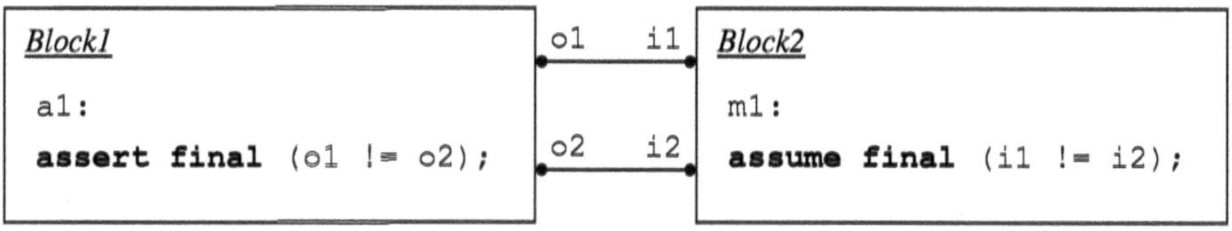

Fig. 20.5 Assume-guarantee paradigm

20.5 Assume-Guarantee Paradigm

As we have seen, the correctness and completeness of FV of a design block depends very much on the specification of assumptions. To check assumptions for the block, it is necessary to prove them as assertions on the parts of the design that drive the block, as shown in Fig. 20.5.

In this example, we verify *Block2* using assumption `m1` stating that the two inputs `i1` and `i2` are complements of each other. This assumption should be proven as an assertion `a1` on the outputs `o1` and `o2` when verifying *Block1*.

Unfortunately, it is not always possible formally to verify assumptions as assertions on another block:

- the other block may be more complex,
- the drivers of the signals participating in the assumption may belong to different blocks which then must be taken together, thus increasing the complexity of the model,
- the signals in the assumptions may be generated by an Intellectual Property (IP) block, etc.

In all these cases, the assumptions should at least be checked in simulation. Furthermore, the assumptions should also be checked in simulation of larger models in which the design is integrated.

20.6 Formal Verification Efficiency

For any FV flow, and especially for exhaustive verification, the quality of assertions is critical. If in simulation inefficient assertions increase the simulation time, in FV assertion efficiency may be a question of life and death: an FV session with an inefficient assertion may not produce a conclusive result. Of course, assertion efficiency in FV, as in simulation, is a matter of specific algorithms and tools, but there are common principles that should be understood in order to write efficient assertions.

Unfortunately, the requirements for assertion efficiency imposed by simulation and FV are often different, sometimes even contradictory. The good news is that in many cases a reasonable compromise can be found. When no compromise exists, one should go after intended assertion usage—simulation or FV. If an assertion is targeted for both modes, then efficiency in FV should be preferred. In rare critical cases the same assertion can have different implementations for simulation and for FV. These cases should be avoided whenever possible because it may be difficult to ensure assertion equivalence.

The assertion efficiency requirements in emulation are usually more aligned with FV than with simulation since both emulation and FV require assertion synthesis, while assertion simulation algorithms may be implemented in a different way that does not require synthesis. Note, however, that assertions synthesized for FV can be nondeterministic, while for emulation they must be deterministic.

20.7 Hybrid Verification

There exist also hybrid methods combining simulation and FV for checking assertions. These methods provide better coverage than simulation but are less exhaustive than FV. The methods can usually handle much bigger designs than

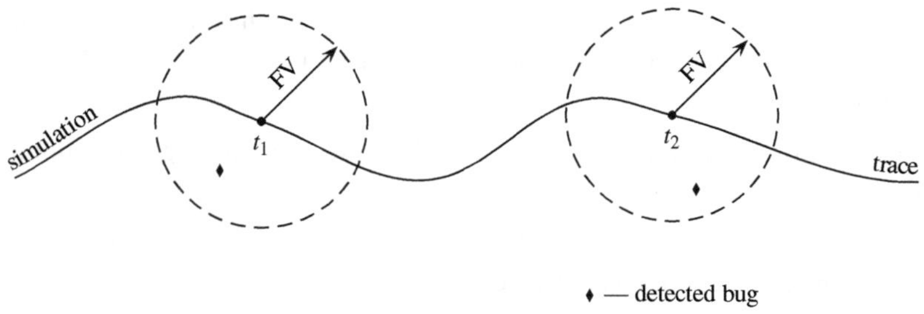

Fig. 20.6 Hybrid verification

pure FV. Actually, hybrid verification is a special case of underapproximation. There are many variations of hybrid verification methods, but usually their main idea is to interleave conventional or random simulation and FV [17,42].

The concept of hybrid verification is illustrated in Fig. 20.6. The design is simulated until simulation time t_1, and starting from the system state in time t_1 bounded FV is performed. Then the system is simulated until another time moment t_2, and bounded FV is performed again starting from the new state of the system, and so on. Figure 20.6 shows a case when no bugs have been found during the proper simulation, but two bugs are detected by bounded FV in neighborhoods of states of the simulation trace.

Hybrid verification requires that state information from the design and assertions can be mapped between the simulation and formal model. In turn this may require that the design and the assertions be synthesizable.

Exercises

20.1. What are counterexamples and witnesses? What is the purpose of reporting counterexamples and witnesses in formal verification?

20.2. Assume that in Fig. 1.7 Line 9 is omitted. Provide a counterexample exhibiting violation of `check_shift`.

20.3. Based on the shift register shown in Fig. 1.7, provide a witness for the following cover statement:

```
cover property (@(posedge clk) disable iff (rst)
  shift_reg == 8'b10000000 && ##1 8'b000000010);
```

20.4. What are complete and incomplete methods in formal verification?

20.5. What is the meaning of abstraction as used in formal verification?

20.6. What kinds of approximation do you know? When are they used? What kind of approximation can result in false positives? In false negatives? What kind of approximation is simulation?

20.7. What kind of approximation is produced because of a missing assumption? A redundant assumption?

20.8. What result will be produced by FV in the case of contradictory assumptions? In the case when an assumption contradicts the model?

20.9. What is an empty model?

20.10. What is pruning? Why is it used in FV?

20.11. One of the additional pruning methods is *black-boxing*, in which a sub-model is considered to be a black box. What kind of approximation is introduced by black-boxing?

20.12. What is the easiest way to debug specification correctness?

20.13. Why are spurious counterexamples produced by FV tools? How can one check whether a counterexample is spurious?

20.14. What should be done if the FV result is inconclusive for an assertion?

20.15. How can assumption correctness be checked?

20.16. What is hybrid verification and when could it be used?

Chapter 21
Formal Verification and Models

Hope is a great falsifier. Let good judgment keep her in check.

— Baltasar Gracian

The scope of this book does not admit detailed explanation of formal verification algorithms. Instead, we provide hints about the way formal verification is conducted and explain how assertions are interpreted in formal verification (FV). The material in this chapter is primarily useful to people who deal with formal verification. If you are interested in assertion simulation only you may skip the chapter, but we recommended to read the chapter if you want to obtain a deeper understanding of SystemVerilog assertions.

The DUT is represented as a set of states and a set of transitions between these states. In FV, all DUT transitions are synchronized by the *global clock*, which is the fastest clock in the particular part of the design hierarchy. See discussion about possible multiple global clocks within a design in Sect. 4.4.2. All other clocks are synchronized with it.

Throughout this chapter, we make the following assumptions unless otherwise stated:

- We use the following conventions: the letters `a`, `b`, and `e` designate Boolean expressions; the letters `r` and `s` designate sequences; and the letters `p` and `q` designate properties.
- We freely switch between the abstract research notation and SystemVerilog notation depending on the context. For example, when explaining theoretical background we denote the disjunction of Boolean variables as $a \vee b$. Illustrating the same formula in SystemVerilog, we write it as `a || b`. Analogously, in the research notation we use $\neg$ for negation (Boolean NOT), $\wedge$ for conjunction (Boolean AND), $\rightarrow$ for implication, and $\oplus$ for exclusive disjunction (modulo two addition, or XOR).

E. Cerny et al., *SVA: The Power of Assertions in SystemVerilog*,
DOI 10.1007/978-3-319-07139-8_21

- For temporal logic operators we may also use the research notation: G for **`always`**, F for **`s_eventually`**, and X for **`nexttime`**.
- All variables and expressions are 2-state, even if their type is explicitly specified as **`logic`**. The values `x` and `z` are interpreted as 0.
- All properties are clocked by the global clock.

21.1 Auxiliary Notions

In this section, we briefly describe logical and mathematical notions that will be used later in this chapter. The reader familiar with them may skip this section.

21.1.1 Relations

The *Cartesian product* of two sets, A and B, is the set $A \times B$ consisting of all ordered pairs (a, b) such that $a \in A$ and $b \in B$.

Example 21.1. If $A = \{x, y\}$, and $B = \{0, 1, 2\}$, then $A \times B = \{(x, 0), (x, 1), (x, 2), (y, 0), (y, 1), (y, 2)\}$. □

A Cartesian product of an arbitrary number of sets $A_1 \times A_2 \times \ldots \times A_n$ is defined as the set of all tuples $(a_1, \ldots, a_n)$, where $a_1 \in A_1, \ldots, a_n \in A_n$.

A *binary relation* R between two sets A and B is any set of ordered pairs (a, b) such that $a \in A$ and $b \in B$. In other words, R is a binary relation iff $R \subseteq A \times B$.

Example 21.2. The order $\leq$ of integer numbers $\mathbb{Z}$ is a binary relation: $\leq \subseteq \mathbb{Z} \times \mathbb{Z}$. For example, $(3, 5) \in \leq$, but $(6, 4) \notin \leq$. Of course, we are accustomed to writing $3 \leq 5$ instead of $(3, 5) \in \leq$ and $6 \not\leq 4$ instead of $(6, 4) \notin \leq$. □

It is possible to define relations of an arbitrary arity: an n-ary relation R between the sets $A_1, \ldots, A_n$ is any set of tuples $(a_1, \ldots, a_n)$ such that $a_1 \in A_1, \ldots, a_n \in A_n$. In other words, R is an n-ary relation between $A_1, \ldots, A_n$ iff $R \subseteq A_1 \times \cdots \times A_n$.

21.1.2 Logic Notation and Quantifiers

Boolean Logic: In logical formulas, we are using the following notation for Boolean operators: $\neg$ for negation (Boolean NOT), $\wedge$ for conjunction (Boolean AND), $\vee$ for disjunction (Boolean OR), $\rightarrow$ for implication, and $\oplus$ for XOR.

Quantifiers: In mathematical logic, there are two quantifiers: a *universal quantifier* $\forall$ and an *existential quantifier* $\exists$. If $P(x)$ is a formula dependent on some variable x then $\forall x\ P(x)$ is true iff $P(x)$ is true for all values of x. $\exists x\ P(x)$ is true iff $P(x)$ is true for some value of x. Of course, the result of the quantification depends on the variable domain.

Example 21.3. Suppose that the domain of x is the set of integers. The formula $\forall x \exists y\ x = 2y$ is true if the domain of y is the set of real or rational numbers, but it is false if the domain of y is the set of integers or natural numbers. □

21.1.3 Languages

We call a finite set $\Sigma = \{\sigma_1, \ldots, \sigma_k\}$ an *alphabet*, and its elements $\sigma_1, \ldots, \sigma_k$ *letters*. Any sequence of letters is called a *word*, or a *trace*, and we use these terms interchangeably. If the word does not contain any letters, it is called the *empty word* and is denoted ε. We distinguish between *finite* and *infinite* words (traces).

Any set L of words is called a *language*. If all the words of the language are finite, the language is called *finitary*, if all the words of the language are infinite, the language is called *infinitary*.

Example 21.4. The words of written English form a finitary language according to our definition. Its alphabet consists of 26 Latin letters "a" through "z",[1] and for every sequence of Latin letters we can say whether it is an English word or not. For example, *building* is an English word, whereas *buildign* is not. □

21.1.4 Finite Automaton

A *finite automaton* $\mathscr{A}$ is a tuple $\langle \Sigma, S, S_0, \rho, F \rangle$, where Σ is an alphabet, $S = \{s_1, \ldots, s_n\}$ is a finite set of *states*, $S_0 \subseteq S$ is the set of initial states, $\rho \subseteq S \times \Sigma \times S$ is the *transition relation*, and $F \subseteq S$ is the set of the *accepting states*.

It is convenient to represent a finite automaton as a directed graph in which vertices are automaton states and edges are labeled with the alphabet letters to represent the transition relation. If $s_i, s_j \in S$ and $\sigma \in \Sigma$, then there is a labeled edge $s_i \xrightarrow{\sigma} s_j$ in the graph iff $(s_i, \sigma, s_j) \in \rho$. We mark the initial states with a double incoming arrow and the final states with a double circle.

[1]For our purpose, there is no need to distinguish between small and capital letters. We also ignore the fact that there exist words with spaces, hyphens, etc.

Fig. 21.1 Finite automaton

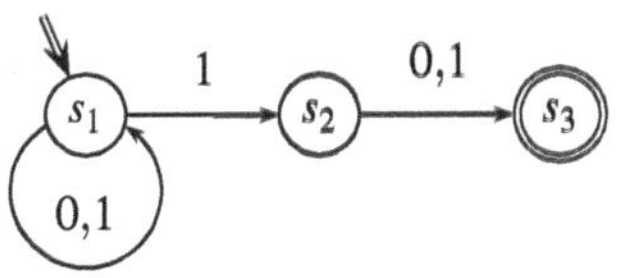

Example 21.5. The alphabet of the automaton depicted in Fig. 21.1 consists of two letters: 0 and 1. The automaton has three states: s_1, s_2, and s_3. There is one initial state, s_1, and one final state, s_3. The transition relation ρ consists of the following triples: $(s_1, 0, s_1)$, $(s_1, 1, s_1)$, $(s_1, 1, s_2)$, $(s_2, 0, s_3)$, and $(s_2, 1, s_3)$.

This automaton is nondeterministic in the following sense: from state s_1 on letter 1, the automaton can transition either to s_1 or to s_2. □

A finite automaton $\mathscr{A}$ *accepts a word* w on Σ iff there is a path from one of its initial states to one of its final states such that the successive transitions are labeled by the consecutive letters from w. The set of the words accepted by the automaton $\mathscr{A}$ forms the language $L(\mathscr{A})$, called the *language of the automaton* $\mathscr{A}$.

Example 21.6. The automaton $\mathscr{A}$ defined in Example 21.5 accepts all words over the alphabet $\{0, 1\}$ that have length of at least two and 1 as the penultimate letter. □

21.2 Formal Verification Model

The DUT in FV is represented as a *formal verification model*, also known as a *Kripke structure*.

The formal verification model M is a tuple $\langle Q, I, V, R\rangle$, where Q is a set of states,[2] $I \subseteq Q$ is the set of initial states, V is a finite set of Boolean variables, and $R \subseteq Q \times Q$ is the transition relation. We also assume that the transition relation R is *total*, i.e., for any $q \in Q$ there exists $q' \in Q$ such that $(q, q') \in R$. In other words, the relation is such that from any state there is at least one transition (possibly to the same state).[3]

Each state is characterized by the set of variables that are true in it. If two different states have exactly the same set of variables true in them, then these states may be merged into one. R contains all state pairs such that it is possible to transition from the first state of the pair to the second one.

[2]The definition does not require having a finite number of states, but we assume that their number is finite since this is true for any RTL model.

[3]In general, one includes in the definition of a Kripke structure a *labeling function* $L : Q \to 2^V$, which defines which Boolean variables are true in each state. In our case we just identify Q with 2^V, assuming that each state corresponds to some specific valuation of the variables.

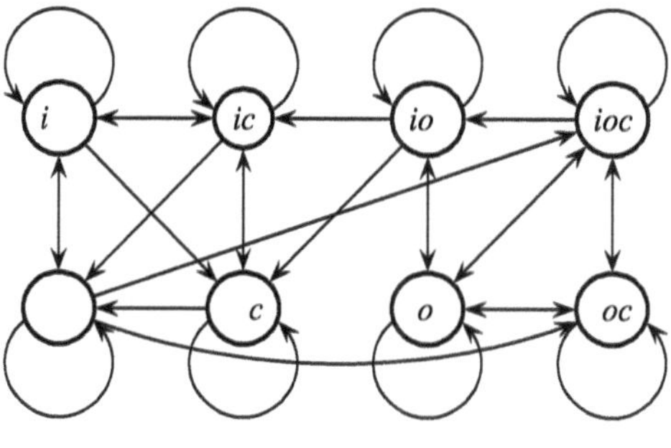

Fig. 21.2 Formal verification model corresponding to module m

Example 21.7. Consider the module m defined by

```
module m(input logic i, c, output o);
  wire a = !i;
  always @(posedge c)
    o <= a;
endmodule : m
```

This module can be represented as an FV model. It seems at first that V should consist of the four variables i, a, c, and o. But it is easy to see that the value of a is uniquely determined by the value of i. Therefore, the variable a is redundant. This situation can be generalized:

Only state variables and primary inputs should be included in the variable set V of an FV model. The signals that are Boolean functions of other signals should not be included in the variable set.

For the FV model of m, we thus have $V = \{i, c, o\}$ and Q consists of the eight states $\emptyset, \{i\}, \{c\}, \{o\}, \dots, \{i, c, o\}$. The transitions of this model are depicted in Fig. 21.2.

Note that the input signals i and c may have any initial values, and that the initial value of o is unknown. In simulation, this unknown value is designated as x, while in the FV model the unknown value just means that both initial values 0 and 1 for o are possible. All the eight combinations of these variables may occur at system initialization. It follows that all the states of this model are also initial states, i.e., $I = Q$.

We now explain the transitions on several examples. Note that the output o can change only when c changes from 0 to 1. Therefore, if no input changes then neither does the output change. Consequently, from any state there is a transition to the same state in Fig. 21.2, a self-loop. The input i may change independently of c and o. Therefore, transitions for all values of i are possible. This explains why there are vertical bidirectional edges in Fig. 21.2. When c changes from 0 to 1, the next value of o must be equal to the negation of the current value of i. This yields the transitions $\emptyset \to \{o, c\}$, $\emptyset \to \{i, o, c\}$, $\{i\} \to \{c\}$, and $\{i\} \to \{i, c\}$. We leave the explanation of the other transitions as an exercise to the reader. □

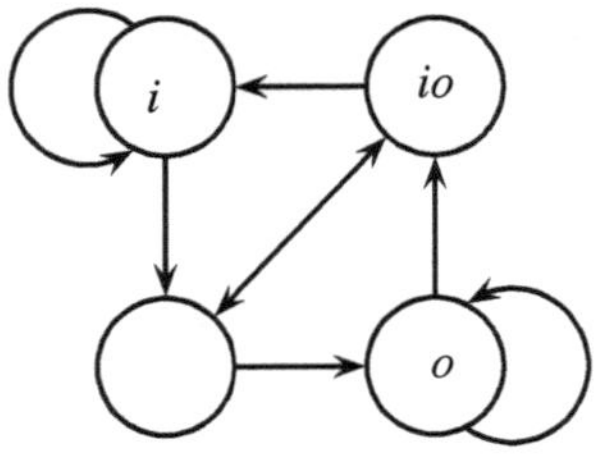

Fig. 21.3 Formal verification model explicitly clocked by `$global_clock`

21.2.1 Time

To complete the picture, we need to introduce the notion of time. We assume that RTL models are *synchronous*, i.e., that there is a *global clock*, called also the *primary system clock* that synchronizes all the system transitions. Any signal change may happen only at a tick of the global clock. This assumption is applicable even if the system has several clock domains. For example, if there are two clock domains controlled by clocks `clk1` and `clk2`, then the global clock should be defined as an event `clk1 or clk2`. In SystemVerilog, the global clock is introduced with **global clocking**, and it can be referenced as `$global_clock` (Sect. 4.4.2). The mapping of the global clock to events using **global clocking** is important in simulation, but in FV this mapping is not necessary. In this chapter, we refer to the global clock even if **global clocking** has not been defined in the design.

One might wonder why ticks of the global clock are not identified with simulation ticks. Sometimes this is a good idea, but if all interesting signals are synchronized by a relatively slow clock, then this mapping is too inefficient. Another reason may be that we need to map the system clock to some clock in simulation to ignore transitions that happen between ticks of the global clock.

The time in FV of RTL is discrete, and it is defined in terms of the ticks of the global clock. Time 0 corresponds to the initial tick of the global clock, time 1 corresponds to the next one, etc. It is also assumed that the global clock never stops ticking. Therefore, unlike in simulation, the time in FV is infinite.

All transitions in formal verification models are synchronized by `$global_clock`.

Example 21.8. The model in Example 21.7 is controlled by an arbitrary clock **posedge** `c`. If **posedge** `c` is the global clock, as in the code snippet below, then the FV model can be simplified: $V = \{i, o\}$, $Q = \{\emptyset, \{i\}, \{o\}, \{i, o\}\}$, $I = Q$. The transitions of this model are depicted in Fig. 21.3:

```
module m(input logic i, c, output o);
  wire a = !i;
  global clocking @(posedge c); endclocking
```

```
  always @($global_clock)
      o <= a;
endmodule : m
```

This model is much simpler than that in Example 21.7. While the simulation model remains exactly the same, c is no longer a variable of the FV model. To understand the transition diagram, consider the state $\{i\}$ as an example. In this state, $i = 1$ and $o = 0$. In the next tick of the system clock o must be 0, while i may assume any value. Thus, we have two transitions: $\{i\} \to \emptyset$, and $\{i\} \to \{i\}$. □

21.2.2 Model Language

We can interpret each state of an FV model as a letter in the alphabet $\Sigma = 2^V$. The notation 2^V is used to designate the set of all subsets of the set V. For instance, the alphabet of the FV model from Example 21.8 is $\Sigma = \{\emptyset, \{i\}, \{o\}, \{i, o\}\}$.[4]

According to its transition relation, the FV model accepts some sequences, or *paths*, of its states, while it prohibits others. For instance, the model from Example 21.8 accepts the sequence of states $\{i\}, \{i\}, \ldots$ because it has the transition $\{i\} \to \{i\}$", whereas the model forbids the sequence $\emptyset, \emptyset, \ldots$ because it does not have the transition $\emptyset \to \emptyset$. Therefore, an FV model defines an infinitary language (see Sect. 21.1.3) over Σ consisting of all the paths it accepts. The path $q_0, q_1, \ldots$ is accepted by the FV model if it starts in an initial state: $q_0 \in I$, and each pair of consecutive states q_i and q_{i+1} is connected by the transition relation: $(q_i, q_{i+1}) \in R$. These paths are exactly the words of the model language $L(M)$. According to the terminology introduced in Sect. 21.1.3, the words are also called traces, and they are traces as also understood by hardware engineers. Indeed, a trace may be considered as a dump of all variable values at each tick of the global clock. The main difference between these traces and conventional simulation traces is that the simulation traces are finite, whereas the FV model defines infinite traces.

21.2.3 Symbolic Representation

In the above examples, we built explicit representations of FV models: each state and each state transition appeared separately. This approach is feasible only when the number of variables is very small. Even a modest design containing 300 state elements (latches and flip-flops) may have more states than there are atoms in the universe! To address this problem, a symbolic state representation is used.

[4]We distinguish between the alphabet Σ and the set of states Q since in the general Q does not need to contain all combinations of variables, whereas Σ does.

It is possible to represent each state as a Boolean function that has the value 1 in this state and 0 in all other states. Such a function identifying a single state from 2^V can be written as a conjunction of literals, one for each variable: if a variable is false in the state then the corresponding literal is the negation of the variable; otherwise the corresponding literal is the variable itself. A conjunction of this form is called a *minterm* over V. It is easy to see that the minterms over V are in one-to-one correspondence with the elements of 2^V.

Example 21.9. The states $\emptyset, \{i\}, \{o\}$, and $\{i, o\}$ from Example 21.8 may be represented by the minterms $\neg i \wedge \neg o, i \wedge \neg o, \neg i \wedge o$, and $i \wedge o$, respectively. □

Boolean functions may also be used to represent sets of states. Given a set $S \subseteq 2^V$, each element of S can be represented symbolically by its corresponding minterm, and the Boolean function representing S itself is then just the disjunction of these minterms. This function returns the value 1 on the elements of S and the value 0 on elements not in S. For this reason, the function is called the *characteristic function* of the set S, and it is denoted as χ_S.

Example 21.10. The set of states $S = \{\{o\}, \{i, o\}\}$ from Example 21.8 has the characteristic function $\chi_S = (\neg i \wedge o) \vee (i \wedge o) = o$. The empty set has the characteristic function $\chi_\emptyset = 0$, while the set of all states has the characteristic function $\chi_Q = 1$. □

The same principle may be applied to symbolically represent transition relations with their characteristic functions. Recall that the transition relation R is a binary relation, that is, a set of pairs (sect. 21.1.1): $R \subseteq Q \times Q$. To build the characteristic function of a pair, it is necessary to distinguish between the variables describing the first state of the pair and those describing the second state. This is achieved by duplicating the set of variables for the second set and distinguishing the variables with a prime. We call the original variables *current state variables* and the primed variables *next state variables*. We also expand this notation to variable expressions: if e is an expression built from the current state variables, then e' is the expression obtained from e by replacing each current state variable by the corresponding next state variable.

If a pair $p \in R$, then $p = (c(p), n(p))$. $c(p)$ is the current state of p, and $n(p)$ is the next state of p. Let $\chi_{c(p)}$ and $\chi_{n(p)}$ be the characteristic functions of $c(p)$ and $n(p)$, respectively. Using our convention of primes for next state variables, the characteristic function of p is $\chi_{c(p)} \wedge \chi'_{n(p)}$ and the characteristic function of the transition relation R as a whole is

$$\bigvee_{p \in R} \chi_{c(p)} \wedge \chi'_{n(p)}.$$

Example 21.11. The symbolic representation of the transition relation R from Example 21.7, Fig. 21.3 is

$$\begin{aligned}
&\neg i \wedge \neg o \wedge \neg i' \wedge \ \ o' \vee \\
&\neg i \wedge \neg o \wedge \ \ i' \wedge \ \ o' \vee \\
&\ \ i \wedge \neg o \wedge \ \ i' \wedge \neg o' \vee \\
&\ \ i \wedge \neg o \wedge \neg i' \wedge \neg o' \vee \\
&\neg i \wedge \ \ o \wedge \neg i' \wedge \ \ o' \vee \\
&\neg i \wedge \ \ o \wedge \ \ i' \wedge \ \ o' \vee \\
&\ \ i \wedge \ \ o \wedge \ \ i' \wedge \neg o' \vee \\
&\ \ i \wedge \ \ o \wedge \neg i' \wedge \neg o' =
\end{aligned}$$

$$\begin{aligned}
&\neg i \wedge \neg o \wedge \ \ o' \vee \\
&\ \ i \wedge \neg o \wedge \neg o' \vee \\
&\neg i \wedge \ \ o \wedge \ \ o' \vee \\
&\ \ i \wedge \ \ o \wedge \neg o' =
\end{aligned}$$

$$(\neg i \wedge o') \vee (i \wedge \neg o') =$$

$$i \oplus o'$$

To understand how this symbolic representation is obtained, consider the state $\{i\}$ as an example. From Fig. 21.3, there are two transitions from $\{i\}$: $\{i\} \rightarrow \{i\}$, and $\{i\} \rightarrow \emptyset$. To symbolically represent the transition $\{i\} \rightarrow \{i\}$, we encode $\{i\}$ from the source of the transition with the current state variables and $\{i\}$ from the target of the transition with the next state variables, and then form a conjunction. This yields $i \wedge \neg o \wedge i' \wedge \neg o'$. Similarly, the symbolic encoding of the transition $\{i\} \rightarrow \emptyset$ is $i \wedge \neg o \wedge \neg i' \wedge \neg o'$. The set of these two transitions is represented as a disjunction $(i \wedge \neg o \wedge i' \wedge \neg o') \vee (i \wedge \neg o \wedge \neg i' \wedge \neg o')$. Although the further transformations simplifying the characteristic function of the transition relation are routine, it is instructive to understand an interpretation of these transformations. For example, $i \wedge \neg o \wedge i' \wedge \neg o' \vee i \wedge \neg o \wedge \neg i' \wedge \neg o' = i \wedge \neg o \wedge \neg o'$ means that all the transitions from state $\{i\}$ lead to states where o is false. This agrees with the behavior of module `m`: if the current value of i is high then the next value of o is low.

The series of simplifications yields a compact formula for the entire transition relation: $i \oplus o'$. This result expresses the essence of the whole model: the next value of o is the negation of the current value of i. □

21.2.3.1 Sampled Value Functions

The next state v' of a variable $v \in V$ may be interpreted as the value of v in the next tick of the global clock, and therefore in SystemVerilog v' corresponds to `$future_gclk(v)`. Using global clocking future sampled value functions (Sect. 7.2.2.2), it is possible to explicitly express a transition relation in SystemVerilog.

Example 21.12. The transition relation $i \oplus o'$ of the module `m` from Example 21.8 may be represented in SystemVerilog as `i != $future_gclk(o)`, and the module `m` is equivalent to the following assumption `m1`

```
m1: assume property (@$global_clock i != $future_gclk(o));
```

This equivalence should be understood in the following sense: the set of infinite traces (relative to the global clock) consistent with the module `m` is identical to the set of the traces satisfying the assumption `m1`. □

The global clocking future sampled value functions do not require any additional modeling. They are already built into the FV model. On the contrary, the past sampled value functions do require additional modeling. For example, the function `$past(a,,,@(posedge clk))`, is represented as follows:

```
type(a) pa;
always @(posedge clk) pa <= a;
```

As we have seen, each past value of a one-bit expression effectively adds a new variable to the FV model, and thus *doubles* the total number of the model states!

Efficiency Tip. In FV, future value functions are more efficient than their past value counterparts.[5]

Note, however, that using future value functions in simulation may have the opposite effect. It may increase the simulation load. Therefore, when writing assertions keep in mind the target verification tool.

21.3 Properties

In Sect. 21.2, we saw that an FV model M defines an infinitary language $L(M)$, the set of infinite traces it accepts. Each property p also defines an infinitary language $L(p)$, the set of infinite traces that satisfy it.

Example 21.13. If $V = \{a, b, c\}$, the property $p =$ **always** a defines a language $L(p) = \{a^\omega\}$, that is, a set of all infinite traces $aa\ldots$. The notation a^ω means repetition of a infinitely many times.

The alphabet of this language is $\Sigma = 2^V$, the set of all subsets of the set $\{a, b, c\}$. $aa\ldots$ is therefore not a single trace, but a family of traces, as a is a symbolic representation of the set of the letters $\{\{a\}, \{a, b\}, \{a, c\}, \{a, b, c\}\}$. As a disjunction

[5]Some FV tools represent the transition relations not between current and next variables, but between past and current variables. However, this approach is inconsistent with the SystemVerilog semantics requiring the past value of a bit variable be 0 at the initial moment.

of minterms, $a = (a \wedge \neg b \wedge \neg c) \vee (a \wedge b \wedge \neg c) \vee (a \wedge \neg b \wedge c) \vee (a \wedge b \wedge c)$. The property p requires a to be true at each position of the trace, whereas the variables b and c may have arbitrary values. □

We say that M *satisfies* the property p, or that M is a *model* for p, or that p is *valid* on M, and write $M \models p$, iff all traces that M accepts also satisfy p.

Example 21.14. Let p be defined as the property `always i |=> !o`. The FV model M corresponding to the module `m` from Example 21.8 satisfies p, i.e., $M \models p$, since every trace accepted by M satisfies p. □

There are two important special cases of properties: **true** and **false** (corresponding to SystemVerilog `1'b1` and `1'b0`, respectively). The property **true** is satisfied on any trace. Therefore, its language consists of all possible traces defined by a given set of variables. The property **false** is not satisfied on any trace. Therefore, its language is empty.

In Chap. 4, we introduced SystemVerilog assertion statements: assertions, assumptions, and cover. From the formal point of view, all these statements are properties used in different contexts. Below we discuss their formal meaning.

21.3.1 Asserts

When we check the validity of a property, it plays the role of an *assert*, as explained in Chap. 4. From the definition of $M \models p$, it follows that the property is valid iff the language defined by this property contains the language defined by the model, i.e., iff $L(M) \subseteq L(p)$. For any model $M \models$ **true**, and $M \not\models$ **false**.

21.3.2 Assumes

As a property defines its own infinitary language, it is possible to check validity of an assertion relative to this property instead of checking its validity on a model. We say that the property p is valid assuming the property q, and write $q \models p$, iff all traces satisfying q also satisfy p. When the validity of a property is assumed, it plays the role of an *assumption*, as explained in Chap. 4. Note that for any assertion p, **false** $\models p$ since $\emptyset \subseteq L(p)$.

$q \models p$ is equivalent to $L(q) \subseteq L(p)$. As we have seen in Example 21.8, an assumption may replace a model in the sense that it defines the same language (i.e., set of traces) as the model. More formally, the FV model $M = \langle V, Q, I, R \rangle$ is equivalent to the following assumption:

```
initial assume property (I and always R);
```

If all states of the model are initial, i.e., $I = Q$, then this assumption is just an invariant represented by the transition relation:

```
assume property (R);
```

An assumption may be composed with a model, which is written as $M \,||\, q$. This composition defines the set of traces that are common for both M and q. In other words, $L(M \,||\, q) = L(M) \cap L(q)$. The assertion p is valid on the composition $M \,||\, q$, written as $M \,||\, q \models p$, iff $L(M \,||\, q) \subseteq L(p)$.[6]

Example 21.15. The assertion `always nexttime o` is valid on the composition of the module `m` from Example 21.8 and the assumption `always !i`; □

If assumption q and model M are contradictory, that is, there is no common trace satisfying them, then $L(M) \cap L(q) = \emptyset$. In this case, $M \,||\, q \models p$ for any assertion p. In this situation, the composite model $M \,||\, q$ is *empty* (Sect. 20.3.2). An empty composite model also results when several assumptions are contradictory. It is a common mistake to think that formal verification fails when a model is empty, in fact the opposite is true:

> If a model is empty then formal verification declares success for any assertion.

Example 21.16. Adding the assumption `always i |=> o` to Example 21.8, both assertions `always o |=> !o` and `always i != o` hold (as do any other assertions) because the model is empty. □

FV tools may report when an contradictory set of assumptions is detected. If not, deploying coverage on some key properties of the design can provide useful information.

21.3.3 Coverage

If a property p is used as an assertion, we verify its validity on a model M, that is, *every* trace accepted by M satisfies p. If we replace the requirement of validity by the requirement of satisfiability, we obtain the notion of *coverage* (Sect. 4.7).

We say that a property p is *satisfiable*, or that p is *covered*, on M if *there exists* a trace accepted by M that satisfies p. In other words, p is satisfiable, or covered, on M iff $L(M) \cap L(p) \neq \emptyset$. The notions of satisfiability and of validity are dual: p is satisfiable on M iff $M \not\models \neg p$.

---

[6]In this case $M \,||\, q$ represents the so-called parallel composition of M and q and not the Verilog Boolean OR.

It follows that a contradictory model will make cover properties unsatisfiable, indicating to the user that something is not correct in the model if the expectation is that the cover should be reachable.

Example 21.17. Property `s_eventually o` is satisfiable on model M from Example 21.8 because the trace $\neg i, \ldots$ satisfies p (in this trace the value of i at time 0 is 0, and thus the value of o at time 1 is 1). The same property is not valid on M because it is not satisfied on any trace of the form i^{ω}, which are legal traces of M. The failure of the assertion on i^{ω} indicates that some assumption on the behavior of the input i is needed, such that i may not stay constant 1 from some point in time on. □

21.3.4 Constraining a Model with Assumptions

If assertion p holds on model M then it also holds on M composed with assumption q. This is because the composition $M \,||\, q$ may not accept more traces than model M alone does.

The situation with coverage is the opposite: if property p is covered on a composition $M \,||\, q$, then it is also covered on model M alone, since M accepts all traces that $M||q$ does.

Informally speaking, adding assumptions increases the chances of a property to become valid, but decreases its chances of being covered. This does not necessarily mean that adding assumptions makes FV easier for assertions and more difficult for coverage goals. For example, adding assumptions may make the formulas used in FV more complicated, and thus make the work of FV tools much harder. See Sect. 20.3 for a discussion on approximation and abstraction.

21.4 Safety and Liveness

Compare the following two assertions `a1` and `a2`:

```
a1: assert property (always a);
a2: assert property (s_eventually a);
```

Can these assertions fail in simulation? Assertion `a1` certainly can if in some global clock tick the value of `a` is 0. We can even say that if this assertion is violated on some infinite trace then there is a finite prefix of this trace where this assertion is violated, namely, the trace fragment until (including) the first occurrence of `a == 0`. But what about assertion `a2`? We can simulate this assertion during millions of ticks and never see `a` taking on the value 1. Does it mean that assertion `a2` is violated in simulation? Obviously not: had we simulated `a2` a few more ticks we might discover that `a` holds. Assertion `a2` cannot fail in simulation except at the end as discussed in Chap. 10.

We can formulate our findings in terms of the counterexamples: every counterexample of `a1` has a finite prefix such that all its extensions are counterexamples. On the contrary, in the case of `a2`, every finite trace has an extension that is not a counterexample. Informally speaking, all the counterexamples of the assertion `a1` are finite, whereas all the counterexamples of the assertion `a2` are infinite. In fact all the counterexample of the assertion `a2` have the form $w = \neg a^{\omega}$.

This leads us to the following property classification. Property p is *safety* if each counterexample has a finite prefix, and all its extensions are counterexamples. Such a prefix is a *bad prefix*. The property p is *(pure) liveness* if every finite trace has an extension that is not a counterexample. Speaking informally, we can say that safety properties check that something bad never happens, whereas liveness properties check that something good eventually happens. Safety properties may fail in simulation, whereas pure liveness properties cannot. Only FV can fully check liveness properties as FV can deal with infinite traces.[7] The vast majority of all properties used in practice are safety.

Are there properties that are neither safety nor liveness? Yes, such properties exist. As an example, consider the property `a` **`s_until`** `b`. The trace $w_1 = (\neg a \wedge \neg b)b \ldots$ has a bad prefix $(\neg a \wedge \neg b)$; all the extensions are counterexamples, while the trace $w_2 = (a \wedge \neg b)^{\omega}$ provides an infinite counterexample, such that each finite prefix has an extension that is not a counterexample. Indeed, on w_1 there is no `a` before the first occurrence of `b`, and on w_2 `b` never happens. Such properties are called *hybrid* or *general liveness* properties. These properties have both a liveness and a safety component.

For our purpose, the distinction between the pure and general liveness properties is not that important, and we will reserve the term *liveness* for both pure and general liveness properties. We explicitly use the terms "pure" or "general" when we need to distinguish between these two types of liveness.

The notions of safety and liveness are fundamental in FV. The users of FV tools should be able to determine for any property whether it is safety or liveness, and try to use safety properties as much as possible. This is necessary for FV to be efficient, as checking safety is generally more efficient than checking liveness. To help the user write efficient properties, some FV tools report the type of each property, safety, or liveness.

21.4.1 Safety Properties

Algorithms for formal verification of assertions are called *model checking*. A systematic description of model checking is beyond the scope of this book, and we

---

[7]Some liveness properties may pass in simulation, like property **`s_eventually`** `a` if `a` is observed to hold at one point in time. However, there are liveness properties that can neither fail nor pass in simulation, like the property **`s_eventually always`** `a`.

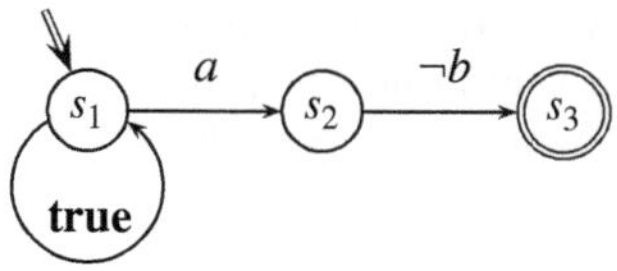

Fig. 21.4 Automaton for complement of property **always** a |=> b

discuss here only a simple special case to give an idea of how an FV tool works for safety properties. Some basic understanding will give insight into assertion efficiency.

The basic idea of model checking of safety properties is simple. All states of the FV model are partitioned into good and bad. Entering a bad state signifies a failure of the property. Model checking of a safety property can thus be formulated as a reachability problem: is there a path from one of the initial states of the FV model to one of its bad states? If such a path exists, the property fails, if not, it holds. This path, if it exists, represents a bad prefix of the property, or, informally speaking, its finite counterexample.

Property **always** a provides the simplest example. All states of the FV model containing a are good, while those not satisfying a are bad. The good states are those in the set $\{v \in V | \mathtt{a} \in v\}$, while the bad states are those in the set $\{v \in V | \mathtt{a} \notin v\}$.

An arbitrary safety property may be reduced to a property of this form. The common algorithm is to build a finite nondeterministic automaton $\mathscr{A}$ (Sect. 21.1.4) corresponding to the complement of the safety property. The language of this automaton consists of bad prefixes of the safety properties, or, in other words, this automaton recognizes bad prefixes of the safety property[8]. The accepting states of $\mathscr{A}$ are called the *bad states*. It can be shown [63] that it is always possible to construct $\mathscr{A}$ with a single bad state. The automaton $\mathscr{A}$ may be synthesized into RTL [14] such that each automaton state becomes a new RTL variable. This results in an augmented FV model M' containing both variables of the original model M and the variables corresponding to the states of $\mathscr{A}$. If the bad state is represented by variable b then the original safety property is equivalent to the property **always** b of the augmented model.

The exact description of building the automaton of the complement of a safety property is beyond the scope of this book. Instead, we illustrate the idea on the following examples.

Example 21.18. Build an automaton for the complement of the property **always** a |=> b.

Solution: The automaton for the complement is shown in Fig. 21.4.

Discussion: The automaton in this example is nondeterministic. It remains in the initial state s_1 until it follows one of the evaluation attempts of a |=> b. The attempt

[8]It is not required that the language $L(\mathscr{A})$ of this automaton coincide with *all* bad prefixes of the safety property, it is sufficient if $L(\mathscr{A})$ contains *some* of its bad prefixes [46].

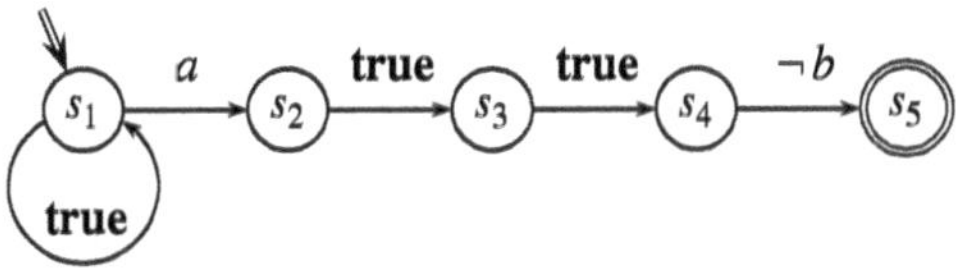

Fig. 21.5 Automaton for complement of property `always a |=> ##2 b`

may start whenever. It will be non-vacuous only if `a` holds when it starts. Then, the automaton moves to state s_2. If in the next tick of the global clock `b` does not hold, the automaton moves to state s_3, which is the bad state of the automaton. □

Example 21.19. Synthesize the automaton from Fig. 21.4 (Example 21.18) into RTL.

Solution:

```
logic s1 = 1'b1, s2 = 1'b0, s3 = 1'b0;
always @$global_clock begin
  s1 <= s1;
  s2 <= s1 && a;
  s3 <= s2 && !b;
end
```

Discussion: This code implies that `s1` always equals to `1'b1` and therefore its computation is redundant. The optimized code is:

```
logic s2 = 1'b0, s3 = 1'b0;
always @$global_clock begin
  s2 <= a;
  s3 <= s2 && !b;
end
```

`s3 == 1` is a bad state: if `s3` becomes `1'b1`, the assertion fails . □

The fact that a safety property may be synthesized into RTL reflects the well-known practice of implementing assertions in RTL as an instrumented code. This practice was common before the assertion specification languages became widely accepted (Chap. 1).

Example 21.20. Build an automaton for the complement of the property `always a |=> ##2 b`.

Solution: Its automaton is depicted in Fig. 21.5. □

As we can see from the comparison of Examples 21.18 and 21.20, the more complex the property is and the bigger its bounded time windows are, the more states its automaton has. Interestingly, infinite time windows do not introduce many additional states as illustrated in the following example.

Example 21.21. Build an automaton of the complement of the property `always a[+] ##1 b |=> c`.

Fig. 21.6 Automaton for complement of property **always** a[+] ##1 b |=> c

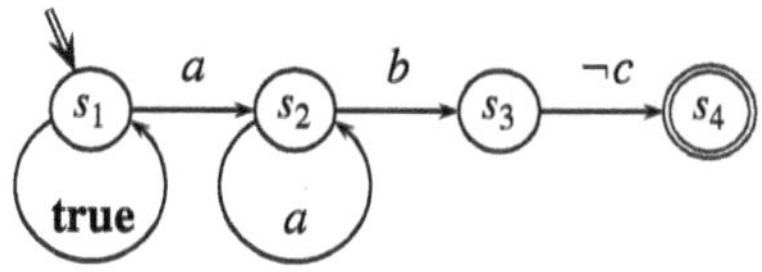

Solution: See Fig. 21.6.

Discussion: This automaton contains only 4 states. Compare it with the automaton for the complement of the property a[*3] ##1 b |=> c (Exercise 21.2). □

Efficiency Tip. The general trend is as follows: The more complex the automaton for a property, the less efficient it is in FV.

How can a safety property be recognized? In the case the property does not contain negations, the property is safety if it uses the following operators only:

- weak sequence
- suffix implications
- **nexttime**
- **always**
- **until**, **until_with**
- Boolean connectives **and** and **or**.

If the above operators are in the scope of a negation (or of an odd number of negations) the resulting property is usually not safety. However, operators such as **strong** sequences, **s_eventually**, **s_until**, and **s_until_with** become safety properties *when negated*.

Example 21.22. Property **nexttime** (a **until_with** b) is a safety property because it consists of the operators **nexttime** and **until_with** that are not in the scope of any negation.

Property **not strong** (a[*] ##1 b) is a safety property since the strong sequence is negated.

Property **strong** (a[*] ##1 b) **implies weak** (c ##[+] d) is a safety property because it contains a strong sequence under negation in the antecedent, and a weak sequence without negation in the consequent. Recall that A **implies** B is equivalent to **not** A **or** B.

Property **not** (**not weak**(a[*] ##1 b) **or not weak**(c ##[+] d)) is a safety property because both weak sequences are in the scope of two negations. □

Note that most FV tools recognize safety and liveness properties syntactically. For example, even though a **until** b is by definition the same thing as (a **s_until** b) **or always** a, it is likely that most FV tools report the former property as safety, while the latter as liveness (see [33, 46]).

21.4.2 Liveness Properties

Model checking of liveness properties is significantly more complex than model checking of safety properties. Liveness properties cannot be verified using direct reachability analysis. This is because for safety properties any path to a bad state is finite, and all counterexamples found this way are necessarily finite, while liveness properties have infinite counterexamples. Checking liveness properties is much less efficient than checking safety properties because the tool must search (explicitly or implicitly) for infinite cycles in the model that do not satisfy the property.

Efficiency Tip. Avoid writing liveness properties unless they are absolutely necessary.

Example 21.23. To check that some condition `c` holds between two events `ev1` and `ev2` implemented as Boolean expressions, we can write the following property: `always ev1 |-> c s_until_with ev2`. Before we proceed further, we should ask ourselves whether this property is one that we really need to check. It verifies two things:

1. `c` holds between `ev1` and `ev2`, and
2. Each time that `ev1` holds, `ev2` also *eventually* holds.

Do we really want to check the second condition which is liveness? Usually, the answer is "no". If not, we should rewrite the property as `always ev1 |-> c until_with ev2`. This property is safety, and it only checks for the first condition.[9] □

Example 21.24. To check that each request is granted, we can write the following property: `always req |=> s_eventually gnt`. This property is liveness, and in this case liveness is what we actually want. Although this property is expensive, we are prepared to pay for it. □

21.4.2.1 Why Write Liveness Properties?

Why should we write liveness properties if they are expensive? There may be several reasons for doing so. One common reason is to check that there is no starvation in the system. For example, if a CPU needs to access some resource, this resource should be eventually available.

Another common reason to use liveness properties is specification abstraction. For example, we might need to ensure that each request is eventually granted. The following liveness property is suitable for this purpose:

```
always req |-> s_eventually gnt
```

[9]By a chance the operator **`s_until_with`** has a clumsier syntax in SVA than the operator **`until_with`**: this was done to make the safety property **`until_with`** a natural choice.

Of course, in a real system there is always an upper bound on the request service time, but if the system implementation changes or if the upper bound is large, then a liveness property can be more appropriate.

For example, suppose we know that the request service time is bounded by 600 clock ticks, and also that the system implementation may be modified in which case the upper bound may grow to 700. To make it a safety property, we could have written the property as

```
always req |-> ##[1:701] gnt
```

However, we are not really interested in the exact time bound, and the efficiency of this bounded safety property in FV is very poor. We thus have a tradeoff between the safety property, with an automaton comprising more than 700 states and with about twice as many edges, and the liveness property, with only a couple of states. Checking the liveness property is likely to be more efficient in this case.

21.4.2.2 Counterexamples for Liveness Properties

There is an important question about liveness property checking: How can infinite counterexamples be found and reported? Obviously, tools and humans can deal with finite representations only. Fortunately, it can be shown [29] that if a property has an infinite counterexample, then it also has a lasso-shaped counterexample, that is, a counterexample of the form $w_1 w_2^{\omega}$. Here, w_1 and w_2 are finite words, w_1 is a prefix, and w_2 is an infinitely repeated part. For instance, $(a \wedge b)ba(ab)^{\omega}$ has the form $w_1 w_2^{\omega}$ for $w_1 = (a \wedge b)ba$ and $w_2 = ab$. The lasso-shaped counterexample stems from the fact that the model has a finite number of states, hence it must in a bounded number of transitions enter some previously visited state thus forming the lasso loop.

21.4.2.3 Assumptions and Liveness

So far we have considered property classification into safety and liveness from the point of view of assertions. It turns out that even the simplest assumptions introduce liveness into the system.

Example 21.25. Given assertion `a1` and assumption `m1` not in the scope of an **`initial`** procedure, they are equivalent to the assertion `a2`

```
m1: assume property (a);
a1: assert property (b);
initial a2: assert property ((always a) implies (always b));
```

as explained in Example 22.1.

For conciseness, we switch now to the notation of the research literature, where operator **`always`** is designated as G, and operator **`s_eventually`** as F. Using this notation, the property corresponding to the assertion `a2` may be written as $Ga \to Gb$. This property is equivalent to $(\neg Ga) \vee Gb$, which is, in its turn, equivalent to $(F\neg a) \vee Gb$. The latter form explicitly contains an eventuality, and shows that `a2` is a general liveness assertion. Hence, the combination of `m1` and `a1` introduces liveness condition into the system. □

Example 21.25 shows that all assumptions of the form **`always`** `q`, that is, the vast majority of all assumptions, introduce liveness.[10] To understand why, consider some assertion p and an assumption of the form Ga stating that a is an invariant of the system. Assume, for simplicity, that p is a safety property. Suppose that at some time i assertion p fails on trace w. In other words, the trace prefix $w^{0,i}$ is a counterexample of p: $w^{0,i} \not\models p$. We are ignoring the assumption Ga for a moment. Assume now that the assumption Ga holds on the trace prefix $w^{0,i}$, that is, a holds at each position of w from 0 to i (inclusive). Does it mean that $w^{0,i}$ is a real counterexample of p? Strictly speaking, no: if a = **false** at some point in time $j > i$, then assumption Ga does not hold, and therefore assertion p holds (vacuously).

In practice, invariant assumptions are always used, which means that the model checking of safety properties is not relevant at all—if we take assumptions into account, all our properties become liveness! In spite of this, the FV tools usually ignore the liveness component introduced by assumptions when checking safety properties and only check that until the time of an assertion failure no assumptions have been violated. Some tools even check that the assumptions are not violated a few more global clock ticks after the assertion violation. Strictly speaking, this is incorrect, but it is a reasonable compromise. Also, this method is safe: if the assertion passes then the subsequent behavior of an assumption is not important.

21.4.2.4 Automata of Clocked Properties

In Sect. 21.4.1, we showed that safety properties may be represented as finite automata. How will the property automaton change if we take the property's clock into account?

Consider property `@c a |=> b`. We assume that occurrence of clocking event `@c` has been resolved to testing of a Boolean expression `c` with respect to the global clock (see Sect. 20.4). The automaton for the property complement is shown in Fig. 21.7 (the corresponding unclocked version is shown in Fig. 21.4).

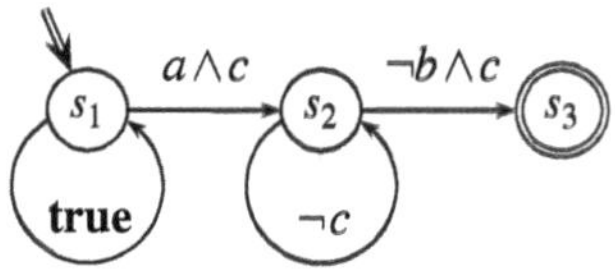

Fig. 21.7 Automaton for complement of property `@c` **`always`** `a |=> b`

[10] We are not talking about corner cases such as when the same property is used both as an assumption and as an assertion.

The automaton in Fig. 21.7 differs from the automaton in Fig. 21.4 in several points:

- There is a self-loop in the state s_2 labeled with $\neg c$, which reflects the automaton waiting for the clock c to transition to the state s_3.
- There is an additional guard c on the edges $s_1 \rightarrow s_2$ and $s_2 \rightarrow s_3$ allowing the state transitions only at a clock tick.

The automaton reaches its accepting state s_3 iff the property fails. For the property to fail, the clock must tick enough times to allow the transition from the initial state to the accepting state.

It is quite natural that the automaton of the complement of the clocked property is more complex than that of the unclocked property, and therefore FV of clocked properties is more expensive than FV of the unclocked ones. The main penalty of the verification of the clocked properties is due to the fact that their clock is considered to be an arbitrary signal. In practice, however, the waveform of the clock is often well known, and can be expressed through the system clock in a regular manner. For example, the clock may tick each second tick of the global clock. In such a case, FV of clocked properties can be made much more efficient. Therefore, many FV tools ask for a clock pattern. Some of them may also derive this information from the **global clocking** statement and RTL. However, some FV tools may ignore the **global clocking** statement.

21.5 Weak and Strong Operators

Temporal operators behave differently relative to their clock: some of them require their clock to tick enough times to witness success, whereas others just require no evidence of failure while the clock is ticking. The operators belonging to the first group are called *strong*, and the operators from the second group are called *weak*.

The weak operators considered so far are **weak**, `|->`, `|=>`, **nexttime**, **always**, **until**, and **until_with**. The operators **strong**, **s_nexttime**, **s_eventually**, **s_until**, and **s_until_with** are strong. The suffix conjunction operators `#-#` and `#=#` are also strong. All strong operators denoted by keywords have a special mnemonics: their names begin with the prefix **s_**, except for the operator **strong**. The weak operators, when they are not in the scope of a negation, yield only safety properties.

Example 21.26. Consider the following properties:

1. `@clk s_nexttime a`
2. `@$global_clock s_nexttime a`
3. `@clk a s_until b`
4. `@$global_clock a s_until b`
5. `@clk (a s_until b)or always a.`

Property (1) is a liveness property, as it checks that `clk` ticks at least twice. Property (2) is a safety property even though it has a strong operator. As the global clock is fair[11] according to its definition, the operators **`nexttime`** and **`s_nexttime`** mean the same thing when they are controlled by the global clock.

Property (3) is a liveness property as it checks that `b` eventually happens and that `clk` ticks enough times to witness the occurrence of `b`. Property (4) is also liveness, even though it is controlled by the global clock, as it checks that `b` eventually happens. Property (5) is a safety property since it is equivalent to `a` **`until`** `b`. However, it is likely that most FV tools will not recognize it as safety, as it contains a **`s_until`** operator, and it looks syntactically like a general liveness property. □

Most property operators have both weak and strong versions, such as **`until`** and **`s_until`**. However, the unbounded operator **`always`** has only a weak form, and the unbounded operator **`s_eventually`** has only a strong form. When we say that eventually `a` happens, our intent is that `a` happens in some clock tick, and therefore the clock cannot stop ticking before `a` has been detected. When we say that `a` always happens, our intent is that `a` happens at each clock tick. If there are no clock ticks, the value of `a` is not checked. Therefore, in this case there is no requirement that the clock ticks.

Suffix Implication

Why is suffix implication a weak operator? Suffix implication requires its consequent to be true for each match of its antecedent. There is no requirement that the antecedent match, and therefore there is no requirement that the clock ticks enough times to witness a match of the antecedent.

Negation

Negation reverses the strength of an operator.

Example 21.27. Property `@clk` **`not weak`**`(a)` is equivalent to `@clk` **`strong`**`(!a)`. Indeed, according to the clock rewriting rules, explained in Sect. 20.4, `@clk` **`not weak`**`(a)` is equivalent to **`not weak`**`(@clk a)`. The latter property holds iff property **`weak`**`(@clk a)` fails. This property fails iff either `clk` does not tick at least once or `a` is false at the first clock tick `clk`. This is in its turn equivalent to `@clk` **`strong`**`(!a)`. □

[11] I.e., the clock never stops.

Example 21.28. Property `not (a until b)` is strong. It checks that the clock ticks enough times to witness that `a until b` does not hold. Property `not (a s_until b)` is weak. If the clock stops ticking early, property `a s_until b` will fail, and its negation will pass. □

Since property implication `p implies q` is equivalent to `(not p) or q`, the strength of the antecedent is reversed and the strength of the consequent is preserved. Therefore, for the implication to be weak, property `p` should be strong and property `q` should be weak.

Example 21.29. Consider the following assertions:

```
a1: assert property (@clk !a);
a2: assert property (@clk not a);
initial a3: assert property (@clk !a);
initial a4: assert property (@clk not a);
```

Assertions `a1` and `a2` are equivalent. Property `@clk not a` is equivalent to property `@clk strong(!a)` (Example 21.27). In the context of assertion `a2`, properties `@clk strong(!a)` and `@clk weak(!a)` are equivalent as we check them only at all the ticks of `clk` due to the implicit top-level `always`.

Assertions `a3` and `a4` are not equivalent: `a3` is satisfied if `clk` never ticks, whereas assertion `a4` fails in this case. □

> The implicit always in continuously monitored assertions causes the clock to be treated weakly at the top-level.

Operator Composition

We know that strong operators require their clock to tick enough times, while there is no such restriction in weak operators. To interpret the semantics of a property containing both weak and strong operators, one should refer to the formal semantics of these operators and to their clock rewriting rules.

Example 21.30. Which requirements are imposed on the clock in property `@clk always s_eventually p`? Property `@clk always s_eventually p` holds if in each tick of `clk` property `@clk s_eventually p` holds. This means that if there is an ith tick of `clk`, then there must exist a jth tick of `clk`, $j \geq i$, at which property `p` holds. To summarize, property `@clk always s_eventually p` is satisfied if one of the following conditions holds:

- Clock `clk` does not tick at all.
- Clock `clk` ticks finitely many times and `p` holds when the clock ticks for the last time.

- Clock `clk` ticks infinitely many times, and `p` holds in infinitely many ticks of the clock.

□

Sometimes mixing weak and strong operators in the same property is unavoidable, as in the case of **always s_eventually** p, or **s_eventually always** p. However, when there is no special reason to mix weak and strong operators, it is more efficient and more intuitive to use operators of the same strength in the property. For example, instead of p **until s_nexttime** q, use either p **until nexttime** q or p **s_until s_nexttime** q, depending on whether or not it is important to prove that q eventually happens.

Do not mix weak and strong operators in the same property unless it is unavoidable.

21.6 Embedded Assertions

Thus far we have considered the formal semantics of stand-alone assertions (i.e., concurrent assertions placed outside procedural code) and properties. In this section, we discuss how embedding concurrent assertions in procedural code modifies their formal semantics. The simulation semantics of embedded assertions is described in Chap. 14. In FV, the support of embedded assertions is restricted, but it covers most cases of importance.

As explained in Chap. 5, if an assertion is in the scope of an **initial** procedure, it is monitored only once, and if it is in the scope of an **always** procedure, it is monitored continuously. In other words, if an assertion or an assumption is embedded in an **always** procedure, an outermost **always** property operator is implicitly added to this assertion.[12]

If an assertion or an assumption is placed in the scope of one or several procedural conditional statements, such as **if** or **case**, in the FV model an implicit suffix implication is added to this assertion with the overall condition in the antecedent.[12]

Example 21.31. Consider the following code fragment:

```
always @(posedge clk) begin : b1
  if (en) begin : b2
    // ...
    if (cond) begin : b3
      // ...
```

[12]See Chap. 18 for the semantics of embedded coverage statements.

```
    end : b3
    else begin : b4
      // ...
      a1: assert property (p);
    end : b4
    // ...
  end : b2
  // ...
end : b1
```

Assertion a1 is equivalent to the following stand-alone assertion:

```
assert property (@(posedge clk) en && !cond |-> p);
```

This equivalent assertion has an additional suffix implication operator |-> with the antecedent en && !cond. The condition en && !cond is the entry condition of the block b4 to which the assertion a4 belongs. This is quite natural since we expect the assertion attempt to be triggered only when the block b4 is activated. □

As we know, assertions may also be placed in the scope of a looping statement. To be tractable for FV, the looping statement must be statically unrollable, and the assertion is replicated according to the iterations.

Example 21.32. Consider the following code fragment:

```
property p(int i, j);
  a[i] |-> ##[1:2] b[j];
endproperty : p

always @(posedge clk) begin : b1
  // ...
  for (int i = 0; i < 5; i++) begin : b2
    // ...
    for (int j = 0; j < i; j++) begin : b3
      // ...
      a1: assert property (p(i, j));
    end : b3:
  end : b2
end : b1
```

The assertion a1 is equivalent to the following 10 stand-alone assertions:

```
assert property (@(posedge clk) p(1, 0));
assert property (@(posedge clk) p(2, 0));
...
assert property (@(posedge clk) p(4, 3));
```

Note that the loops b2 and b3 are statically unrollable in spite of the fact that the upper bound of the loop b3 is not a constant. □

21.7 Immediate and Deferred Assertions

So far in this chapter we have dealt only with concurrent assertions. Immediate and deferred assertions (both observed and final) are also meaningful for FV when they are instantiated in synthesizable code and when the data they reference are synthesizable. For FV, there is no difference between immediate and deferred assertions (observed and final). Therefore, we limit our discussion to deferred assertions.

For FV, a possible implementation of deferred assertions is to transform them into (possibly embedded) concurrent assertions using the following scheme: assertion `assert final (e);` is transformed into assertion `assert property (@clk e);` for some clock `clk`. Assumptions and cover statements are transformed similarly. It remains to define the clock for the transformed assertions.

If the deferred assertion is not in the scope of an **always** procedure such that the clock is inferable from it for concurrent assertions, the global clock is inferred. The global clock is inferred even if there is no **global clocking** definition in the design: recall that **global clocking** is only required to map the global clock into an event for simulation, and in FV models the global clock is inherently defined .

Example 21.33. Consider the following code fragment:

```
a1: assert #0 (e1);

always_comb begin
  // ...
  if (en) begin
    // ...
    a2: assert #0 (e2);
  end else // ...
end
```

For FV, this code is rewritten using concurrent assertions as follows:

```
a1: assert property (@$global_clock e1);

always_comb begin
  // ...
  if (en) begin
    // ...
    a2: assert property (@$global_clock e2);
  end else // ...
end
```

□

If the deferred assertion is in the scope of an **always** procedure with an inferable clock then this deferred assertion is treated in FV as a concurrent assertion.

Example 21.34. Consider the following code fragment:

```
always @(posedge clk) begin
  // ...
  if (en) begin
    // ...
    a1: assert #0 ($onehot0({a, b});
  end
end
```

In FV, the deferred assertion `a1` is interpreted as a concurrent assertion:

```
always @(posedge clk) begin
  // ...
  if (en) begin
    // ...
    a1: assert property ($onehot0({a, b});
  end
end
```

□

Exercises

21.1. Write a module implementing a counter modulo 4 and build the corresponding FV model. Write a symbolic representation of its transition relation.

21.2. Build an automaton for the complement of the property `always a[*3] ##1 b |=> c`. Compare it with the automaton for the complement of the property `always a[+] ##1 b |=> c`

21.3. What do the following properties mean?

(1) `@clk s_eventually always p`.
(2) `@clk s_nexttime nexttime p`.
(3) `@clk nexttime s_nexttime p`.
(4) `@clk p until s_nexttime q`.
(5) `@clk p s_until nexttime q`.

21.4. Rewrite the following properties without using negation:

(a) `@clk not weak(a ##1 b)`
(b) `@clk not strong(s[->2])`.

21.5. What is the meaning of the property `@clk p s_until_with always q`?

21.6. What is the meaning of the embedded assertion `a1`?

```
always @(posedge clk) begin
  // ...
  for (int i = 0; i < 4; i++) begin
    // ...
    if (en[i]) begin
      // ...
      a1: assert property (a[i][*2]);
    end
  end
end
```

21.7. Suppose that assertion `a1` in Exercise 21.6 is changed to **`assert final (a[i])`**. What could be its possible implementation in FV?

Chapter 22
Formal Semantics

I don't want to get bogged down in semantics causing problems.

—Pervez Musharraf

To enable formal verification of assertions, the assertions must be formally defined: it must be possible to unambiguously tell whether a specific DUT behavior satisfies a given assertion or not. The formal definition of the meaning of an assertion is called its *formal semantics*. Our intuition works well for simple assertions, but to understand the exact meaning of complex assertions requires knowledge of the formal semantics. This chapter is dedicated to the description of the formal semantics of sequences, properties, and assertions.

This chapter can be skipped on the first reading, but those who wish to obtain deep insight into SystemVerilog assertions need to carefully study it. Some of the aspects of this chapter are useful primarily to people who deal with formal verification, but others are important also for simulation. Throughout this chapter, unless otherwise specified, we make the same notational assumptions as in Chap. 21. We also continue to assume that variables and expressions are 2-state and that global clocking applies to all properties unless otherwise specified.

22.1 Formal Semantics of Properties

We have seen that a property defines its own language, we now need to describe this language precisely for every property that can be expressed in SVA. The formal definition of the language is called the *formal semantics* of the property. Understanding the formal semantics of each SVA property is important to understanding the exact meaning of each temporal formula. In Chaps. 5 and 10 we described the semantics of the properties informally. In this chapter, we provide the formal description.

E. Cerny et al., *SVA: The Power of Assertions in SystemVerilog*,
DOI 10.1007/978-3-319-07139-8_22

We use the following notation:

- w is a (finite or infinite) word, or trace, over alphabet $\Sigma = 2^V$.
- Each letter in the word w is numbered by a nonnegative integer number; the ith letter is denoted as w^i. The numbering starts from 0, so that the first letter is w^0.
- If w is a finite word, its length is denoted as $|w|$.
- The empty word is denoted as ε. The empty word does not contain any letters, and its length is 0.
- $w^{i..}$ is the suffix of the word w starting from the letter w^i. More precisely, $w^{i..}$ is the word obtained from w by deleting its first i letters. If $|w| \leq i$ then $w^{i..} = \varepsilon$.
- $w^{i,j}$, where $i \leq j$ is the finite segment of the word w starting from the letter w^i and ending at the letter w^j. More precisely, $w^{i,j}$ is the finite word obtained from the word w by deleting its i first letters and also deleting all its letters after the $j + 1^{\text{st}}$ (i.e., after the letter j).

We also use symbolic notation for the letters. For example, assuming that $V = \{a, b\}$, then $w^0 = a \wedge \neg b$ means that in the first position of the word (trace) a is 1 and b is 0.

To denote that the word w satisfies the property p, we write $w \models p$. In the context of property satisfaction, w is assumed to be infinite. Finally, we assume that all the properties in this section are unclocked (or controlled by the global clock).

22.1.1 Basic Property Forms

We consider first basic SVA property operators: Boolean property, negation, conjunction, **nexttime**, and **s_until**. Their formal semantics is defined recursively, starting from the Boolean property.

22.1.1.1 Boolean Property

The Boolean property e, where e is a Boolean expression over variables in V is satisfied on all words such that e holds on their first letter:

$$w \models e \text{ iff } w^0 \to e.$$

$w^0 \to e$ means that if the values of the variables of V specified in w^0 are substituted into e, then the result is true. Equivalently, $w^0 \to e$ means that this Boolean formula is a tautology over valuations of the variables in V. In this case, we also write $w^0 \models e$.

For example, if $w^0 = a \wedge \neg b$ then $w \models a$, and $w \models a \vee b$, but $w \not\models a \wedge b$.

In what follows, we assume that we know the formal semantics of each subproperty, and thus we define the semantics of compound properties recursively in terms of the semantics of their components.

22.1.1.2 Negation Property

The property **not** p holds on w iff the property p does not hold on w:

$$w \models \texttt{not}\ p \text{ iff } w \not\models p.$$

In the research literature, **not** p is usually denoted as $\neg p$. We continue the discussion on the formal semantics of negation in Sect. 22.3.4.

22.1.1.3 Conjunction Property

The property p **and** q holds on w iff both properties p and q hold on w:

$$w \models p\ \texttt{and}\ q \text{ iff } w \models p \text{ and } w \models q.$$

In the research literature, p **and** q is usually denoted as $p \wedge q$.

22.1.1.4 Nexttime Property

The property **nexttime** p holds on w iff the property p holds on $w^{1..}$:

$$w \models \texttt{nexttime}\ p \text{ iff } w^{1..} \models p$$

In the research literature, **nexttime** p is usually denoted either as Xp or as $\bigcirc p$. Derived forms of this property are described in Sect. 10.5.

22.1.1.5 Strong Until Property

The property p **s_until** q holds on w iff there exists $i \geq 0$ such that p holds in all positions of w up until, but not including, position i, and q holds in position i:

$$w \models p\ \texttt{s\_until}\ q \text{ iff there exists } i \text{ so that } w^{i} \models q \text{ and for every } 0 \leq j < i,\ w^{j..} \models p.$$

In the research literature, p **s_until** q is usually denoted[1] as pUq.

[1] In the research literature, the operator **s_until** is called simply *until*, and the operator **until** is called *weak until*.

22.1.2 Derived Properties

The rest of the SVA property operators that do not operate on sequences are derived. They can be expressed through the basic operators defined in Sect. 22.1.1. In this section, we provide the list of these operators, except for the operators **`if`** and **`case`** because their derivation was explained in Sect. 10.2.

22.1.2.1 Boolean Connectives

In addition to **`and`**, SVA defines the Boolean property connectives **`or`**, **`implies`**, and **`iff`**.

Disjunction Property: $w \models p$ **`or`** q iff either $w \models p$ or $w \models q$. p **`or`** q is a shortcut notation for **`not`**`(`**`not`** p **`and not`** q`)`. In the research literature, p **`or`** q is usually denoted as $p \vee q$.

Implication Property: $w \models p$ **`implies`** q iff either $w \not\models p$ or $w \models q$. p **`implies`** q is a shortcut notation for **`not`**$(p$ **`or not`** $q)$. In the research literature, p **`implies`** q is usually denoted as $p \rightarrow q$.

Given an assertion p, an assumption q, and a model M, it is possible to rewrite the relation $M||q \models p$[2] using an implication property as $M \models q \rightarrow p$. Indeed, the first formula means that all words satisfying both M and q also satisfy p, and the second formula means that all words satisfying M either satisfy p or do not satisfy q.

Example 22.1. Consider the following assertion statements:

```
m1: assume property (a);
a1: assert property (b);
a2: assert property (a implies b);
```

Assertion `a1` together with assumption `m1` are *not* equivalent to assertion `a2` if they are not in the scope of an **`initial`** procedure. The reason is that assertions and assumptions out of scope of an **`initial`** procedure are interpreted as having an implicit outermost **`always`** operator. At module level, for example, `a1` together with `m1` is equivalent to the following:

```
initial a3: assert property ((always a) implies (always b));
```

Assertion `a3` is weaker than `a2`. □

Equivalence Property: $w \models p$ **`iff`** q iff either both $w \models p$ and $w \models q$, or $w \not\models p$ and $w \not\models q$. p **`iff`** q is a shortcut notation for $(p$ **`implies`** $q)$ **`and`** $(q$ **`implies`** $p)$. In the research literature, p **`iff`** q is usually denoted as $p \leftrightarrow q$ or as $p \equiv q$.

[2] Recall that in this context || represents parallel composition of the two models.

22.1.2.2 Eventually Property

The property **s_eventually** p holds on w iff the property p holds on $w^{i..}$ for some $i \geq 0$. **s_eventually** p is a shortcut for **true s_until** p. Indeed, since the property **true** holds on any trace, **true s_until** p only checks that p eventually holds. In the research literature, **s_eventually** p is usually denoted as Fp or as $\Diamond p$. Bounded forms are described in Sect. 10.5.

22.1.2.3 Always Property

The property **always** p holds on w iff the property p holds on $w^{i..}$ for every $i \geq 0$. **always** p is a shortcut for **not s_eventually not** p. Indeed, p always holds iff the fact that **not** p holds at some time is false. In the research literature, **always** p is usually denoted as Gp or as $\Box p$. Bounded forms are described in Sect. 10.5.

22.1.2.4 Until Properties

The property p **until** q holds on w iff either (1) there exists $i \geq 0$ such that p holds in all positions of w up until, but not including, position i, and q holds in position i, or (2) p holds in all positions of w. p **until** q is a shortcut for (p **s_until** q) **or always** p. The remaining two properties from **until** family are also simple shortcuts:

- p **until_with** q is a shortcut for p **until** (p **and** q).
- p **s_until_with** q is a shortcut for p **s_until** (p **and** q).

The property **always** p can also be directly expressed through **until** as p **until false**. Since **false** never holds, p must always hold.

22.2 Formal Semantics of Sequences

Like a property, a sequence defines a language, namely, the set of words (traces) that match the sequence. Unlike the language of a property, the language of a sequence is finitary.

Example 22.2. Given the set of variables $V = \{a, b\}$, the sequence `a[*2] ##1 b` defines the language $\{aab\}$ over the alphabet $\Sigma = 2^V$. aab is a shortcut for the following traces: $\{a\}\{a\}\{b\}$, $\{ab\}\{a\}\{b\}$, $\{a\}\{ab\}\{b\}$, $\{a\}\{ba\}\{ab\}$, $\{ab\}\{ab\}\{b\}$, $\{ab\}\{ba\}\{ab\}$, $\{ba\}\{ab\}\{ab\}$, $\{ab\}\{ab\}\{ab\}$. □

We say that sequence s is *tightly satisfied* on word w, and write $w \models\equiv s$, iff w matches s. In Chaps. 6 and 11, we have informally defined a match for each type of sequence. Here, we provide the formal semantics of sequence match. As in the case of properties, the formal semantics of sequences is defined recursively from the base case of a Boolean sequence. We describe here only the formal semantics of the basic sequence forms. The derived sequence forms have been defined as shortcuts in Chaps. 6 and 11.

In the context of sequence tight satisfaction, we assume that w is a finite word. By Σ^*, we understand the language consisting from all finite words over the alphabet Σ.

Boolean Sequence: The Boolean sequence e is tightly satisfied on w iff $|w| = 1$ and e is true on w:

$$w \models\equiv e \text{ iff } (|w| = 1) \text{ and } (w^0 \models e).$$

Concatenation: The concatenation r `##1` s of the sequences r and s is tightly satisfied on the word w iff it is possible to break w into two words x and y such that r is tightly satisfied on x and s is tightly satisfied on y:

$$w \models\equiv r \texttt{ ##1 } s \text{ iff there exist } x, y \text{ so that } w = xy, \text{ and } x \models\equiv r \text{ and } y \models\equiv s.$$

Fusion: The fusion r `##0` s of the sequences r and s is tightly satisfied on the word w iff it is possible to break w into three words x, y, and z, where the size of y is 1, such that r is tightly satisfied on xy and s is tightly satisfied on yz:

$w \models\equiv r$ `##0` s iff there exist x, y, z so that $w = xyz$ and $|y| = 1$, and $xy \models\equiv r$ and $yz \models\equiv s$.

Disjunction: The disjunction r **`or`** s of the sequences r and s is tightly satisfied on the word w iff either sequence is tightly satisfied on w:

$$w \models\equiv r \textbf{ or } s \text{ iff } (w \models\equiv r) \text{ or } (w \models\equiv s).$$

Intersection:

The intersection `r` **`intersect`** `s` of the sequences r and s is tightly satisfied on the word w iff both sequences are tightly satisfied on w:

$$w \models\equiv r \textbf{ intersect } s \text{ iff } (w \models\equiv r) \text{ and } (w \models\equiv s).$$

Empty Sequence: The empty sequence s `[*0]` is tightly satisfied on the word w iff w is empty:

$$w \models\equiv s\texttt{[*0]} \text{ iff } |w| = 0.$$

Iteration: The sequence s [+] (also written as s[*1:$]) is tightly satisfied on the word w iff it is possible to break w into one or more words so that each of them tightly satisfies s:

$w \models s$ [+] iff there exist words $w_1, w_2, \ldots, w_j$ $(j \geq 1)$ so that $w = w_1 w_2 \ldots w_j$ and for every i so that $0 < i \leq j$ $w_i \models s$.

First Match: The first match **first_match**(s) of the sequence s is tightly satisfied on the word w iff s is tightly satisfied on some prefix x of w then the suffix y of w must be empty:

$w \models$ **first_match**(s) iff $w \models s$ and there exist x, y so that $w = xy$ and $\bar{x} \models s$ then $y = \varepsilon$ ($\bar{x}$ is explained in 22.3.2).

22.3 Formal Semantics: Sequences and Properties

In Sect. 22.1, we defined how to build up the semantics of properties recursively, starting from Boolean properties. However, from Chap. 6 we know that properties are built on top of sequences and that the sequential property serves as the basic building block for other properties. In this section, we define the formal semantics of the properties built on top of sequences.

22.3.1 Strong Sequential Property

$w \models$ **strong**(s) iff there exists $i \geq 0$ so that $w^{0,i} \models s$.

Note the condition $i \geq 0$—tight satisfaction on the empty word does not make the strong sequential property hold. In fact, the LRM [8] states that a sequential property admitting empty match is illegal.

Example 22.3. **strong**(e[*]) is illegal according to the LRM [8] because $e[*]$ admits empty match. Even so, consider applying the rule above for its semantics on the trace $w = \neg e^{\omega}$ where at each position e evaluates to 0. The only match of $e[*]$ on a prefix of w is empty, but the rule above requires nonempty match for property satisfaction. □

22.3.2 Extension of Alphabet

Before we proceed to the definition of the formal semantics of weak sequential properties, we need to revisit the alphabet definition. In Sect. 21.2.2, we defined the model alphabet as the set of all variable valuations: $\Sigma = 2^V$. For the purposes of the weak sequence definition discussed in Sect. 22.3.3, and of the reset definition discussed in Chap. 13, we need to extend the alphabet Σ with two special letters:

$\top$ and $\bot$. The letter $\top$ satisfies every Boolean expression, even **false**, and the letter $\bot$ does not satisfy any Boolean expression, not even **true**. In other words, for any Boolean e, $\top \models e$ and $\bot \not\models e$. These letters are mathematical devices useful only for defining the formal semantics; there are no corresponding SystemVerilog constructs to express them directly.

Example 22.4. Let s denote the sequence `##1 a[*2]`. s is tightly satisfied on the trace $w_1 = \neg aa\top$, but it is not tightly satisfied on the traces $w_2 = \neg a\neg a\top$ and $w_3 = \bot aa$.

Note that $w_1^0 = \neg a \models$ **true**, as **true** is satisfied by every letter except $\bot$. Also, $w_1^1 = a \models a$. And $w_1^2 = \top \models a$ since $\top$ satisfies every Boolean expression. Therefore, $w_1 \models\equiv s$.

$w_2 \not\models\equiv s$ since $w_2^1 = \neg a \not\models a$. The $\top$ at time 2 does not help match the sequence. At time 1, we already have the evidence that the sequence s cannot be matched regardless the trace content at future time moments.

$w_3 \not\models\equiv s$ since $w_3^0 = \bot \not\models$ **true**: $\bot$ does not satisfy any Boolean, not even **true**. The sequence started with unary `##1`. The important point is that this does not simply mean to skip the first letter. Rather, the first letter must satisfy **true**. □

Example 22.5. Let s denote the sequence `a ##1 0`. Sequence s is tightly satisfied on the trace $w = a\top$, i.e., $w \models\equiv s$, since $w^0 = a \models a$ and $w^1 = \top \models 0$. The latter holds because $\top$ satisfies every Boolean expression, even 0 (or, **false**). Note that the sequence s cannot be tightly satisfied on any trace that does not contain $\top$ since no other letter satisfies **false**. □

Example 22.6. The property `a until b` is satisfied on the trace $a \wedge \neg b\ \top\ \neg a \wedge \neg b \ldots$. Indeed, at time 0 `a` is true, and at time 1 all Boolean expressions are satisfied. □

Example 22.7. The property `always a` holds on the trace $w = a\top\neg a \ldots$ because `always a` is a shortcut for `a until 0` and 0 (or **false**) is satisfied by $\top$ (see Example 22.6), even though `a` does not hold in time 2. □

22.3.3 Weak Sequential Property

Now that we have defined the extension of the alphabet, we are ready to define the formal semantics of weak sequences:

$$w \models\equiv \texttt{weak}(s) \text{ iff for every } i \geq 0\ w^{0,i}\top^\omega \models \texttt{strong}(s).$$

Informally, this definition means that no finite prefix of the trace can be evidence that the sequence cannot match. As in the case of a strong sequential property the sequence must not admit an empty match.

Example 22.8. `weak(a[*2])` is satisfied on any trace of the form $aa\ldots$. Indeed, the sequence `a[*2]` is tightly satisfied on the traces $a\top$ and aa where the match of the sequence `a[*2]` can be witnessed.

However, `weak(a[*2])` is not satisfied on the trace $a\neg a\ldots$, as the sequence `a[*2]` is not tightly satisfied on the trace $a\neg a$. □

Consider an infinite trace and suppose that there is some time T such that any prefix matching the sequence is of length at most T. Then there is no difference between the weak and the strong satisfaction of the sequence on the trace. This is the case of bounded sequences described in Sect. 6.2.

For instance, the sequence `a[*2]` from Example 22.8 satisfies these conditions, provided it is controlled by the global clock since only traces of length 2 may match it. Therefore, there is no difference between the properties `weak(a[*2])` and `strong(a[*2])` when they are controlled by the global clock.

Example 22.9. The property `weak(##[*] a)` holds on any trace without any special letter $\top$ or $\bot$. Indeed, since $\top \models a$, $w^{0,i}\top \models$ `##[*] a` for any $w \in \Sigma^\omega$ and for any i (recall that Σ^ω is an infinite sequence of the letters from the alphabet of the language defined by the FV model, see Sect. 21.2.2). However, property `strong(##[*] a)` is equivalent to `s_eventually a`.

The property `weak(##[*] a ##1 0)` also holds on any infinite trace without any special letter $\top$ or $\bot$. Since $\top \models a$ and $\top \models$ **false**$(=$ `0`$)$, $w^{0,i}\top\top \models$ `##[*]` a `##1 0` for any $w \in \Sigma^*$ and any i. On the contrary, the property `strong(##[*] a ##1 0)` is a contradiction, as **false** does not match any letter from the original unextended alphabet. □

22.3.4 Property Negation

In Sect. 22.1.1.2, we gave the following definition of property negation:

$$w \models \texttt{not}\ p \text{ iff } w \not\models p.$$

This definition is only correct if w does not contain special letters $\top$ or $\bot$. According to that definition, if $w = \top^\omega$ then, for any a, $w \models a$, and thus $w \not\models$ `not` a. This is counterintuitive since we expect `not` a to behave the same way as `!`a when the property is controlled by the global clock. `!`a is satisfied on w since $\top$ satisfies every Boolean expression.

We need to modify the formal semantics of negation to eliminate this counterintuitive behavior. Given a word w we build the word $\overline{w}$ by interchanging the letters $\top$ and $\bot$ in w and leaving all other letters unmodified. Then we can define the formal semantics of property negation as follows:

$$w \models \texttt{not}\ p \text{ iff } \overline{w} \not\models p.$$

Example 22.10. If $w = \top^{\omega}$, then $\overline{w} = \bot^{\omega} \not\models a$ for any a. Therefore, according to this modified definition, $w \models$ **not** a. □

Example 22.11. $w \models$ `not weak(##[*] a)` iff $\overline{w} \not\models$ `weak(##[*] a)`. The latter means that there exists $i \geq 0$ such that $\overline{w}^{0,i}\top^{\omega} \not\models$ `strong(##[*] a)`. This is impossible if w does not contain special letters. Therefore, we have that `not weak(##[*] a)` cannot hold on a trace without special letters, which agrees with our intuition.

$w \models$ `not strong(##[*] a)` iff $\overline{w} \not\models$ `strong(##[*] a)`. The latter means that the sequence `##[*] a` does not match any finite prefix of $\overline{w}$. Assuming that w has no special letter, this is equivalent to $w^i \not\models a$ for all i, or, equivalently, to $w^i \models\ !a$ for all i. This, in turn, means that `always !a` holds on w. This is not surprising, since `strong(##[*] a)` is equivalent to `s_eventually a`, and the negation of `s_eventually a` is `always !a` for Boolean `a` (Chap. 5). □

22.3.5 Suffix Implication

The formal semantics of suffix implication is as follows:

$$w \models s\,|\text{->}\,p \text{ iff for every } i \geq 0,\ \left(\overline{w}^{0,i} \equiv\!\!\models s\right) \rightarrow \left(w^{i..} \models p\right).$$

This definition means that for each finite prefix of trace w matching sequence s, the suffix of the trace satisfies property p. The prefix and the suffix overlap at the letter i. This definition agrees with the informal definition of the overlapping suffix implication from Sect. 6.4.

Note the interchange of the letters $\top$ and $\bot$ in the antecedent. This is necessary to make the formal semantics intuitive, as illustrated in the following examples.

Example 22.12. It is natural to expect that a |-> b behaves the same as a **implies** b when a and b are Booleans. Assume that $|w| \geq 1$. Then $w \models a$ **implies** b iff $w \models$ **not** a **or** b iff $\overline{w} \not\models a$ or $w \models b$ iff $\overline{w}^{0,0} \equiv\!\!\models a$ implies $w^{0..} \models b$. □

Example 22.13. Given the trace $w = a \wedge \neg b \top a \wedge \neg b \ldots$ and the property `a[*3] |-> b` we can see that `b` does not belong to the first three letters of the trace. However, the property holds on w, as $\overline{w} = a \wedge \neg b \bot a \wedge \neg b \ldots$, and `a[*3]` does not match any finite prefix of $\overline{w}$.

This example is important for understanding the behavior of implication under reset discussed in Chap. 13. □

Is the sequence in the antecedent of a suffix implication strong or weak? The question is malformed since weak and strong sequences are *properties*, whereas the antecedent of a suffix implication is a *sequence*.

As explained in Sect. 6.4, nonoverlapping implication is a shortcut:

$$(s\ \texttt{|=>}\ p) \equiv (s\ \texttt{\#\#1 1 |->}\ p).$$

22.3.6 *Suffix Conjunction: Followed-by*

The formal semantics of followed-by (also called suffix conjunction) is as follows:

$$w \models s\ \texttt{\#-\#}\ p \text{ iff for some } i \geq 0,\ w^{0,i} \models\!\!\equiv s \text{ and } w^{i..} \models p.$$

Followed-by and suffix implication are dual operations (Chap. 10.3):

$$(s\ \texttt{\#-\#}\ p) \equiv (\texttt{not}(\ s\ \texttt{|->}\ \texttt{not}\ p)).$$

Indeed, property s `|->` p is false iff at some tight satisfaction point of sequence s property p does not hold.

Nonoverlapping followed-by is defined as a shortcut:

$$(s\ \texttt{\#=\#}\ p) \equiv (s\ \texttt{\#\#1 1 \#-\#}\ p).$$

22.4 Formal Semantics of Clocks

The following notations, with and without subscripts, are used throughout this section: b, c, d, x, y, and z denote Boolean expressions; e denotes an event expression; r denotes a sequence; p denotes a property; v denotes a local variable (see Chap. 15); h denotes an expression.

The formal semantics treats clocks like Boolean expressions. If e is a clocking event, then $\tau(e)$ denotes the associated Boolean expression. The semantics of $\tau(e)$ is that $\ell \models \tau(e)$ iff there is a tick of e in letter ℓ. In this way, the question of whether e occurs at a particular tick of the global clock is transformed into the question of whether the Boolean expression $\tau(e)$ is true at that tick of the global clock. τ is defined as follows:

C1 If e is a named event, then $\tau(e)$ is assumed to be understood as a Boolean expression at the granularity of the global clock.[3]

C2 $\tau($`$global_clock`$) =$ `1'b1`.

C3 $\tau(b) =$ `$changing_gclk(`b`)`, where b is not `$global_clock`.

[3]The formal semantics does not attempt to expand the meaning of a named event in terms of the various code that may trigger the event.

C4 τ(**posedge** b) = `$rising_gclk(`b`)`.
C5 τ(**negedge** b) = `$falling_gclk(`b`)`.
C6 τ(**edge** b) = τ(**posedge** b) `||` τ(**negedge** b).
C7 τ(e **iff** b) = $\tau(e)$ `&&` b.
C8 $\tau(e_1$ **or** $e_2) = \tau(e_1)$ `||` $\tau(e_2)$.
C9 $\tau(e_1$ `,` $e_2) = \tau(e_1)$ `||` $\tau(e_2)$.

For example,

τ(**posedge** `clk` **iff** `enable`) = `$rising_gclk(clk) && enable`

After transforming clocking events into Boolean expressions, the formal semantics eliminates the clocking event controls by folding the associated Boolean expressions into the underlying sequences and properties. The resulting *unclocked* sequences and properties are then interpreted over traces (i.e., finite and infinite words). The elimination of the clocking event controls is accomplished by a system of *clock rewrite rules*. The rewrite rules define a function $\mathcal{T}^s$ that transforms a sequence and the Boolean expression for a clock into an unclocked sequence. They also define a function $\mathcal{T}^p$ for transforming properties similarly. The clock rewrite rules are as follows, where c denotes a clocking event:

SCR0 $\mathcal{T}^s$(`(`r`)`, c) = `(`$\mathcal{T}^s(r, c)$`)`.
SCR1 $\mathcal{T}^s(b, c)$ = `!`c`[*] ##1` c `&&` b.
SCR2 $\mathcal{T}^s$(`(1,` v `=` h`)`,c) = $\mathcal{T}^s$(`1`,c) `##0 (1,` v `=` h`)`.
SCR3 $\mathcal{T}^s$(`@(`c_2`)` r, c_1) = $\mathcal{T}^s(r, c_2)$.[4]
SCR4 $\mathcal{T}^s(r_1$ *op* $r_2, c)$ = $\mathcal{T}^s(r_1, c)$ *op* $\mathcal{T}^s(r_2, c)$, where *op* can be any of the operators `##0`, `##1`, **or**, and **intersect**.
SCR5 $\mathcal{T}^s$(**first_match**`(`r`)`, c) = **first_match**`(`$\mathcal{T}^s(r, c)$`)`.
SCR6 $\mathcal{T}^s$(r`[*`n`]`, c) = $\mathcal{T}^s(r, c)$`[*`n`]`. The same rule applies to ranged repetition operators.

PCR0 $\mathcal{T}^p$(`(`p`)`, c) = `(`$\mathcal{T}^p(p, c)$`)`.
PCR1 $\mathcal{T}^p$(*op*`(`r`)`, c) = *op*`(`$\mathcal{T}^s(r, c)$`)`, where *op* is either **strong** or **weak**.
PCR2 $\mathcal{T}^p$(`@(`c_2`)` p, c_1) = $\mathcal{T}^p(p, c_2)$.[5]
PCR3 $\mathcal{T}^p$(*op* p, c) = *op* $\mathcal{T}^p(p, c)$, where *op* is any of **not**, **disable iff** `(`b`)`, **accept_on**`(`b`)`, and **reject_on**`(`b`)`.
PCR4 $\mathcal{T}^p(p_1$ *op* $p_2, c)$ = $\mathcal{T}^p(p_1, c)$ *op* $\mathcal{T}^p(p_2, c)$, where *op* can be any of the operators **and**, **or**, **implies**, and **iff**.
PCR5 $\mathcal{T}^p$(**sync_accept_on**`(`b`)` p, c) = **accept_on**`(`b `&&` c`)` $\mathcal{T}^p(p, c)$. An analogous rule applies to rewrite the synchronous reject in terms of the asynchronous reject.

[4]This rule can be applied only after applying the clock flow rules in Sect. 12.2.4.1.
[5]See footnote 4.

PCR6 $\mathcal{T}^p(r\ op\ p, c) = \mathcal{T}^s(r, c)\ op\ \mathcal{T}^p(p, c)$, where *op* is any of `|->`, `|=>`, `#-#`, and `#=#`.[6]

PCR7 $\mathcal{T}^p($**`nexttime`** $p, c) =$
`!`c **`until`** `(`c **`and nexttime`** `(!`c **`until`** `(`c **`and`** $\mathcal{T}^p(p, c)$`)))`.

PCR8 $\mathcal{T}^p(p_1$ **`until`** $p_2, c) =$
`(`c **`implies`** $\mathcal{T}^p(p_1, c)$`)` **`until`** `(`c **`and`** $\mathcal{T}^p(p_2, c)$`)`.

Rule *PCR7* is the most complicated. It contains two alignments to c, performed by the idiomatic subproperty `!`c **`until`** c. The first brings evaluation to alignment with a tick of c, and the second performs the advance to the next tick of c, as specified by **`nexttime`** (see Sect. 12.2.5.1). The first alignment is needed to get the proper semantics when the clock is changing. If evaluation of **`nexttime`** p begins in a tick of c, then the first alignment does not advance time.

The clock rewrite rules above cover all the primitive operators and a number of derived operators. For other derived operators, rewrite rules are obtained by applying the rules above to the definition of the derived operator in terms of more primitive operators.

Example 22.14. Calculate a clock rewrite rule for b`[->1]`.

Solution: b`[->1]` is defined to be equivalent to `!`b`[*] ##1` b. Therefore,

$\mathcal{T}^s(b$`[->1]`$, c)$
$= \mathcal{T}^s($`!`b`[*] ##1` $b, c)$
$= \mathcal{T}^s($`!`b`[*]`$, c)$ `##1` $\mathcal{T}^s(b, c)$ [*SCR4*]
$= \mathcal{T}^s($`!`$b, c)$`[*] ##1` $\mathcal{T}^s(b, c)$ [*SCR6*]
$=$ `(!`c`[*] ##1` c `&&` `!`b`)[*] ##1 !`c`[*] ##1` c `&&` b [*SCR1*]

□

Example 22.15. Calculate a clock rewrite rule for r_1 **`within`** r_2.

Solution: r_1 **`within`** r_2 is defined to be equivalent to

`(1[*] ##1` r_1 `##1 1[*])` **`intersect`** r_2

Therefore,

$\mathcal{T}^s(r_1$ **`within`** $r_2, c)$
$= \mathcal{T}^s($`(1[*] ##1` r_1 `##1 1[*])` **`intersect`** $r_2, c)$
$= \mathcal{T}^s($`(1[*] ##1` r_1 `##1 1[*])`$, c)$ **`intersect`** $\mathcal{T}^s(r_2, c)$ [*SCR4*]
$=$ `(`$\mathcal{T}^s($`1[*] ##1` r_1 `##1 1[*]`$, c)$`)` **`intersect`** $\mathcal{T}^s(r_2, c)$ [*SCR0*]
$=$ `(`$\mathcal{T}^s($`1[*]`$, c)$ `##1` $\mathcal{T}^s(r_1, c)$ `##1` $\mathcal{T}^s($`1[*]`$, c)$`)`
`intersect` $\mathcal{T}^s(r_2, c)$ [*SCR4*]
$=$ `(`$\mathcal{T}^s($`1`$, c)$`[*] ##1` $\mathcal{T}^s(r_1, c)$ `##1` $\mathcal{T}^s($`1`$, c)$`[*])`
`intersect` $\mathcal{T}^s(r_2, c)$ [*SCR6*]

[6] See footnote 4.

$$
\begin{aligned}
= \ & \texttt{(} \\
& \quad \texttt{(!}c\texttt{[*] ##1 } c \texttt{ \&\& 1)[*]} \\
& \quad \texttt{##1 } \mathcal{T}^s(r_1, c) \\
& \quad \texttt{##1 (!}c\texttt{[*] ##1 } c \texttt{ \&\& 1)[*]} \\
& \texttt{)} \\
& \quad \textbf{intersect } \mathcal{T}^s(r_2, c) && [SCR1] \\
\equiv \ & \texttt{((}c\texttt{[->1])[*] ##1 } \mathcal{T}^s(r_1, c) \texttt{ ##1 (}c\texttt{[->1])[*])} \\
& \quad \textbf{intersect } \mathcal{T}^s(r_2, c)
\end{aligned}
$$

□

Example 22.16. Calculate a clock rewrite rule for **always** p.

Solution: **always** p is defined to be equivalent to p **until** `1'b0`. Boolean operands of a weak operator like **until** are automatically weakened, as though they were within implicit instances of **weak**. Therefore

$$
\begin{aligned}
& \mathcal{T}^p(\textbf{always } p, c) \\
= \ & \mathcal{T}^p(p \textbf{ until } \texttt{1'b0}, c) \\
= \ & (c \textbf{ implies } \mathcal{T}^p(p, c)) \textbf{ until } (c \textbf{ and } \mathcal{T}^s(\texttt{1'b0}, c)) && [PCR8] \\
= \ & (c \textbf{ implies } \mathcal{T}^p(p, c)) \textbf{ until} \\
& \quad (c \textbf{ and } \texttt{(!}c\texttt{[*] ##1 } c \texttt{ \&\& 1'b0))} && [SCR1] \\
\equiv \ & (c \textbf{ implies } \mathcal{T}^p(p, c)) \textbf{ until } (c \texttt{ \&\& 1'b0)} \\
\equiv \ & (c \textbf{ implies } \mathcal{T}^p(p, c)) \textbf{ until } \texttt{1'b0}
\end{aligned}
$$

□

Example 22.17. Show that the following equivalence is preserved under $\mathcal{T}^p(\cdot, c)$:

$$r \texttt{ #-# } p \equiv \textbf{not}(r \texttt{ |-> } \textbf{not } p) \qquad (*)$$

Solution: Compute:

$$
\begin{aligned}
& \mathcal{T}^p(\textbf{not}(r \texttt{ |-> } \textbf{not } p), c) \\
= \ & \textbf{not}(\mathcal{T}^p(r \texttt{ |-> } \textbf{not } p, c)) && [PCR3] \\
= \ & \textbf{not}(\mathcal{T}^s(r, c) \texttt{ |-> } \mathcal{T}^p(\textbf{not } p, c)) && [PCR6] \\
= \ & \textbf{not}(\mathcal{T}^s(r, c) \texttt{ |-> } \textbf{not } \mathcal{T}^p(p, c)) && [PCR3] \\
\equiv \ & \mathcal{T}^s(r, c) \texttt{ #-# } \mathcal{T}^p(p, c) && [(*)] \\
= \ & \mathcal{T}^p(r \texttt{ #-# } p, c) && [PCR6]
\end{aligned}
$$

□

It is important to understand that the clock rewrite rules do not account for scoping of clocks as defined by the clock flow rules. The addition of clocking event controls consistent with the clock flow rules may be necessary before the clock rewrite rules may be applied.

Example 22.18. Assume that c is the incoming clock to x `##1 @(`d`)` y `|=>` z. Show that application of $\mathcal{T}^p(\cdot, c)$ to this property does not correctly account for clock flow. Add clocking event controls to the property so that application of $\mathcal{T}^p(\cdot, c)$ yields the correct rewrite.

Solution: By *CF3*, *CF7*, and *CF8*, d governs y and z, while the incoming clock c governs x. Compute

$$\begin{aligned}
&\mathcal{T}^p(x \texttt{ ##1 @(}d\texttt{) } y \texttt{ |=> } z, c) \\
&= \mathcal{T}^s(x \texttt{ ##1 @(}d\texttt{) } y, c) \texttt{ |=> } \mathcal{T}^p(z, c) && [PCR6] \\
&= \mathcal{T}^s(x, c) \texttt{ ##1 } \mathcal{T}^s(\texttt{@(}d\texttt{) } y, c) \texttt{ |=> } \mathcal{T}^p(z, c) && [SCR4] \\
&= \mathcal{T}^s(x, c) \texttt{ ##1 } \mathcal{T}^s(y, d) \texttt{ |=> } \mathcal{T}^p(z, c) && [SCR3]
\end{aligned}$$

Thus, according to the rewrite, d governs only y, while c governs x and z. To fix the problem, add the clocking event control `@(`d`)` after `|=>`:

$$\begin{aligned}
&\mathcal{T}^p(x \texttt{ ##1 @(}d\texttt{) } y \texttt{ |=> @(}d\texttt{) } z, c) \\
&= \mathcal{T}^s(x \texttt{ ##1 @(}d\texttt{) } y, c) \texttt{ |=> } \mathcal{T}^p(\texttt{@(}d\texttt{) } z, c) && [PCR6] \\
&= \mathcal{T}^s(x, c) \texttt{ ##1 } \mathcal{T}^s(\texttt{@(}d\texttt{) } y, c) \texttt{ |=> } \mathcal{T}^p(\texttt{@(}d\texttt{) } z, c) && [SCR4] \\
&= \mathcal{T}^s(x, c) \texttt{ ##1 } \mathcal{T}^s(y, d) \texttt{ |=> } \mathcal{T}^p(z, d) && [SCR3, PCR2]
\end{aligned}$$

□

22.5 Formal Semantics of Resets

The special letters $\top$ and $\bot$ play an important role in the formal semantics of resets. If a reset condition occurs at some point in a trace w, then the disposition of the evaluation depends on replacement of the suffix of w beginning at that point by $\top^\omega$, $\bot^\omega$, or both, depending on the particular reset. The intuition is that all properties hold on $\top^\omega$, while no properties hold on $\bot^\omega$.[7] Therefore, replacing a suffix of w by $\top^\omega$ precludes failure after the point of replacement, while replacing the suffix by $\bot^\omega$ forces success to be complete prior to the point of replacement. Abort operators of the "accept" kind replace the suffix beginning at the occurrence of the abort condition by $\top^\omega$, whereas abort operators of the "reject" kind replace the suffix by $\bot^\omega$. For **`disable iff`**, the disposition on occurrence of the disable condition is neither "pass" nor "fail", but "disabled", so the formal semantics uses both replacements: the replacement of the suffix by $\top^\omega$ confirms that failure did not occur before the disable condition, whereas the replacement of the suffix by $\bot^\omega$ confirms that success also did not occur before the disable condition.

[7]There are some theoretical subtleties involved. It is possible to write a sequence that does not match any word, even when interpreted over the alphabet extended with $\top$ and $\bot$. An example is a sequence involving a length mismatch, such as $r =$ `1'b1` **`intersect`** `(1'b1 ##1 1'b1)`. From r, one can create a property, such as **`weak`**`(`r`)`, that is not satisfiable, even by the word $\top^\omega$. However, by restricting each maximal sequence appearing in a concurrent assertion to be *non-degenerate*, i.e., to match at least one nonempty finite word when interpreted over the extended alphabet, this problem is avoided. The SystemVerilog LRM requires maximal sequences to be nondegenerate in the appropriate circumstances to avoid such problematic cases.

Exposition of the semantics is simplified by the fact that only **disable iff** and **accept_on** need be treated as primitive operators. The other resets are derived by the following rules:

- **reject_on** (b) p $\equiv$ **not accept_on** (b) **not** p.
- **sync_reject_on** (b) p $\equiv$ **not sync_accept_on** (b) **not** p.
- $\mathcal{T}^p($**sync_accept_on** (b) $p, c)$ $=$ **accept_on** $(b$ && $c)$ $\mathcal{T}^p(p, c)$.

The formal semantics of **accept_on** is simpler than that of **disable iff** because it involves only the two dispositions "pass" and "fail". It is defined as follows:

$w \models$ **accept_on** (b) p
iff either
$w \models p$
or
there exists $0 \leq i < |w|$ such that $w^i \models b$ and $w^{0,i-1}\top^\omega \models p$.

The first case allows w to satisfy **accept_on** (b) p if w satisfies p itself, regardless of the abort condition. The second case accounts for an occurrence of the abort condition and replaces the suffix of w beginning at the letter where the abort condition occurs by $\top^\omega$ before checking for satisfaction of the underlying property p.

Example 22.19. Derive direct semantics for **reject_on** (b) p.

Solution:

$w \models$ **reject_on** (b) p
iff $w \models$ **not accept_on** (b) **not** p
iff $\bar{w} \not\models$ **accept_on** (b) **not** p
iff both
$\bar{w} \not\models$ **not** p
and
for all $0 \leq i < |w|$ such that $\bar{w}^i \models b$: $\bar{w}^{0,i-1}\top^\omega \not\models$ **not** p
iff both
$w \models p$
and
for all $0 \leq i < |w|$ such that $\bar{w}^i \models b$: $w^{0,i-1}\bot^\omega \models p$

□

It thus follows that $reject\_on(b)p$ succeeds when w satisfies p and this happens before the earliest occurrence of b.

The "pass" and "fail" dispositions of a **disable iff** are defined formally as follows:

$w \models$ **disable iff** (b) p
iff either
$w \models p$ and no letter of w satisfies b
or
for some i, $w^i \models b$, and $w^{0,i-1}\bot^\omega \models p$ for the least such i.

The "disabled" disposition of a **disable iff** is defined formally as follows:

$w \models^d$ **disable iff** $(b)\ p$
iff for some i, $w^i \models b$, and for the least such i, both
$$w^{0,i-1}\top^\omega \models p$$
and
$$w^{0,i-1}\bot^\omega \not\models p.$$

Intuitively, the requirement that $w^{0,i-1}\top^\omega \models p$ ensures that p has not failed prior to the occurrence of the disable condition, while the requirement that $w^{0,i-1}\bot^\omega \not\models p$ ensures that p has not succeeded prior to the occurrence of the disable condition.

22.6 Formal Semantics of Local Variables

This section gives an overview of the formal semantics of local variables. For more details, see the LRM [8]. The technical report [41] is another good reference, covering details of theoretical results that are not presented in the LRM. [23] presents complexity results for local variables. Only the unclocked semantics is discussed here. As usual, the clocked semantics is obtained by first transforming to unclocked forms using the clock rewrite rules and then applying the unclocked semantics.

22.6.1 Formalizing Local Variable Flow

If X is a set of local variables and r is a sequence, let $flow(X,r)$ denote the set of local variables that flow out of r given that the local variables in X flow into r. From the local variable flow rules in Sect. 16.4.1, one obtains the following recursive representation of $flow(X,r)$:

SF1 $flow(X,b) = X$. Analogous equalities hold for Boolean repetitions b`[->`n`]`, b`[=`n`]`, etc.
SF2 $flow(X,$ `(`r`,` v `=` e`)`$) = flow(X,r) \cup \{v\}$.
SF3 $flow(X,r$ `##`n $s) = flow(flow(X,r),s)$. Analogous equalities hold for variants of the binary concatenation operator.
SF4 $flow(X,r$ **or** $s) = flow(X,r) \cap flow(X,s)$.
SF5 $flow(X,r$ **and** $s) = (flow(X,r) - assign(s)) \cup (flow(X,s) - assign(r))$. Here $assign(r)$ (resp., $assign(s)$) denotes the set of local variables assigned anywhere within r (resp., s). Analogous equalities hold for **intersect** and **within**.
SF6 $flow(X,b$ **throughout** $r) = flow(X,r)$.
SF7 $flow(X,$ **first_match**`(`r`)`$) = flow(X,r)$.
SF8 $flow(X,r$`[*0]`$) = X$.

SF9 $flow(X, r\texttt{[+]}) = flow(X, r\texttt{[*}n\texttt{]}) = flow(X, r)$, provided n is positive. Analogous equalities apply to ranged forms of the repetition operator, provided the lower range is positive. If the lower range is zero, then the flow rule is obtained by decomposing $r\texttt{[*0:}n\texttt{]}$ as $r\texttt{[*0]}$ **or** $r\texttt{[*1:}n\texttt{]}$ (n positive or `$`).

22.6.2 Local Variable Contexts

The principal technical device introduced to define the semantics of local variables is the local variable context. A *local variable context* is a partial function mapping local variables to values. It can be written as a set of ordered pairs:

$$L = \{(v_1, a_1), (v_2, a_2), \ldots, (v_n, a_n)\}$$

L specifies that for each i, $1 \leq i \leq n$, the local variable v_i is assigned and has the value a_i. It is customary to say that $\{v_1, v_2, \ldots, v_n\}$ is the *domain* of L, written $\mathrm{dom}(L)$. Thus, the domain of a local variable context is the set of local variables that are assigned in that context. Any local variable not in the domain of the local variable context is unassigned in that context. A local variable context may be empty, in which case its domain is also empty and all local variables are unassigned in that context.

If $D \subseteq \mathrm{dom}(L)$, then $L|_D$ denotes the local variable context obtained by restricting L to the domain D. For example, with L as above and $D = \{v_1, v_3\}$,

$$L|_D = \{(v_1, a_1), (v_3, a_3)\}$$

$L \setminus v$ denotes $L|_{\mathrm{dom}(L)-\{v\}}$. It is the local variable context that results from L by removing v from the domain, if it was there to begin with.

22.6.3 Sequence Semantics with Local Variables

The semantics of matching (i.e., tight satisfaction) of unclocked sequences can now be defined as a four-way relation:

$$w, L_0, L_1 \models\!\equiv r$$

The relation holds if r matches the finite word w starting with incoming local variable context L_0 and resulting in outgoing local variable context L_1. Whenever the relation holds, $\mathrm{dom}(L_1) = flow(\mathrm{dom}(L_0), r)$. This fact is proven in [41]. The substance of defining the relation is to capture how the local variable contexts evolve, as a result of assignments to local variables, local variables becoming unassigned according to the flow rules, and inductively through intermediate local variable contexts. Here are the definitions for basic sequence operators:

- $w, L_0, L_1 \models b$ iff $|w| = 1$ and $w^0 \models b[L_0]$ and $L_1 = L_0$. Here, $b[L_0]$ denotes the expression obtained by substituting values from L_0 for any references in b to local variables in dom(L_0).
- $w, L_0, L_1 \models$ `(`r`, `v` = `e`)` iff there exists L such that $w, L_0, L \models r$ and $L_1 = L \backslash v \cup \{v, e[L, w^0]\}$. Here, $e[L, w^0]$ denotes the expression obtained by substituting values from L_0 for any references in e to local variables in dom(L_0) and then using values from w^0 to evaluate remaining references in e. If w^0 is a special letter $\top$ or $\bot$, then $e[L, w^0]$ can be any value.
- $w, L_0, L_1 \models r$ `##1` s iff there exist x, y, L such that $w = xy$ and $x, L_0, L \models r$ and $y, L, L_1 \models s$.
- $w, L_0, L_1 \models r$ `##0` s iff there exist x, y, z, L such that $w = xyz$ and $|y| = 1$ and $xy, L_0, L \models r$ and $yz, L, L_1 \models s$.
- $w, L_0, L_1 \models r$ **`or`** s iff there exists L such that both
  - either $w, L_0, L \models r$ or $w, L_0, L \models s$, and
  - $L_1 = L|_D$, where $D = \mathit{flow}(\mathit{dom}(L_0), r$ **`or`** $s)$.
- $w, L_0, L_1 \models r$ **`intersect`** s iff there exist L, L' such that $w, L_0, L \models r$ and $w, L_0, L' \models s$ and $L_1 = L|_D \cup L'|_{D'}$, where
  - $D = \mathit{flow}(\mathrm{dom}(L_0), r) - \mathit{assign}(s)$
  - $D' = \mathit{flow}(\mathrm{dom}(L_0), s) - \mathit{assign}(r)$
- $w, L_0, L_1 \models$ **`first_match`**`(`r`)` iff both
  - $w, L_0, L_1 \models r$, and
  - if there exist x, y, L such that $w = xy$ and $\bar{x}, L_0, L \models r$, then y is empty.
- $w, L_0, L_1 \models r$ `[*0]` iff $|w| = 0$ and $L_1 = L_0$.
- $w, L_0, L_1 \models r$ `[+]` iff there exist $L_{(0)} = L_0, w_1, L_{(1)}, \ldots, w_k, L_{(k)} = L_1, k \geq 1$, such that $w = w_1 \cdots w_k$ and $w_i, L_{(i-1)}, L_{(i)} \models r$ for every i, $1 \leq i \leq k$.

22.6.4 Property Semantics with Local Variables

Local variables do not flow out of properties, so the semantics of property satisfaction with local variables is simpler than that of sequence matching. The relation is three-way:

$$w, L_0 \models p$$

The relation holds if p is satisfied by word w starting with the incoming local variable context L_0. Here are the definitions for the basic property operators:

- $w, L_0 \models$ **`not`** p iff $\bar{w}, L_0 \not\models p$.
- $w, L_0 \models$ **`strong`**`(`r`)` iff there exists $0 \leq j < |w|$ and L such that $w^{0,j}, L_0, L \models r$. r must not admit an empty match.

- $w, L_0 \models$ `weak(r)` iff for every $0 \leq j < |w|$, $w^{0,j}\top^\omega, L_0 \models$ `strong(r)`. r must not admit an empty match.
- $w, L_0 \models r$ `|->` p iff for every $0 \leq j < |w|$ and L such that $\bar{w}^{0,j}, L_0, L \models\!\equiv r$, $w^{j..}, L \models p$.
- $w, L_0 \models p$ `or` q iff either $w, L_0 \models p$ or $w, L_0 \models q$.
- $w, L_0 \models p$ `and` q iff both $w, L_0 \models p$ and $w, L_0 \models q$.
- $w, L_0 \models$ `nexttime` p iff either $|w| = 0$ or $w^{1..}, L_0 \models p$.
- $w, L_0 \models p$ `s_until` q iff there exists $0 \leq j < |w|$ such that $w^{j..}, L_0 \models q$ and for every $0 \leq i < j$, $w^{i..}, L_0 \models p$.
- $w, L_0 \models$ `accept_on(b)` p iff either
  - $w, L_0 \models p$ and no letter of w satisfies b, or
  - for some $0 \leq i < |w|$, $w^i \models b$ and $w^{0,i-1}\top^\omega, L_0 \models p$.[8]

It is worth noting that in the rule for `|->`, the intermediate local variable context L is universally quantified. This is how the formal semantics specifies that multiple matches over the same interval that result in distinct values of the local variables must be treated as separate matches, each obligating a check of the consequent with the corresponding local variable values.

22.7 Formal Semantics of Recursive Properties

This section gives a brief description of the formal semantics of recursive properties. The main idea is to define the semantics of a recursive property in terms of the semantics of associated non-recursive properties. Intuitively, these non-recursive properties are approximations to greater and greater depth of the unrolling of the recursion. There is some subtlety to the definition because recursive properties may instantiate non-recursive properties and vice versa. This section presumes that the semantics of instantiation of non-recursive properties is understood, at least at the intuitive level of substituting actual arguments for formal arguments, while avoiding aliasing that is contrary to the rules of name resolution. For further discussion, see Annex F.4 of the LRM.

First, let us say more precisely what constitutes a recursive property and a recursive instance. Consider the source code for a SystemVerilog model. Within it are finitely many declarations of named properties and finitely many instances of named properties. The *dependency digraph* is the directed graph $\langle V, E \rangle$ formed as follows. The node set V consists of the named properties that appear in the source code. Each named property has a unique name, where the name may be expanded as a hierarchical name, e.g., as needed for disambiguation. If p and q are two named properties, then there is a directed edge from p to q in the edge set E iff there is an instance of q within the declaration of p. A named property is said

[8] In case $i = 0$, $w^{0,-1}$ is understood to denote the empty word.

to be *recursive* iff it belongs to a nontrivial strongly connected component of the dependency digraph. An instance of a named property p is said to be *recursive* iff it appears within the declaration of a named property q such that p and q belong to the same nontrivial strongly connected component of the dependency digraph. This condition is satisfied iff there is a loop in the dependency digraph that has at least one arc and that passes through both p and q.[9]

Example 22.20. Compute the dependency digraph for the following named property declarations:

```
property p1;
   a and nexttime p2(p2(p3));
endproperty
property p2(property p)
   eventually[0:1] p;
endproperty
property p3;
   b and nexttime p1;
endproperty
```

Identify which named properties are recursive and which instances are recursive.

Solution: The node set of the dependency digraph is $V = \{$`p1`,`p2`,`p3`$\}$. The declaration of `p1` instantiates `p2` (twice) and `p3`, so there are directed edges `p1` $\rightarrow$ `p2` and `p1` $\rightarrow$ `p3` in the edge set E. Similarly, there is a directed edge `p3` $\rightarrow$ `p1` in E. Below is a graphical representation of the dependency digraph. There is a single

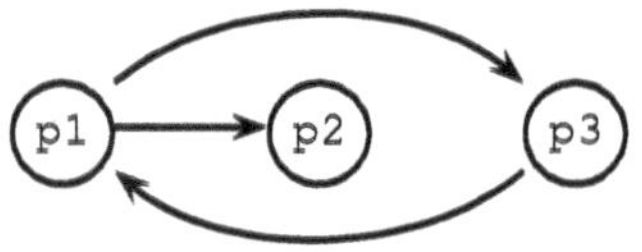

nontrivial strongly connected component in the digraph consisting of the nodes $\{$`p1`,`p3`$\}$. Therefore, `p1` and `p3` are recursive properties. The instance of `p3` in the declaration of `p1` and the instance of `p1` in the declaration of `p3` are both recursive instances.`p2` is a non-recursive property. It is worth noting that, under rewriting, the instance `p2(p2(p3))` expands to

```
eventually[0:1] eventually[0:1] p3,
```

which is equivalent to `eventually[0:2] p3`. □

For $k \geq 0$, the k-fold approximation to a named property p is denoted $p[k]$ and is defined as follows:

- $p[0]$ is a named property whose declaration is obtained from that of p by replacing the body property by `1'b1`.

[9]In case $p = q$, the loop can be a self-loop on p.

- For $k > 0$, $p[k]$ is a named property whose declaration is obtained from that of p by the following replacements:
  - Each instance of a recursive property q in the declaration of p is replaced by the instance of $q[k-1]$ obtained by passing the same actual arguments.
  - Each instance of a non-recursive property q in the declaration of p is replaced by the instance of $q[k]$ obtained by passing the same actual arguments.

For any property π, the k-fold approximation to π is denoted $\pi[k]$ and is obtained from π by replacing each instance of a named property p by the instance of $p[k]$ obtained by passing the same actual arguments. For a trace w and a local variable context L, unclocked satisfaction of π is defined by $w, L \models \pi$ iff $w, L \models \pi[k]$ for all $k > 0$.

Several examples are given below to illustrate how these definitions work.

Example 22.21. Show that for `k` > 0, `my_always[k]`(*q*) is equivalent to

```
(always[0:k-1] q) and nexttime[k] 1'b1,
```

where `my_always` is as declared in Fig. 17.1.

Solution: The definitions above imply the following declarations:

```
property my_always[0] (property p);
   1'b1;
endproperty
property my_always[k] (property p);
   (nexttime[0] p) and nexttime my_always[k-1](p);
endproperty
```

for `k` positive. Then we get

```
my_always[1](q) ≡ (nexttime[0] q) and nexttime 1'b1
                ≡ (always[0:0] q) and nexttime[1] 1'b1
```

Suppose inductively that

```
my_always[k](q) ≡ (always[0:k-1] q) and nexttime[k] 1'b1
```

Then

```
my_always[k+1](q) ≡ (nexttime[0] q) and nexttime my_always[k](q)
                  ≡ (nexttime[0] q) and
                    nexttime(
                      (always[0:k-1] q) and nexttime[k] 1'b1
                    )
                  ≡ (always[0:k] q) and nexttime[k+1] 1'b1
```

□

Example 22.22. Let π = `my_always(a)`, where `a` is a Boolean. Show that for w a word without any special letter $\top$ or $\bot$, $w \models \pi$ iff $w \models$ **`always`** `a` in the unclocked semantics.

Solution: By definition, $w \models \pi$ iff for all `k` > 0, $w \models \pi$`[k]`. Also by definition,

$$\pi\texttt{[k]} = \texttt{my\_always[k](a)}$$

By the preceding example, for all `k` > 0,

$$\texttt{my\_always[k](a)} \equiv \texttt{(}\textbf{\texttt{always}}\texttt{[0:k-1] a) } \textbf{\texttt{and nexttime}}\texttt{[k] 1'b1}$$

Since w has no special letter, $w \models$ **`nexttime`**`[k] 1'b1` for all `k` > 0, and so

$$w \models \pi \quad \text{iff} \quad \text{for all } 0 \le i < |w|, w^i \models \texttt{a}$$

On the contrary, **`always`** `a` is formally defined as `a` **`until`** `0` in Annex F of the LRM. Since w has no special letter, $w^{i..} \not\models$ `0` for all $0 \le i < |w|$. Therefore,

$$w \models \textbf{\texttt{always}} \texttt{ a} \quad \text{iff} \quad \text{for all } 0 \le i < |w|, w^i \models \texttt{a}$$

□

Example 22.23. Let `a` and `b` be Booleans, and let `q` be declared by

```
property q(property p);
   a |-> p;
endproperty
```

Determine the unclocked formal semantics of the property

$$\pi = \texttt{q(my\_always(q(b)))}$$

Solution: By definition, $w \models \pi$ iff for all `k` > 0, $w \models \pi$`[k]`. Also by definition,

$$\pi\texttt{[k]} = \texttt{q[k](my\_always[k](q[k](b)))}$$

The declaration of `q` instantiates no named property, so for all `k` > 0, `q[k]` $=$ `q`. Therefore, for all `k` > 0,

$$\begin{aligned}\pi\texttt{[k]} &\equiv \texttt{q(my\_always[k](q(b)))}\\ &\equiv \texttt{a |-> my\_always[k](a |-> b)}\\ &\equiv \texttt{a |-> ((}\textbf{\texttt{always}}\texttt{[0:k-1] a |-> b) }\textbf{\texttt{and nexttime}}\texttt{[k] 1'b1)}\end{aligned}$$

$w \models$ **`nexttime`**`[k] 1'b1` iff either $|w| <$ `k` or $w^{\texttt{k}} \neq \bot$. Therefore, $w \models \pi$ iff

either

1. $|w| = 0$, or
2. $\bar{w}^0 \not\models$ `a`, or
3. both

a. for all $0 \le i < |w|$ such that $\bar{w}^i \models$ `a`, $w^i \models$ `b`, and
b. for all $0 < i < |w|$, $w^i \ne \bot$.

□

Exercises

22.1. Write the explicit formal semantics for the operators **`or`**, **`always`**, **`s_eventually`**, **`until`**, **`until_with`**, and **`s_until_with`**.

22.2. Prove the following semantic equivalences:

(a) `not (p until q)` $\equiv$ `(not q) s_until_with (not p)`
(b) `not (p s_until q)` $\equiv$ `(not q) until_with (not p)`.

22.3. Write the explicit formal semantics for the following sequences: `s[*n]`, `s[*]`, `##[*] s`, `r ##[*] s`. Here `r` and `s` are sequences, and `n` is a positive integer number.

22.4. Prove that `c[->1] ##0 c[->1]` $\equiv$ `c[->1]` as unclocked sequences.

22.5. Prove that `s_nexttime p` $\equiv$ `not nexttime not p` as unclocked properties.

22.6. Define the clock rewriting rules for the operators **`until_with`** and **`s_until_with`**.

22.7. For each of the event expressions below, compute the Boolean expression that results from applying τ.

1. `$global_clock iff` b.
2. `(posedge` c`) or (negedge` d`)`.
3. `(edge` c`),` b.
4. `(negedge` c`) or (negedge` d `iff` b`)`.

22.8. Compute clock rewrite rules for the following derived forms:

1. b`[->`n`]` $\equiv$ `(`b`[->1])[*`n`]`.
2. b`[=`n`]` $\equiv$ `(`b`[->`n`] ##1 !`b`[*]`.
3. `(1,` v_1 `=` h_1`,` v_2 `=` h_2`)` $\equiv$ `(1,` v_1 `=` h_1`) ##0 (1,` v_2 `=` h_2`)`.
4. b `throughout` r $\equiv$ b`[*] intersect` r.
5. r_1 `and` r_2 $\equiv$
   `((`r_1 `##1 1[*]) intersect` r_2`) or (`r_1 `intersect (`r_2 `##1 1[*]))`.
6. `s_eventually` p $\equiv$ `not always not` p.
7. `s_nexttime` p $\equiv$ `not nexttime not` p.
8. `if (`b`)` p_1 `else` p_2 $\equiv$ `(`b `|->` p_1`) and (weak(`b`) or` p_2`)`.

22.9. Show that the following equivalences are preserved under $\mathcal{T}^p(\cdot, c)$:

1. r `#=#` p $\equiv$ `not(`r `|=> not` p`)`.
2. `reject_on(`b`)` p $\equiv$ `not accept_on(`b`) not` p.
3. p_1 `iff` p_2 $\equiv$ `(`p_1 `implies` p_2`) and (`p_2 `implies` p_1`)`.

22.10. The following facts about the SVA formal semantics can be proved:

a. If $w \models p$ and w' results from w by changing zero or more letters to $\top$, then $w' \models p$.
b. If $w \models p$ and w' results from w by changing zero or more letters away from $\bot$, then $w' \models p$.

Use these facts to prove the following:

1. If $0 \leq i < j < |w|$ and $w^{0,j-1}\top^{\omega} \models p$, then $w^{0,i-1}\top^{\omega} \models p$.
2. If $0 \leq i < j < |w|$ and $w^{0,i-1}\bot^{\omega} \models p$, then $w^{0,j-1}\bot^{\omega} \models p$.

How do these implications relate to the formal semantics of resets?

22.11. Let a, b, and c be Booleans. Determine the dependency digraph for the following declarations.

```
property p1;
   case(a)
      1'b0:  p2;
      1'b1:  p3;
   endcase
endproperty
property p2;
   p4 or (b and nexttime p1);
endproperty
property p3;
   p4 and nexttime(a |-> p2);
endproperty
property p4;
   a |=> c;
endproperty
```

Which properties are recursive and which are non-recursive?

22.12. Let my_until be as declared in Fig. 17.3. Let a and b be Booleans. Show that for k > 0, my_until[k](a,b) is equivalent to

```
weak( (a[*0:k-1] ##1 b) or (a[*k] ##1 1) )
```

Show that my_until(a,b) is equivalent to

```
weak(a[*] ##1 b)
```

22.13. Let a be a Boolean. Show that for any word w, including special letters $\top$ and $\bot$, $w \models$ my_always(a) implies $w \models$ a **until** 0. Show that the converse can fail if $\top$ is a letter in w. (cf. Example 22.22.)

Part VI
Advanced Checkers

Chapter 23
Checkers in Formal Verification

Hope is a great falsifier. Let good judgment keep her in check.

— Baltasar Gracian

This chapter discusses applications of checkers to formal verification. Checkers were introduced in Chap. 9, where we described their primary use as an instantiable container for assertions and associated modeling code. This chapter presents an additional feature of checkers that is important for formal verification—*free variables*. Free variables support the construction of abstract nondeterministic models and reasoning about them. Recall from Chap. 20 that an abstract model is an over-approximation of the DUT. If an assertion has been proven for an abstract model, then it also holds for the DUT. Good abstraction hides irrelevant details and yields a simpler model on which to conduct formal verification. For example, to reason about a pipeline latency, the exact data passed through the pipeline and the exact operations at specific pipeline stages may be immaterial and may be abstracted away. Checkers provide natural building blocks for these abstract models, and their usage in FV modeling is similar to the usage of modules in RTL.

Another important application of checkers with free variables is to build abstract models of the DUT environment. Such models may be supported both in simulation and in FV, in contrast to traditional software testbenches that are not FV-friendly. Careful modeling of the environment, avoiding false negatives and false positives (see Sect. 20.3), is a major and effort-consuming problem in FV. Of course, carefully crafting the environment and writing tests is also important for simulation. Usually there is no reuse of the environment between simulation and FV. Checkers with free variables provide an approach to environment modeling that can, in many situations, be used by both engines and eliminate redundant effort.

Checker free variables, including constant free variables (also called *rigid variables*), can be used in an alternative coding style instead of assertion local variables. Each style has advantages and disadvantages.

E. Cerny et al., *SVA: The Power of Assertions in SystemVerilog*,
DOI 10.1007/978-3-319-07139-8_23

This chapter is mainly intended for people interested in FV. However, it is also useful for people interested in creating testbenches that can be processed by FV tools.

23.1 Free Variables

Free variables in checkers have the same syntax as regular checker variables, but they are prefixed with the keyword **rand**. If a free variable is unrestricted, it may assume *any* value at *any* time. Essentially free variables are similar to primary inputs to the model.

Free variables may only be declared in a checker body. They cannot be used as checker formal arguments (ports).

23.1.1 Free Variables in Assertions

We will study first the behavior of free variables in an assertion context, introducing the main ideas with examples. The following statement characterizes the semantics of free variables in the absence of assumptions:

> The behavior of a free variable in an assertion context corresponds to universal quantification over its domain in each clock tick.

Example 23.1. Consider the following checker:

```
checker failure;
  default clocking @$global_clock; endclocking
  rand bit v;
  a1: assert property(v);
endchecker : failure
```

Since v may assume any value in any clock tick, it may assume value 0 (i.e., false) in clock tick 0, and therefore assertion a1 fails.

Here is a more formal explanation. The formal semantics of the assertion **always** a, where a is a regular variable, is $\forall i\ w^i \models a$, which means that a holds in each clock tick on the trace w. If we want to stress that the value of a depends on the clock tick *i*, we can rewrite this formula as $\forall i\ w^i \models a^i$. In this formula, we assume that a is well defined, i.e., the values of a are given in each clock tick. In our case, when v is a free variable, it may assume any value in any clock tick, and therefore the formal semantics of the assertion a1 is $\forall i \forall v^i\ w^i \models v^i$, which is apparently false. The important consideration here is that the formal semantics of the free variable is obtained from a universal quantification over its domain in each clock tick. □

Example 23.2. Assertion `a2` in the following checker passes:

```
checker success;
  default clocking @$global_clock; endclocking
  rand bit v;
  a1: assert property(v || !v);
endchecker : success
```

The reason for the success of this assertion is that its underlying expression is a tautology: it holds for any values of `v`. The formal semantics of this assertion is $\forall i \forall v^i\; w^i \models v^i \vee \neg v^i$, and this formula is true. □

This example illustrates the general principle that at any given time, all references to the same free variable within the same assertion have consistent values. In FV, references to the same free variable in different assertions need not have consistent values, even at the same time. Put another way, referencing the same free variable in two assertions does not introduce any relation between the assertions.[1] We have the same situation as in first-order logic: the names of bound variables are not important.[2] For example, both statements $\forall x\; x \leq 0$ and $\forall x\; x > 0$ are false, though there is no single x that falsifies both $x \leq 0$ and $x > 0$.

Example 23.3. In the following checker

```
checker check;
  default clocking @$global_clock; endclocking
  rand bit v;
  a1: assert property (s_eventually v);
  a2: assert property (s_eventually !v);
endchecker : check
```

both assertions `a1` and `a2` fail. A counterexample for assertion `a1` is `v` = `0, 0, ...`, whereas a counterexample for `a2` is `v` = `1, 1, ...`. These two assertions do not have a common counterexample: on an infinite trace either `v` or `!v` must occur infinitely often. This fact does not prevent failure of both assertions because the references to `v` in different assertions need not have consistent values. □

Formal Semantics. We consider here only assertions without assumptions. The effect of assumptions is discussed in the next section. For simplicity we limit our discussion to variables of type **`bit`**.

We partition the set V of all model variables into the subset A of regular variables and the subset F of free variables: $V = A \cup F$ and $A \cap F = \emptyset$. Since we have assumed all the variables to be of type **`bit`**, a *valuation* of the variables is just an

[1] This is not true in simulation, where free variables are randomized and hold consistent values across all assertion statements. See Sect. 23.3.

[2] Ironically enough, bound variables in first-order logic correspond to free variables in SVA.

assignment to each variable in V of either 0 or 1 as a value. We can identify the valuations of V naturally with the power set 2^V, where $U \in 2^V$ corresponds to the valuation that assigns 1 to each element in U and 0 to each element in $V \setminus U$. Since $V = A \cup F$ is a partition, $2^V = 2^A \times 2^F$ and there is a projection $\pi : 2^V \to 2^A$ defined by $\pi(U) = U \cap A$. In terms of valuations, π retains the values assigned to the regular variables A and forgets the values assigned to the free variables F. For example, suppose $A = \{a_1, a_2\}$ and $F = \{f_1, f_2\}$. Let $U = \{a_1, f_2\}$. As a valuation, U assigns $a_1 = 1, a_2 = 0, f_1 = 0, f_2 = 1$. Then $\pi(U) = \{a_1, f_2\} \cap A = \{a_1\}$. As a valuation, $\pi(U)$ assigns $a_1 = 1$, $a_2 = 0$, and we simply drop the values assigned to f_1 and f_2.

We consider two kinds of traces, reduced and extended. A *reduced trace* is a trace w whose letters are valuations of regular variables only: $w^i \in 2^A$, $i = 0, 1, \ldots$. An *extended trace* is a trace W whose letters are valuations of all variables, both regular and free: $W^i \in 2^V, i = 0, 1, \ldots$. The projection $\pi : 2^V \to 2^A$ can be applied letter-wise to define *projection* of an extended trace to a reduced trace. We use the same notation π for this projection of traces: if W is an extended trace, then $\pi(W)$ is the reduced trace obtained from W by forgetting the values assigned to the free variables in each letter. In other words, $\pi(W) = w$ iff $\pi(W^i) = w^i$ for all $i = 0, 1, \ldots$.

Assertion satisfaction is naturally defined in terms of extended traces, where all variables that may be referenced by the assertion are included (see Sect. 21.3). To define formal semantics for free variables, we need to define assertion satisfaction in terms of reduced traces, which contain regular variables only. The definition is as follows. Assertion p is *satisfied* on reduced trace w, written $w \models p$, iff p is satisfied in the usual sense on all extended traces W that project to w. In other words,

$$w \models p \quad \text{iff} \quad \forall\, W \text{ s.t. } \pi(W) = w : W \models p$$

This universal quantification over extended traces makes precise the sense in which free variables are universally quantified over their domains.

23.1.2 Free Variables in Assumptions

What happens if free variables are referenced within assumptions? To keep the notation clear, we will assume that there is one free variable v, two assumptions q_1 and q_2 and two assertions p_1 and p_2. We will talk only about the meaning of the evaluation attempt starting at the beginning of the trace. Property p_1 in the presence of the assumptions is equivalent to property $q_1 \wedge q_2 \to p_1$. Analogously, p_2 is equivalent to property $q_1 \wedge q_2 \to p_2$. As we discussed in Sect. 23.1.1, the values of v in formulas $q_1 \wedge q_2 \to p_1$ and $q_1 \wedge q_2 \to p_2$ are taken independently. In order for $q_1 \wedge q_2 \to p_j$ to hold, it is sufficient to check on a given trace that p_j holds for all values of v that satisfy both assumptions q_1 and q_2.

Assumptions with free variables constrain the free variable domains for all assertions using these free variables.

Example 23.4. Consider the following checker:

```
checker check;
  default clocking @$global_clock; endclocking
  rand bit[5:0] v;
  m1: assume property (v > 2);
  m2: assume property (v < 7);
  a1: assert property (v > 1);
  a2: assert property (s_eventually v <= 5);
  a3: assert property (s_eventually v > 5);
endchecker : check
```

Assumptions `m1` and `m2` constrain free variable `v` such that `v` may only assume values 3, 4, 5 and 6 in all global clock ticks for all assertions. For all other values of `v`, all the assertions are understood to hold vacuously under the assumptions. This statement means that if `v` assumes some other value violating the assumptions, then we don't even need to check the assertions for that value.

Assertion `a1` holds since it holds for the constrained values of `v`. Assertions `a2` and `a3` fail, as explained in Example 23.3. □

Formal Semantics. The formal semantics of an assertion relative to a set of assumptions in the presence of free variables is as follows. Let $\mathcal{Q}$ be the set of assumptions, and let $Q = \bigwedge \mathcal{Q}$ be their conjunction. Reduced trace w *satisfies* assertion p *relative to* $\mathcal{Q}$, written $w, \mathcal{Q} \models p$, iff $w \models Q$ **implies** p, where Q **implies** p is treated as an assertion without assumptions. From the preceding section, $w \models Q$ **implies** p iff for every extended trace W such that $\pi(W) = w$, $W \models Q$ **implies** p. Now, $W \models Q$ **implies** p means that if $W \models Q$, then $W \models p$. We will say that an extended trace W is $\mathcal{Q}$-*feasible* iff $W \models Q$, i.e., iff $W \models q$ for all $q \in \mathcal{Q}$. Then we can say that

$$w, \mathcal{Q} \models p \quad \text{iff} \quad \forall\ \mathcal{Q}\text{-feasible } W \text{ s.t. } \pi(W) = w : W \models p$$

If the set $\mathcal{Q}$ of assumptions is understood, it may be dropped from the notation, and we may say more briefly that in the presence of assumptions,

$$w \models p \quad \text{iff} \quad \forall \text{ feasible } W \text{ s.t. } \pi(W) = w : W \models p$$

23.1.3 Free Variables in Cover Statements

As defined in Sect. 21.3.3, the property p is covered iff there is a trace that satisfies p. If p depends on a free variable v then in order for p to be covered it should be possible to choose values of v in each position on the trace so that p becomes satisfied. Therefore, we come to the following characterization of free variables in the coverage context:

> The behavior of a free variable in a coverage context is obtained from existential quantification over its domain in each clock tick.

Example 23.5. Consider the following checker:

```
checker check(bit a);
  default clocking @$global_clock; endclocking
  rand bit v;
  c1: cover property ((a && v)[*2]);
  c2: cover property ((a || v)[*2]);
endchecker: check
```

If `a` is always low, `c1` is not covered, while `c2` is. There are no values of `v` that could cause sequence `(a && v)[*2]` to match. If `v` is `1` in clock ticks 0 and 1, sequence `(a || v)[*2]` has a match in clock tick 1. □

If free variables are constrained with assumptions, these assumptions should be taken into account when evaluating the coverage.

Example 23.6. Consider the following checker:

```
checker check;
  default clocking @$global_clock; endclocking
  rand bit v;
  m1: assume property ($steady_gclk(v));
  c1: cover property (v ##1 !v);
endchecker : check
```

Cover statement `c1` cannot be satisfied because of assumption `m1` imposed on free variable `v`, which requires it to keep the same value all the time. Without `m1`, `c1` is covered in clock tick 1: it is sufficient to choose the value of `v` to be `1` in clock tick 0 and `0` in clock tick 1. □

Formal Semantics. Cover statement c is *satisfied* on reduced trace w iff there exists a feasible extended trace W that projects to w and satisfies c. In other words,

$$w \models c \quad \text{iff} \quad \exists \text{ feasible } W : \pi(W) = w \text{ and } W \models p$$

This existential quantification over extended traces makes precise the sense in which free variables are existentially quantified over their domains in the semantics of a cover statement.

23.2 Checker Modeling with Free Variables

Free variables may participate in regular checker modeling, though some limitations apply. In this section we discuss the semantics of free variables in the context of checker modeling and the specifics of their usage.

23.2.1 *Free Variable Initialization*

Free variables may be initialized with a declaration assignment. Initialization constrains only the initial value of a free variable; its values at other times are not affected by the initialization.

Example 23.7. Consider the following checker:

```
checker check1;
  rand bit rst = 1'b1;
  bit a = 1'b0;
  bit b = 1'b0;

  default clocking @$global_clock; endclocking
  default disable iff rst;

  always_ff @$global_clock
    a <= 1'b1;

  a1: assert property (a);
  a2: assert property (b);
endchecker : check1
```

The first tick of the global clock happens at time 0, and both assertions `a1` and `a2` are disabled since the value of `rst` at time 0 is `1`. At the next tick of the global clock, `a1` either is disabled or passes because the value of `a` has been set to `1`. On the contrary, the assertion `a2` fails because `b` is always `0` (it is not a free variable, so it keeps its value until it is again assigned), while `rst` is constrained only in global clock tick 0. For example, the trace `rst` = `1, 0, ...` violates assertion `a2`.

Now, let us modify our checker to be controlled by some general clock:

```
checker check2(event clk);
  rand bit rst = 1'b1;
  bit a = 1'b0;
  bit b = 1'b0;

  default clocking @clk; endclocking
  default disable iff rst;

  always_ff @clk a <= 1'b1;

  a1: assert property (a);
  a2: assert property (b);
endchecker : check2
```

In checker `check2`, assertion `a2` fails for the same reason as in checker `check1`. The behavior of assertion `a1` now depends on the waveform of `clk`: if `clk` ticks for the first time at time 0 then assertion `a1` will not fail, as in the checker `check1`. Otherwise, it will also fail since the value of `rst` is constrained only in global clock tick 0. □

Assignment to checker variables, including free variables, is forbidden in **initial** procedures to avoid contention. For example, if the code

```
bit v; // or rand bit v
initial v = 1'b0;
initial v = 1'b1;
```

were legal, what would the resulting initial value of `v` be?

Formal Semantics. A free variable initialization

```
rand some_type v = e;
```

is equivalent to the following assumption:

```
initial
  assume property (@$global_clock v === e);
```

Essentially, there is nothing specific for free variables here. The same definition is applicable to the initialization of any variable.

Using notation from research literature, we will also write the initialization as

$$\text{init}\, v = e,$$

and the initial assumption is just the Boolean property $v = e$.

23.2.2 *Free Variable Assignment*

A free variable may be written as the left-hand side of a nonblocking assignment (NBA). It is also possible to assign expressions containing free variables to regular variables in blocking, nonblocking and continuous assignments. A free variable may not be written as the left-hand side of a blocking or continuous assignment.

> It is illegal to assign to a free variable in a blocking or continuous assignment.

23.2.2.1 Unconditional Assignment to Free Variables

We consider first the case when the free variable assignment does not belong to the scope of any conditional or looping statement. Let us start with the following example.

Example 23.8. The following checker fragment defines a periodic clock, `myclk`:

```
rand bit myclk;
always_ff @$global_clock
   myclk <= !myclk;
```

In this example, `myclk` is assigned, but uninitialized. `myclk` changes its value with each tick of the global clock. Nevertheless, `myclk` is nondeterministic. Since `myclk` is uninitialized, it has two possible patterns: `0101...` and `1010....` □

Formal Semantics (Global Clock). The formal semantics of a free variable NBA controlled by the global clock

```
rand some_type v;
always_ff @$global_clock v <= e;
```

is defined as:

```
assume property (@$global_clock $future_gclk(v) === e);
```

For readability, we extend the formal notation from Chap. 21. Free variable assignment will be designated as $v' \leftarrow e$, meaning that the next state variable v' is assigned the value e. This is just a shortcut notation for the invariant defined by the assumption $G(v' = e)$. This invariant defines a transition relation that happens to be a function: the value of next state variable v' is uniquely determined from the values of the current state variables forming the expression e. Such a transition relation is called a *next state function*. Because of their explicit unidirectionality, free variable assignments may be handled more efficiently by FV tools than an equivalent set of assumptions.[3]

Example 23.9. The assignment form Example 23.8 can be rewritten as

```
assume property (@$global_clock $changing_gclk(myclk));
```

but this version might be less efficient since the explicit unidirectionality of the assignment of the free variable `myclk` has been lost. □

Formal Semantics (Arbitrary Clock). The semantics of free variable assignment $v' \leftarrow$ `@(`**`edge`** clk`)` e, which is controlled by clocking event `@(`**`edge`** clk`)`, is defined as $v' \leftarrow (clk \neq clk')$ `?` e `:` v. This means that v gets the value e only when the clock ticks, and it retains its old value when the clock does not tick. In this

[3]Note that NBAs in a checker may conflict with assumptions, but not with other assignments. In a checker, NBAs are allowed only within **`always_ff`** procedures (Sect. 9.2.2.2), and thus their targets cannot be overridden by other processes (Sect. 2.2.1.3).

case, the transition is $v' \leftarrow f(e, v, clk, clk')$, where the next state function is $f(e, v, clk, clk') = (clk \neq clk')$? e : v. This is a more general form than in the case of the global clock. There the right-hand side of the free variable assignment depended only on the current state variables, while here it depends also on the next state variable clk'. From a theoretical point of view, there is no problem allowing next state variables on the right-hand side of an NBA. SystemVerilog does not, however, allow explicit specification of next state variables on the right-hand side, and therefore they can be referenced there only implicitly, as through an event control.

Free variable assignments controlled by other forms of clocking event are treated in similar ways (see Exercise 23.1).

Example 23.10. We revisit Example 23.8 to define `myclk` to be synchronized by some specific clock `clk`:

```
rand bit myclk;
always_ff @clk  myclk <= !myclk;
```

Here, `myclk` ticks every tick of `clk`. Since the initial value of `myclk` is undefined, the code allows two patterns: `0101...` and `1010....` Note that according to the formal semantics of the free variable assignment, `myclk` holds a stable value between two consecutive ticks of the clock `clk`.

It is possible to replace the assignment with the following equivalent, but less efficient, assumption:

```
m1: assume property (@$global_clock
    $future_gclk(myclk) == ($changing_gclk(clk) ? !myclk : myclk)
);
```

One might be tempted to write the following assumption instead:

```
m2: assume property (@clk ##1 $changed(myclk));
```

`m2` states that `myclk` changes from every tick of `clk` to the next,[4] but assumption `m2` *is not equivalent* to the free variable assignment. `m2` only states that `myclk` has different values at consecutive ticks of `clk`, imposing no restrictions on the free variable behavior between the ticks of `clk`. The assignment semantics requires keeping the free variable stable between the ticks of `clk`. □

Efficiency Tip. It is preferable to assign free variables than to constrain them with assumptions.

What happens if there are several unconditional assignments to the same free variable within the same **`always_ff`** procedure? We assume that the free variable is one bit wide, because we can treat different bits of a free variable as different free variables. Recall that in checker **`always_ff`** procedures only nonblocking assignments are allowed (see Sect. 9.2.2.2). Therefore we have the following situation for free variable `v`:

---

[4]Recall (Sect. 7.2.2) that it is illegal to use future sampled value functions with arbitrary clocks, so we use a past value function here delayed by one clock tick.

```
v <= ...;
...
v <= ...;
```

Since nonblocking assignments are executed in order, only the last assignment counts and the preceding assignments to v are simply ignored.

23.2.2.2 Conditional Assignment to Free Variable

The formal semantics of a free variable assignment within a conditional or looping statement is not elaborated in the LRM. The description here is according to our understanding. We limit our consideration to the case when the **always_ff** procedure in which the free variable is assigned is controlled by the global clock.[5] The case of an arbitrary clocking event is obtained using the generalization described in Sect. 23.2.2.1 and is left to the reader (see Exercise 23.2). We limit our discussion to plain **if** and **if...else** statements. Other conditional statements of SystemVerilog can be rewritten using these forms. If a free variable assignment is within the scope of a looping statement, directly or otherwise, we assume that the looping statement is statically unrollable. We also assume that the loop management does not cause side effects.

Formal Semantics. The formal semantics is obtained by the recursive application of the following rewriting rules for free variable v.

1. Unroll all loops containing a free variable assignment (directly or otherwise). As a result, the free variable is not within the scope of any looping statement, directly or otherwise.
2. Here and in the remaining rules we consider direct scopes only. If v has multiple assignments within the same clause of a conditional statement, eliminate all the assignments except the last one (see Sect. 23.2.2.1). Recall that we consider different bits of a free variable as different free variables.
   For example,

   ```
   if (cond) begin
      ... // sequence of statements 1
      v <= expr1;
      ... // sequence of statements 2
      v <= expr2;
   end
   ```

   is rewritten as

   ```
   if (cond) begin
      ... // sequence of statements 1
   ```

---

[5]Since free variables may be assigned only within **always_ff** procedures and since a variable assigned in an **always_ff** procedure cannot be assigned in any other process (see Sect. 2.2.1.3), we can focus on a single **always_ff** procedure.

```
        ... // sequence of statements 2
        v <= expr2;
    end
```

3. Uniformize all conditionals to be of the form **if...else** by adding trivial **else begin end** to any **if** statement with no *else* clause. Further uniformize so that if v is assigned in a conditional statement, then v is assigned in both the *then* and the *else* clauses. Do this by adding the trivial assignment `v <= v;` to any clause where assignment to v is missing but needed.
4. If v is assigned in both clauses of a conditional statement

```
    if (cond) begin
        ... // sequence of statements 1
        v <= expr1;
        ... // sequence of statements 2
    end
    else begin
        ... // sequence of statements 3
        v <= expr2;
        ... // sequence of statements 4
    end
```

   extract it out of the scope of the conditional operator as follows:

```
    v <= cond ? expr1 : expr2;
    if (cond) begin
        ... // sequence of statements 1
        ... // sequence of statements 2
    end
    else begin
        ... // sequence of statements 3
        ... // sequence of statements 4
    end
```

   The semantic soundness of this extraction follows from the fact that only NBA can be written in a checker **always_ff**, and so the statements preceding the NBA in the conditional do not affect the value of its RHS.

23.2.2.3 Fully Assigned Free Variables

As we have seen, assignment to a free variable leaves it little freedom: only its initial value contributes nondeterminism.[6] If we also initialize it, there remains no freedom introduced by this variable. Such a free variable is said to be *fully assigned*. Essentially, there is no difference between a fully assigned free variable and a regular checker variable.

[6] If we assign one free variable to a second free variable, the second one may still be completely nondeterministic by virtue of the nondeterminism of the first. The point is that a free variable that is assigned loses its "own" nondeterminism. See also Sect. 23.2.2.4.

Example 23.11. Consider the following code:

```
checker check1(a, b, event clk = $inferred_clock);
  rand var type(b) v = a;
  always_ff @clk v <= b;
endchecker : check1
```

In checker `check1` free variable `v` is fully assigned, and it may be replaced with a regular checker variable, as in the checker `check2`:

```
checker check2(a, b, event clk = $inferred_clock);
  var type(b) v = a;
  always_ff @clk v <= b;
endchecker : check2
```

□

Thus, regular checker variables may be considered a special case of free variables, and the formal semantics of their NBA is a special case of the formal semantics of free variable NBA provided that the same limitations are imposed (e.g., they are not assigned in tasks, etc.)

23.2.2.4 Assigning Free Variables to Checker Variables

It is legal to assign an expression containing free variables to a regular checker variable; it is also possible to initialize a regular variable with an expression containing free variables. This means that the regular variables may also be nondeterministic. However, the regular variables do not introduce any new nondeterminism to the system; their values are completely defined by the values of the variables on which they depend.

Example 23.12. Consider the following checker fragment:

```
bit [3:0] a;
rand bit [3:0] v;
always_ff  @clk a <= v;
```

If the value of `v` is nondeterministic, then starting from clock tick 1 the value of `a` is also nondeterministic. The initial value of `a` is `0` by default—only a free variable can introduce nondeterminism through its initial value, and then only if it has not been explicitly initialized. Nevertheless, `a` is deterministic in the following sense: its value is uniquely defined by the value of `v`; `a` does not introduce any new nondeterminism. □

Example 23.13. Nondeterminism can also be introduced by an event control, as illustrated by the following checker fragment:

```
rand bit clk;
bit [3:0] a;
always_ff @clk
   a <= e; // e - deterministic expression defined elsewhere
```

In this example, the clocking event `@clk` is nondeterministic, and so also is the value of `a` because the assignment is performed at nondeterministic time moments. However, `a` is deterministic in the sense that it does not introduce any new nondeterminism: its value is fully defined by the values of `e` and `clk`.

The nondeterminism associated with the clocking event is not of an essentially different kind because the effect of the clocking event could be pushed into the right-hand side of a checker variable assignment:

```
always_ff @clk  a <= e;
```

has the same formal semantics as

```
always_ff @$global_clock  a <= $changing_gclk(clk) ? e : a;
```

had the latter statement been legal in SystemVerilog. □

A situation similar to assignment of a free variable to a checker variable arises when a free variable of one checker is passed to a function or to another checker as an input argument. Formal arguments of a checker can never be declared as free variables.

Example 23.14. Checker `check1` below defines a free variable `v` and passes it to checker `check2`.

```
checker check1;
  rand bit v;
  check2 c2(v, $global_clock);
endchecker : check1

checker check2(a, clk = $inferred_clock);
  a1: assert property (@clk not always a iff s_eventually !a);
endchecker : check2
```

In checker, `check2` formal argument `a` is untyped. We could define it to be of type **`bit`**, but *not* of type **`rand bit`** since formal checker arguments cannot be free variables. □

Though it is illegal to assign free variables by continuous or blocking assignments, they may be referenced on the right-hand sides.

Example 23.15. In FV modeling, it may be useful to define an expression that may assume only values from a specific set, for example, only values 2 and 5. A straightforward way to do this is to define a three-bit free variable and add an assumption constraining its value, such as:

```
rand bit [2:0] v;
m1: assume property(@$global_clock v == 3'd2 || v == 3'd5);
```

A better solution is to define a one-bit free variable and to use a continuous assignment to a regular variable instead:

```
bit [2:0] v;
rand bit x;
assign v = x ? 3'd2 : 3'd5;
```

The latter implementation is usually more efficient because it replaces a three-bit free variable with a one-bit free variable. FV efficiency significantly decreases with each free bit added. Another advantage is that using free variables in assignments is more efficient than in assumptions □

The situation with passing free variables to functions is similar. Though it is illegal to declare free variables within a function, either in a function body or among function ports, it is legal to pass free variables to a function, as illustrated in the following example.

Example 23.16. The following code may be used to find a maximum among two free variables of size N, where N is an elaboration-time constant.

```
rand bit[N] v1, v2;

function bit[N] max(bit[N] a, b);
   if (a >= b) return a;
   return b;
endfunction : max
```

Though function max operates with regular variables only, it is legal to pass free variables v1 and v2 to it. □

Why is it forbidden to assign free variables by continuous or blocking assignment? The main reason is that it is difficult to define acceptable simulation semantics in this case (see Sect. 23.3). There is also a problem of combinational loops. Consider, for example, the following illegal code:

```
rand bit a, b;
bit c;
// c assigned elsewhere
assign a = ~b & c;
assign b =  a & c;
```

If c changes from 0 to 1 then it is completely unclear what should happen with a and b.

Note that the same problem exists if a and b are defined as regular variables. In simulation this causes an infinite iteration. This is why in SystemVerilog 2009 it was illegal to use any continuous or blocking assignments in checkers. However, this problem also exists in modules, interfaces and programs. It is usually solved at a tool level. Simulators may detect an infinite iteration, abort it and issue an error message. Synthesis and FV tools detect combinational loops statically.[7] Because the problem is not specific to checkers, SystemVerilog 2012 relaxes the restrictions and allows continuous and blocking assignments to regular variables in checkers.

A contention problem, when the same variable is assigned in different processes, does not exist in checkers. If a variable is assigned in one checker process it cannot

---

[7] Simulators may also do this.

be assigned in another (see Sect. 2.2.1). Nor can a variable be assigned by two continuous assignments, according to general SystemVerilog rules.[8]

Formal Semantics. It is difficult to give a general definition of the formal semantics of continuous and blocking assignments to regular checker variables because of various constructs, such as task calls in procedures and general looping statements. If we avoid these constructs and assume that there are no side effects for the variables in question, that all the loops are statically unrollable, and that there are no combinational loops, we can provide a formal semantics for continuous and blocking assignments.

Consider the continuous assignment

```
assign v = e;
```

where v is a regular variable and expression e does not depend on v. Its formal semantics is given by the assumption

```
assume property (@$global_clock v === e);
```

Expression e may contain free variables.

In the notation of research literature, the assignment is written $v \leftarrow e$, and the equivalent assumption is the invariant $G(v = e)$. The transition relation for v is in this case a function. This function is called the *current state function*, stressing that it assigns the current, rather than the next, state value of v (cf. Sect. 23.2.2.1).

It remains to provide a formal semantics for blocking assignments to regular checker variables. This task is much more complicated and requires a thorough consideration of many details beyond the scope of this book. We provide only a sketch. In the theory of compilation it is proven that any synthesizable program can be represented in a Static Single Assignment (SSA) form [31] (see also [51]).[9] In SSA form each **always_comb** and **always_latch** procedure is represented equivalently using only continuous assignments, whose semantics has been given. But the situation is complicated by latch modeling, which introduces incomplete assignments and NBAs. Detailed description of latch modeling is beyond the scope of this book. We will just illustrate the concept in the following example.

Example 23.17. Consider the variable declarations

```
bit c, i;
bit q = 1'b0;
```

A simple latch may be defined either using an NBA, as in

```
always_latch
   if (c) q <= i;
```

or using a blocking assignment:

---

[8]Multiple continuous assignments are allowed only for nets.

[9]The program does not have to be synthesizable to be represented in SSA form. The synthesizability is required for an efficient representation of the ϕ-function.

```
always_latch
    if (c) q = i;
```

Both representations are equivalent from the perspective of formal semantics, and we will consider the latter representation. The assignment to `q` is *incomplete* in the sense that it is executed only when condition `c` is true. The rewriting produces a continuous assignment with a combinational self-loop: the assignment right-hand side depends on its target:

```
assign q = c ? i : q;
```

To reflect the dependency of the new value of `q` on its previous value, this assignment should be modeled in terms of its next state variables, and therefore it is represented with the following next state function:

$$q' \leftarrow c' \ ? \ i' : q$$

Representation of the initialization of `q` is straightforward:

$$\text{init } q \leftarrow 0$$

□

23.2.2.5 Checker Data Model

We give here a summary of the FV model build from the modeling code of a synthesizable checker. We assume that all variables are split into individual bits. Each variable v introduces one and only one of the following options:

- v has neither assignment nor initialization.
- Initialization init $v \leftarrow e_1$, corresponding to Boolean assumption $v = e_1$. Here e_1 is an expression containing current state variables only.
- Next state function $v' \leftarrow e_2$, corresponding to assumption $G(v' = e_2)$. Here e_2 is an expression that may contain both current and next state variables.
- Both initialization and next state function.
- Current state function $v \leftarrow e_1$, corresponding to invariant assumption $G(v = e_1)$. Here e_1 is an expression containing current state variables only.

The current state function is actually redundant, as it is equivalent to the combination of an initialization and a next state function:

$$\text{init } v \leftarrow e_1; \ v' \leftarrow e_1'$$

Essentially the same transition relation is created to build an FV model for general SystemVerilog modules and interfaces, provided they conform to restrictions for "FV synthesizability". The details of such a build are much more complicated, involving multiple assignments to nets, general purpose **always** procedures, **initial** procedures, etc. Support for an FV model build has been tacitly assumed in Sect. 21.2.

23.2.3 Example: Building Abstract Models with Checkers

Checkers with free variables can be used to build nondeterministic models for FV.

Example 23.18. Consider a pipeline consisting of `n` stages. Each stage has its own enabler signal `en[i]`, where `i` is the number of the stage. This signal `en[i]` is asserted when the `i`-th stage is ready to process the new data. When the data is ready at the pipeline input, the `data_ready` flag is asserted. We want to write a checker verifying that the pipeline latency (the number of clock ticks required to pass through the pipeline) does not exceed `max_latency`.

The RTL implementing this model may be very complex, but the exact data and exact operations performed in the pipeline are not important for our purpose. We can build an abstract model as a checker that will only take into account the progress of the data in the pipeline. This checker is shown in Fig. 23.1.

At the beginning of the checker, there are definitions of the checker variables and several **`let`**-definitions provided for convenience. `ub(x)` (Line 7) defines an upper bound of a vector that can store the value `x`. The lower bound of the vector is understood to be `0`. For example, if `x = 5` then 3 bits are needed to store `x`. The upper bound of this vector will be `2` assuming that its lower bound is `0`.

The variable `stage` (Line 14) contains the number of the active stage, the first active stage having number `1`. The `stage` can also assume two dummy values: stage `0`, meaning that the data has not been sent to the pipeline yet, and stage `nstages` (Line 8), which exceeds the number of actual stages by one, signaling that data has been fully processed by the pipeline.

The variable `latency` (Line 15) keeps the current latency of the data. It is limited by `max_latency + 1` (`big_latency`, Line 10)—in our implementation, if the `latency` reaches this value, it is not incremented anymore, as the exact value of the latency is not important, only the fact that the latency has exceeded the maximum allowed value.

The main point in this checker is to understand when the transaction starts. There may be many simultaneous transactions in the system, different data may be simultaneously processed by different stages of the pipeline, and therefore, checking the condition `data_ready` alone is not sufficient for counting the stage and latency. The checker needs to be able to track any of these transactions, but by taking advantage of nondeterminism it does not have to allocate and manage resources to track them all at once. It only needs resources to track one!

The key here is to introduce free variable `start` (Line 13) and to start counting the first time that both `start` and `data_ready` are simultaneously true. The role of `start` is to choose nondeterministically which among all the possible transactions the checker will track. Since `start` is free to take any value in each clock tick, we effectively consider all combinations of possible attempts, and among them also all possible single attempts. For convenience, we introduce variable `go` (Line 18) for the simultaneous occurrence of `data_ready` and `start`.

For example, assume that `data_ready` is high at clock ticks 10 and 20. The following scenarios are possible:

```
checker check_latency(en, data_ready, max_latency,
                      event clk = $inferred_clock,
                      untyped rst = $inferred_disable);
  default clocking @clk; endclocking
  default disable iff rst;

  let ub(x) = $clog2(x + 1) - 1;
  let nstages = $bits(en) + 1;
  let stage_ub = ub(nstages);
  let big_latency = max_latency + 1;
  let latency_ub = ub(big_latency);

  rand bit start;
  bit [stage_ub:0] stage = '0;
  bit [latency_ub:0] latency = '0;

  bit go;
  assign go = data_ready && start;

  function type(stage) next_stage;
    if (rst) return 0;
    if (stage == 0 && go) return 1;
    if (stage != 0 && stage != nstages && en[stage])
      return stage + 1;
    return stage;
  endfunction : next_stage

  function type(latency) next_latency;
    if (rst || latency == 0 && !go) return 0;
    if (latency == big_latency) return big_latency;
    return latency + 1;
  endfunction : next_latency

  always @clk begin
    stage <= next_stage();
    latency <= next_latency();
  end

  a1: assert property (
    $rose(stage == nstages) -> latency != big_latency);
endchecker: check_latency
```

Fig. 23.1 Checking pipeline latency

1. `start` is low at both clock ticks 10 and 20.
2. `start` is low at clock tick 10 and high at clock tick 20.
3. `start` is high at clock tick 10 and low at clock tick 20.
4. `start` is high at both clock ticks 10 and 20.

In the first, we never start counting and do not check anything; in the second, we check only the transaction starting at clock tick 20; in the third, we check only the transaction starting at clock tick 10; and in the fourth, we check the transaction starting at clock tick 10, and when the second transaction comes at clock tick 20, it is ignored. Why? The checker tracks only one transaction! Essentially, in the fourth case we check the same scenario as in the third one, since a subsequent `go` does not restart counting. FV considers all four scenarios together when it universally quantifies the free variable, so we will be able to detect a maximal latency violation for both transactions (if it happens): for the first transaction in the third and in the fourth scenarios, and for the second transaction in the second scenario.

Lines 20–26. Initially, `stage` is 0, and it is not incremented until `go` becomes 1, which may happen in an arbitrary clock tick when the `data_ready` is high, as `start` is a free variable. Otherwise, the `stage` number is incremented when the `en` control of the corresponding stage is asserted, provided the data has not passed through the entire pipe, i.e., provided `stage != nstages`.

Lines 28–32. Initially, the value of `latency` is 0, and it is not incremented until free variable `start` is asserted and `data_ready` is high. Then it is incremented each clock tick until the `max_latency` is exceeded (the value of `big_latency` is reached). Then it remains stuck at the value of `big_latency` forever.

Assertion `a1` (Line 39) checks that when data leaves the pipeline, that is when `stage` becomes `nstages` for the first time, `latency` should not exceed `max_latency`. Note that in this case it is safe to use the sampled value function `$rose` without an initial delay in the antecedent (see Sect. 7.2.1.3), since by construction, `stage` cannot assume the value `nstage` at the beginning or immediately after the `rst` deactivation. Note also that we chose `latency != big_latency`, and not `latency < big_latency`, since the synthesis model of the former expression is more efficient than that of the latter, and therefore the former expression is also more efficient in FV than the latter.

Discussion: In our implementation, free variable `start` is unconstrained, and hence it can freely oscillate on the trace. Although, as we have seen, this oscillation does not affect FV correctness, it may affect FV performance,[10] as many redundant traces are considered. For instance, the fourth scenario is redundant, as was explained above. We can use the following assumption to get rid of these redundant scenarios:

```
m1: assume property (start |=> always !start);
```

This assumption says that `start` may be high at most in one clock tick.

We might want to get rid of other useless behaviors, those in which `start` does not happen at all or in which it is not synchronized with `data_ready`. This can be remedied with the following assumption:

```
initial
  m2: assume property (s_eventually(start && data_ready));
```

However, assumption `m2` is likely to affect negatively the performance of some FV tools. □

[10] This is highly dependent on the specific tool.

23.3 Free Variables in Simulation

So far we have discussed the formal semantics of free variables. In this section, we discuss the behavior of free variables in simulation.

As their syntax suggests, free variables are randomized in simulation. All uninitialized free variables are randomized initially. Otherwise, only unassigned free variables are randomized. The following rules are applicable to all randomized free variables:

- A randomized free variable may change its value at most once during each simulation tick.
- If a randomized free variable gets a new random value in some simulation tick, this value becomes ready before the Observed region.
- A sampled value of a free variables is its current value.

The sampled value of a free variable coincides with its current value.

23.3.1 *Unconstrained Free Variables*

If there are no assumptions in a checker or any of its child checkers, and a checker variable is neither assigned nor initialized then it can get a new value at any time step.

Example 23.19. In the original code of Example 23.18, there are no assumptions, and so the simulator may assign free variable `start` a new random value at any simulation step. One extreme case would be assigning it a random value only once, at its initialization, and then leaving it unchanged. The other extreme case would be assigning it a new random value at each simulation step. Both behaviors are legal. However, the first behavior is too "boring", while the second one is an overkill: there is no advantage in changing the value of `start` every simulation tick since all the assignments and the assertions where it is used are controlled by `clk`. Therefore, most simulators will assign `start` a new random value at every tick of `clk`. □

23.3.2 *Assigned Free Variables*

If a free variable is assigned a value, then it is only randomized at its initialization, provided it is not initialized. If a free variable is unassigned, but initialized, it is randomized, but not at its initialization. If a free variable is both initialized and assigned, it is not randomized.[11]

Example 23.20. Consider the following code:

```
checker check(..., clk = $inferred_clock);
  rand bit v;
  rand bit w = 1'b0;

  always_ff @clk
   v <= ...;
   ...
endchecker : check
```

In checker `check`, free variable `v` is randomized at its initialization only, whereas free variable `w` may be randomized at every simulation step except for step 0, where the value of `w` is 0. □

23.3.3 Checkers with Assumptions

The set of all assumptions contained in a checker instance and in its child checkers is called the *assume set* of the checker instance.[12] If the assume set of a checker instance is nonempty, then the unassigned free variables are randomized in each tick of any clock event used in any assumption of the assume set. The random values of the free variables should be chosen in a way to satisfy all the assumptions in the assume set, if possible—this means that no assumption from the set should fail at the simulation step when the new randomization values have been assigned if such values exist. However, lookahead analysis is not required in constraint solving. In time steps that are not ticks of any clock in the assume set, the unassigned free variables may be randomized arbitrarily.

Example 23.21. The assume set of checker `check1` consists of single assumption `m1`. This assumption is clocked by two clocks, `clk1` and `clk2`.

```
checker check1(sequence s, ..., event clk1, clk2);
  rand bit [3:0] v, w, r;
  ...
  m1: assume property (@clk1 s |-> @clk2 v + w < 7);
endchecker : check1
```

[11]If a free variable is of an aggregate data type—array or structure, some of its elements may be assigned, while others remain unassigned. The LRM defines that all elements of the singular data types (i.e., of all data types except unpacked ones) should be randomized monolithically, whereas different elements of the unpacked data types may be randomized independently: some elements may be randomized, while other may not.

[12]There are several subtleties concerning assumption set definition for procedural checker instantiations and for instances involving **const** cast of checker arguments. These are not covered here. Refer to the LRM [8] for the exact definitions.

There are three free variables v, w, and r in the checker, all of them uninitialized and unassigned. All three variables are randomized every tick of clk1 and of clk2, even though r does not enter assumption m1. The randomized values of r are unconstrained, whereas the randomized values of v and of w are constrained by assumption m1. □

Example 23.22. In the code fragment below, checker check1 from Example 23.21 is rewritten so that assumption m1 has been moved to separate checker c2. Although there are no more assumptions in checker check1, the assume set of checker check1 consists of assumption mycheck.m1, and the simulation of free variables v, w, and r is exactly the same as in Example 23.21.

```
checker check1(sequence s, ..., event clk1, clk2);
  rand bit [3:0] v, w, r;
  ...
  check2 mycheck(s, v, w, clk1, clk2);
endchecker : check1

checker check2(sequence s, untyped a, b, event clk1, clk2);
  m1: assume property (@clk1 s |-> @clk2 a + b < 7);
endchecker : check2
```

□

Example 23.23. In the following checker

```
checker check1(sequence s, ..., event clk1, clk2);
  rand bit [3:0] v, w, r;
  ...
  always @clk1
    w <= v + 1;
  m1: assume property (@clk1 s |-> @clk2 v + w < 7);
endchecker : check1
```

free variable w is randomized *only* at its initialization, as it is assigned a value. Free variables v and r are randomized every tick of clocks clk1 and clk2. The random constraint imposed by assumption m1 is used to randomize the value of v. □

If the randomization constraint imposed by the checker assumptions cannot be solved, the free variables may be assigned arbitrary values, and the corresponding assumptions fail.

Example 23.24. In the following checker

```
checker check(..., event clk);
  default clocking @clk; endclocking
  rand bit [3:0] v, w;
  ...
  m1: assume property (v + w < 2);
  m2: assume property (v + w > 3);
endchecker : check
```

assumptions m1 and m2 cannot be satisfied together. Arbitrary values of v and w will be chosen in each clock tick, and the failure of an appropriate assumption will be reported. For example, if in clock tick 0 the value 0 was assigned to both v and w, assumption m1 would pass and assumption m2 would be violated. If in clock tick 1, v gets the value of 1 and w gets the value of 2, then both assumptions will be violated. □

Example 23.25. In the following checker

```
checker check(..., event clk);
  rand bit v;
  ...
m1: assume property (@clk v |-> ##3 1'b0);
endchecker : check
```

the value of free variable v must always be low to satisfy assumption m1: if in some clock tick v is high, assumption m1 will fail in three clock ticks. However, in simulation there is no obligation to satisfy the assumptions in future clock ticks, and your simulator can choose the value 1 for v in some clock tick, as it does not result in violation of the assumption in the same clock tick. Of course, in this case, assumption m1 will fail in three clock ticks. □

If there are deferred (or final) assumptions in the assume set,[13] all checker free variables are randomized at every tick of a clock from a concurrent assumption from the assume set and, in addition, they may also be randomized in any other simulation time step.[14]

Example 23.26. In the following checker

```
checker check(..., event clk);
  rand bit v, w;
  ...
  m1: assume final ($onehot0({v,w}));
  m2: assume property (@clk v |=> w);
endchecker : check
```

free variables v and w must be randomized every tick of clk and, in addition, they may be randomized in any simulation tick. □

Efficiency Tip. Constraining free variables by assumptions significantly slows down the simulation and sometimes even leads to bogus assumption violations. It is always preferable to use unconstrained free variables and their assignments whenever possible. However, from a methodological point of view assumptions on the model interface may be desirable for assume-guarantee reasoning.

---

[13]One should avoid using free variables in all forms of assertion statements but concurrent, see Sect. 23.3.4.

[14]This is our interpretation of the standard. The standard is not explicit about free variable randomization with deferred assumptions.

Free variable support in simulation does not provide correctness confidence even for a given simulation trace, but just shows one arbitrary realization of the free variables in time. In the presence of temporal assumptions, an assumption failure in simulation does not necessarily signify incorrectness of the implementation.

23.3.4 Limitations Imposed on Free Variables

Why are continuous and blocking assignments to free variables illegal? One reason was explained in Sect. 23.2.2.4. There is another, more compelling reason related to the simulation semantics of free variables. Had continuous assignments to free variables been permitted it would be reasonable to expect the following statements to behave similarly for a free variable `v` and an expression `e`:

```
assign v = e;
assume property (@global_clock v === e);
```

From the assumption we know that the sampled values of `v` and `e` coincide (at the ticks of the global clock). This requires a right-hand side of a continuous assignment in checkers to be also sampled. However, such a definition would not work for a continuous assignment to regular checker variables. The same problem exists also for blocking assignments.

An attempt to define the simulation semantics of free variable assignment in a different manner, and require their right-hand side to be sampled, would not work, either: Since continuous and blocking assignments in checkers are executed in the Reactive region, assertions would miss new values of free variables, as shown in the following example.

Example 23.27. Consider the following code fragment assuming that continuous assignments to free variables are legal and their right-hand side is sampled:

```
1 rand bit v;
2 bit a;
3 ...
4 assign v = a;
5 a1: assert property(@$global_clock v == a);
```

The assignment in Line 4 would execute in the Reactive region, and the value of `a` would be sampled. Let regular variable `a` change from 0 to 1. Then the value of `v` would only change in the Reactive region, whereas assertion `a1` executes in the Observed region, and it would see the stale value of `v`, and therefore, would fail, which is counterintuitive (cf. Exercise 23.8). □

Unfortunately, there is one more source of inconsistency related to the simulation semantics of free variables: SystemVerilog allows using free variables in deferred assertions. Their simulation semantics looks innocent, and Example 23.26 shows a reasonable usage. Consider now a different example:

```
rand bit v;
bit a;
...
m1: assume final(v == a);
a1: assert property (@$global_clock v == a);
```

Suppose that `a` transitions from 0 to 1. According to Sect 23.3.3, `v` receives the value 1 *before* the Observed region. Concurrent assertion `a1` is executed in the Observed region. The sampled value of `v` is 1 (its current value), and the sampled value of `a` is 0 (its preponed value), so that the assertion fails. Therefore, using free variables in deferred (or final) assertion statements is a bad idea.

> Never use free variables in deferred (or final) assertion statements.

23.4 Rigid Variables

23.4.1 Rigid Variables in Formal Verification

Free variables defined in Sect. 23.1 can assume any value at any simulation step (unless constrained by assumptions). *Rigid variables* may assume any value initially, but their value does not change in time. In other words, rigid variables are constant free variables, which fact is reflected in their syntax. For example,

```
rand const bit [3:0] r;
```

is a declaration of rigid variable `r`. This variable may assume any 4-bit value, but this value remains unchanged all the time. Note that the **rand** qualifier must appear first; **const rand** is illegal.

It is straightforward to model rigid variables with free variables. For example, the above declaration of rigid variable `r` is equivalent to:

```
rand bit [3:0] r;
assume property(@$global_clock $steady_gclk(r));
```

This is, essentially, the formal semantics of free variables: rigid variable r is a free variable with additional assumption $G(r = r')$ imposed (we use research literature notation here).

Rigid variable initialization is legal, but it renders the rigid variable equivalent to a non-free constant variable:

```
rand const bit [3:0] r = 4'd5;
```

is equivalent to

```
const bit [3:0] r = 4'd5;
```

We will always assume that rigid variables are uninitialized. Of course, it is illegal to assign a rigid variable, as it is illegal to assign any constant variable.

Example 23.28. Checker `same1` verifies that two large data words contain the same value:

```
checker same1(bit [127:0] word1, word2,
  event clk = $inferred_clock);
  a1: assert property (@clk word1 == word2);
endchecker : same1
```

This equality check is expensive as the size of the words is big. For some FV tools,[15] it might be more efficient to choose an arbitrary bit and check that its values in both words coincide. This may be achieved by introducing a rigid variable `i` that indicates which bit in the words to compare, as implemented in checker `same2`:

```
checker same2(bit [127:0] word1, word2,
  event clk = $inferred_clock);
  rand const bit [6:0] i;
  a2: assert property (@clk word1[i] == word2[i]);
endchecker : same2
```

Assertion `a2` compares only one bit in two words instead of comparing 128 bits as in assertion `a1`. This comes, however, at a price of introducing a 7-bit rigid variable `i`. □

Example 23.29. Checker `data_consistency1` verifies that `out_data` at the end of a transaction (`end_t` asserted) has the same value as `in_data` at the beginning of the transaction (`start_t` asserted).

```
checker data_consistency1(start_t, end_t, in_data, out_data,
                          event clk = $inferred_clock,
                          untyped rst = $inferred_disable);
  default clocking @clk; endclocking
  default disable iff rst;
  rand const var type(in_data) data;

  a1: assert property (start_t && data == in_data ##1 end_t[->1]
                       |-> out_data == data);
endchecker : data_consistency1
```

Assume for simplicity that both `in_data` and `out_data` are of type **bit** `[1:0]`. Then assertion `a1` is equivalent to the set of the four assertions:

```
a10: assert property (start_t && 2'b00 == in_data ##1 end_t[->1]
                      |-> out_data == 2'b00);
a11: assert property (start_t && 2'b01 == in_data ##1 end_t[->1]
                      |-> out_data == 2'b01);
```

[15] It highly depends on the specific tool.

```
a12: assert property (start_t && 2'b10 == in_data ##1 end_t[->1]
                      |-> out_data == 2'b10);
a13: assert property (start_t && 2'b11 == in_data ##1 end_t[->1]
                      |-> out_data == 2'b11);
```

Thus, specifying rigid variable `data` is equivalent to checking the data correspondence for all possible `in_data` values.[16]

We can rewrite assertion `a1` using local variables instead of rigid variables:

```
checker data_consistency2(start_t, end_t, in_data, out_data,
                          event clk = $inferred_clock,
                          untyped rst = $inferred_disable);
  default clocking @clk; endclocking
  default disable iff rst;

  property data_consistent;
    var type(in_data) data;
    (start_t, data = in_data) ##1 end_t[->1]
      |-> out_data == data;
  endproperty : data_consistent

  a2: assert property (data_consistent);
endchecker : data_consistency2
```

Using rigid variables in checker `data_consistency1` is even syntactically similar to local variables, only instead of assigning to a local variable, there is a comparison with a rigid variable. In this sense, local and rigid variables are interchangeable when in the beginning of the transaction they store a value of some data and then check this value later during the transaction. □

23.4.2 Rigid Variable Support in Simulation

In simulation, rigid variables are randomly initialized, and then their values remain unchanged. If there are assumptions imposed on rigid variables, a simulator does not have any obligation to take these assumptions into account when assigning initial random values to rigid variables. Therefore, there is no guarantee to fulfill even straightforward assumptions imposed on rigid variables.

The simulation of rigid variables does not fully reflect their nature. Instead of checking an assertion for all possible values of rigid variables, only one random value is checked. For instance, if in simulation of checker `same2` from Example 23.28, variable `i` was initialized with 47, only equality of the 47-th bit in two words would be checked.

[16] In PSL, the construct similar to rigid variables is even called *forall*.

23.4.3 Rigid and Free Variables Versus Local Variables

Rigid variables are typically used for latching: they effectively store a value at the beginning of a transaction, and then check it at the transaction end, as shown in Example 23.29. This usage is also common for local variables. The advantage of rigid variables is their straightforward implementation in FV, while not all assertions with local variables are supported by FV tools. On the contrary, local variables have full support in simulation, while the rigid variable support is rather poor. Both assertion implementations look very similar.

> It is recommended to use local variables rather than rigid variables, unless there is a restriction on local variables imposed by an FV tool.

Efficiency Tip. Rigid variables are expensive in FV, but they are significantly less expensive than the free variables of the same size. When rigid variables may be used interchangeably with local variables, the efficiency of the rigid variables and of the local variables in FV is about the same, provided that the FV tool can handle local variables efficiently.

Sometimes it is possible to achieve the same goal by using local variables instead of free variables. As opposed to rigid variables, here we are talking about free variables of a *smaller* size than corresponding local variables. For instance, the checker described in Example 23.18 can be implemented as an assertion with local variables, as discussed in Exercise 23.3, but in this case one-bit free variable `start` would be replaced by local variables of a larger size. Both approaches have their own advantages and drawbacks:

- Processing of unconstrained free variables is straightforward for FV tools. Although free variables impose a heavy burden on FV, they are efficiently synthesized. Sophisticated assertions with local variables may be difficult to synthesize, and their synthesis may introduce substantial penalty *in addition* to the penalty introduced by free variables.
- If an abstract model requires complex modeling, it is difficult to keep assertions with local variables of manageable size. Some modularity may be achieved by partitioning complex properties into subproperties. Splitting these assertions into several smaller assertions is problematic, since different assertions cannot share their local variables, and one will have to duplicate many common parts of the modeling in different assertions. On the contrary, when using free variables the modeling can be shared.
- For manageable models, using local variables is more intuitive and readable than using free variables.
- Local variables can easily be checked in simulation, whereas free variables have only partial simulation support.

23.5 Checkers as Generators

Checkers may be used as input stimuli generators for a DUT, i.e., as a synthesizable testbench. The major advantage of such a testbench is its natural support both in simulation and in FV. Writing synthesizable testbenches saves the effort spent on creation of two different environments. A synthesizable testbench is useful even when created only for FV because it provides an excellent tool for debugging the FV model correctness in simulation.

Let us start with a simple example in which two inputs of a DUT are restricted to be mutually exclusive. For simplicity we assume that these inputs are of type **`bit`**.

Example 23.30. Conventional implementation of mutually exclusive stimuli.

Using conventional SystemVerilog testbench constructs — programs, classes, randomization and constraints, — this testbench may be implemented as shown in Fig. 23.2. The program implementing this testbench declares class `Mutex` (Lines 2–7) to represent a data structure for two random bits. These bits are defined on Line 3. Here **`rand`** qualifier specifies only randomness; it is not to be confused with the checker specification of free variables. These random bits are constrained using the SystemVerilog **`constraint`** construct (Lines 4–6). This constraint has name `c`, and its body contains a condition for the mutual exclusion of `x` and `y` (Line 5).

The stimuli are generated in the **`initial`** procedure of the program (Lines 11–20). Line 9 instantiates class `Mutex` and names the created object `vals`. The value generation is done in the loop (Lines 13–19). Lines 14–15 generate the

```
1  program gen_mutex(output bit out1, out2, clk);
2    class Mutex;
3      rand bit x, y;
4      constraint c {
5        $onehot0({x, y});
6      }
7    endclass : Mutex
8
9    Mutex vals = new;
10
11   initial begin
12     clk = 1'b1;
13     for (int i = 0; i < 100; i ++) begin
14       #5 clk = !clk;
15       #5 clk = !clk;
16       vals.randomize();
17       out1 = vals.x;
18       out2 = vals.y;
19     end
20   end
21 endprogram : gen_mutex
```

Fig. 23.2 Conventional testbench generating mutually exclusive stimuli

```
1 checker gen_mutex(output bit out1, out2);
2   default clocking @$global_clock; endclocking
3   rand bit x, y;
4   m1: assume property($onehot0({x, y}));
5   assign {out1, out1} = {x, y};
6 endchecker : gen_mutex
```

Fig. 23.3 Synthesizable testbench generating mutually exclusive stimuli

clock and advance the simulation time. Function `randomize` (Line 16) randomizes values of class member variables `x` and `y` taking mutual exclusion constraint `c` into account. Lines 17–18 assign the generated values of `x` and `y` to output ports `out1` and `out2` of the program. □

Compare now the implementation of Example 23.30 with the checker-based implementation shown in Example 23.31.

Example 23.31. Checker-based implementation of mutually exclusive stimuli.
This implementation is shown in Fig. 23.3. Here, `x` and `y` are free variables (Line 3), and they are randomized by assumption `m1` (Line 4). What remains is to copy the values of the free variables to the checker output ports, which is done by the continuous assignment statement on Line 5. We have to copy the free variables to the output ports because the checker ports cannot be free variables. □

We conclude this section with a more meaningful example of a synthesizable testbench. This testbench implements a packet generator for the 1-bit serial interface described in Sect. 1.2.1. The output of the testbench is the input of a DUT that we need to verify. The packet generator produces packets and transmits them as transactions bit by bit to the DUT. The verbal specification of the serialization protocol is as follows.

TX1 The following packet (or transaction) types are supported:

- Data
- Control
- Void

These types are encoded using the following enumeration:

```
typedef enum logic[1:0] {
   txa_data    = 2'b10,
   txa_control = 2'b01,
   txa_void    = 2'b00,
   txa_forbid  = 2'b11
} txa_t;
```

`txa_forbid` indicates the forbidden combination of bits.

TX2 All packets have the same size. Each packet consists of a type field, tag field and data field. The corresponding SystemVerilog type definitions are as follows:

```
typedef bit [5:0] tag_t;
typedef bit [23:0] data_t;
typedef struct packed {
   txa_t txa;
   tag_t tag;
   data_t data;
} packet_t;
```

TX3 The serialized data is sent to `tx_data` input pin of the DUT. The validity of the data bit is indicated by the assertion of `tx_valid` DUT input pin. There are two versions of the serialization protocol: *normal* and *quiet*. In the quiet mode it is required that `tx_data` be set to zero when `tx_valid` is deasserted.

TX4 Each packet is transmitted bit by bit, from LSB to MSB, one bit per clock cycle.

TX5 During reset an active transaction, if any, is aborted.

TX6 Packets are transmitted sequentially. Transmission of a different packet is allowed only when the current packet has been fully transmitted, or when its transmission has been aborted because of a reset.

TX7 Each packet my be retransmitted according to the internal logic of the transmitter.

TX8 There are no timing constraints imposed on the transmission of two consecutive packets or on the retransmission of the same packet. Back-to-back packet transmission is allowed.

Checker `tx` implementing this testbench is shown in Figs. 23.4 and 23.5. Fig 23.4 shows the checker interface and several internal declarations. Input checker

```
checker tx(QUIET = 1'b0,
           event clk = $inferred_clock,
           untyped rst = $inferred_disable,
           output bit tx_valid, tx_data
);
  default clocking @clk; endclocking
  default disable iff rst;

  let legal_txa(txa) =  txa != txa_forbid;

  rand bit valid;
  rand bit data;
  rand bit packet_start;
  rand bit new_packet;
  rand packet_t packet;

  assign tx_valid = valid;
  assign tx_data  = data;
  ...
endchecker : tx
```

Fig. 23.4 Packet generation for 1-bit serial interface. Checker skeleton

```
let LENGTH = $bits(packet_t);
typedef bit[$clog2(LENGTH):0] tx_ptr_t;
m_rst_vs_packet_start:
  assume property (disable iff (1'b0) rst |-> !packet_start);
m_rst_vs_valid:
  assume property (disable iff (1'b0) rst |-> !valid);
m_non_overlap_start: assume property(
  packet_start |=> !packet_start[*LENGTH - 1]);

tx_ptr_t ptr;
function tx_ptr_t next_tx_ptr();
  if (rst) return 0;
  if (packet_start) return 1;
  if (ptr > 0 && ptr < LENGTH - 1)
    return ptr + 1;
  return 0;
endfunction : next_tx_ptr
always_ff @clk  ptr <= next_tx_ptr();

m_def_valid: assume property (
  valid == (packet_start || ptr != 0));
m_def_data: assume property (
  valid |-> data == packet[LENGTH-1-ptr]);
m_new_only_if_start: assume property (
  new_packet |-> packet_start);

bit packet_seen = 1'b0;
always_ff @clk
  packet_seen <= !rst && (packet_start || packet_seen);

m_new_if_start_not_seen: assume property (
  packet_start && !packet_seen |-> new_packet);
m_tx_packet_new: assume property (
  new_packet |-> legal_txa(packet.txa));
m_tx_packet_stable: assume property (
  ##1 !new_packet |-> $stable(packet));
if (QUIET)
  m_quiet_data: assume property (!tx_valid_i |-> !tx_data_i);
a_sanity_valid_0: assert property (
  !packet_start && !packet_seen |-> !valid);
a_sanity_valid_1: assert property (
  packet_start |-> valid[*LENGTH]);
```

Fig. 23.5 Packet generation for 1-bit serial interface. Checker contents

argument `QUIET` (Line 1) is an elaboration time constant, and its non-zero value indicates the quiet mode. By default the serialization protocol is executed in the normal mode. The other input arguments `clk` and `rst` have their standard meaning. Output arguments `tx_valid` and `tx_data` (Line 4) are defined in accordance with Rule *TX3*.

Let-declaration `legal_txa` on Line 9 is aimed to check the validity of the packet type, as specified by Rule *TX1*.

Lines 11–15 contain free variable declarations. Free bit `valid` (Line 11) indicates when the transmitted data bit is valid. Free bit `data` is a data bit—the result of the serialization. Both these free variables are directly copied to the checker outputs (Lines 17–18). The necessity of this copy is explained in Example 23.31.

Free bit `packet_start` (Line 13) indicates the beginning of a packet transmission: either a transmission of a new packet or a retransmission of a recently transmitted one. Free bit `new_packet` (Line 14) indicates the beginning of a new packet transmission. Finally, free variable `packet` corresponds to packet contents (Rule *TX2*). The described free variables are constrained by the assumptions shown in Fig. 23.5. Let us now move to that figure.

Assumptions `m_rst_vs_packet_start` and `m_rst_vs_valid` restrict the generator behavior during the reset (Rule *TX5*). Assumption `m_rst_vs_packet_start` (Line 3) forbids a new transmission when the reset is active. Assumption `m_rst_vs_valid` (Line 5) enforces `valid` bit deassertion during the reset. Both assumptions have an explicitly specified disable condition to override the default. Note the problem with the inconsistent sampling of a reset condition. In **`disable iff`** clauses `rst` is not sampled, whereas in these two assumptions it is, because it is part of their body. The same problem exists for the assignment on Line 18 which implicitly uses the sampled value of `rst` (Line 12). It remains to hope that the value of `rst` does not change at the same time steps when the clock does.[17]

Assumption `m_non_overlap_start` (Line 7) follows Rule *TX4*: No new transaction is allowed until the completion of the current one. The transaction length `LENGTH` is defined on Line 1.

Line 10 declares `ptr`, a pointer to a currently transmitted packet bit. This variable occupies the number of bits necessary for referencing any bit in a packet, the bit numeration starting from 0 (Line 2). The pointer is updated at each clock tick (Line 18); the next value of the pointer is calculated by function `next_tx_ptr` (Lines 11–17). Each time the reset is active the processing starts anew, and the pointer is set to 0 (Line 12; Rule *TX5*). When `packet_start` is active, i.e., when the transmission of a new packet or the retransmission of a transmitted one begins, the pointer value is set to 1 (Line 13; see also Exercise 23.9). Whenever a packet transmission continues, the pointer is incremented (Lines 14–15; Rule *TX4*). Finally, when the packet transmission has been completed, the pointer is reset to 0 (Line 16). See Exercise 23.10 regarding an alternative implementation of packet transmission.

Assumption `m_def_valid` (Line 20) defines the conditions of data validity: at the beginning of the packet transmission (`packet_start` asserted) or in its middle (`ptr != 0`). We could eliminate free variable `valid` and use output argument `tx_valid` directly, because this assumption may be replaced with assignment

```
assign tx_valid = packet_start || ptr != 0;
```

---

[17] Remember that this problem exists only in simulation. In FV the reset value is sampled even in **`disable iff`** clauses.

However, we prefer to keep free variable `valid` because it improves the checker readability and makes its code more robust for potential modifications of the protocol.

Assumption `m_def_data` (Line 22) requires `data` to contain the currently transmitted bit when the result is valid. Assumption `m_new_only_if_start` (Line 24) takes care that `packet_start` is set for a new packet.

Lines 27–29 define an indication (`packet_seen`) that at least one packet has arrived. `packet_seen` is zeroed at a reset. Assumption `m_new_if_start_not_seen` (Line 31) takes care to set `new_packet` when the first packet has arrived. Assumption `m_tx_packet_new` (Line 33) limits packet generation to packets of a legal type (Fig. 23.4, Line 9). It is sufficient to require this for new packets only. Assumption `m_tx_packet_stable` (Line 35) guarantees that the contents of `packet` may change only when a new packet arrives.

Lines 37–38 handle the quiet mode. Assumption `m_quiet_data` (Line 38) requires the data to be zero when the result is invalid (Rule *TX3*). Note that the **`if`** statement on Line 37 is a generate **`if`** as it belongs directly to the checker scope.

The checker also contains two sanity assertions that check its implementation. Assertion `a_sanity_valid_0` (Line 39) makes sure that the result is invalid until the first packet starts transmitting. Assertion `a_sanity_valid_1` verifies that the packet is transmitted continuously from the beginning to the end (Rule *TX4*).

As explained in Sect. 23.3.3 checker free variables are randomized in simulation subject to the constraints imposed by the assumptions. Contemporary simulators have strong abilities of solving combinatorial constraints. But our checker contains two temporal assumptions: `m_non_overlap_start` (Line 7) and `m_tx_packet_stable` (Line 35). This may become an obstacle to generate a viable test. Note, however, that the implementation of assumption `m_non_overlap_start` in simulation should be straightforward as the actual constraint on `packet_start` at each clock is unambiguously defined by the values of this variable in the preceding `LENGTH - 1` clock cycles. As for assumption `m_tx_packet_stable`, it may be rewritten as

```
##1 $changed(packet) |-> !new_packet
```

which makes it, essentially, combinational.

Exercises

23.1. Define the formal semantics of unconditional free variable nonblocking assignment controlled by a clocking event of each of the following forms:

(a) `@(`**`posedge`**` clk)`,
(b) `@(`**`negedge`**` clk)`,
(c) `@(`**`edge`**` clk `**`iff`**` en)`,
(d) `@(`**`posedge`**` clk `**`iff`**` en)`,
(e) `@(`**`negedge`**` clk `**`iff`**` en)`.

23.2. Define the formal semantics of a conditional free variable assignment controlled by a clocking event of the form of `@(edge clk)` and of each of the forms listed in Exercise 23.1.

23.3. Implement the checker from Example 23.18 using local variables instead of free variables.

23.4. What is the difference between the behavior of checkers `check1` and `check2` (dotted parts are the same and do not assign v in both checkers)?

```
checker check1(...);
  bit [3:0] v = ...;
  // ...
endchecker : check1

checker check2(...);
  rand bit [3:0] v = ...;
  // ...
endchecker : check2
```

How should the checker `check2` be modified to have the same behavior as the checker `check1`? The variable v should remain free in `check2`.

23.5. Rewrite the following fragment of a checker **`always_ff`** procedure to eliminate all conditional statements.

```
if (c1)
  if (c2) begin
    a <= x; b <= y; a <= z;
  end
  else a <= x;
  else begin
    a <= y;
    if (c2); else a <= x;
  end
```

23.6. Implement checker `data_consistency` from Example 23.29 without using local, free, or rigid variables. Assume that the transactions do not overlap.

23.7. The output data `out_data` at the end of transaction (`end_t` asserted) has the same value as the input data `in_data` at the beginning of the transaction (`start_t` asserted). With each transaction is associated a tag: the tag at the beginning of the transaction is contained in `stag`, and the tag at the end of the transaction is contained in `etag`. For the same transaction, `stag` and `etag` have the same values.

Implement this specification as a checker

(a) using rigid variables, and
(b) as an assertion with local variables.

23.8. What is the problem with the following definition of simulation semantics of a continuous assignment to a free variable: Continuous assignment to a free variable executes in the Active region? Consider both the case when the right-hand side is sampled, and when it is not. Hint: consider an assignment where the right-hand side is a checker input argument whose actual value is generated in a module.

23.9. Why is a separate processing of a packet start (Line 13) in Fig. 23.5 required? Why is this processing not subsumed by Lines 14–15?

23.10. Implement a packet transmission in checker `tx` from Sect. 23.5 using a shift register instead of a pointer to a current bit (Fig. 23.5, Lines 10–17).

23.11. Modify checker `tx` (Figs. 23.4 and 23.5) to support packet reception acknowledgment: the generator should receive an acknowledgment (`ack` asserted) from the DUT within 4 clock cycles upon the packet transmission. In the absence of the acknowledgment the same packet must be retransmitted. No new packet, but the first one, may start transmission until the acknowledgment of the previous packet reception has been obtained.

23.12. Limit the number of retries described in Exercise 23.11 to 3.

Chapter 24
Checker Libraries

A room without books is like a body without a soul.

— G.K. Chesterton

The enhancements to the IEEE SystemVerilog language in the 2009 and 2012 standards and, in particular, to the SystemVerilog Assertions (SVA) allow us to create much more useful and versatile checker libraries. In this chapter, we first identify the weaknesses of the current checker libraries by examining an example from the OVL library. We then provide a classification of *checkers*, and show how various forms of effective checker libraries can be created using the new constructs. We use the term *checker* and *checker library* in a broad sense to denote a verification unit and library, possibly assertion based. We refer to the SystemVerilog checker construct using `checker`.

There are many functional properties common to any design that are reusable modulo some expression changes. Therefore, to speed up the deployment of assertions without requiring extensive knowledge of the syntax and semantics of the SystemVerilog assertions language, it is essential to create libraries of checkers. Such checker libraries have been around for some time, such as the Accellera Open Verification Library (OVL) [10], and other checker libraries from EDA vendors. A similar approach was used even before the arrival of assertion languages by hiding procedural or RTL implementation of assertions in modules used as checkers. The initial implementation as well as the Verilog'95 implementation of OVL is in this form.

The enhancements to the SystemVerilog language in the 2009 and 2012 standards and, in particular to the assertion features allow us to create much more useful and versatile checker libraries. They benefit primarily from the following features: New encapsulation, `let` declarations, clock and disable inference, deferred assertions, elaboration error tasks, and enhanced property operators. The new `checker` encapsulation can be used to replace the module. These enhancements

E. Cerny et al., *SVA: The Power of Assertions in SystemVerilog*,
DOI 10.1007/978-3-319-07139-8_24

in SVA provide a solution to many problems when designing a checker library. Let us recall the main new SVA features that help checker library development and deployment [26]:

- **checker** encapsulation is versatile for assertion libraries.
  - Argument specification is similar to that of properties.
  - **checker** can be instantiated in procedural code.
  - Inference of clocking event, disable condition on the ports of the checker is possible.
  - In an **always** and **initial** procedure, evaluation is triggered by control reaching the checker instance.
- Inference functions `$inferred_clock` and `$inferred_disable` can be used as default values on formal ports of checker, sequence and property declarations.
- **global clocking** and **default disable iff** declarations are possible.
- Free variables
- Modeling code in checker
- **let** construct allows for making abstractions from expressions.
- Checking of configuration parameters at elaboration time.

24.1 Weaknesses of Existing Checker Libraries

To explain the current weaknesses, let us consider a simple checker `assert_handshake` inspired by its equivalent in the OVL library [10]. The checker is reduced to include only its important excerpts. Details of included files are omitted. The user may wish to consult the OVL library for further details if necessary. First, let us consider the checker interface.

Example 24.1. **module**-based `assert_handshake` checker interface

```
// Accellera Standard V2.8.1 Open Verification Library (OVL).
// Accellera Copyright (c) 2005--2014. All rights reserved.
'module ovl_handshake (
  clock, reset, enable, req, ack, fire);
  parameter severity_level = 'OVL_SEVERITY_DEFAULT;
  parameter min_ack_cycle = 0;
  parameter max_ack_cycle = 0;
  parameter req_drop = 0;
  parameter deassert_count = 0;
  parameter max_ack_length = 0;
  parameter property_type = 'OVL_PROPERTY_DEFAULT;
  parameter msg ='OVL_MSG_DEFAULT;
  parameter coverage_level= 'OVL_COVER_DEFAULT;
  parameter clock_edge = 'OVL_CLOCK_EDGE_DEFAULT;
  parameter reset_polarity =' OVL_RESET_POLARITY_DEFAULT;
  parameter gating_type = 'OVL_GATING_TYPE_DEFAULT;
```

```
  input clock, reset, enable;
  input req;
  input ack;
  output ['OVL_FIRE_WIDTH-1:0] fire;
  //...
'endmodule // ovl_handshake
```
□

The macros `'module` and `'endmodule` resolve to either **module** and **endmodule** or **interface** and **endinterface**. This distinction is made so that the checker could also be instantiated in SV interfaces. In either case, the kinds of ports such checkers are allowed to have impose severe constraints on the deployment of the checker in a design:

- Clock port `clock` cannot be an event such as **edge** `clk` **iff** `en`.
- Clock, disabling condition `reset`, and the enabling condition cannot be inferred from the instantiation context.
- The checker cannot be instantiated inside a procedure.
- The ports `req` and `ack` must be expressions of type **logic**, they cannot be of type **sequence** or **property**.

The restrictions make the usage of the checker tedious. In particular, the last item makes the checker less flexible to use because if either the requests or the acknowledgments are more complex temporal sequences of signal values, additional modeling code must be added on the outside of the checker instance to detect such sequences. This code and the checker instance are usually not to be included in the synthesized code, hence enclosing them between **'ifndef**—**'endif** compilation controls becomes necessary.

Before transforming the `assert_handshake` checker to its **checker**-based form, let us review the kinds of checkers and the main characteristics a library should possess.

24.2 Kinds of Checkers and Their Characteristics

Checkers can be classified according to four criteria:

1. Temporality: combinational (has no clock) vs. concurrent (requires a clock).
2. Encapsulation: **checker** (or **module** or **interface**) vs. **property** (or **let**) based.
3. Packaging: in a Verilog library vs. in a SystemVerilog package.
4. Configurability: Local per-instance vs. global for all instances.

24.2.1 Temporality

Many interesting checkers can be stated as unclocked Boolean expressions. Often clock is not needed and the user may be interested in instantiating the checker in procedures or design modules that do not have access to any clock. Such checkers cannot use concurrent assertions because they would require a clocking event. For this purpose, deferred assertions (Sect. 4.3) are the best candidates. When it is required to verify behaviors that are synchronous to some clock, concurrent assertions need to be used. They can be Boolean expressions evaluating each attempt at a single clock tick or temporal properties evaluating their attempts over several clock ticks.

24.2.2 Encapsulation

`property`-based encapsulation for temporal checkers, and **`let`**-based encapsulation for combinational checkers are the simplest ones. They are easy to use, but they allow no modeling code and can encompass only a single assertion. They are usually part of relatively simple checker libraries. More complex checkers that may consist of several assertions, modeling code, and coverage items need encapsulation in **`module`**, **`interface`** or more importantly now in **`checker`** constructs.

24.2.3 Packaging

Packaging checker libraries as a series of files, one per checker, in a "library" directory that is included automatically during compilation is the most typical usage. This mechanism has been used with the various existing **`module`**-based checker libraries. **`checker`**, **`property`** and **`let`** encapsulations allow for a more robust use model by packaging them in the SystemVerilog **`package`** enclosure. In this way, the appropriate library can be "imported" only where it is needed. Therefore, even different checkers with the same names can be deployed in different parts of the design. Of course, there is always the third possibility by accessing checker definitions that are brought into the source code using the **`` `include``** directive. However, this method provides the least flexibility and we do not consider it further.

24.2.4 Configurability

Global configuration is achieved best by macros, for example, to accomplish the following:

- Enabling all assertions or all functional coverage or both.
- Exclusion of nonsynthesizable code like action block reporting tasks, covergroups or testbench related items.

Local configuration on a per-instance basis is best achieved by elaboration-time constants and conditional generate blocks. This includes:

- Selection of specific functional coverage items or levels, from a combination of **`cover property`** and **`covergroup`** constructs.
- Selection of **`assert`**, **`restrict`**, or **`assume`** forms of assertions.
- Configuration and selection of subsets of assertions that should be active in a checker instance.
- Specification of minimal and maximal delay latencies and repetition counts in clock cycles.
- Specific user failure and success messages.
- Specification of severity level of assertion failure.

Elaboration-time constants must provide default values. For example,

- Most typical assertion usage (assertion kind, temporal delays and repetitions).
- Default failure message.
- Minimal useful functional coverage.

Functional coverage should provide several levels of detail whenever practically useful. Some may or may not be suitable for formal and synthesis tools and these should also be under global control. Here is a typical gradation:

- Minimal—"did the checked behavior ever happen?"
- More detailed—"which specific delay or data values were observed?"
- Corner cases—"were min and max delays, and boundary data points ever encountered?"

It is often desirable to perform `x/z` value checks on signals used in a checker. There may be separate checkers that perform just that task and report a failure when an `x` or `z` is detected. However, even "regular" assertions may have to include such checks to disable the assertion from failing or forcing a failure. The choice depends on whether separate checks for these values are used. If yes, then there is no point in reporting a failure in regular assertions, they should just report a vacuous or disabled success. Usually, the detection of `x/z` is done using the system function `$isunknown` that returns true if an `x` or `z` is detected in the argument expression value. For more detailed testing of `x` and `z` values, system function `$countbits` (Sect. 7.1.1) can be used as well.

The ease of using configuration capability is important when different test environments are used. For example, control may be provided over the following features:

- Choice of failure, success and information reporting integrated with the SystemVerilog testbench verification methodologies (e.g., UVM [9], VMM [19] or OVM [38]), or only reporting using `$display`, or using run-time error tasks, such as `$error`, etc.

- Macro encapsulation over the checker such that it would automatically provide some of the keywords, thus simplifying instantiation (e.g., Example 24.6). It may also hide differences between **checker**, **property**, and **module**-based checkers.
- Validation of values of arguments used as elaboration-time constants at elaboration time. Using conditional generate statements to test the constant arguments, elaboration tasks can issue error messages at elaboration, rather than at a later time during simulation (possibly many hours after the start of the compilation of the design).

The checker instances should also be easily identifiable by synthesis and formal tools, without the need of **`ifndef**—**`endif** enclosures around the checker instances. It is then left up to the tool to specify whether checkers should or should not be included in the process.

24.3 Examples of Typical Checker Kinds

We now examine examples of different forms of checkers, illustrating the various characteristics and limitations.

24.3.1 *Simple Combinational Checker*

The combinational checker shown in the following example is defined in a **let** declaration, which is then used in a deferred assertion. Notice the configuration mechanism using **`ifdef** SYNTHESIS for selecting a form that is suitable for synthesis and formal tools.

Example 24.2. **let**-based checker (in a package)

```
let onehot0 (sig, reset = 1'b1) =
`ifdef SYNTHESIS
 // Selected for synthesis or formal
   !|reset || $onehot0(sig);
`else
// Selected for 4-valued simulation
   |reset === 0 ||
     ($onehot0(sig) && !$isunknown(reset_n);
`endif
```
□

Such a checker can be instantiated in a module (program or interface), and procedural scope as follows.

Example 24.3. **let**-based checker instantiation

```
module m(input logic [3:0] r1,
         output logic [3:0] r2);
  a1: assert final (onehot0(r1))
```

```
  else $error("a1 failed"); //check input
  always_comb begin
    r2 = r1;
    a2: assert final (onehot0(.sig(r2)))
    else $error("a2 failed"); //check output
  end
endmodule
```
□

Note the following features:

- A macro definition `SYNTHESIS` selects between two forms of let, one suitable for formal tools and synthesizable checkers, and the other one for four-valued simulation. In the latter case, the assertion is enabled when `reset` is `1`, disabled (success) when `reset` is `0`, and it is forced to fail when `reset` is `x`, or `z`.
- Both positional (`a1`) and named (`a2`) argument association can be used.
- The system function `$onehot0` could be used directly in the assertion, however, it would not provide for a disabling condition.
- The `reset` argument has a default actual argument `1'b1`, meaning that when the actual is not provided in an instance, the resetting condition is false by default as shown in both `a1` and `a2`.
- The assertion is in the deferred form, hence it filters out 0-width glitches on both `reset` and `sig` actual arguments.
- The assertions can be instantiated in the module scope like `a1`, or in a procedure like `a2`.
- The disabling condition cannot be inferred in **`let`** instances. That is, `$inferred_disable` may not be used as a default actual argument.

24.3.2 A Checker-Based Combinational Checker

Next we examine a more flexible combinational checker. We assume that the convention is that **`default disable iff`** provides an active low reset.

Example 24.4. Combinational **`checker`**

```
typedef enum {ASSERT, ASSUME, NONE} assert_type;
typedef bit [15:0] cover_type;

checker onehot0
  (sig,
   assert_type usage_kind = ASSERT,
   cover_type cover_level = 1,
   reset = $inferred_disable,
   string msg = "", synthesis = `SYNTHESIS);

if (cover_level<16'b0 || cover_level>16'b11)
// check valid coverage selection
  $error("Coverage level is invalid %d",
    cover_level,
```

```
    "\nonly 1(level 0), 2(level 2), 3(both)",
    "or 0 (disabled) are allowed");
if (usage_kind != ASSERT || usage_kind != ASSUME)
  $warning("No assert or assume selected");

if (synthesis) begin : SYNTH
  let check_onehot0 (sig, reset) =
    ((!|reset_n) || $onehot0(sig));
  let cover_onehot0 (sig, reset) =
    ((|reset) && $onehot0(sig));
end : SYNTH
else begin : NO_SYNTH
  let check_onehot0 (sig, reset) =
    ((|reset === 0) || $onehot0(sig) && !$isunknown(reset));
  let cover_onehot0 (sig, reset) =
    ((|reset === 1) && $onehot0(sig));
end : NO_SYNTH

`ifdef ASSERT_ON
if (usage_kind == ASSERT) begin : ASSERT
  Assert_onehot0:
    assert final (check_onehot0(sig, reset))
      else $error(msg);
end : ASSERT
else if (usage_kind == ASSUME) begin : ASSUME
  Assume_onehot0:
    assume final (check_onehot0(sig, reset))
      else $error(msg);
end : ASSUME
`endif

`ifdef COVER_ON
if (cover_level & 1) begin : COVER_L1
  Cover_onehot0_1:
    cover final (cover_onehot0(sig, reset));
end : COVER_L1

if (!synthesis && (cover_level & 2))
  begin : COVER_L2
  function int position(logic $bits(sig) arg);
    for (int i = 0; i < $bits(sig); i++)
          if (sig[i] === 1) return i;
    return 0;
  endfunction // position

  covergroup cg_onehot0_2 with
    function sample(int index);
    coverpoint index;
  endgroup
  cg_onehot0_2 onehot0_2_index = new();
  Cover_onehot0_2:
    cover #0 (cover_onehot0(sig, reset))
      onehot0_2_index.sample(position(sig));
```

```
end : COVER_L2
`endif

endchecker : onehot0
```

□

This combinational checker illustrates many of the features that the **checker** encapsulation provides over the simpler `let`-based form:

- Coverage can be enabled globally for all checker instances by defining the symbol `COVER_ON`. Similarly, verification statements (**assert** or **assume**) can be globally enabled by defining `ASSERT_ON`.
- Synthesizable form is selected by a conditional generate block controlled by the argument `synthesis` that has as default actual value the macro symbol `SYNTHESIS`. This allows overriding the global selection if so required.
- The reset condition may be inferred from a **default disable iff** declaration.
- Deferred **assert** (`usage_kind == ASSERT`) or **assume** (`usage_kind == ASSUME`) statement or none can be selected using the argument `usage_kind`.
- When an invalid value is provided for `cover_level`, no coverage is enabled in this instance and an error message is issued.
- When an invalid value (or `NONE`) is provided for `usage_kind`, no verification statement (**assert** or **assume**) is enabled in this instance and a warning message is issued.
- For `Cover_onehot0_2`, we use deferred `#0` because **final** cannot be used when a user defined function or task is invoked in the action block. Otherwise, **final** would have been preferred.
- Two levels of functional coverage are provided, they can be individually enabled or disabled:
  - Level 1—when `cover_level == 1` is selected, it collects information on how many times a one hot or 0 condition was encountered while not disabled by reset.
  - Level 2—when `cover_level == 2` is selected, the **covergroup** classifies the bit positions that are set to 1 when the one hot condition holds. A deferred **cover** statement is used to trigger sampling of the bit position index by calling the `sample` method of the **covergroup** in the pass action statement of the deferred **cover** statement.
  - Both levels can be selected by setting `cover_level == 3`.

The **checker** can be instantiated in a simpler way than the one using a **let** declaration because the disable condition can be inferred:

Example 24.5. Combinational **checker** instantiation

```
`define ASSERT_ON
module m(input logic [3:0] r1,
         output logic [3:0] r2,
         input logic rst_n);
```

```
  default disable iff rst_n;
  onehot0 A1(r1); // check input
  always_comb begin
    r2 = r1;
    onehot0 A2(r1); // check output
  end
endmodule
```
□

As with the previous simple combinational checker (Sect. 24.3.1), this more complex checker can still be instantiated both inside (instance a1) and outside (instance a2) a procedure. Both checker instances use default values for configuration constants and enable verification statement **assert final** because ASSERT_ON is defined and usage_kind default value of ASSERT is used. Coverage is globally disabled because COVER_ON is not defined.

24.3.3 A Simple Property-Based Temporal Checker

We now turn our attention to checkers that verify behavior over time–temporal checkers.

Similarly as with the simple combinational checker and **let** declarations, we can define a simple temporal, clocked, checker using **property** declarations. As before, we assume that **default disable iff** defines an active low reset.

Example 24.6. **property**-based temporal checker

```
property time_interval_p
  (sequence trig, property cond,
   start_tick = 1, end_tick = 1,
   event clk = $inferred_clock,
   untyped rst_n = $inferred_disable);
     @clk disable iff (!bit'(|rst_n))
     trig |->
        always [start_tick:end_tick] cond;
endproperty : time_interval_p
```
□

Note that in the consequent of |-> we used the **always** operator instead of using the consecutive repetition cond[*start_tick:end_tick]. The reason is that we obtain maximum generality as to the actual argument for the formal cond. It can not only be a Boolean or a **sequence**, but also any **property** expression.

The property verifies that when trig occurs cond holds true in the interval start_tick to end_tick clock ticks, unless it is disabled by rst_n being 1'b0. The property has the following characteristics:

- The actual argument for trig is restricted to the type **sequence** because it is used in the antecedent of |->. The actual argument for cond can be any **property** expression (Boolean, sequence or property). The actual argument for clk must be a clocking event, while rst_n is left **untyped** for the user to be able to pass any valid expression.

- `trig` and `cond` do not have default actual arguments, hence the user must supply valid arguments there.
- Both `clk` and `rst` can be inferred from the context because the inference functions are used as default actual arguments.
- The arguments `start_tick` and `end_tick` have the typical default value of `1`. If used as in the following instantiation example, the property will check that `cond` holds true at the next clock tick after `trig` holds true.

A simple instantiation of property `time_interval_p` is illustrated in the next example.

Example 24.7. **`property`**-based checker instantiation

```
module m(input logic clk, reset_n, load,
         input logic [3:0] r1,
         output logic [3:0] r2);
default disable iff reset_n;
always @(posedge clk) begin
  if (!reset_n) r2 <= 'b0;
  else if (load) r2 <= r1;
  loaded_r2: assert property(time_interval_p(
    $past(load), r2 == $past(r1))) else
    $error("r2 not loaded correctly by r1");
end
endmodule
```

□

Except for the arguments that are used in the actual verification, all other ones use default values. The clock and the disabling condition are inferred from the **`always`** procedure and from the **`default disable iff`** declaration, respectively.

24.3.4 A Checker-Based Temporal Checker

The final example illustrates the full power of a **`checker`**-based temporal checker definition. We show a modified form of checker `assert_handshake` discussed at the beginning of this chapter (Sect. 24.1), but for reasons of brevity we include only those portions of the code that illustrate the differences.

The interface of the new checker is now as follows[1]:

Example 24.8. `assert_handshake` **`checker`** definition

```
import std_ovl_defines::*;
checker assert_handshake (
  sequence req, sequence ack,
  event clk          = $inferred_clock,
  untyped reset      = $inferred_disable,
  //elaboration-time constants:
```

[1]Port `enable` is not used in the OVL checker, so it is not included in this description.

```
  int severity_level = `OVL_SEVERITY_DEFAULT,
  int min_ack_cycle  = 0,
  int max_ack_cycle  = 0,
  int req_drop       = 0, // these three arguments
  int deassert_count = 0, // may not be needed
  int max_ack_length = 0, // since req is a sequence
  int property_type  = `OVL_PROPERTY_DEFAULT,
  string msg         = `OVL_MSG_DEFAULT,
  int coverage_level = `OVL_COVER_DEFAULT,
  int synthesis      = `SYNTHESIS,
  output [`OVL_FIRE_WIDTH-1:0]   fire
);
//...
generate // elaboration-time constant checks at compile time
  if (min_ack_cycle < 0)
    $error("min_ack_cycle is negative");
  if (max_ack_cycle < min_ack_cycle) $error(
    "max_ack_cycle is less than min_ack_cycle");
  if (req_drop < 0 || req_drop > 1) $warning(
    "req_drop \%0d is not 0 or 1",
    req_drop, "positive assumed 1,"
    "anything less than 1 assumed 0");
  // ...  checks for other arguments ...
endgenerate

default clocking checker_clk @clk; endclocking
default disable iff (reset);
assign fire = {`OVL_FIRE_WIDTH{1'b0}}; // Tied low in V2.3
//... Body of the checker ...

endchecker : assert_handshake
```
□

The parameters from the original checker became regular arguments of the **checker**-based checker. It simplifies instantiation, although the user should be aware that these arguments must be elaboration-time constants. The argument values of constants are verified at elaboration time using a conditional **generate** and elaboration time error tasks. If the values are illegal, then an error message is issued, or if a reasonable alternative exists, then that value is used and a warning is issued.

The following parameters from the original checker are missing:

```
parameter clock_edge =`OVL_CLOCK_EDGE_DEFAULT;
parameter reset_polarity = `OVL_RESET_POLARITY_DEFAULT;
parameter gating_type = `OVL_GATING_TYPE_DEFAULT;
```

This is because

- `clock_edge` is not needed as the argument `clk` can be an event expression.
- `reset_polarity` is not needed because we can pass any expression to the checker and it can infer the appropriate default expression from the contextual **default disable iff** declaration.

- `gating_type` is omitted for the same reason as `clock_edge`—the actual clocking event provided for `clk` can contain **`iff`** enabling condition.

The formal arguments `clk` and `reset` were placed after the arguments that do not have defaults. This simplifies instantiation of the checker when all arguments use default values. The type of `reset` is left unspecified (the keyword **`untyped`**) to provide more flexibility as to the kind of the actual reset expression.

The default value constants for the arguments are no more `` `defines``, but instead they are constants picked up from package `std_ovl_defines` as **`enum`** type values.

The type for `req` and `ack` is specified as **`sequence`** to allow Booleans and sequences, but prohibit supplying a property expression as the actual argument. This makes the checker more general, eliminating the need for modeling code to reduce a complex temporal behavior to a Boolean expression.

The body of the checker has to be modified to comply with restrictions on modeling code in **`checker`** constructs, and to use all the new features that help implementing and using checkers. The body of the **`checker`**-based checker is shown next. Refer to the OVL library to compare with the original checker body [10].

Only those portions as in the example of the original checker are shown that illustrate the differences with the original checker. The following piece of code shows the transformation needed in the modeling code of the checker.

Example 24.9. Body of `assert_handshake` **`checker`**

```
`ifdef ASSERT_ON
bit first_req = 1'b0;

function logic setFirstReq();
  if (!reset_n) return 1'b0;
  if((first_req ^ first_req) == 1'b0)
    return s_req.triggered;
  return 1'b0;
endfunction : setFirstReq

always @(clk) first_req <= setFirstReq();
```

Variable `first_req` is used in a **`property`** in the following code fragment:

```
property ASSERT_HANDSHAKE_ACK_MIN_CYCLE_P;
  req |-> not s_eventually [0:min_ack_cycle] ack;
endproperty
property
  ASSERT_HANDSHAKE_ACK_WITHOUT_REQ_FIRST_REQ_P;
  (##1 ack) implies
    (first_req or req.triggered);
endproperty

// other properties ...

// this remains as before
case (property_type)
```

```
    OVL_ASSERT_2STATE, // defined as enum types
    OVL_ASSERT: begin : ovl_assert
      if (min_ack_cycle > 0)
      begin : a_assert_handshake_ack_min_cycle
        A_ASSERT_HANDSHAKE_ACK_MIN_CYCLE_P:
          assert property (ASSERT_HANDSHAKE_ACK_MIN_CYCLE_P)
        else ovl_error_t("...as before...");
      end
// other assert and assume statements
endcase
`endif //ASSERT_ON

`ifdef COVER_ON
generate
  if (coverage_level != OVL_COVER_NONE) begin : ovl_cover
    if (OVL_COVER_BASIC_ON)
    begin : ovl_cover_basic
      cover_req_asserted:
      cover property
        (reset throughout req))
          ovl_cover_t("req_asserted covered");
    end
    //... other cover statement ...
  end
endgenerate
`endif // COVER_ON
```
□

Notice the following differences:

- Case default values on parameters are removed since constant argument values are checked at compile time.
- Case item labels are predefined **enum** types rather than **`define** symbols.
- The property expressions use property operators to allow sequences as the arguments and to make the assertions more efficient for formal tools. For example, in property `ASSERT_HANDSHAKE_ACK_MIN_CYCLE_P`, the sequence repetition is replaced by **not s_eventually** ... to accept a sequence expression for `ack`.
- Since both `req` and `ack` can be sequences, `$rose` had to be removed from both of these operands of the property. Thus, if the user wishes to use `$rose` on a Boolean, the appropriate expression has to be passed as the actual argument.[2]

Note that by extending the type of the arguments to **sequence**, it is now impossible to include checks for the presence of `x/z` values in the variables involved in the actual arguments. If such checking is required, the best approach is to create specific checkers just for the purpose of verifying `x/z` on variables. An open question remains how to disable existing assertions within the checker in such cases. It requires either an enhancement to the SystemVerilog language to provide

[2]If a system function existed that allowed to distinguish Boolean expressions from temporal sequences and properties, a conditional generate could be used to construct different forms of properties depending on the actual argument.

a function that detects `x/z` in sequences and properties, or an enhancement in the simulator to evaluate assertion in a pessimistic fashion.

The following is an example of instantiation of the new checker. For simplicity, all elaboration-time arguments take on default values.

Example 24.10. `assert_handshake` **`checker`** instantiation

```
module m;
bit clk;
logic rst_n, request,
      acknowledgment, endtrans;

default clocking @(posedge clk iff enabled);
endclocking

default disable iff rst_n;

//... some design code ...

always @(posedge clk) begin
  assert_handshake chk_handshake_inst(
    .req($rose(request)),
    .ack(acknowledgment ##1 endtrans));
  if (!rst_n) begin
    //... some design procedure ...
  end
end

//... some design code ...

endmodule
```
□

The main points are:

- The checker instantiation syntax is similar to that of a module, except that there is no parameter section.
- It can be instantiated in an **`always`** procedure.
- The `reset` argument is inferred from **`default disable`** declaration.
- The clocking event is inferred from the **`always`** procedure, hence even though **`default clocking`** is defined, the clock from the **`always`** procedure takes precedence.
- The actual argument for the formal argument `req` is `$rose(request)`; its clock is obtained from the default clocking defined in the module.
- The actual argument for `ack` is a sequence expression.

In the next and final section, we summarize the transformations to consider when converting the old-style module-based checkers into the new format based on checker encapsulation. This may be of interest when it is not desired to support two different formats of a checker library.

24.4 Converting Module-Based Checkers to the New Format

The set of transformations include the following items:

- Replace `` `define `` for various constants by **`typedef`** declarations using an **`enum`** type whenever possible.
- Replace **`always`** with **`always_ff`**, **`always_comb`** or **`always_latch`**.
- Create compile-time checks on elaboration-time constant values.
- Use initial procedures only for indicating that the enclosed assertions should have only one evaluation attempt. Initialize variables in their declaration.
- Change interface definition to include original parameters as regular arguments.
- Provide inference functions as default arguments to clock and reset.
- Provide default actual arguments wherever appropriate.
- Generalize the type of arguments to **`sequence`** or **`property`** wherever the checker properties can admit such operands.
- Checker instance identification task calls in initial procedures should be replaced by initial and an immediate assert statement on true, with a pass action statement displaying the required identification message.
- Consider using **`covergroup`** statements to provide more detailed coverage, selectable by an argument.
- Add **`default clocking`** and **`disable iff`** declarations and simplify assertions.
- Place the new checkers in a package for easy and controlled access from a design unit.

Exercises

24.1. Suppose that your design contains some legacy code with **`module`**-based checker instances while new parts of the design should use a **`checker`**-based version of the same checkers. The latter have the same names as the old module based ones, but are enclosed in a package. How can you use both of these checkers but in different parts of the design without name clashes?

24.2. The OVL checkers can be obtained from Accellera at [10]. Modify the OVL SVA checker `assert_proposition` into the **`checker`** form. What kind of assertion should it use, concurrent, immediate, or deferred?

24.3. When transforming the OVL `assert_handshake` checker into the checker form, we omitted any discussion on `x/z` checking on the arguments `req` and `ack`. This is because the actual arguments can be temporal sequences in which case we cannot use `$isunknown` on the argument to check for the presence of `x/z`. What would you provide as a solution to the user? Are any extensions to the SystemVerilog language necessary?

24.4. List some extensions to the SystemVerilog language that would be useful to have for creating effective checker libraries.

Appendix A
Expression Sampling

In concurrent assertions, in the scope of **always_ff** procedures in checkers and in several other SVA constructs *sampled* values of expressions are used. In most cases this sampling is done in the Preponed region, but there are exceptions for different corner cases. In this appendix we provide an exhaustive formal definition of sampling borrowed from the LRM with some modifications.

A.1 Default Sampled Value

As the first step we define the *default sampled value*—the sampled value that an expression has at the beginning of simulation. In the examples below we will use notation Φ for the default sampled value; this notation, of course, is not a part of SystemVerilog.

The default sampled value of an expression is defined as follows:

- The default sampled value of a static variable is the value assigned at its declaration.

  Example A.1. Given the declaration:

  ```
  logic [7:0] a = 8'h15;
  ```

  Φ(a) = `8'h15`. □

- If no value is assigned to the static variable at its declaration then the default sampled value is the default value of the corresponding type.

  Example A.2. Given the declaration:

  ```
  logic [7:0] b;
  ```

  Φ(a) = `8'hxx`, because the variables of type **logic** are initialized by default with `x`. □

E. Cerny et al., *SVA: The Power of Assertions in SystemVerilog*,
DOI 10.1007/978-3-319-07139-8

- The default sampled value of any other variable or net is the default value of the corresponding type.

  Example A.3. Given the declaration:

  ```
  wire w = a;
  ```

  Φ(a) =`1'bz`, because the default sampled value of a net is the default value of its type. □

- If `s` is a sequence then the default sampled values of `s.triggered` and `s.matched` are false (`1'b0`).
- The default sampled value of an expression is defined recursively.

  Example A.4. Given the declarations:

  ```
  bit a;
  logic [3:0] b = 7, c = 4;
  ```

  Φ(`a || b > c`) = Φ(a)`||`Φ(`b > c`) = `1'b0 ||` Φ(b) `>` Φ(c) = `7 > 4` = `1'b1`. □

A default sampled value is used in the definition of sampled value functions when there is need to reference a sampled value of an expression before time 0 (see Sect. 7.2.1).

A.2 Sampled Value of Variable

The general rule for variable sampling is as follows:

- The sampled value of a variable in time slot 0 is its default sampled value defined in Sect. A.1.
- The sampled value of a variable in any other time slot is the value of the variable in the Preponed region of this time slot.

This rule has the following exceptions:

- The sampled value of an automatic variable is its current value. When a past or a future value of an automatic variable is referenced by a sampled value function, the current value of the automatic variable is taken instead. The indication of past of future is thus disregarded.

  Example A.5. The sampled value of `i` in the following loop:

  ```
  for (int i = 0; i < 7; i++) begin
      ...
  end
  ```

  is its current value. If `a` is a vector, then in both `$sampled(a[i])` and `$past(a[i])` the current value of `i` in the current time slot is assumed. □

- The sampled value of a local variable (see Chaps. 15 and 16) is its current value.
- The sampled value of a free checker variable (see Sect. 23.1) is its current value. If a free checker variable is referred to by a sampled value function (see Sect. 7.2), then sampling takes place in the Postponed region of the corresponding clock tick.

  Example A.6. Given the following declaration:

  ```
  rand bit v;
  ```

  `$sampled(v)` is the current value of `v`. In the expression `$past(v)` the value of `v` is taken from the Postponed region of the previous clock tick. The rationale of this definition for past or future sampled value functions is to take the final value of the variable in the corresponding clock tick: the notion of the current value makes sense for the current clock tick only. □

- An input variable of a clocking block must be sampled by the clocking block with `#1step` sampling.[1] This is the sampled value of a such variable (see Sect. 2.3 and the LRM).

A.3 Sampled Value of Expression

Having defined the sampled value of a variable, we now can define the sampled value of an expression recursively:

- The sampled value of an expression consisting of a single variable is the sampled value of this variable.
- The sampled value of a const cast expression is the current value of its argument.

  Example A.7. `$sampled(const'(a))` is the current value of `a`. □

- When a past or a future value of a const cast expression is referenced by a sampled value function, the current value of this expression is taken instead.

  Example A.8. `$past(const'(a))` is the current value of `a`. □

- The sampled value of the sequence methods `triggered` and `matched` is defined as the current value returned by the sequence method.
- When a past or a future value of a sequence method is referenced by a sampled value function, this value is sampled in the Postponed region of the corresponding past or future clock tick

  Example A.9. If `s` is a sequence then in `$past(s.triggered)` the value of `s.triggered` is sampled in the Postponed region of the previous tick of the corresponding clock. The rationale of this definition is to take the final value

[1]The clocking block sampling may be defined other than `#1step`, but in this case the input variable cannot be used in contexts requiring variable sampling in SVA sense, such as a body of a concurrent assertion.

of `s.triggered` at a past or future time slot, because the notion of the current value makes sense for the current time slot only. □

- The sampled value of any other expression is defined recursively using the values of its operands.

Example A.10. If `a` is a static variable and `s` is a sequence then `$sampled(a && s.triggered)` = `$sampled(a) && $sampled(s.triggered)`, i.e., the value of `a` is taken from the Preponed region, and the value of `s.triggered` is the current value. □

Example A.11. If `a` and `s` are variables and `f` is a function then `$sampled(f(a, b))` = `f($sampled(a), $sampled(b))`.

Discussion: The LRM is not clear whether the global variables accessed by the function are sampled or not. Our interpretation is that these variables should be sampled. Otherwise, it would cause sampling inconsistency when a function is invoked from a right-hand side of a checker NBA. See, for example, function `next_tx_ptr` in Fig. 23.5, Lines 11–18. Note that in the concurrent assertion context functions must be automatic and have no side effects (see Sect. 5.1), therefore the definition is accurate in this case. □

References

1. IEEE Std. 1364–2001, IEEE Standard Verilog Hardware Description Language (2001)
2. IEEE Std. 1364–2005, IEEE Standard Verilog Hardware Description Language (2005)
3. IEEE Std. 1800–2005, IEEE Standard for SystemVerilog—Unified Hardware Design, Specification, and Verification Language (2005)
4. IEEE Std. 1076–2008, IEEE Standard VHDL Language Reference Manual (2008)
5. IEEE Std. 1800–2009, IEEE Standard for SystemVerilog—Unified Hardware Design, Specification, and Verification Language (2009)
6. IEEE Std. 1850–2010, IEEE Standard for Property Specification Language (PSL) (2010)
7. IEEE Std. 1666–2011, IEEE Standard SystemC®Language Reference Manual (2011)
8. IEEE Std. 1800–2012, IEEE Standard for SystemVerilog—Unified Hardware Design, Specification, and Verification Language (2012)
9. Accellera. *Universal Verification Methodology (UVM) 1.1* (2011)
10. Accellera. *Accellera Standard Open Verification Library (OVL) V2.8* (2013)
11. A. Adir, S. Copty, S. Landa, A. Nahir, G. Shurek, A. Ziv, C. Meissner, J. Schumann, A unified methodology for pre-silicon verification and post-silicon validation, in *Proceedings of Design, Automation and Test in Europe Conference and Exhibition (DATE), 2011* (IEEE, 2011), pp. 1–6
12. R. Armoni, L. Fix, A. Flaisher, R. Gerth, B. Ginsburg, T. Kanza, A. Landver, S. Mador-Haim, E. Singerman, A. Tiemeyer, M.Y. Vardi, Y. Zbar, The ForSpec temporal logic: a new temporal property-specification language, in *TACAS'02: Proceedings of the 8th International Conference on Tools and Algorithms for the Construction and Analysis of Systems* (Springer, London, 2002), pp. 296–211
13. R. Armoni, L. Fix, A. Flaisher, O. Grumberg, N. Piterman, A. Tiemeyer, M.Y. Vardi, Enhanced vacuity detection in linear temporal logic, in *Proceeding of International Conference on Computer-Aided Verification*, Lecture Notes in Computer Science. ISBN 3-540-40524-0, pp. 368–380 (2003)
14. R. Armoni, S. Egorov, R. Fraer, D. Korchemny, M. Vardi, Efficient LTL compilation for SAT-based model checking, in *IEEE/ACM International Conference on Computer-Aided Design* (IEEE, 2005)
15. P. Ashar, S. Dey, S. Malik, Exploiting multicycle false paths in the performance optimization of sequential logic circuits. IEEE Trans. Comput. Aided Des. Integrated Circ. Syst. I **14**(9), 1067–1075 (1995)
16. M.A. Azadpour, *SystemVerilog for Design and Verification Using UVM: From RTL to Synthesis* (Springer, New York, 2013)
17. A. Aziz, J. Kukula, T. Shiple, Hybrid verification using saturated simulation, in *Proceedings of the Design Automation Conference*, pp. 615–618 (IEEE and ACM, 1998)

E. Cerny et al., *SVA: The Power of Assertions in SystemVerilog*,
DOI 10.1007/978-3-319-07139-8

18. C. Baier, J.-P. Katoen, *Principles of Model Checking* (The MIT Press, Cambridge, 2008)
19. J. Bergeron, E. Cerny, A. Hunter, A. Nightingale, *Verification Methodology Manual for SystemVerilog* (Springer, New York, 2006)
20. M. Bernardo, A. Cimatti, *Formal Methods for Hardware Verification: 6th International School on Formal Methods for the Design of Computer, Communication, and Software Systems, SFM.* Lecture Notes in Computer Science (Springer, New York, 2006)
21. R. Bloem, B. Jobstmann, N. Piterman, Y. Saár. Synthesis of reactive (1) designs. J. Comput. Syst. Sci. **78**, 911–938 (2011)
22. D. Bustan, Mantis Item 290: Recursive properties can define non-regular languages, Erratum submitted to the IEEE 1800 SV-AC as part of the development and revision of IEEE 1800–2005, November 2004
23. D. Bustan, J. Havlicek, Some complexity results for SystemVerilog assertions, in *Proceedings of Computer Aided Verification*, Lecture Notes in Computer Science. ISBN 3-540-37406-X, pp. 205–218 (Springer, 2006)
24. D. Bustan, A. Flaisher, O. Grumberg, O. Kupferman, M.Y. Vardi. Regular vacuity. Lect. Notes Comput. Sci. **3725**, 191–206 (2005) ISBN 3-540-37406-X
25. D. Bustan, D. Korchemny, E. Seligman, J. Yang, SystemVerilog Assertions: past, present, and future SVA standardization experience. IEEE Des. Test Comput. **29**(2), 23–31 (2012)
26. E. Cerny, D. Korchemny, L. Piper, E. Selingman, S. Dudani, Verification case studies: evolution from sva 2005 to sva 2009, in *Proceedings of Design Verification Conference, DVCon* (Accellera System Initiative, 2009)
27. R. Chadha. *Static Timing Analysis for Nanometer Designs* (Springer, New York, 2009)
28. L. Claesen, J.-P. Schupp, P. Das, P. Johannes, S. Perremans, H. De Man, Efficient false path elimination algorithms for timing verification by event graph preprocessing. Integr. VLSI J. **8**(2), 173–187 (1989)
29. E.M. Clarke, O. Grumberg, D. Peled, *Model Checking*, 6th edn. (MIT Press, Cambridge, 2008)
30. V.R. Cooper, *Getting Started with UVM: A Beginner's Guide* (Verilab Publishing, Austin, 2013)
31. R. Cytron, J. Ferrante, B.K. Rosen, M.N. Wegman, F.K. Zadeck, Efficiently computing static single assignment form and the control dependence graph. ACM Trans. Program. Lang. Syst. **13**(4), 451–490 (1991)
32. A. Das, P. Basu, A. Banerjee, P. Dasgupta, P.P. Chakrabarti, C. Rama Mohan, L. Fix, R. Armoni, Formal verification coverage: computing the coverage gap between temporal specifications, in *ICCAD '04: Proceedings of the 2004 IEEE/ACM International conference on Computer-aided design*, (IEEE Computer Society, Washington, 2004), pp. 198–203
33. C. Eisner, D. Fisman, J. Havlicek, A topological characterization of weakness, in *Proceedings of Principles of Distributed Computing*, pp. 1–8 (ACM, 2005)
34. E.A. Emerson, Temporal and modal logic, in *Handbook of Theoretical Computer Science*, ed. by J. van Leeuwen (Elsevier Sience Publishers B.V., Amsterdam, 1990), pp. 996–1072
35. E.A. Emerson, J.Y. Halpern, Decision procedures and expressiveness in the temporal logic of branching time, in *STOC '82: Proceedings of the 14th Annual ACM Symposium on Theory of Computing* (ACM, New York, 1982), pp. 169–180
36. D. Fisman, O. Kupferman, S. Seinvald, M. Vardi, A framework for inherent vacuity, in *Hardware and Software: Verification and Testing*, vol. 5394. Lecture Notes in Computer Science (Springer, New York, 2008), pp. 7–22
37. H. Foster, Assertion-based verification: Industry myths to realities (invited tutorial), in *Proceedings of Computer Aided Verification*, Lecture Notes in Computer Science. ISBN 978-3-540-70543-7, pp. 5–10 (2008)
38. M. Glasser. *Open Verification Methodology Cookbook* (Springer, New York, 2009)
39. K. Gulati, S.P. Khatri, *Hardware Acceleration of EDA Algorithms. Custom ICs, FPGAs and GPUs* (Springer, New York, 2010)
40. J. Havlicek, N. Levi, H. Miller, K. Shultz, Extended CBV statement semantics. Part of a proposal presented to the Accellera Formal Verification Technical Committee, April 2002

41. J. Havlicek, K. Shultz, R. Armoni, S. Dudani, E. Cerny, Accellera Technical Report 2004.01: Notes on the Semantics of Local Variables in Accellera SystemVerilog 3.1 Concurrent Assertions, May 2004
42. S. Hazelhurst, O. Weissberg, G. Kamhi, L. Fix, A hybrid verification approach: getting deep into the design, in *DAC '02: Proceedings of the 39th Conference on Design Automation* (ACM, New York, 2002), pp. 111–116
43. J.L. Hennessy, D.A. Patterson, *Computer Architecture, Fourth Edition: A Quantitative Approach*. (Morgan Kaufmann Publishers Inc., San Francisco, 2011)
44. C.R. Ho, M. Theobald, B. Batson, J. Grossman, S.C. Wang, J. Gagliardo, M.M. Deneroff, R.O. Dror, D.E. Shaw, Four pillars of assertion-based verification, in *Proceedings of the Design and Verification Conference and Exhibition* (San Jose, 2009)
45. A. Kuehlmann, C.A.J. van Eijk, Combinational and sequential equivalence checking, in *Logic Synthesis and Verification* (Kluwer Academic Publishers, Norwell, 2002), pp. 343–372
46. O. Kupferman, M.Y. Vardi, Model checking of safety properties. Form. Methods Syst. Des. **19**(3), 291–314 (2001)
47. O. Kupferman, M.Y. Vardi, Vacuity detection in temporal model checking. Int. J. Softw. Tools Technol. Transfer **4**(2), 224–233 (2003)
48. L. Lamport, *Specifying Systems, The TLA+ Language and Tools for Hardware and Software Engineers*. (Addison-Wesley, Boston, 2002)
49. S. Malik, A case for runtime validation of hardware, in *Haifa Verification Conference*, Lecture Notes in Computer Science. ISBN 978-3-540-32604-5, pp. 30–42 (Springer, 2006)
50. Microsoft Research, AsmL: Abstract State Machine Language (2001), http://research.microsoft.com/en-us/projects/asml
51. S.S. Muchnick, *Advanced Compiler Design and Implementation*. (Morgan Kaufmann, San Francisco, 1973)
52. R. Mukhopadhyay, S.K. Panda, P. Dasgupta, J. Gough, Instrumenting ams assertion verification on commercial platforms. ACM Trans. Des. Autom. Electron. Syst. **14**(2), 1–47 (2009)
53. The Open Group, Assertion Definition Language (2009), http://adl.opengroup.org/about/index.html
54. R.H. Parker. Caution: clock crossing. a prescription for uncontaminated data across clock domains. *Chip Design Magazine* 5 April 2004
55. M. Pellauer, M. Lis, D. Baltus, R. Nikhil, Synthesis of synchronous assertions with guarded atomic actions, in *2nd ACM/IEEE International Conference on Formal Methods and Models for Co-Design*, (IEEE Computer Society, Washington, 2005), pp. 15–24
56. R.B. Reese, M.A. Thornton. *Introduction to Logic Synthesis Using Verilog HDL (Synthesis Lectures on Digital Circuits and Systems)* (Morgan and Claypool, San Rafael, 2006)
57. H. Rotithor, Postsilicon validation methodology for microprocessors. IEEE Des. Test, **17**(4), 77–88 (2000)
58. T. Schubert, High-level formal verification of next-generation microprocessors, in *Proceedings of the Design Automation Conference* (IEEE/ACM, 2003), pp. 1–6
59. C. Spear, G. Tumbush. *SystemVerilog for Verification: A Guide to Learning the Testbench Language Features*. (Springer, New York, 2012)
60. Sun Developer Network, Java Programming Language, Java Standard Edition 6 (2010), http://java.sun.com/
61. S. Sutherland, S. Davidmann, P. Flake, *SystemVerilog for Design: A Guide to Using SystemVerilog for Hardware Design and Modeling*, 2nd edn. (Springer, New York, 2006)
62. D. Tabakov, M.Y. Vardi, G. Kamhi, E. Singerman, A temporal language for SystemC, in *FMCAD '08: Proceedings of the 2008 International Conference on Formal Methods in Computer-Aided Design*, (IEEE Press, Piscataway, 2008), pp. 1–9
63. B.A. Trakhtenbrot, B.I.A. *Finite Automata: Behaviour and Synthesis* (North-Holland Publishing Co., Amsterdam, 1973)
64. F. Vahid, *Digital System Design with SystemVerilog*, 2nd edn. (Wiley (1000), New York, 2010)
65. I. Wagner, V. Bertacco, Reversi: Post-silicon validation system for modern microprocessors, in *IEEE International Conference on Computer Design, 2008* (IEEE, 2008), pp. 307–314

66. Y. Wolfsthal, Abstract for the ISOLA special session on "Industrial use of tools for formal analysis", in *ISoLA (Preliminary proceedings)*, Lecture Notes in Computer Science. ISBN 3-540-48928-2, pp. 190–190 (Springer, 2004)
67. P. Yeung, Four pillars of assertion-based verification, in *Euro DesignCon*, International Engineering Consortium (2004)
68. J. Yuan, K. Albin, A. Aziz, C. Pixley, Constraint synthesis for environment modeling in functional verification, in *Proceedings of Design Automation Conference* (IEEE and ACM, 2003), pp. 296–299

Index

E. Cerny et al., *SVA: The Power of Assertions in SystemVerilog*,
DOI 10.1007/978-3-319-07139-8

MIX
Papier aus verantwortungsvollen Quellen
Paper from responsible sources
FSC® C105338

If you have any concerns about our products,
you can contact us on
ProductSafety@springernature.com

In case Publisher is established outside the EU,
the EU authorized representative is:
Springer Nature Customer Service Center GmbH
Europaplatz 3, 69115 Heidelberg, Germany

Printed by Libri Plureos GmbH
in Hamburg, Germany